The Electric Power Engineering Handbook

POWER SYSTEMS

THIRD EDITION

T0204015

The Electric Power Engineering Handbook
Third Edition

Edited by
Leonard L. Grigsby

Electric Power Generation, Transmission, and Distribution, Third Edition
Edited by Leonard L. Grigsby

Electric Power Transformer Engineering, Third Edition
Edited by James H. Harlow

Electric Power Substations Engineering, Third Edition
Edited by John D. McDonald

Power Systems, Third Edition
Edited by Leonard L. Grigsby

Power System Stability and Control, Third Edition
Edited by Leonard L. Grigsby

The Electric Power Engineering Handbook

POWER SYSTEMS

THIRD EDITION

EDITED BY

LEONARD L. GRIGSBY

CRC Press
Taylor & Francis Group
Boca Raton London New York

CRC Press is an imprint of the
Taylor & Francis Group, an **informa** business

CRC Press
Taylor & Francis Group
6000 Broken Sound Parkway NW, Suite 300
Boca Raton, FL 33487-2742

First issued in paperback 2019

ISBN-13: 978-1-4398-5633-8 (hbk)
ISBN-13: 978-0-367-38148-6 (pbk)

Library of Congress Cataloging-in-Publication Data

Power systems / editor, Leonard L. Grigsby. -- 3rd ed.
    p. cm.
  Includes bibliographical references and index.
  ISBN 978-1-4398-5633-8 (hardback)
  1. Electric power systems. I. Grigsby, Leonard L.

TK1001.P65 2013
621.31--dc23
                                                2011044126

Visit the Taylor & Francis Web site at
http://www.taylorandfrancis.com

and the CRC Press Web site at
http://www.crcpress.com

Contents

PART III   Power System Planning (Reliability)

PART IV Power Electronics

Preface

The generation, delivery, and utilization of electric power and energy remain one of the most challenging and exciting fields of electrical engineering. The astounding technological developments of our age are highly dependent upon a safe, reliable, and economic supply of electric power. The objective of the Electric Power Engineering Handbook is to provide a contemporary overview of this far-reaching field as well as a useful guide and educational resource for its study. It is intended to define electric power engineering by bringing together the core of knowledge from all of the many topics encompassed by the field. The chapters are written primarily for the electric power engineering professional who seeks factual information, and secondarily for the professional from other engineering disciplines who wants an overview of the entire field or specific information on one aspect of it.

The first and second editions of this handbook were well received by readers worldwide. Based upon this reception and the many recent advances in electric power engineering technology and applications, it was decided that the time was right to produce a third edition. Because of the efforts of many individuals, the result is a major revision. There are completely new chapters covering such topics as FACTS, smart grid, energy harvesting, distribution system protection, electricity pricing, linear machines. In addition, the majority of the existing chapters have been revised and updated. Many of these are major revisions.

The handbook consists of a set of five books. Each is organized into topical parts and chapters in an attempt to provide comprehensive coverage of the generation, transformation, transmission, distribution, and utilization of electric power and energy as well as the modeling, analysis, planning, design, monitoring, and control of electric power systems. The individual chapters are different from most technical publications. They are not journal-type articles nor are they textbooks in nature. They are intended to be tutorials or overviews providing ready access to needed information while at the same time providing sufficient references for more in-depth coverage of the topic.

This book is devoted to the subjects of power system protection, power system dynamics and stability, and power system operation and control. If your particular topic of interest is not included in this list, please refer to the list of companion books referred to at the beginning.

In reading the individual chapters of this handbook, I have been most favorably impressed by how well the authors have accomplished the goals that were set. Their contributions are, of course, key to the success of the book. I gratefully acknowledge their outstanding efforts. Likewise, the expertise and dedication of the editorial board and section editors have been critical in making this handbook possible. To all of them I express my profound thanks.

They are as follows:

- Nonconventional Power Generation   Saifur Rahman
- Conventional Power Generation   Rama Ramakumar
- Transmission Systems   George G. Karady
- Distribution Systems   William H. Kersting

- Electric Power Utilization   Andrew P. Hanson
- Power Quality   S. Mark Halpin
- *Transformer Engineering* (a complete book)   James H. Harlow
- *Substations Engineering* (a complete book)   John D. McDonald
- Power System Analysis and Simulation   Andrew P. Hanson
- Power System Transients   Pritindra Chowdhuri
- Power System Planning (Reliability)   Gerry Sheblé
- Power Electronics   R. Mark Nelms
- Power System Protection   Miroslav M. Begovic[*]
- Power System Dynamics and Stability   Prabha S. Kundur[†]
- Power System Operation and Control   Bruce Wollenberg

I wish to say a special thank-you to Nora Konopka, engineering publisher for CRC Press/Taylor & Francis, whose dedication and diligence literally gave this edition life. I also express my gratitude to the other personnel at Taylor & Francis who have been involved in the production of this book, with a special word of thanks to Jessica Vakili. Their patience and perseverance have made this task most pleasant.

Finally, I thank my longtime friend and colleague—Mel Olken, editor, the *Power and Energy Magazine*—for graciously providing the picture for the cover of this book.

---

[*] Arun Phadke for the first and second editions.
[†] Richard Farmer for the first and second editions.

Editor

Leonard L. ("Leo") Grigsby received his BS and MS in electrical engineering from Texas Tech University, Lubbock, Texas and his PhD from Oklahoma State University, Stillwater, Oklahoma. He has taught electrical engineering at Texas Tech University, Oklahoma State University, and Virginia Polytechnic Institute and University. He has been at Auburn University since 1984, first as the Georgia power distinguished professor, later as the Alabama power distinguished professor, and currently as professor emeritus of electrical engineering. He also spent nine months during 1990 at the University of Tokyo as the Tokyo Electric Power Company endowed chair of electrical engineering. His teaching interests are in network analysis, control systems, and power engineering.

During his teaching career, Professor Grigsby received 13 awards for teaching excellence. These include his selection for the university-wide William E. Wine Award for Teaching Excellence at Virginia Polytechnic Institute and University in 1980, the ASEE AT&T Award for Teaching Excellence in 1986, the 1988 Edison Electric Institute Power Engineering Educator Award, the 1990–1991 Distinguished Graduate Lectureship at Auburn University, the 1995 IEEE Region 3 Joseph M. Beidenbach Outstanding Engineering Educator Award, the 1996 Birdsong Superior Teaching Award at Auburn University, and the IEEE Power Engineering Society Outstanding Power Engineering Educator Award in 2003.

Professor Grigsby is a fellow of the Institute of Electrical and Electronics Engineers (IEEE). During 1998–1999, he was a member of the board of directors of IEEE as the director of Division VII for power and energy. He has served the institute in 30 different offices at the chapter, section, regional, and international levels. For this service, he has received seven distinguished service awards, such as the IEEE Centennial Medal in 1984, the Power Engineering Society Meritorious Service Award in 1994, and the IEEE Millennium Medal in 2000.

During his academic career, Professor Grigsby has conducted research in a variety of projects related to the application of network and control theory to modeling, simulation, optimization, and control of electric power systems. He has been the major advisor for 35 MS and 21 PhD graduates. With his students and colleagues, he has published over 120 technical papers and a textbook on introductory network theory. He is currently the series editor for the Electrical Engineering Handbook Series published by CRC Press. In 1993, he was inducted into the Electrical Engineering Academy at Texas Tech University for distinguished contributions to electrical engineering.

Contributors

Hirofumi Akagi
Department of Electrical and Electronic
Engineering
Tokyo Institute of Technology
Tokyo, Japan

Richard E. Brown
Quanta Technology
Cary, North Carolina

William A. Chisholm
Kinectrics/Université du Quebec à Chicoutimi
Toronto, Ontario, Canada

Pritindra Chowdhuri
Tennessee Technological University
Cookeville, Tennessee

Mariesa L. Crow
Department of Electrical and Computer
Engineering
Missouri University of Science and Technology
Rolla, Missouri

Francisco De la Rosa
Electric Power Systems
Fremont, California

Juan Dixon
Department of Electrical Engineering
Pontificia Universidad Católica de Chile
Santiago, Chile

M. José Espinoza
Department of Electrical Engineering
Universidad de Concepción
Concepción, Chile

James W. Feltes
Siemens Power Technologies International
Schenectady, New York

Michael G. Giesselmann
Department of Electrical and Computer
Engineering
Center for Pulsed Power & Power Electronics
Texas Tech University
Lubbock, Texas

Leonard L. Grigsby
Department of Electrical and Computer
Engineering
Auburn University
Auburn, Alabama

Charles A. Gross
Department of Electrical and Computer
Engineering
Auburn University
Auburn, Alabama

Andrew P. Hanson
The Structure Group
Raleigh, North Carolina

Gerald T. Heydt
School of Electrical, Computer, and Energy
Engineering
Arizona State University
Tempe, Arizona

Alireza Khotanzad
Department of Electrical
Engineering
Southern Methodist University
Dallas, Texas

Stephen R. Lambert
Shawnee Power Consulting, LLC
Williamsberg, Virginia

Juan A. Martinez-Velasco
Department of Electrical Engineering
Universitat Politecnica de Catalunya
Barcelona, Spain

Thomas E. McDermott
Meltran, Inc.
Pittsburgh, Pennsylvania

Hyde M. Merrill
Merrill Energy, LLC
Salt Lake City, Utah

Luis Morán
Department of Electrical Engineering
Universidad de Concepción
Concepción, Chile

R. Mark Nelms
Department of Electrical and Computer
  Engineering
Auburn University
Auburn, Alabama

Wei Qiao
Department of Electrical Engineering
University of Nebraska-Lincoln
Lincoln, Nebraska

Kaushik Rajashekara
Rolls-Royce Corporation
Indianapolis, Indiana

N. Dag Reppen
Niskayuna Power Consultants, LLC
Niskayuna, New York

José Rodríguez
Department of Electronics Engineering
Universidad Técnica Federico Santa María
Valparaiso, Chile

Peter W. Sauer
Department of Electrical and Computer
  Engineering
University of Illinois at Urbana-Champaign
Champaign, Illinois

Gerald B. Sheblé
Quanta Technology, LLC
Raleigh, North Carolina

Z. John Shen
Department of Electrical Engineering and
  Computer Science
University of Central Florida
Orlando, Florida

Anthony F. Sleva
Altran Solutions
Cranbury, New Jersey

Mahesh M. Swamy
Yaskawa America Incorporated
Waukegan, Illinois

Lawrence J. Vogt
Mississippi Power Company
Gulfport, Mississippi

I

Power System Analysis and Simulation

Andrew P. Hanson

Andrew Hanson is a senior manager with The Structure Group, Raleigh, North Carolina. He has 20 years of experience in power system engineering, operations, and consulting, having worked at Tampa Electric, Siemens, and ABB and having led the development of an office for a small privately held engineering firm. His expertise is focused on power delivery system operations and planning, having led the development of processes, forecasts, plans, and distribution automation implementations for a number of utilities. Dr. Hanson has a BEE from The Georgia Institute of Technology and an MEE and a PhD in electrical engineering from Auburn University.

The Per-Unit System

In many engineering situations, it is useful to scale or normalize quantities. This is commonly done in power system analysis, and the standard method used is referred to as the per-unit system. Historically, this was done to simplify numerical calculations that were made by hand. Although this advantage has been eliminated by using the computer, other advantages remain:

- Device parameters tend to fall into a relatively narrow range, making erroneous values conspicuous.
- The method is defined in order to eliminate ideal transformers as circuit components.
- The voltage throughout the power system is normally close to unity.

Some disadvantages are that component equivalent circuits are somewhat more abstract. Sometimes phase shifts that are clearly present in the unscaled circuit are eliminated in the per-unit circuit.

It is necessary for power system engineers to become familiar with the system because of its wide industrial acceptance and use and also to take advantage of its analytical simplifications. This discussion is limited to traditional AC analysis, with voltages and currents represented as complex phasor values. Per-unit is sometimes extended to transient analysis and may include quantities other than voltage, power, current, and impedance.

The basic per-unit scaling equation is

$$\text{Per-unit value} = \frac{\text{actual value}}{\text{base value}}. \tag{1.1}$$

The base value always has the same units as the actual value, forcing the per-unit value to be dimensionless. Also, the base value is always a real number, whereas the actual value may be complex. Representing a complex value in polar form, the angle of the per-unit value is the same as that of the actual value.

Consider complex power

$$S = VI^* \tag{1.2}$$

or

$$S\angle\theta = V\angle\alpha I\angle-\beta$$

where

V = phasor voltage, in volts

I = phasor current, in amperes

Suppose we arbitrarily pick a value S_{base}, a real number with the units of volt-amperes. Dividing through by S_{base},

$$\frac{S\angle\theta}{S_{base}} = \frac{V\angle\alpha I\angle -\beta}{S_{base}}.$$

We further define

$$V_{base}\, I_{base} = S_{base}. \tag{1.3}$$

Either V_{base} or I_{base} may be selected arbitrarily, but not both. Substituting Equation 1.3 into Equation 1.2, we obtain

$$\frac{S\angle\theta}{S_{base}} = \frac{V\angle\alpha(I\angle -\beta)}{V_{base}\, I_{base}}$$

$$S_{pu}\angle\theta = \left(\frac{V\angle\alpha}{V_{base}}\right)\left(\frac{I\angle -\beta}{I_{base}}\right) \tag{1.4}$$

$$\mathbf{S}_{pu} = V_{pu}\angle\alpha(I_{pu}\angle -\beta)$$

$$\mathbf{S}_{pu} = \mathbf{V}_{pu}\,\mathbf{I}_{pu}^{*}$$

The subscript pu indicates per-unit values. Note that the form of Equation 1.4 is identical to Equation 1.2. This was not inevitable, but resulted from our decision to relate V_{base} I_{base} and S_{base} through Equation 1.3. If we select Z_{base} by

$$Z_{base} = \frac{V_{base}}{I_{base}} = \frac{V_{base}^2}{S_{base}}. \tag{1.5}$$

Convert Ohm's law:

$$\mathbf{Z} = \frac{\mathbf{V}}{\mathbf{I}} \tag{1.6}$$

into per-unit by dividing by Z_{base}.

$$\frac{\mathbf{Z}}{Z_{base}} = \frac{\mathbf{V}/\mathbf{I}}{Z_{base}}$$

$$\mathbf{Z}_{pu} = \frac{\mathbf{V}/V_{base}}{\mathbf{I}/I_{base}} = \frac{\mathbf{V}_{pu}}{\mathbf{I}_{pu}}.$$

Observe that

$$\mathbf{Z}_{pu} = \frac{\mathbf{Z}}{Z_{base}} = \frac{R + jX}{Z_{base}} = \left(\frac{R}{Z_{base}}\right) + j\left(\frac{X}{Z_{base}}\right)$$
$$Z_{pu} = R_{pu} + jX_{pu} \tag{1.7}$$

Thus, separate bases for R and X are not necessary:

$$Z_{base} = R_{base} = X_{base}$$

By the same logic,

$$S_{base} = P_{base} = Q_{base}$$

Example 1.1

(a) Solve for **Z**, **I**, and **S** at Port ab in Figure 1.1a.
(b) Repeat (a) in per-unit on bases of $V_{base} = 100$ V and $S_{base} = 1000$ V. Draw the corresponding per-unit circuit.

Solution

(a)

$$\mathbf{Z_{ab}} = 8 + j12 - j6 = 8 + j6 = 10 \angle 36.9° \ \Omega$$

$$\mathbf{I} = \frac{\mathbf{V_{ab}}}{\mathbf{Z_{ab}}} = \frac{100\angle 0°}{10\angle 36.9°} = 10\angle -36.9° \text{ A}$$

$$\mathbf{S} = \mathbf{VI^*} = (100\angle 0°)(10\angle -36.9°)^*$$

$$= 1000\angle 36.9° = 800 + j600 \text{ VA}$$

$$P = 800 \text{ W} \quad Q = 600 \text{ var}$$

(b) On bases V_{base} and $S_{base} = 1000$ VA:

$$Z_{base} = \frac{V_{base}^2}{S_{base}} = \frac{(100)^2}{1000} = 10 \ \Omega$$

$$I_{base} = \frac{S_{base}}{V_{base}} = \frac{1000}{100} = 10 \text{ A}$$

$$\mathbf{V_{pu}} = \frac{100\angle 0°}{100} = 1\angle 0° \text{ pu}$$

$$\mathbf{Z_{pu}} = \frac{8 + j12 - j6}{10} = 0.8 + j0.6 \text{ pu}$$

$$= 1.0\angle 36.9° \text{ pu}$$

$$\mathbf{I_{pu}} = \frac{\mathbf{V_{pu}}}{\mathbf{Z_{pu}}} = \frac{1\angle 0°}{1\angle 36.9°} = 1\angle -36.9° \text{ pu}$$

$$\mathbf{S_{pu}} = \mathbf{V_{pu}} \mathbf{I_{pu}}^* = (1\angle 0°)(1\angle -36.9°)^* = 1\angle 36.9° \text{ pu}$$

$$= 0.8 + j0.6 \text{ pu}$$

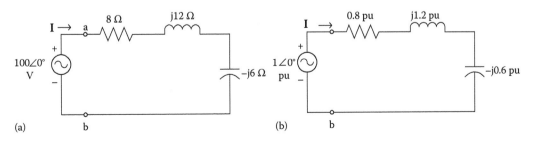

FIGURE 1.1 (a) Circuit with elements in SI units. (b) Circuit with elements in per-unit.

Converting results in (b) to SI units:

$$\mathbf{I} = (\mathbf{I}_{pu})I_{base} = (1\angle{-36.9°})(10) = 10\angle{-36.9°}\,\text{A}$$

$$\mathbf{Z} = (\mathbf{Z}_{pu})Z_{base} = (0.8 + j0.6)(10) = 8 + j6\,\Omega$$

$$\mathbf{S} = (\mathbf{S}_{pu})S_{base} = (0.8 + j0.6)(1000) = 800 + j600\,\text{W, var}$$

The results of (a) and (b) are identical.

For power system applications, base values for S_{base} and V_{base} are arbitrarily selected. Actually, in practice, values are selected that force results into certain ranges. Thus, for V_{base}, a value is chosen such that the normal system operating voltage is close to unity. Popular power bases used are 1, 10, 100, and 1000 MVA, depending on system size.

1.1 Impact on Transformers

To understand the impact of pu scaling on transformer, consider the three-winding ideal device (see Figure 1.2).

For sinusoidal steady-state performance:

$$\mathbf{V}_1 = \frac{N_1}{N_2}\mathbf{V}_2 \tag{1.8a}$$

$$\mathbf{V}_2 = \frac{N_2}{N_3}\mathbf{V}_3 \tag{1.8b}$$

$$\mathbf{V}_3 = \frac{N_3}{N_1}\mathbf{V}_1 \tag{1.8c}$$

and

$$N_1\mathbf{I}_1 + N_2\mathbf{I}_2 + N_3\mathbf{I}_3 = 0 \tag{1.9}$$

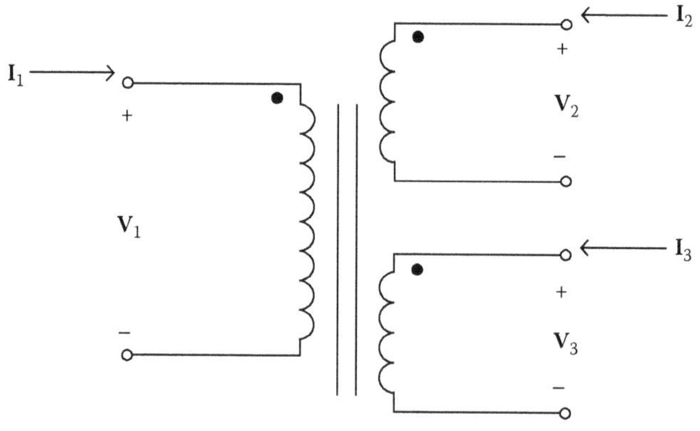

$N_1:N_2:N_3$

Ideal

FIGURE 1.2   The three-winding ideal transformer.

Consider the total input complex power **S**.

$$\mathbf{S} = \mathbf{V}_1\mathbf{I}_1{}^* + \mathbf{V}_2\mathbf{I}_2{}^* + \mathbf{V}_3\mathbf{I}_3{}^*$$

$$= \mathbf{V}_1\mathbf{I}_1{}^* + \frac{N_2}{N_1}\mathbf{V}_1\mathbf{I}_2{}^* + \frac{N_3}{N_1}\mathbf{V}_1\mathbf{I}_3{}^*$$

$$= \frac{\mathbf{V}_1}{N_1}[N_1\mathbf{I}_1 + N_2\mathbf{I}_2 + N_3\mathbf{I}_3]^*$$

$$= 0 \tag{1.10}$$

The interpretation to be made here is that the ideal transformer can neither absorb real nor reactive power. An example should clarify these properties.

Arbitrarily select two base values V_{1base} and S_{1base}. Require base values for windings 2 and 3 to be

$$V_{2base} = \frac{N_2}{N_1} V_{1base} \tag{1.11a}$$

$$V_{3base} = \frac{N_3}{N_1} V_{1base} \tag{1.11b}$$

and

$$S_{1base} = S_{2base} = S_{3base} = S_{base} \tag{1.12}$$

By definition,

$$I_{1base} = \frac{S_{base}}{V_{1base}} \tag{1.13a}$$

$$I_{2base} = \frac{S_{base}}{V_{2base}} \tag{1.13b}$$

$$I_{3base} = \frac{S_{base}}{V_{3base}} \tag{1.13c}$$

It follows that

$$I_{2base} = \frac{N_1}{N_2} I_{1base} \tag{1.14a}$$

$$I_{3base} = \frac{N_1}{N_3} I_{1base} \tag{1.14b}$$

Recall that a per-unit value is the actual value divided by its appropriate base. Therefore:

$$\frac{\mathbf{V}_1}{V_{1base}} = \frac{(N_1/N_2)\mathbf{V}_2}{V_{1base}} \tag{1.15a}$$

and

$$\frac{\mathbf{V}_1}{V_{1base}} = \frac{(N_1/N_2)\mathbf{V}_2}{(N_1/N_2)V_{2base}} \tag{1.15b}$$

or

$$\mathbf{V}_{1pu} = \mathbf{V}_{2pu} \tag{1.15c}$$

indicates per-unit values. Similarly,

$$\frac{\mathbf{V}_1}{V_{1base}} = \frac{(N_1/N_3)\mathbf{V}_3}{(N_1/N_3)V_{3base}} \tag{1.16a}$$

or

$$\mathbf{V}_{1pu} = \mathbf{V}_{3pu} \tag{1.16b}$$

Summarizing:

$$V_{1pu} = V_{2pu} = V_{3pu} \tag{1.17}$$

Divide Equation 1.9 by N_1

$$I_1 + \frac{N_2}{N_1}I_2 + \frac{N_3}{N_1}I_3 = 0$$

Now divide through by I_{1base}

$$\frac{\mathbf{I}_1}{I_{1base}} + \frac{(N_2/N_1)\mathbf{I}_2}{I_{1base}} + \frac{(N_3/N_1)\mathbf{I}_3}{I_{1base}} = 0$$

$$\frac{\mathbf{I}_1}{I_{1base}} + \frac{(N_2/N_1)\mathbf{I}_2}{(N_2/N_1)I_{2base}} + \frac{(N_3/N_1)\mathbf{I}_3}{(N_3/N_1)I_{3base}} = 0$$

Simplifying to

$$\mathbf{I}_{1pu} + \mathbf{I}_{2pu} + \mathbf{I}_{3pu} = 0 \tag{1.18}$$

Equations 1.17 and 1.18 suggest the basic scaled equivalent circuit, shown in Figure 1.3. It is cumbersome to carry the pu in the subscript past this point: no confusion should result, since all quantities will show units, including pu.

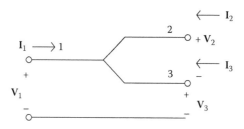

FIGURE 1.3   Single-phase ideal transformer.

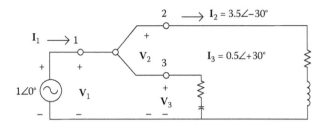

FIGURE 1.4 Per-unit circuit.

Example 1.2

The three-winding single-phase transformer of Figure 1.1 is rated at 13.8 kV/138 kV/4.157 kV and 50 MVA/40 MVA/10 MVA. Terminations are as follows:

  13.8 kV winding: 13.8 kV Source
  138 kV winding: 35 MVA load, pf = 0.866 lagging
  4.157 kV winding: 5 MVA load, pf = 0.866 leading

Using S_{base} = 10 MVA, and voltage ratings as bases,

  (a) Draw the pu equivalent circuit.
  (b) Solve for the primary current, power, and power, and power factor.

Solution

  (a) See Figure 1.4.
  (b)

$$S_2 = \frac{35}{10} = 3.5\,\text{pu} \quad S_2 = 3.5\angle+30°\,\text{pu}$$

$$S_3 = \frac{5}{10} = 0.5\,\text{pu} \quad S_3 = 0.5\angle-30°\,\text{pu}$$

$$V_1 = \frac{13.8}{13.8} = 1.0\,\text{pu} \quad V_1 = V_2 = V_3 = 1.0\angle0°\,\text{pu}$$

$$I_2 = \left(\frac{S_2}{V_2}\right)^* = 3.5\angle-30°\,\text{pu}$$

$$I_3 = \left(\frac{S_3}{V_3}\right)^* = 0.5\angle+30°\,\text{pu}$$

All values in Per-Unit Equivalent Circuit:

$$I_1 = I_2 + I_3 = 3.5\angle-30° + 0.5\angle+30° = 3.464 - j1.5 = 3.775\angle-23.4°\,\text{pu}$$

$$S_1 = V_1 I_1^* = 3.775\angle+23.4°\,\text{pu}$$

$$S_1 = 3.775(10) = 37.75\,\text{MVA}; \text{pf} = 0.9177 \text{ lagging}$$

$$I_1 = 3.775\left(\frac{10}{0.0138}\right) = 2736\,\text{A}$$

1.2 Per-Unit Scaling Extended to Three-Phase Systems

The extension to three-phase systems has been complicated to some extent by the use of traditional terminology and jargon, and a desire to normalize phase-to-phase and phase-to-neutral voltage simultaneously. The problem with this practice is that it renders Kirchhoff's voltage and current laws invalid in some circuits. Consider the general three-phase situation in Figure 1.5, with all quantities in SI units.

Define the complex operator:

$$\mathbf{a} = 1\angle 120°$$

The system is said to be balanced, with sequence abc, if

$$\mathbf{V}_{bn} = \mathbf{a}^2\mathbf{V}_{an}$$

$$\mathbf{V}_{cn} = \mathbf{a}\mathbf{V}_{an}$$

and

$$\mathbf{I}_b = \mathbf{a}^2\mathbf{I}_a$$

$$\mathbf{I}_c = \mathbf{a}\mathbf{I}_a$$

$$-\mathbf{I}_n = \mathbf{I}_a + \mathbf{I}_b + \mathbf{I}_c = 0$$

Likewise:

$$\mathbf{V}_{ab} = \mathbf{V}_{an} - \mathbf{V}_{bn}$$

$$\mathbf{V}_{bc} = \mathbf{V}_{bn} - \mathbf{V}_{cn} = \mathbf{a}^2\mathbf{V}_{ab}$$

$$\mathbf{V}_{ca} = \mathbf{V}_{cn} - \mathbf{V}_{an} = \mathbf{a}\mathbf{V}_{ab}$$

If the load consists of wye-connected impedance:

$$\mathbf{Z}_Y = \frac{\mathbf{V}_{an}}{\mathbf{I}_a} = \frac{\mathbf{V}_{bn}}{\mathbf{I}_b} = \frac{\mathbf{V}_{cn}}{\mathbf{I}_c}$$

The equivalent delta element is

$$\mathbf{Z}_\Delta = 3\mathbf{Z}_Y$$

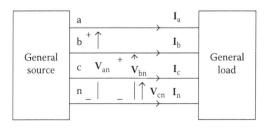

FIGURE 1.5   General three-phase system.

To convert to per-unit, define the following bases:

$S_{3\phi base}$ = The three-phase apparent base at a specific location in a three-phase system, in VA.
V_{Lbase} = The line (phase-to-phase) rms voltage base at a specific location in a three-phase system, in V.

From the above, define:

$$S_{base} = \frac{S_{3\phi base}}{3} \quad (1.19)$$

$$V_{base} = \frac{V_{Lbase}}{\sqrt{3}} \quad (1.20)$$

It follows that

$$I_{base} = \frac{S_{base}}{V_{base}} \quad (1.21)$$

$$Z_{base} = \frac{V_{base}}{I_{base}} \quad (1.22)$$

An example will be useful.

Example 1.3

Consider a balanced three-phase 60 MVA 0.8 pf lagging load, sequence abc operating from a 13.8 kV (line voltage) bus. On bases of $S_{3\phi base}$ = 100 MVA and V_{Lbase} = 13.8 kV:

(a) Determine all bases.
(b) Determine all voltages, currents, and impedances in SI units and per-unit.

Solution

(a)

$$S_{base} = \frac{S_{3\phi base}}{3} = \frac{100}{3} = 33.33\,MVA$$

$$V_{base} = \frac{V_{Lbase}}{\sqrt{3}} = \frac{13.8}{\sqrt{3}} = 7.967\,kV$$

$$I_{base} = \frac{S_{base}}{V_{base}} = 4.184\,kA$$

$$Z_{base} = \frac{V_{base}}{I_{base}} = 1.904\,\Omega$$

(b)

$$V_{an} = 7.967\angle 0°\,kV \quad (1.000\angle 0°\,pu)$$

$$V_{bn} = 7.967\angle -120°\,kV \quad (1.000\angle -120°\,pu)$$

$$V_{cn} = 7.967\angle +120°\,kV \quad (1.000\angle +120°\,pu)$$

$$S_a = S_b = S_c = \frac{S_{3\phi}}{3} = \frac{60}{3} = 20\,\text{MVA}\,(0.60\,\text{pu})$$

$$S_a = S_b = S_c = 16 + j12\,\text{MVA}\,(0.48 + j0.36\,\text{pu})$$

$$I_a = \left(\frac{S_a}{V_{an}}\right) = 2.510\angle{-36.9°}\,\text{kA}(0.6000\angle{-36.9°}\,\text{pu})$$

$$I_b = 2.510\angle{-156.9°}\,\text{kA}(0.6000\angle{-156.9°}\,\text{pu})$$

$$I_c = 2.510\angle{83.1°}\,\text{kA}(0.6000\angle{83.1°}\,\text{pu})$$

$$Z_Y = \frac{V_{an}}{I_a} = 3.174\angle{+36.9°} = 2.539 + j1.904\,\Omega(1.33 + j1.000\,\text{pu})$$

$$Z_\Delta = 3Z_Y = 7.618 + j5.713\,\Omega(4 + j3\,\text{pu})$$

$$V_{ab} = V_{an} - V_{bn} = 13.8\angle{30°}\,\text{kV}(1.732\angle{30°}\,\text{pu})$$

$$V_{bc} = 13.8\angle{-90°}\,\text{kV}(1.732\angle{90°}\,\text{pu})$$

$$V_{ca} = 13.8\angle{150°}\,\text{kV}(1.732\angle{150°}\,\text{pu})$$

Converting voltages and currents to symmetrical components:

$$\begin{bmatrix} V_0 \\ V_1 \\ V_2 \end{bmatrix} = \frac{1}{3}\begin{bmatrix} 1 & 1 & 1 \\ 1 & a & a^2 \\ 1 & a^2 & a \end{bmatrix}\begin{bmatrix} V_{an} \\ V_{bn} \\ V_{cn} \end{bmatrix} = \begin{bmatrix} 0\,\text{kV} & (0\,\text{pu}) \\ 7.967\angle{0°}\,\text{kV} & (1\angle{0°}\,\text{pu}) \\ 0\,\text{kV} & (0\,\text{pu}) \end{bmatrix}$$

$$I_0 = 0\,\text{kA}(0\,\text{pu})$$

$$I_1 = 2.510\angle{-36.9°}\,\text{kA}(0.6\angle{-36.9°}\,\text{pu})$$

$$I_2 = 0\,\text{kA}(0\,\text{pu})$$

Inclusion of transformers demonstrates the advantages of per-unit scaling.

Example 1.4

A 3ϕ 240 kV <Insert Symbol1>:15 kV <Insert Symbol2> transformer supplies a 13.8 kV 60 MVA pf = 0.8 lagging load, and is connected to a 230 kV source on the high-voltage (HV) side, as shown in Figure 1.6.

(a) Determine all base values on both sides for $S_{3\phi base}$ = 100 MVA. At the low-voltage (LV) bus, V_{Lbase} = 13.8 kV.
(b) Draw the positive sequence circuit in per-unit, modeling the transformer as ideal.
(c) Determine all currents and voltages in SI and per-unit.

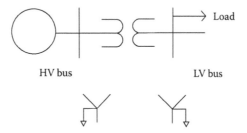

FIGURE 1.6   A three-phase transformer situation.

Solution

(a) Base values on the LV side are the same as in Example 1.3. The turns ratio may be derived from the voltage ratings ratios:

$$\frac{N_1}{N_2} = \frac{240/\sqrt{3}}{15/\sqrt{3}} = 16$$

$$\therefore (V_{base})_{HV\,side} = \frac{N_1}{N_2}(V_{base})_{LV\,side} = 16.00(7.967) = 127.5\,kV$$

$$(I_{base})_{HV\,side} = \frac{S_{base}}{(V_{base})_{HV\,side}} = \frac{33.33}{0.1275} = 261.5\,A$$

Results are presented in the following chart.

Bus	$S_{3\phi base}$ MVA	$V_{L\,base}$ kV	S_{base} MVA	I_{base} kA	V_{base} kV	Z_{base} Ohm
LV	100	13.8	33.33	4.184	7.967	1.904
HV	100	220.8	33.33	0.2615	127.5	487.5

(b)

$$V_{LV} = \frac{7.967\angle 0°}{7.967} = 1\angle 0°\,pu$$

$$S_{1\phi} = \frac{60}{3} = 20\,MVA$$

$$S_{1\phi} = \frac{20}{33.33} = 0.6\,pu$$

The positive sequence circuit is shown as Figure 1.7.

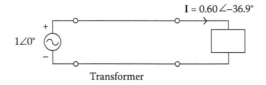

FIGURE 1.7   Positive sequence circuit.

(c) All values determined in pu are valid on both sides of the transformer! To determine SI values on the HV side, use HV bases. For example:

$$\mathbf{V}_{an} = (1\angle 0°)127.5 = 127.5\angle 0° \,\text{kV}$$

$$\mathbf{V}_{ab} = (1.732\angle 30°)(127.5) = 220.8\angle 30° \,\text{kV}$$

$$\mathbf{I}_a = (0.6\angle -36.9°)(261.5) = 156.9\angle -36.9° \,\text{A}$$

Example 1.5

Repeat the previous example using a 3ϕ 240 kV:15 kV <Insert Symbol1> Δ

Solution

All results are the same as before. The reasoning is as follows.

The voltage ratings are interpreted as line (phase-to-phase) values *independent of connection* (wye or delta). Therefore the turns ratio remains:

$$\frac{N_1}{N_2} = \frac{240/\sqrt{3}}{15/\sqrt{3}} = 16$$

As before:

$$(V_{an})_{LV\,side} = 7.967 \,\text{kV}$$

$$(V_{an})_{HV\,side} = 127.5 \,\text{kV}$$

However, $\mathbf{V}_{an}$ is no longer in phase on both sides. This is a consequence of the transformer model, and not due to the scaling procedure. Whether this is important depends on the details of the analysis.

1.3 Per-Unit Scaling Extended to a General Three-Phase System

The ideas presented are extended to a three-phase system using the following procedure.

1. Select a three-phase apparent power base ($S_{3ph\,base}$), which is typically 1, 10, 100, or 1000 MVA. This base is valid at every bus in the system.
2. Select a line voltage base ($V_{L\,base}$), user defined, but usually the nominal rms line-to-line voltage at a user-defined bus (call this the "reference bus").
3. Compute

$$S_{base} = \frac{(S_{3ph\,base})}{3} \quad \text{(Valid at every bus)} \tag{1.23}$$

4. At the reference bus:

$$V_{base} = \frac{V_{L\,base}}{\sqrt{3}} \tag{1.24}$$

$$I_{base} = \frac{S_{base}}{V_{base}} \tag{1.25}$$

$$Z_{base} = \frac{V_{base}}{I_{base}} = \frac{V_{base}^2}{S_{base}}$$ (1.26)

5. To determine the bases at the remaining busses in the system, start at the reference bus, which we will call the "from" bus, and execute the following procedure: Trace a path to the next nearest bus, called the "to" bus. You reach the "to" bus by either passing over (a) a line or (b) a transformer.
   a. The "line" case: $V_{L\,base}$ is the same at the "to" bus as it was at the "from" bus. Use Equations 1.2 through 1.4 to compute the "to" bus bases.
   b. The "transformer" case: Apply $V_{L\,base}$ at the "from" bus, and treat the transformer as ideal. Calculate the line voltage that appears at the "to" bus. This is now the new $V_{L\,base}$ at the "to" bus. Use Equations 1.2 through 1.4 to compute the "to" bus bases.
   Rename the bus at which you are located, the "from" bus. Repeat the above procedure until you have processed every bus in the system.
6. We now have a set of bases for every bus in the system, which are to be used for every element terminated at that corresponding bus. Values are scaled according to

per-unit value = actual value/base value

where actual value = the actual complex value of S, V, Z, or I, in SI units (VA, V, Ω, A); base value = the (user-defined) base value (real) of S, V, Z, or I, in SI units (VA, V, Ω, A); per-unit value = the per-unit complex value of S, V, Z, or I, in per-unit (dimensionless).

Finally, the reader is advised that there are many scaling systems used in engineering analysis, and, in fact, several variations of per-unit scaling have been used in electric power engineering applications. There is no standard system to which everyone conforms in every detail. The key to successfully using any scaling procedure is to understand how all base values are selected at every location within the power system. If one receives data in per-unit, one must be in a position to convert all quantities to SI units. If this cannot be done, the analyst must return to the data source for clarification on what base values were used.

2

Symmetrical Components for Power System Analysis

Anthony F. Sleva
Altran Solutions

2.1 Introduction

Three-phase power systems are difficult to model and analyze using traditional methods, that is, using loop equations (Kirchoff's laws) and node equations (Norton's laws) when transformer turns ratios and transformer winding connections are included in the calculations. In this chapter, an alternate approach using "symmetrical components" will be introduced. The concept was first proposed by C.L. Fortescue in 1918 in a classic paper devoted to consideration of the general N-phase case.

Per phase analysis can be developed for the simple three-phase system illustrated in Figure 2.1 by solving Equations 2.1a through 2.1d. After line and residual currents are calculated, voltage can be calculated:

$$V_A - I_A \times (Z_A \text{ Source} + Z_A \text{ System} + Z_A \text{ Load}) - I_R \times Z_R \text{ System} = 0 \qquad (2.1a)$$

$$V_B - I_B \times (Z_B \text{ Source} + Z_B \text{ System} + Z_B \text{ Load}) - I_R \times Z_R \text{ System} = 0 \qquad (2.1b)$$

$$V_C - I_C \times (Z_C \text{ Source} + Z_C \text{ System} + Z_C \text{ Load}) - I_R \times Z_R \text{ System} = 0 \qquad (2.1c)$$

$$I_A + I_B + I_C - I_R = 0 \qquad (2.1d)$$

Solving Equations 2.1 is tedious because all values are vectorial, that is, voltages are represented as a voltage magnitude with an associated angle and impedances are represented as complex impedances that include resistance and reactance components. In addition, the residual system impedance (Z_R) is the parallel combination of neutral and earth impedance.

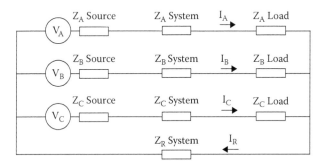

FIGURE 2.1   General three-phase circuit.

Using symmetrical components, analysis is greatly simplified because calculations are a function of load or fault condition rather than a function of source voltage. Conversion from phase quantities to symmetrical quantities or from symmetrical quantities to phase quantities is easily accomplished using rules established by Dr. Fortescue. The basic premise underlying symmetrical components is that calculations, for any load or fault condition, can be developed using single phase methods. Figures 2.2 through 2.5 illustrate impedance networks used to calculate three-phase faults, phase to phase faults, single phase to ground faults, and double phase to ground faults.

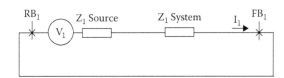

FIGURE 2.2   Three-phase fault—symmetrical component equivalent circuit.

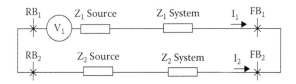

FIGURE 2.3   Phase to phase fault—symmetrical component equivalent circuit.

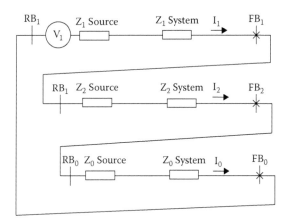

FIGURE 2.4   Single phase to ground fault—symmetrical component equivalent circuit.

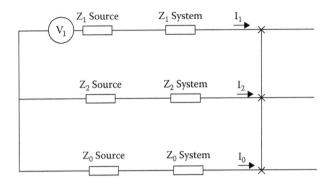

FIGURE 2.5 Double phase to ground fault—symmetrical component equivalent circuit.

Generally, two simplifying assumptions are made, namely, source voltages and system impedances are balanced. Although symmetrical components can be used to develop calculations for unbalanced source voltages and/or unbalanced system impedances, such calculations are beyond the scope of this book.

2.2 Discussion of Per Unit Quantities

Calculations are developed using per unit quantities for voltage, current, and impedance as per unit quantities eliminate the need to use transformer turns ratios and conversions throughout calculations. The use of per unit quantities, however, introduces the need to convert all impedances to a common MVA base before developing calculations and the need to convert per unit quantities to volts and amps after calculations have been developed.

Common practice is to develop calculations using nominal system voltages as base voltages. Calculations with transformers operating with off-nominal tap positions are only developed when special considerations arise.

Transformer impedances, which are listed in percent on transformer nameplates, are determined by test during factory acceptance tests. Per unit values are equal to percent values divided by 100. Transformer winding connections need to be included as essential system data.

Generator impedances, which are listed on data sheets in per unit quantities, are a function of fault duration. For short circuit analysis, X_d'' (generator subtransient reactance) is used. X_d' (generator transient reactance a few cycles after a three-phase fault occurs on the terminals of a generator) and X_d (generator synchronous reactance when a three-phase fault persists on the terminals of a generator) are used when developing specific calculations to analyze generator parameters, generating station circuit breaker interrupting capability, stuck circuit breaker conditions at generating stations, etc.

Line (feeder or circuit) impedances are a function of conductor length, size, type, spacing, phase relationship, construction, bundling, etc., are application specific, and need to be calculated for each specific line. (In many applications, standardized line impedances are available and the calculation becomes a function of line length only.)

Motor impedance is a function of the condition being analyzed—running, starting, or backfeeding a fault. The duration of motor backfeed is a function of inertia of the driven load. When specific motor data is not available, a common practice is to assume running power factor is 85%, starting power factor is 30%, and motor impedance during backfeed conditions is 0.10 per unit.

When developing calculations using per unit quantities, any MVA base can be used. Common practice is to use 1 MVA as the base for low-voltage systems, 10 MVA for medium voltage systems, and 100 MVA for high-voltage systems. After base MVA and base voltage are selected, base amps and base ohms are calculated using the Equations 2.2a and 2.2b for three-phase systems and Equations 2.3a and 2.3b for single phase systems:

$$\text{Base amps} = \frac{\text{base MVA}(3\varphi) \times 1000}{\sqrt{3} \times \text{base kV}(\varphi - \varphi)} \quad \text{(three phase)} \tag{2.2a}$$

$$\text{Base ohms} = \frac{(\text{base kV}(\varphi - \varphi))^2}{\text{MVA}(3\varphi)} \quad \text{(three phase)} \tag{2.2b}$$

$$\text{Base amps} = \frac{\text{base MVA}(1\varphi) \times 1000}{\text{base kV}(\varphi - n)} \quad \text{(single phase)} \tag{2.3a}$$

$$\text{Base ohms} = \frac{(\text{base kV}(\varphi - n))^2}{\text{MVA}(1\varphi)} \quad \text{(single phase)} \tag{2.3b}$$

Calculated values for base amps and base ohms, when base MVA and base voltage are selected, are shown in Table 2.1.

Per unit impedances for components are converted from nameplate values to base values using Equations 2.4a and 2.4b:

$$\text{Impedance at base MVA} = \text{Impedance at given MVA} \times \frac{\text{base MVA}}{\text{given MVA}} \tag{2.4a}$$

$$\text{Impedance at base kV} = \text{Impedance at given kV} \times \frac{\text{given kV}^2}{\text{base kV}^2} \tag{2.4b}$$

Table 2.2 Lists per unit impedances that are converted from given MVA to base MVA.
   Table 2.3 Lists per unit impedance that are converted from given kV to base kV.

TABLE 2.1   Example Base Values

Base MVA (MVA)	Base Voltage	Base Amps (A)	Base Ohms (Ω)
1	120 V, 1φ	8333	0.014
1	277 V, 1φ	3610	0.077
1	208 V, 3φ	2775	0.043
1	480 V, 3φ	1202	0.230
10	12.47 kV, 3φ	463	15.55
10	13.8 kV, 3φ	418	19.0
10	34.5 V, 3φ	167	119
100	69 kV, 3φ	837	47.6
100	138 kV, 3φ	418	190
100	230 kV, 3φ	251	529
100	345 kV, 3φ	167	1190
100	500 kV, 3φ	115	2500

TABLE 2.2 Impedance Conversion to Common MVA Base

Given MVA (MVA)	Base MVA (MVA)	Given Impedance (%)	Impedance at Base MVA (%)
0.25	1	1	4.00
0.5	1	1	2.00
1.0	1	1	1.00
2.0	1	1	0.50
2.5	10	5	20.00
5.0	10	5	10.00
10	10	5	5.00
20	10	5	2.50
25	100	10	40.00
50	100	10	20.00
100	100	10	10.00
200	100	10	5.00

TABLE 2.3 Impedance Conversion to Common Voltage Base

Given kV (kV)	Base kV (kV)	Given Impedance (%)	Impedance at Base kV (%)
13.2	13.8	5	4.57
13.8	13.8	5	5.00
14.1	13.8	5	5.22
220	230	10	9.15
230	230	10	10.00
245	230	10	11.35

Per unit values can be converted to actual values using the following equations:

$$\text{Circuit amps} = \text{per unit amps} \times \text{base amps} \qquad (2.5a)$$

$$\text{Circuit volts} = \text{per unit volts} \times \text{base volts} \qquad (2.5b)$$

$$\text{Circuit ohms} = \text{per unit ohms} \times \text{base ohms} \qquad (2.5c)$$

Impedance conversions to common voltage and MVA bases are necessary when transformers with different voltage ratings, but similar turns ratios, are connected in parallel. This is illustrated in Table 2.4.

TABLE 2.4 Transformer Data Converted to Common Base

Transformer 1 Data		Transformer 2 Data	
25	MVA	25	MVA
69	kV	66	kV
13.2	kV	12.47	kV
7%	Impedance	7%	Impedance
5.227	Transformer turns ratio	5.293	Transformer turns ratio
190.44	Base ohms, 25 MVA, 69 kV	174.24	Base ohms, 25 MVA, 66 kV
13.33	Transformer ohms	12.20	Transformer ohms
28.0%	At 100 MVA, 69 kV	25.6%	At 100 MVA, 69 kV

2.3 Fundamental Principles

The fundamental principles of symmetrical components are as follows:

1. Single phase methods can be used to solve problems on three-phase systems.
2. Sequence networks have been developed to calculate sequence voltages and sequence currents. These networks are designated the positive sequence network, the negative sequence network, and the zero sequence network.
3. Each sequence has its own line conductors, transformer equivalent circuits, generator equivalent circuits, etc.
4. Each sequence is independent of the other sequences (for balanced fault conditions).
5. Per unit quantities are used in calculations.
6. Single phase to ground faults are calculated as Aφ to ground faults.
7. Phase to phase faults are calculated as Bφ to Cφ faults.
8. Sequence currents are calculated at a fault bus. Current dividers are used to calculate sequence currents at other points in each sequence network.
9. Sequence voltages are calculated at a fault bus. Voltage dividers are used to calculate sequence voltages at other points in each sequence network.
10. Positive sequence voltage is maximum at a generator and minimum at a fault bus.
11. Negative sequence voltage is maximum at a fault bus and zero at the reference bus.
12. Zero sequence voltage is maximum at a fault bus and zero at the reference bus.
13. Phase voltage can be calculated when sequence voltages are known.
14. Phase current can be calculated when sequence currents are known.
15. Sequence voltages can be calculated if phase voltages are known.
16. Sequence currents can be calculated if phase currents are known.

2.4 "a" Operator

Symmetrical components utilize a unit vector, designated "a," to transform phase quantities to sequence quantities and to transform sequence quantities to phase quantities.

The properties of the "a" vector are as follows:

$$a^0 = 1 @ 0° \qquad\qquad (2.6a)$$

$$a^1 = 1 @ 120° = -0.5 + j0.866 \qquad\qquad (2.6b)$$

$$a^2 = 1 @ 240° = -0.5 - j0.866 \qquad\qquad (2.6c)$$

$$a^3 = 1 @ 360° = a \qquad\qquad (2.6d)$$

$$a^2 - a = 1 @ 240° - 1 @ 120° = -j1.732 = -j\sqrt{3} \qquad\qquad (2.6e)$$

$$a - a^2 = 1 @ 120° - 1 @ 240° = +j1.732 = j\sqrt{3} \qquad\qquad (2.6f)$$

$$1 + a + a^2 = 1 + 1 @ 120° + 1 @ 240° = 0 \qquad\qquad (2.6g)$$

2.5 Phase and Sequence Relationships

Power system voltages and currents are represented as the sum of sequence values:

$$E_a = E_{a1} + E_{a2} + E_{a0} \qquad\qquad (2.7a)$$

$$E_b = E_{b1} + E_{b2} + E_{b0} \qquad\qquad (2.7b)$$

$$E_c = E_{c1} + E_{c2} + E_{c0} \qquad\qquad (2.7c)$$

$$I_a = I_{a1} + I_{a2} + I_{a0} \qquad\qquad (2.8a)$$

$$I_b = I_{b1} + I_{b2} + I_{b0} \qquad\qquad (2.8b)$$

$$I_c = I_{c1} + I_{c2} + I_{c0} \qquad\qquad (2.8c)$$

Sequence values are derived from phase voltages and current using the following:

$$E_{a1} = \frac{1}{3}(E_a + aE_b + a^2E_c) \qquad\qquad (2.9a)$$

$$E_{a2} = \frac{1}{3}(E_a + a^2E_b + aE_c) \qquad\qquad (2.9b)$$

$$E_{a0} = \frac{1}{3}(E_a + E_b + E_c) \qquad\qquad (2.9c)$$

$$I_{a1} = \frac{1}{3}(I_a + aI_b + a^2I_c) \qquad\qquad (2.10a)$$

$$I_{a2} = \frac{1}{3}(I_a + a^2I_b + aI_c) \qquad\qquad (2.10b)$$

$$I_{a0} = \frac{1}{3}(I_a + I_b + I_c) \qquad\qquad (2.10c)$$

Positive sequence voltages and currents are represented by vectors that have equal amplitudes and are displaced 120° relative to each other as shown in Figure 2.6:

$$E_{a1} = E_{a1} \qquad\qquad (2.11a)$$

$$E_{b1} = a^2E_{a1} = E_{a1} \ @ \ 240° \qquad\qquad (2.11b)$$

$$E_{c1} = aE_{a1} = E_{a1} \ @ \ 120° \qquad\qquad (2.11c)$$

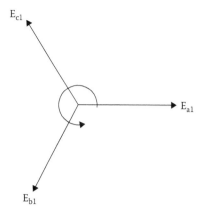

FIGURE 2.6   Positive sequence voltage vectors.

$$I_{a1} = I_{a1} \tag{2.12a}$$

$$I_{b1} = a^2 I_{a1} = I_{a1} \text{ @ } 240° \tag{2.12b}$$

$$I_{c1} = a I_{a1} = I_{a1} \text{ @ } 120° \tag{2.12c}$$

Negative sequence voltages and currents are represented by vectors that have equal amplitudes and are displaced 120° relative to each other, as shown in Figure 2.7. The rotation of the negative sequence vectors is opposite the rotation of the positive sequence vectors:

$$E_{a2} = E_{a2} \tag{2.13a}$$

$$E_{b2} = a E_{a2} = E_{a2} \text{ @ } 120° \tag{2.13b}$$

$$E_{c2} = a^2 E_{a2} = E_{a2} \text{ @ } 240° \tag{2.13c}$$

$$I_{a2} = I_{a2} \tag{2.14a}$$

$$I_{b2} = a I_{a2} = I_{a2} \text{ @ } 120° \tag{2.14b}$$

$$I_{c2} = a^2 I_{a2} = I_{a2} \text{ @ } 240° \tag{2.14c}$$

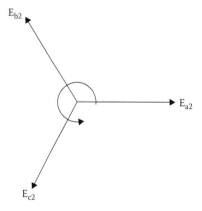

FIGURE 2.7   Negative sequence voltage vectors.

FIGURE 2.8 Zero sequence voltage vectors.

Zero sequence voltages and currents are represented by vectors that have equal amplitudes that are in phase with each other, as shown in Figure 2.8:

$$E_{a0} = E_{a0} \tag{2.15a}$$

$$E_{b0} = E_{a0} \tag{2.15b}$$

$$E_{c0} = E_{a0} \tag{2.15c}$$

$$I_{a0} = I_{a0} \tag{2.16a}$$

$$I_{b0} = I_{a0} \tag{2.16b}$$

$$I_{c0} = I_{a0} \tag{2.16c}$$

When calculations are developed, sequence quantities, that is, I_{a0}, I_{a1}, I_{a2} and E_{a0}, E_{a1}, E_{a2}, are calculated. Then, phase quantities are calculated using previous equations.

The use of symmetrical components and the determination of the magnitude and angle of positive sequence, negative sequence, and zero sequence vectors for various fault conditions will be demonstrated for the power system shown in Figure 2.9. This power system consists of a generator, two transformers, and one transmission line. For this generator X/R = 40 and for these transformers X/R = 25. When X/R is 10 or less, resistance is generally included in calculations.

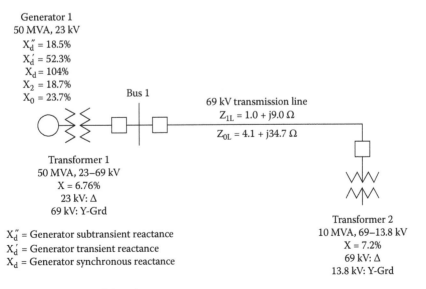

FIGURE 2.9 Power system single line diagram.

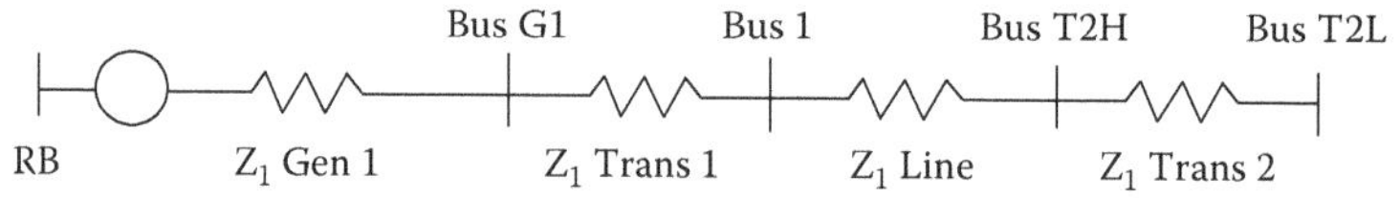

FIGURE 2.10   Positive sequence network for power system shown in Figure 2.9.

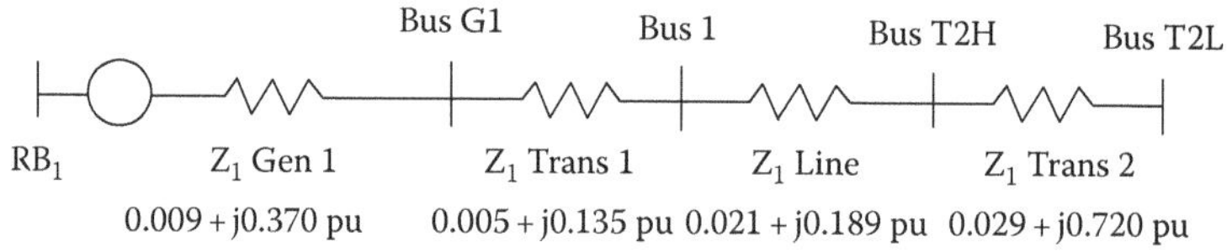

Base MVA = 100 MVA

Base voltages = 23, 69, and 13.8 kV

FIGURE 2.11   Positive sequence network with normalized impedances.

2.6 Positive Sequence Networks

Positive sequence networks are in all respects identical to the usual single line diagram for three-phase networks. The resistances and reactances are the values used in traditional equivalent circuits. Although each machine (generator or motor) is considered to be a voltage (power) source, motors are omitted when fault contributions from motors are negligible.

Figure 2.10 is the positive sequence network for the system shown in Figure 2.9. The positive sequence reference bus, RB_1, the common point to which all power sources are connected, is needed for network analysis. Other buses shown in Figure 2.10 (G1, 1, T2H and T2L) are locations where faults can be analyzed by inserting a jumper from a fault bus (G1, 1, T2H or T2L), to a reference bus, RB_1. Positive sequence jumper connections vary with fault type as shown in Figures 2.2 through 2.5.

Figure 2.11 is the positive sequence network with normalized impedances. To calculate a three-phase fault, a jumper is placed between RB_1 and the fault bus.

2.7 Sample Three-Phase Fault Calculation: Fault at Bus T2H

$$V_1 = 1.00 \text{ pu}$$

$$Z_1 = 0.036 + j0.694 \text{ pu} = 0.695 \text{ pu @ } 87.06°$$

$$I_{a1} = 1.439 \text{ pu @ } -87.06°$$

$$I_{a1} = 1204 \text{ A @ } -87.06° \text{ at } 69 \text{ kV}$$

$$I_A = I_{a1} = 1204 \text{ A @ } -87.06°$$

$$I_B = a^2 I_{a1} = 1 @ 240° \times 1204 \text{ A @ } -87.06° = 1204 \text{ A @ } 152.94°$$

$$I_C = a I_{a1} = 1 @ 120° \times 1204 \text{ A @ } -87.06° = 1204 \text{ A @ } 32.94°$$

$$V_1 \text{ @ Gen 1} = 1.00 \text{ pu} - 1.439 \text{ pu @ } -87.06° \times 0.370 \text{ pu @ } 90°$$

$$V_1 \text{ @ Gen 1} = 0.469 \text{ pu @ } -3.34°$$

$$V_{A-N} \text{ @ Gen 1} = 6.23 \text{ kV @ } -3.34°$$

2.8 Sample Three-Phase Fault Calculation: Fault at Bus T2L

$$V_1 = 1.00 \text{ pu}$$

$$Z_1 = 0.064 + j1.414 \text{ pu} = 1.416 \text{ pu @ } 87.39°$$

$$I_{a1} = 0.706 \text{ pu @ } -87.39°$$

$$I_{a1} = 2955 \text{ A @ } -87.39° \text{ at } 13.8 \text{ kV}$$

$$I_A = I_{a1} = 2955 \text{ A @ } -87.39°$$

$$I_B = a^2 I_{a1} = 1 \text{ @ } 240° \times 2955 \text{ A @ } -87.39° = 2955 \text{ A @ } 152.61°$$

$$I_C = a I_{a1} = 1 \text{ @ } 120° \times 2955 \text{ A @ } -87.39° = 2955 \text{ A @ } 32.61°$$

$$V_1 \text{ @ Gen 1} = 1.00 \text{ pu} - 0.706 \text{ pu @ } -87.39° \times 0.370 \text{ pu @ } 90°$$

$$V_1 \text{ @ Gen 1} = 0.739 \text{ pu @ } -0.92°$$

$$V_{A-N} \text{ @ Gen 1} = 9.81 \text{ kV @ } -0.92°$$

2.9 Negative Sequence Networks

Negative sequence networks are similar to positive sequence networks in that the number of branches is the same. But, negative sequence networks do not contain voltage sources because machines (generators or motors) only generate positive sequence voltage. The other difference between negative sequence networks and positive sequence networks is that different machine impedances may be used in each sequence. For most calculations, machine negative sequence impedance is assumed to be equal to positive sequence impedance. Different positive sequence and negative sequence impedances are used when generator response to short circuit conditions or circuit breaker interrupting capabilities are being evaluated at generating stations.

Figure 2.12 is the negative sequence network for the system shown in Figure 2.9. The negative sequence reference bus, RB_2, the common point to which all power sources are connected, is needed for network analysis. Other buses shown in Figure 2.12 (G1, 1, T2H and T2L) are locations where faults can be analyzed by inserting a jumper from a fault bus to a reference bus. Negative sequence jumper connections vary with fault type, as shown in Figures 2.3 through 2.5.

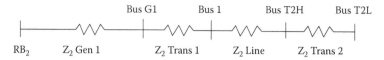

FIGURE 2.12   Negative sequence network for power system shown in Figure 2.9.

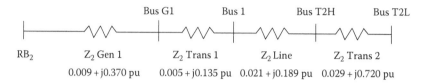

Base MVA = 100 MVA

Base voltages = 23, 69, and 13.8 kV

FIGURE 2.13   Negative sequence network with normalized impedances.

Figure 2.13 is the negative sequence network with normalized impedances. To calculate a phase to phase fault, the positive and negative sequence networks must be connected, as shown in Figure 2.3. This means that one jumper is needed between RB_1 and RB_2 and a second jumper is needed between the positive sequence fault bus and the corresponding negative sequence fault bus.

2.10 Sample Phase to Phase Fault Calculation: Fault at Bus T2L

$$V_1 = 1.00 \text{ pu}$$

$$Z = Z_1 + Z_2 = 2 \times (0.064 + j1.414) \text{ pu} = 2.831 \text{ pu } @ \ 87.39°$$

$$I_{a1} = 0.353 \text{ pu } @ -87.39°$$

$$I_{a2} = -I_{a1} = 0.353 \text{ pu } @ \ 92.61°$$

$$I_{a1} = 1477 \text{ A } @ -87.39° \text{ at } 13.8 \text{ kV}$$

$$I_{a2} = 1477 \text{ A } @ \ 92.61° \text{ at } 13.8 \text{ kV}$$

$$I_A = I_{a1} + I_{a2} = 1477 \text{ A } @ -87.39° + 1477 \text{ A } @ \ 92.61° \text{ at } 13.8 \text{ kV}$$

$$= 0$$

$$I_B = a^2 I_{a1} + a I_{a2} = 1 @ \ 240° \times 1477 \text{ A } @ -87.39° + 1 @ \ 120° \times 1477 \text{ A } @ \ 92.61°$$

$$= 2559 \text{ A } @ -182.61°$$

$$I_C = a I_{a1} + a^2 I_{a2} = 1 @ \ 120° \times 1477 \text{ A } @ -87.39° + 1 @ \ 240° \times 1477 \text{ A } @ -92.61°$$

$$= 2559 \text{ A } @ -2.61°$$

$$V_1 \ @ \ \text{Gen } 1 = 1.00 \ \text{pu} - 0.353 \ \text{pu} \ @ -87.06° \times 0.370 \ \text{pu} \ @ 90°$$

$$V_1 \ @ \ \text{Gen } 1 = 0.869 \ \text{pu} \ @ -0.39°$$

$$V_2 \ @ \ \text{Gen } 1 = -0.353 \ \text{pu} \ @ \ 92.61° \times 0.370 \ \text{pu} \ @ 90°$$

$$V_2 \ @ \ \text{Gen } 1 = 0.131 \ \text{pu} \ @ \ 2.61°$$

$$V_{A-N} \ @ \ \text{Gen } 1 = 0.869 \ \text{pu} \ @ -0.39° + 0.131 \ \text{pu} \ @ \ 2.61°$$

$$V_{A-N} \ @ \ \text{Gen } 1 = 1.0 \ \text{pu} \ @ \ 0°$$

$$V_{A-N} \ @ \ \text{Gen } 1 = 13.28 \ \text{kV} \ @ \ 0°$$

2.11 Zero Sequence Networks

Zero sequence networks, like negative sequence networks, do not contain voltage sources because machines (generators or motors) only generate positive sequence voltages. However, the number and connection of branches will be strongly influenced by transformer winding connections. Most zero sequence networks consist of an array of separate zero sequence networks, as shown in Figure 2.14.

Zero sequence line impedances are obtained by imagining three conductors connected together at the point of fault with the ground forming the return conductor.

Checking transformer winding connections, and properly accounting for them in zero sequence networks, is a necessary requirement when calculating phase to ground fault current and voltage. The connection to the zero sequence reference bus changes the impedance of the zero sequence network and complicates circuit analysis as shown in Figure 2.15.

Transformer equivalent circuits depend on the type of connection for each winding, delta or wye, and if wye, whether the transformer is solidly grounded, resistance grounded, reactance grounded, or ungrounded. Figure 2.16 illustrates transformer winding connections (delta, wye, and wye-grounded)

FIGURE 2.14 Zero sequence network for power system shown in Figure 2.9.

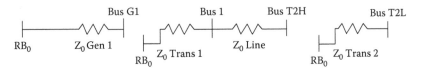

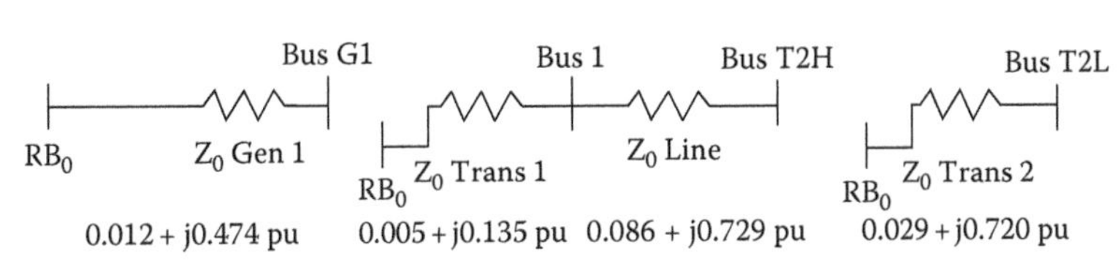

0.012 + j0.474 pu   0.005 + j0.135 pu   0.086 + j0.729 pu   0.029 + j0.720 pu

Base MVA = 100 MVA

Base voltages = 23, 69, and 13.8 kV

FIGURE 2.15 Zero sequence network with normalized impedances.

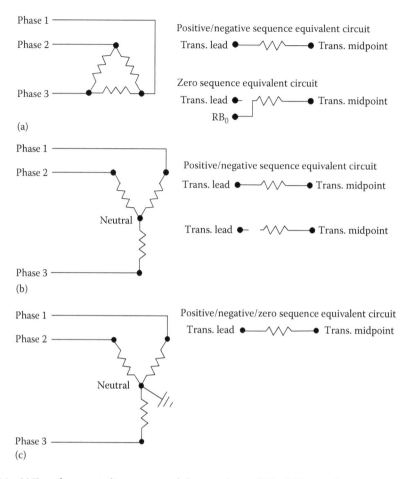

FIGURE 2.16 (a) Transformer windings connected phase to phase—"delta." (b) Transformer windings connected phase to neutral—"wye." (c) Transformer windings connected phase to ground—"wye-grounded."

and the equivalent circuit for each sequence. When using symmetrical components, transformers are represented as "star equivalents." High-voltage terminals and low-voltage terminals are located in the network and then connections from a midpoint to each terminal is established using the equivalents shown in Figure 2.16.

When transformers are connected wye-grounded/wye-grounded, the equivalent circuit is the same in the positive, negative, and zero sequences, as shown in Figure 2.17. For two winding transformers, the impedance listed on the nameplate is the sum of the high-voltage winding impedance and the low voltage winding impedance. For autotransformers and other transformers with more than two windings, impedances must be converted to a "star equivalent."

In wye-grounded/wye-grounded transformers, I_0, zero sequence current can flow in each winding, in each phase, and in each transformer lead. $3I_0$ can flow in the ground circuit.

When transformers are connected delta/wye-grounded, the zero sequence equivalent circuit, shown in Figure 2.18, is different from the positive and negative sequence equivalent circuit. Note that the zero sequence equivalent circuit shows (1) an open circuit between the transformer leads connected to the delta winding and the transformer midpoint and (2) a connection from the transformer midpoint to the zero sequence reference bus.

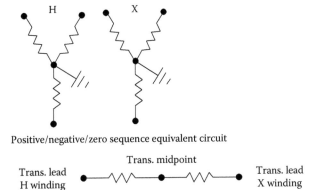

Positive/negative/zero sequence equivalent circuit

FIGURE 2.17   Transformer equivalent circuit, windings connected wye grounded-wye grounded.

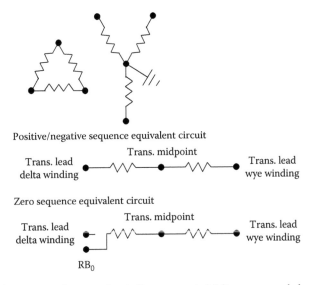

FIGURE 2.18   Transformer equivalent circuit, windings connected delta-wye grounded.

In delta-wye grounded transformers, I_0, can flow in each winding, in each phase, but not in the transformer leads connected to the delta winding. $3I_0$ can flow in the ground circuit connected to the wyegrounded winding.

When transformers are connected delta/wye, wye/wye, or wye/wye-grounded, the equivalent circuit in the zero sequence is an open circuit. The equivalent circuit for a transformer connected delta/wye is shown in Figure 2.19.

Zero sequence impedance of generators, like transformers, depends on the type of winding connection, delta or wye, and if wye, whether the generator is solidly grounded, resistance grounded, reactance grounded, or ungrounded.

Adjacent overhead transmission lines influence the zero sequence impedance of each other. This influence is referred to as mutual impedance.

Figure 2.13 is the negative sequence network with normalized impedances. To calculate a phase to phase fault, the positive and negative sequence networks must be connected as shown in Figure 2.3. This means that one jumper is needed between RB_1 and RB_2 and a second jumper is needed between the positive sequence fault bus and the corresponding negative sequence fault bus.

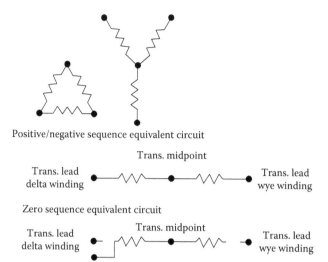

Positive/negative sequence equivalent circuit

Trans. midpoint

Trans. lead
delta winding

Trans. lead
wye winding

Zero sequence equivalent circuit

Trans. midpoint

Trans. lead
delta winding

Trans. lead
wye winding

RB_0

FIGURE 2.19   Transformer equivalent circuits, windings connected delta-wye.

2.12 Sample Phase to Ground Fault Calculation: Fault at Bus T2L

$$V = 1.00 \text{ pu}$$

$$= 2 \times (0.064 + j1.414) + .029 + j0.720 \text{ pu}$$

$$Z = Z_1 + Z_2 + Z_0 = 0.158 + j3.548 \text{ pu} = 3.552 \text{ pu @ } 87.45°$$

$$I_{a1} = 0.282 \text{ pu @ } -87.45°$$

$$I_{a2} = 0.282 \text{ pu @ } -87.45°$$

$$I_{a0} = 0.282 \text{ pu @ } -87.45°$$

$$I_{a1} = 1178 \text{ A @ } -87.45° \text{ at } 13.8 \text{ kV}$$

$$I_{a2} = 1178 \text{ A @ } -87.45° \text{ at } 13.8 \text{ kV}$$

$$I_{a0} = 1178 \text{ A @ } -87.45° \text{ at } 13.8 \text{ kV}$$

$$I_A = I_{a1} + I_{a2} + I_{a0} = 3534 \text{ A @ } -87.45°$$

$$I_B = a^2 I_{a1} + a I_{a2} + I_{a0} = 0$$

$$I_C = aI_{a1} + a^2I_{a2} + I_{a0} = 0$$

$$V_1 \text{ @ Gen } 1 = 1.00 \text{ pu} - 0.282 \text{ pu @ } -87.45° \times 0.370 \text{ pu @ } 90°$$

$$V_1 \text{ @ Gen } 1 = 0.896 \text{ pu @ } -0.30°$$

$$V_2 \text{ @ Gen } 1 = -0.282 \text{ pu @ } -87.45° \times 0.370 \text{ pu @ } 90°$$

$$V_2 \text{ @ Gen } 1 = 0.104 \text{ pu @ } -2.55°$$

$$V_0 \text{ @ Gen } 1 = 0.0 \text{ pu} \times 0.370 \text{ pu @ } 90°$$

$$V_0 \text{ @ Gen } 1 = 0.0$$

$$V_{A-N} \text{ @ Gen } 1 = 0.896 \text{ pu @ } -0.30° + 0.104 \text{ pu @ } -2.55° + 0$$

$$V_{A-N} \text{ @ Gen } 1 = 10.52 \text{ kV @ } -0.67°$$

3

Power Flow Analysis

Leonard L. Grigsby
Auburn University

Andrew P. Hanson
The Structure Group

3.1 Introduction

The equivalent circuit parameters of many power system components are described in other parts of this handbook. The interconnection of the different elements allows development of an overall power system model. The system model provides the basis for computational simulation of the system performance under a wide variety of projected operating conditions. Additionally, "post mortem" studies, performed after system disturbances or equipment failures, often provide valuable insight into contributing system conditions. This chapter discusses one such computational simulation, the power flow problem.

Power systems typically operate under slowly changing conditions, which can be analyzed using steady-state analysis. Further, transmission systems operate under balanced or near-balanced conditions allowing per-phase analysis to be used with a high degree of confidence in the solution. Power flow analysis provides the starting point for most other analyses. For example, the small signal and transient stability effects of a given disturbance are dramatically affected by the "pre-disturbance" operating conditions of the power system. (A disturbance resulting in instability under heavily loaded system conditions may not have any adverse effects under lightly loaded conditions.) Additionally, fault analysis and transient analysis can also be impacted by the pre-disturbance operating point of the power system (although, they are usually affected much less than transient stability and small signal stability analysis).

3.2 Power Flow Problem

Power flow analysis is fundamental to the study of power systems; in fact, power flow forms the core of power system analysis. A power flow study is valuable for many reasons. For example, power flow analyses play a key role in the planning of additions or expansions to transmission and generation facilities. A power flow solution is often the starting point for many other types of power system analyses. In addition, power flow analysis and many of its extensions are an essential ingredient of the studies performed in power system operations. In this latter case, it is at the heart of contingency analysis and the implementation of real-time monitoring systems.

The power flow problem (popularly known as the load flow problem) can be stated as follows:

> For a given power network, with known complex power loads and some set of specifications or restrictions on power generations and voltages, solve for any unknown bus voltages and unspecified generation and finally for the complex power flow in the network components.

Additionally, the losses in individual components and the total network as a whole are usually calculated. Furthermore, the system is often checked for component overloads and voltages outside allowable tolerances.

Balanced operation is assumed for most power flow studies and will be assumed in this chapter. Consequently, the positive sequence network is used for the analysis. In the solution of the power flow problem, the network element values are almost always taken to be in per-unit. Likewise, the calculations within the power flow analysis are typically in per-unit. However, the solution is usually expressed in a mixed format. Solution voltages are usually expressed in per-unit; powers are most often given in kVA or MVA.

The "given network" may be in the form of a system map and accompanying data tables for the network components. More often, however, the network structure is given in the form of a one-line diagram (such as shown in Figure 3.1).

Regardless of the form of the given network and how the network data is given, the steps to be followed in a power flow study can be summarized as follows:

1. Determine element values for passive network components.
2. Determine locations and values of all complex power loads.
3. Determine generation specifications and constraints.
4. Develop a mathematical model describing power flow in the network.
5. Solve for the voltage profile of the network.
6. Solve for the power flows and losses in the network.
7. Check for constraint violations.

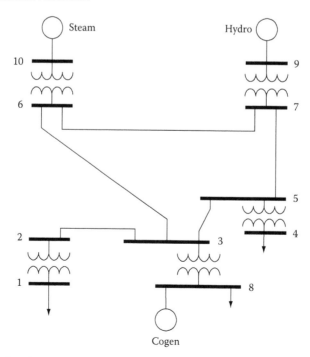

FIGURE 3.1 The one-line diagram of a power system.

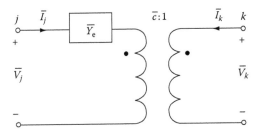

FIGURE 3.2 Off nominal turns ratio transformer.

3.3 Formulation of Bus Admittance Matrix

The first step in developing the mathematical model describing the power flow in the network is the formulation of the bus admittance matrix. The bus admittance matrix is an $n \times n$ matrix (where n is the number of buses in the system) constructed from the admittances of the equivalent circuit elements of the segments making up the power system. Most system segments are represented by a combination of shunt elements (connected between a bus and the reference node) and series elements (connected between two system buses). Formulation of the bus admittance matrix follows two simple rules:

1. The admittance of elements connected between node k and reference is added to the (k, k) entry of the admittance matrix.
2. The admittance of elements connected between nodes j and k is added to the (j, j) and (k, k) entries of the admittance matrix. The negative of the admittance is added to the (j, k) and (k, j) entries of the admittance matrix.

Off nominal transformers (transformers with transformation ratios different from the system voltage bases at the terminals) present some special difficulties. Figure 3.2 shows a representation of an off nominal turns ratio transformer.

The admittance matrix base mathematical model of an isolated off nominal transformer is

$$\begin{bmatrix} \bar{I}_j \\ \bar{I}_k \end{bmatrix} = \begin{bmatrix} \bar{Y}_e & -\bar{c}\,\bar{Y}_e \\ -\bar{c}^*\bar{Y}_e & |\bar{c}|^2\,\bar{Y}_e \end{bmatrix} \begin{bmatrix} \bar{V}_j \\ \bar{V}_k \end{bmatrix} \tag{3.1}$$

where
    $\bar{Y}_e$ is the equivalent series admittance (referred to node j)
    $\bar{c}$ is the complex (off nominal) turns ratio
    $\bar{I}_j$ is the current injected at node j
    $\bar{V}_j$ is the voltage at node j (with respect to reference)

Off nominal transformers are added to the bus admittance matrix by adding the corresponding entry of the isolated off nominal transformer admittance matrix to the system bus admittance matrix.

3.4 Formulation of Power Flow Equations

Considerable insight into the power flow problem and its properties and characteristics can be obtained by consideration of a simple example before proceeding to a general formulation of the problem. This simple case will also serve to establish some notation.

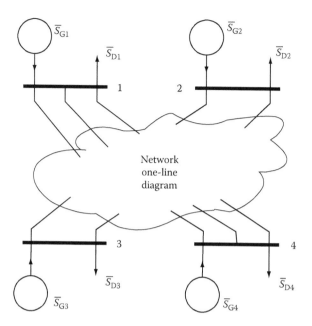

FIGURE 3.3   Conceptual one-line diagram of a four-bus power system.

A conceptual representation of a one-line diagram for a four-bus power system is shown in Figure 3.3. For generality, we have shown a generator and a load connected to each bus. The following notation applies:

$\overline{S}_{GI}$ = Complex power flow into bus 1 from the generator
$\overline{S}_{DI}$ = Complex power flow into the load from bus 1

Comparable quantities for the complex power generations and loads are obvious for each of the three other buses.

The positive sequence network for the power system represented by the one-line diagram of Figure 3.3 is shown in Figure 3.4. The boxes symbolize the combination of generation and load. Network texts refer to this network as a five-node network. (The balanced nature of the system allows analysis using only the positive sequence network; reducing each three-phase bus to a single node. The reference or ground represents the fifth node.) However, in power systems literature it is usually referred to as a four-bus network or power system.

For the network of Figure 3.4, we define the following additional notation:

$\overline{S}_1 = \overline{S}_{G1} - \overline{S}_{D1}$ = Net complex power injected at bus 1
$\overline{I}_1$ = Net positive sequence phasor current injected at bus 1
$\overline{V}_1$ = Positive sequence phasor voltage at bus 1

The standard node voltage equations for the network can be written in terms of the quantities at bus 1 (defined above) and comparable quantities at the other buses:

$$\overline{I}_1 = \overline{Y}_{11}\overline{V}_1 + \overline{Y}_{12}\overline{V}_2 + \overline{Y}_{13}\overline{V}_3 + \overline{Y}_{14}\overline{V}_4 \tag{3.2}$$

$$\overline{I}_2 = \overline{Y}_{21}\overline{V}_1 + \overline{Y}_{22}\overline{V}_2 + \overline{Y}_{23}\overline{V}_3 + \overline{Y}_{24}\overline{V}_4 \tag{3.3}$$

$$\overline{I}_3 = \overline{Y}_{31}\overline{V}_1 + \overline{Y}_{32}\overline{V}_2 + \overline{Y}_{33}\overline{V}_3 + \overline{Y}_{34}\overline{V}_4 \tag{3.4}$$

$$\overline{I}_4 = \overline{Y}_{41}\overline{V}_1 + \overline{Y}_{42}\overline{V}_2 + \overline{Y}_{43}\overline{V}_3 + \overline{Y}_{44}\overline{V}_4 \tag{3.5}$$

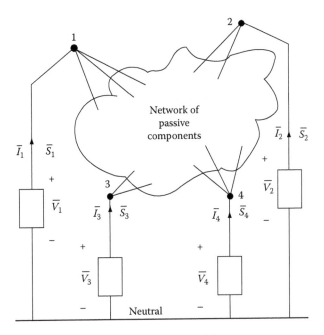

FIGURE 3.4 Positive sequence network for the system in Figure 3.3.

The admittances in Equations 3.2 through 3.5, $\bar{Y}_{ij}$, are the *ij*th entries of the bus admittance matrix for the power system. The unknown voltages could be found using linear algebra if the four currents $\bar{I}_1 \ldots \bar{I}_4$ were known. However, these currents are not known. Rather, something is known about the complex power and voltage, at each bus. The complex power injected into bus *k* of the power system is defined by the relationship between complex power, voltage, and current given by the following equation:

$$\bar{S}_k = \bar{V}_k \bar{I}_k^*$$ (3.6)

Therefore,

$$\bar{I}_k = \frac{\bar{S}_k^*}{\bar{V}_k^*} = \frac{\bar{S}_{Gk}^* - \bar{S}_{Dk}^*}{\bar{V}_k^*}$$ (3.7)

By substituting this result into the nodal equations and rearranging, the basic power flow equations (PFE) for the four-bus system are given as follows:

$$\bar{S}_{G1}^* - \bar{S}_{D1}^* = \bar{V}_1^*[\bar{Y}_{11}\bar{V}_1 + \bar{Y}_{12}\bar{V}_2 + \bar{Y}_{13}\bar{V}_3 + \bar{Y}_{14}\bar{V}_4]$$ (3.8)

$$\bar{S}_{G2}^* - \bar{S}_{D2}^* = \bar{V}_2^*[\bar{Y}_{21}\bar{V}_1 + \bar{Y}_{22}\bar{V}_2 + \bar{Y}_{23}\bar{V}_3 + \bar{Y}_{24}\bar{V}_4]$$ (3.9)

$$\bar{S}_{G3}^* - \bar{S}_{D3}^* = \bar{V}_3^*[\bar{Y}_{31}\bar{V}_1 + \bar{Y}_{32}\bar{V}_2 + \bar{Y}_{33}\bar{V}_3 + \bar{Y}_{34}\bar{V}_4]$$ (3.10)

$$\bar{S}_{G4}^* - \bar{S}_{D4}^* = \bar{V}_4^*[\bar{Y}_{41}\bar{V}_1 + \bar{Y}_{42}\bar{V}_2 + \bar{Y}_{43}\bar{V}_3 + \bar{Y}_{44}\bar{V}_4]$$ (3.11)

Examination of Equations 3.8 through 3.11 reveals that except for the trivial case where the generation equals the load at every bus, the complex power outputs of the generators cannot be arbitrarily selected. In fact, the complex power output of at least one of the generators must be calculated last since it must take up the unknown "slack" due to the, as yet, uncalculated network losses. Further, losses cannot be calculated until the voltages are known. These observations are the result of the principle of conservation of complex power (i.e., the sum of the injected complex powers at the four system buses is equal to the system complex power losses).

Further examination of Equations 3.8 through 3.11 indicates that it is not possible to solve these equations for the absolute phase angles of the phasor voltages. This simply means that the problem can only be solved to some arbitrary phase angle reference.

In order to alleviate the dilemma outlined above, suppose $\overline{S}_{G4}$ is arbitrarily allowed to float or swing (in order to take up the necessary slack caused by the losses) and that $\overline{S}_{G1}$, $\overline{S}_{G2}$, and $\overline{S}_{G3}$ are specified (other cases will be considered shortly). Now, with the loads known, Equations 3.8 through 3.11 are seen as four simultaneous nonlinear equations with complex coefficients in five unknowns $\overline{V}_1$, $\overline{V}_2$, $\overline{V}_3$, $\overline{V}_4$, and $\overline{S}_{G4}$.

The problem of too many unknowns (which would result in an infinite number of solutions) is solved by specifying another variable. Designating bus 4 as the slack bus and specifying the voltage $\overline{V}_4$ reduces the problem to four equations in four unknowns. The slack bus is chosen as the phase reference for all phasor calculations, its magnitude is constrained, and the complex power generation at this bus is free to take up the slack necessary in order to account for the system real and reactive power losses.

The specification of the voltage $\overline{V}_4$ decouples Equation 3.11 from Equations 3.8 through 3.10, allowing calculation of the slack bus complex power after solving the remaining equations. (This property carries over to larger systems with any number of buses.) The example problem is reduced to solving only three equations simultaneously for the unknowns $\overline{V}_1$, $\overline{V}_2$, and $\overline{V}_3$. Similarly, for the case of n buses it is necessary to solve $n-1$ simultaneous, complex coefficient, nonlinear equations.

Systems of nonlinear equations, such as Equations 3.8 through 3.10, cannot (except in rare cases) be solved by closed-form techniques. Direct simulation was used extensively for many years; however, essentially all power flow analyses today are performed using iterative techniques on digital computers.

3.5 *P–V* Buses

In all realistic cases, the voltage magnitude is specified at generator buses to take advantage of the generator's reactive power capability. Specifying the voltage magnitude at a generator bus requires a variable specified in the simple analysis discussed earlier to become an unknown (in order to bring the number of unknowns back into correspondence with the number of equations). Normally, the reactive power injected by the generator becomes a variable, leaving the real power and voltage magnitude as the specified quantities at the generator bus.

It was noted earlier that Equation 3.11 is decoupled and only Equations 3.8 through 3.10 need be solved simultaneously. Although not immediately apparent, specifying the voltage magnitude at a bus and treating the bus reactive power injection as a variable result in retention of, effectively, the same number of complex unknowns. For example, if the voltage magnitude of bus 1 of the earlier four-bus system is specified and the reactive power injection at bus 1 becomes a variable, Equations 3.8 through 3.10 again effectively have three complex unknowns. (The phasor voltages $\overline{V}_2$ and $\overline{V}_3$ at buses 2 and 3 are two complex unknowns and the angle δ_1 of the voltage at bus 1 plus the reactive power generation Q_{G1} at bus 1 result in the equivalent of a third complex unknown.)

Bus 1 is called a *voltage controlled bus*, since it is apparent that the reactive power generation at bus 1 is being used to control the voltage magnitude. This type of bus is also referred to as a *P–V* bus because of the specified quantities. Typically, all generator buses are treated as voltage controlled buses.

3.6 Bus Classifications

There are four quantities of interest associated with each bus:

1. Real power, P
2. Reactive power, Q
3. Voltage magnitude, V
4. Voltage angle, δ

At every bus of the system two of these four quantities will be specified and the remaining two will be unknowns. Each of the system buses may be classified in accordance with which of the two quantities are specified. The following classifications are typical:

Slack bus—The slack bus for the system is a single bus for which the voltage magnitude and angle are specified. The real and reactive power are unknowns. The bus selected as the slack bus must have a source of both real and reactive power, since the injected power at this bus must "swing" to take up the "slack" in the solution. The best choice for the slack bus (since, in most power systems, many buses have real and reactive power sources) requires experience with the particular system under study. The behavior of the solution is often influenced by the bus chosen. (In the earlier discussion, the last bus was selected as the slack bus for convenience.)

Load bus (P–Q bus)—A load bus is defined as any bus of the system for which the real and reactive powers are specified. Load buses may contain generators with specified real and reactive power outputs; however, it is often convenient to designate any bus with specified injected complex power as a load bus.

Voltage-controlled bus (P–V bus)—Any bus for which the voltage magnitude and the injected real power are specified is classified as a voltage controlled (or P–V) bus. The injected reactive power is a variable (with specified upper and lower bounds) in the power flow analysis. (A P–V bus must have a variable source of reactive power such as a generator or a capacitor bank.)

3.7 Generalized Power Flow Development

The more general (n bus) case is developed by extending the results of the simple four-bus example. Consider the case of an n-bus system and the corresponding $n + 1$ node positive sequence network. Assume that the buses are numbered such that the slack bus is numbered last. Direct extension of the earlier equations (writing the node voltage equations and making the same substitutions as in the four-bus case) yields the basic power flow equations in the general form.

3.7.1 Basic Power Flow Equations

$$\bar{S}_k^* = P_k - jQ_k = \bar{V}_k^* \sum_{i=1}^{n} \bar{Y}_{ki}\bar{V}_i \quad \text{for } k = 1, 2, 3, \ldots, n-1 \tag{3.12}$$

and

$$P_n - jQ_n = \bar{V}_n^* \sum_{i=1}^{n} \bar{Y}_{ni}\bar{V}_i \tag{3.13}$$

Equation 3.13 is the equation for the slack bus. Equation 3.12 represents $n-1$ simultaneous equations in $n-1$ complex unknowns if all buses (other than the slack bus) are classified as load buses. Thus, given a set of specified loads, the problem is to solve Equation 3.12 for the $n-1$ complex phasor voltages at the remaining buses. Once the bus voltages are known, Equation 3.13 can be used to calculate the slack bus power.

Bus j is normally treated as a *P–V* bus if it has a directly connected generator. The unknowns at bus j are then the reactive generation Q_{Gj} and δ_j, because the voltage magnitude, V_j, and the real power generation, P_{Gj}, have been specified.

The next step in the analysis is to solve Equation 3.12 for the bus voltages using some iterative method. Once the bus voltages have been found, the complex power flows and complex power losses in all of the network components are calculated.

3.8 Solution Methods

The solution of the simultaneous nonlinear power flow equations requires the use of iterative techniques for even the simplest power systems. Although there are many methods for solving nonlinear equations, only two methods are discussed here.

3.8.1 Newton–Raphson Method

The Newton–Raphson algorithm has been applied in the solution of nonlinear equations in many fields. The algorithm will be developed using a general set of two equations (for simplicity). The results are easily extended to an arbitrary number of equations.

A set of two nonlinear equations are shown in the following equations:

$$f_1(x_1, x_2) = k_1 \tag{3.14}$$

$$f_2(x_1, x_2) = k_2 \tag{3.15}$$

Now, if $x_1^{(0)}$ and $x_2^{(0)}$ are inexact solution estimates and $\Delta x_1^{(0)}$ and $\Delta x_2^{(0)}$ are the corrections to the estimates to achieve an exact solution, Equations 3.14 and 3.15 can be rewritten as

$$f_1(x_1^{(0)} + \Delta x_1^{(0)}, x_2^{(0)} + \Delta x_2^{(0)}) = k_1 \tag{3.16}$$

$$f_2(x_1^{(0)} + \Delta x_1^{(0)}, x_2^{(0)} + \Delta x_2^{(0)}) = k_2 \tag{3.17}$$

Expanding Equations 3.16 and 3.17 in a Taylor series about the estimate yields:

$$f_1\left(x_1^{(0)}, x_2^{(0)}\right) + \left.\frac{\partial f_1}{\partial x_1}\right|^{(0)} \Delta x_1^{(0)} + \left.\frac{\partial f_1}{\partial x_2}\right|^{(0)} \Delta x_2^{(0)} + \text{h.o.t.} = k_1 \tag{3.18}$$

$$f_2\left(x_1^{(0)}, x_2^{(0)}\right) + \left.\frac{\partial f_2}{\partial x_1}\right|^{(0)} \Delta x_1^{(0)} + \left.\frac{\partial f_2}{\partial x_2}\right|^{(0)} \Delta x_2^{(0)} + \text{h.o.t.} = k_2 \tag{3.19}$$

where the subscript, (0), on the partial derivatives indicates evaluation of the partial derivatives at the initial estimate and h.o.t. indicates the higher-order terms.

Neglecting the higher-order terms (an acceptable approximation if $\Delta x_1^{(0)}$ and $\Delta x_2^{(0)}$ are small) Equations 3.18 and 3.19 can be rearranged and written in matrix form:

$$\begin{bmatrix} \left.\dfrac{\partial f_1}{\partial x_1}\right|^{(0)} & \left.\dfrac{\partial f_1}{\partial x_2}\right|^{(0)} \\ \left.\dfrac{\partial f_2}{\partial x_1}\right|^{(0)} & \left.\dfrac{\partial f_2}{\partial x_2}\right|^{(0)} \end{bmatrix} \begin{bmatrix} \Delta x_1^{(0)} \\ \Delta x_2^{(0)} \end{bmatrix} \approx \begin{bmatrix} k_1 - f_1(x_1^{(0)}, x_2^{(0)}) \\ k_2 - f_2(x_1^{(0)}, x_2^{(0)}) \end{bmatrix} \tag{3.20}$$

The matrix of partial derivatives in Equation 3.20 is known as the Jacobian matrix and is evaluated at the initial estimate. Multiplying each side of Equation 3.20 by the inverse of the Jacobian matrix yields an approximation of the required correction to the estimated solution. Since the higher-order terms were neglected, addition of the correction terms to the original estimate will not yield an exact solution, but will often provide an improved estimate. The procedure may be repeated, obtaining successively better estimates until the estimated solution reaches a desired tolerance. Summarizing, correction terms for the ℓth iterate are given in Equation 3.21 and the solution estimate is updated according to Equation 3.22:

$$
\begin{bmatrix} \Delta x_1^{(\ell)} \\ \Delta x_2^{(\ell)} \end{bmatrix} = \begin{bmatrix} \left.\dfrac{\partial f_1}{\partial x_1}\right|^{(\ell)} & \left.\dfrac{\partial f_1}{\partial x_2}\right|^{(\ell)} \\ \left.\dfrac{\partial f_2}{\partial x_1}\right|^{(\ell)} & \left.\dfrac{\partial f_2}{\partial x_2}\right|^{(\ell)} \end{bmatrix}^{-1} \begin{bmatrix} k_1 - f_1(x_1^{(\ell)}, x_2^{(\ell)}) \\ k_2 - f_2(x_1^{(\ell)}, x_2^{(\ell)}) \end{bmatrix}
\tag{3.21}
$$

$$
x^{(\ell+1)} = x^{(\ell)} + \Delta x^{(\ell)}
\tag{3.22}
$$

The solution of the original set of nonlinear equations has been converted to a repeated solution of a system of linear equations. This solution requires evaluation of the Jacobian matrix (at the current solution estimate) in each iteration.

The power flow equations can be placed into the Newton–Raphson framework by separating the power flow equations into their real and imaginary parts and taking the voltage magnitudes and phase angles as the unknowns. Writing Equation 3.21 specifically for the power flow problem:

$$
\begin{bmatrix} \Delta \underline{\delta}^{(\ell)} \\ \Delta \underline{V}^{(\ell)} \end{bmatrix} = \begin{bmatrix} \left.\dfrac{\partial \underline{P}}{\partial \underline{\delta}}\right|^{(\ell)} & \left.\dfrac{\partial \underline{P}}{\partial \underline{V}}\right|^{(\ell)} \\ \left.\dfrac{\partial \underline{Q}}{\partial \underline{\delta}}\right|^{(\ell)} & \left.\dfrac{\partial \underline{Q}}{\partial \underline{V}}\right|^{(\ell)} \end{bmatrix}^{-1} \begin{bmatrix} \underline{P}(\text{sched}) - \underline{P}^{(\ell)} \\ \underline{Q}(\text{sched}) - \underline{Q}^{(\ell)} \end{bmatrix}
\tag{3.23}
$$

The underscored variables in Equation 3.23 indicate vectors (extending the two equation Newton–Raphson development to the general power flow case). The (sched) notation indicates the scheduled real and reactive powers injected into the system. $P^{(\ell)}$ and $Q^{(\ell)}$ represent the calculated real and reactive power injections based on the system model and the ℓth voltage phase angle and voltage magnitude estimates. The bus voltage phase angle and bus voltage magnitude estimates are updated, the Jacobian reevaluated, and the mismatch between the scheduled and calculated real and reactive powers evaluated in each iteration of the Newton–Raphson algorithm. Iterations are performed until the estimated solution reaches an acceptable tolerance or a maximum number of allowable iterations is exceeded. Once a solution (within an acceptable tolerance) is reached, P–V bus reactive power injections and the slack bus complex power injection may be evaluated.

3.8.2 Fast Decoupled Power Flow Solution

The fast decoupled power flow algorithm simplifies the procedure presented for the Newton–Raphson algorithm by exploiting the strong coupling between real power and bus voltage phase angles and reactive power and bus voltage magnitudes commonly seen in power systems. The Jacobian matrix is simplified by approximating as zero the partial derivatives of the real power equations with respect to the bus voltage magnitudes. Similarly, the partial derivatives of the reactive power equations with respect to the bus voltage phase angles are approximated as zero. Further, the remaining partial derivatives are often approximated using only the imaginary portion of the bus admittance matrix. These approximations yield the following correction equations:

$$\Delta\delta^{(\ell)} = [B']^{-1}[\underline{P}(\text{sched}) - \underline{P}^{(\ell)}]$$ (3.24)

$$\Delta V^{(\ell)} = [B'']^{-1}[\underline{Q}(\text{sched}) - \underline{Q}^{(\ell)}]$$ (3.25)

where B' is an approximation of the matrix of partial derivatives of the real power flow equations with respect to the bus voltage phase angles and B'' is an approximation of the matrix of partial derivatives of the reactive power flow equations with respect to the bus voltage magnitudes. B' and B'' are typically held constant during the iterative process, eliminating the necessity of updating the Jacobian matrix (required in the Newton–Raphson solution) in each iteration.

The fast decoupled algorithm has good convergence properties despite the many approximations used during its development. The fast decoupled power flow algorithm has found widespread use, since it is less computationally intensive (requires fewer computational operations) than the Newton–Raphson method.

3.9 Component Power Flows

The positive sequence network for components of interest (connected between buses i and j) will be of the form shown in Figure 3.5.

An admittance description is usually available from earlier construction of the nodal admittance matrix. Thus,

$$\begin{bmatrix} \overline{I}_i \\ \overline{I}_j \end{bmatrix} = \begin{bmatrix} \overline{Y}_a & \overline{Y}_b \\ \overline{Y}_c & \overline{Y}_d \end{bmatrix} \begin{bmatrix} \overline{V}_i \\ \overline{V}_j \end{bmatrix}$$ (3.26)

Therefore the complex power flows and the component loss are

$$\overline{S}_{ij} = \overline{V}_i \overline{I}_i^* = \overline{V}_i [\overline{Y}_a \overline{V}_i + \overline{Y}_b \overline{V}_j]^*$$ (3.27)

$$\overline{S}_{ji} = \overline{V}_j \overline{I}_j^* = \overline{V}_j [\overline{Y}_c \overline{V}_i + \overline{Y}_d \overline{V}_j]^*$$ (3.28)

$$\overline{S}_{\text{loss}} = \overline{S}_{ij} + \overline{S}_{ji}$$ (3.29)

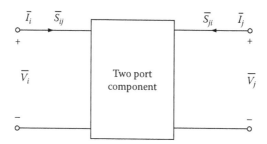

FIGURE 3.5   Typical power system component.

The calculated component flows combined with the bus voltage magnitudes and phase angles provide extensive information about the power systems operating point. The bus voltage magnitudes may be checked to ensure operation within a prescribed range. The segment power flows can be examined to ensure no equipment ratings are exceeded. Additionally, the power flow solution may be used as the starting point for other analyses.

An elementary discussion of the power flow problem and its solution is presented in this section. The power flow problem can be complicated by the addition of further constraints such as generator real and reactive power limits. However, discussion of such complications is beyond the scope of this chapter. The references provide detailed development of power flow formulation and solution under additional constraints. The references also provide some background in the other types of power system analysis.

References

Bergen, A.R. and Vital, V., *Power Systems Analysis*, 2nd edn., Prentice-Hall, Inc., Englewood Cliffs, NJ, 2000.

Elgerd, O.I., *Electric Energy Systems Theory—An Introduction*, 2nd edn., McGraw-Hill, New York, 1982.

Glover, J.D. and Sarma, M., *Power System Analysis and Design*, 3rd edn., Brooks/Cole, Pacific Grove, CA, 2002.

Grainger, J.J. and Stevenson, W.D., *Elements of Power System Analysis*, McGraw-Hill, New York, 1994.

Gross, C.A., *Power System Analysis*, 2nd edn., John Wiley & Sons, New York, 1986.

Further Information

The references provide clear introductions to the analysis of power systems. An excellent review of many issues involving the use of computers for power system analysis is provided in July 1974, *Proceedings of the IEEE* (Special Issue on Computers in the Power Industry). The quarterly journal *IEEE Transactions on Power Systems* provides excellent documentation of more recent research in power system analysis.

4

Fault Analysis
in Power Systems

Charles A. Gross
Auburn University

A *fault* in an electrical power system is the unintentional and undesirable creation of a conducting path (a *short circuit*) or a blockage of current (an *open circuit*). The short-circuit fault is typically the most common and is usually implied when most people use the term *fault*. We restrict our comments to the short-circuit fault.

The causes of faults include lightning, wind damage, trees falling across lines, vehicles colliding with towers or poles, birds shorting out lines, aircraft colliding with lines, vandalism, small animals entering switchgear, and line breaks due to excessive ice loading. Power system faults may be categorized as one of four types: single line-to-ground, line-to-line, double line-to-ground, and balanced three-phase. The first three types constitute severe unbalanced operating conditions.

It is important to determine the values of system voltages and currents during faulted conditions so that protective devices may be set to detect and minimize their harmful effects. The time constants of the associated transients are such that sinusoidal steady-state methods may still be used. The method of symmetrical components is particularly suited to fault analysis.

Our objective is to understand how symmetrical components may be applied specifically to the four general fault types mentioned and how the method can be extended to any unbalanced three-phase system problem.

Note that phase values are indicated by subscripts, *a*, *b*, *c*; sequence (symmetrical component) values are indicated by subscripts 0, 1, 2. The transformation is defined by

$$
\begin{bmatrix} \overline{V}_a \\ \overline{V}_b \\ \overline{V} \end{bmatrix} = \begin{bmatrix} 1 & 1 & 1 \\ 1 & a^2 & a \\ 1 & a & a^2 \end{bmatrix} \begin{bmatrix} \overline{V}_0 \\ \overline{V}_1 \\ \overline{V}_2 \end{bmatrix} = [T] \begin{bmatrix} \overline{V}_0 \\ \overline{V}_1 \\ \overline{V}_2 \end{bmatrix}
$$

4.1 Simplifications in the System Model

Certain simplifications are possible and usually employed in fault analysis.

- Transformer magnetizing current and core loss will be neglected.
- Line shunt capacitance is neglected.
- Sinusoidal steady-state circuit analysis techniques are used. The so-called *DC offset* is accounted for by using correction factors.
- Prefault voltage is assumed to be $1\angle 0°$ per-unit. One per-unit voltage is at its nominal value prior to the application of a fault, which is reasonable. The selection of zero phase is arbitrary and convenient. Prefault load current is neglected.

For hand calculations, neglect series resistance is usually neglected (this approximation will not be necessary for a computer solution). Also, the only difference in the positive and negative sequence networks is introduced by the machine impedances. If we select the subtransient reactance X_d'' for the positive sequence reactance, the difference is slight (in fact, the two are identical for nonsalient machines). The simplification is important, since it reduces computer storage requirements by roughly one-third. Circuit models for generators, lines, and transformers are shown in Figures 4.1 through 4.3, respectively.

Our basic approach to the problem is to consider the general situation suggested in Figure 4.4a. The general terminals brought out are for purposes of external connections that will simulate faults. Note carefully the positive assignments of phase quantities. Particularly note that the currents flow *out of* the system. We can construct general *sequence* equivalent circuits for the system, and such circuits are

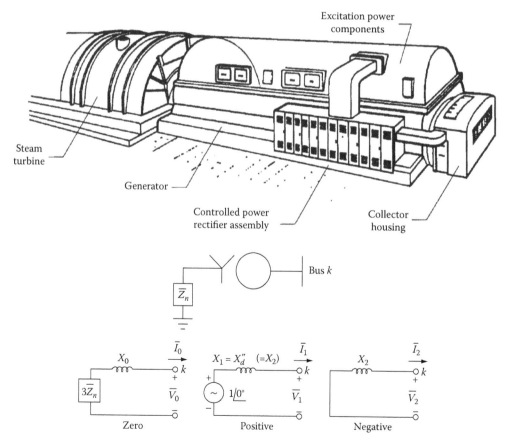

FIGURE 4.1   Generator sequence circuit models.

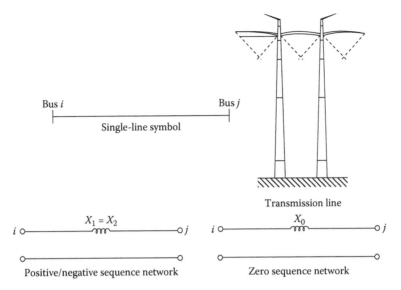

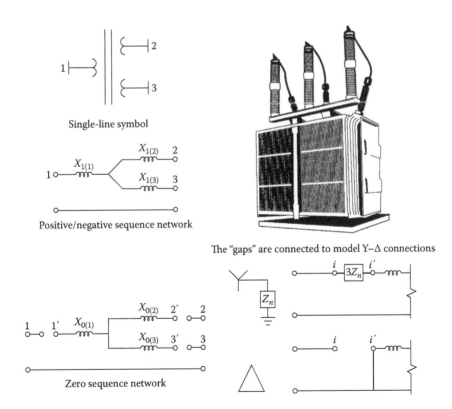

FIGURE 4.2   Line sequence circuit models.

FIGURE 4.3   Transformer sequence circuit models.

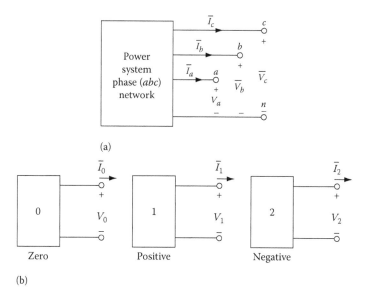

(a)

(b)

FIGURE 4.4   General fault port in an electric power system. (a) General fault port in phase (*abc*) coordinates; (b) corresponding fault ports in sequence (012) coordinates.

indicated in Figure 4.4b. The ports indicated correspond to the general three-phase entry port of Figure 4.4a. The positive sense of sequence values is compatible with that used for phase values.

4.2 The Four Basic Fault Types

4.2.1 The Balanced Three-Phase Fault

Imagine the general three-phase access port terminated in a fault impedance $(\overline{Z}_f)$ as shown in Figure 4.5a. The terminal conditions are

$$\begin{bmatrix} \overline{V}_a \\ \overline{V}_b \\ \overline{V}_c \end{bmatrix} = \begin{bmatrix} \overline{Z}_f & 0 & 0 \\ 0 & \overline{Z}_f & 0 \\ 0 & 0 & \overline{Z}_f \end{bmatrix} \begin{bmatrix} \overline{I}_a \\ \overline{I}_b \\ \overline{I}_c \end{bmatrix}$$

Transforming to $[Z_{012}]$,

$$[Z_{012}] = [T]^{-1} \begin{bmatrix} \overline{Z}_f & 0 & 0 \\ 0 & \overline{Z}_f & 0 \\ 0 & 0 & \overline{Z}_f \end{bmatrix} [T] = \begin{bmatrix} \overline{Z}_f & 0 & 0 \\ 0 & \overline{Z}_f & 0 \\ 0 & 0 & \overline{Z}_f \end{bmatrix}$$

The corresponding network connections are given in Figure 4.6a. Since the zero and negative sequence networks are passive, only the positive sequence network is nontrivial.

$$\overline{V}_0 = \overline{V}_2 = 0 \tag{4.1}$$

$$\overline{I}_0 = \overline{I}_2 = 0 \tag{4.2}$$

$$\overline{V}_1 = \overline{Z}_f \overline{I}_1 \tag{4.3}$$

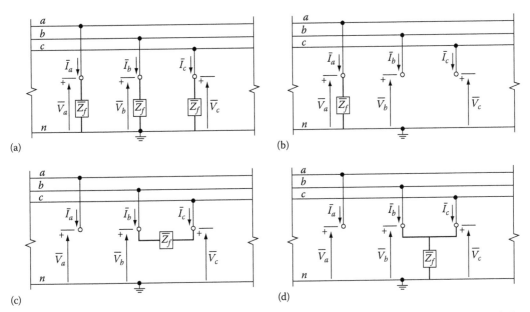

FIGURE 4.5 Fault types. (a) Three-phase fault; (b) single phase-to-ground fault; (c) phase-to-phase fault; (d) double phase-to-ground fault.

4.2.2 The Single Phase-to-Ground Fault

Imagine the general three-phase access port terminated as shown in Figure 4.5b. The terminal conditions are

$$\bar{I}_b = 0 \quad \bar{I}_c = 0 \quad \bar{V}_a = \bar{I}_a \bar{Z}_f$$

Therefore,

$$\bar{I}_0 + a^2 \bar{I}_1 + a \bar{I}_2 = \bar{I}_0 + a \bar{I}_1 + a^2 \bar{I}_2 = 0$$

or

$$\bar{I}_1 = \bar{I}_2$$

Also,

$$\bar{I}_b = \bar{I}_0 + a^2 \bar{I}_1 + a \bar{I}_2 = \bar{I}_0 + (a^2 + a)\bar{I}_1 = 0$$

or

$$\bar{I}_0 = \bar{I}_1 = \bar{I}_2 \tag{4.4}$$

Furthermore, it is required that

$$\bar{V}_a = \bar{Z}_f \bar{I}_a$$

$$\bar{V}_0 + \bar{V}_1 + \bar{V}_2 = 3\bar{Z}_f \bar{I}_1 \tag{4.5}$$

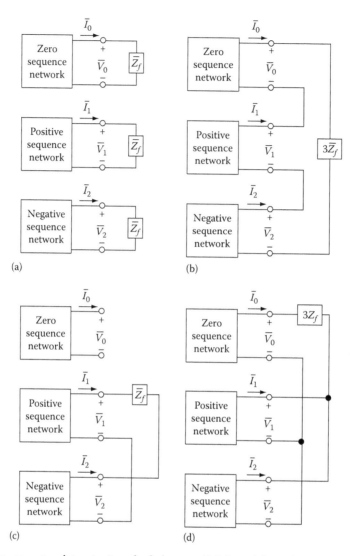

FIGURE 4.6 Sequence network terminations for fault types. (a) Balanced three-phase fault; (b) single phase-to-ground fault; (c) phase-to-phase fault; (d) double phase-to-ground fault.

In general then, Equations 4.4 and 4.5 must be simultaneously satisfied. These conditions can be met by interconnecting the sequence networks as shown in Figure 4.6b.

4.2.3 The Phase-to-Phase Fault

Imagine the general three-phase access port terminated as shown in Figure 4.5c. The terminal conditions are such that we may write

$$\bar{I}_0 = 0 \quad \bar{I}_b = -\bar{I}_c \quad \bar{V}_b = \bar{Z}_f \bar{I}_b + \bar{V}_c$$

It follows that

$$\bar{I}_0 + \bar{I}_1 + \bar{I}_2 = \bar{0} \tag{4.6}$$

$$\overline{I}_0 = 0 \qquad (4.7)$$

$$\overline{I}_1 = -\overline{I}_2 \qquad (4.8)$$

In general then, Equations 4.6 through 4.8 must be simultaneously satisfied. The proper interconnection between sequence networks appears in Figure 4.6c.

4.2.4 The Double Phase-to-Ground Fault

Consider the general three-phase access port terminated as shown in Figure 4.5d. The terminal conditions indicate

$$\overline{I}_a = 0 \quad \overline{V}_b = \overline{V}_c \quad \overline{V}_b = (\overline{I}_b + \overline{I}_c)\overline{Z}_f$$

It follows that

$$\overline{I}_0 + \overline{I}_1 + \overline{I}_2 = \overline{0} \qquad (4.9)$$

$$\overline{V}_1 = \overline{V}_2 \qquad (4.10)$$

and

$$\overline{V}_0 - \overline{V}_1 = 3\overline{Z}_f \overline{I}_0 \qquad (4.11)$$

For the general double phase-to-ground fault, Equations 4.9 through 4.11 must be simultaneously satisfied. The sequence network interconnections appear in Figure 4.6d.

4.3 An Example Fault Study

Case: EXAMPLE SYSTEM

Run:

System has data for two Line(s); two Transformer(s);

four Bus(es); and two Generator(s)

Transmission line data

Line	Bus	Bus	Seq	R	X	B	Srat
1	2	3	pos	0.00000	0.16000	0.00000	1.0000
			zero	0.00000	0.50000	0.00000	
2	2	3	pos	0.00000	0.16000	0.00000	1.0000
			zero	0.00000	0.50000	0.00000	

Transformer data

Transformer	HV Bus	LV Bus	Seq	R	X	C	Srat
1	2	1	pos	0.00000	0.05000	1.00000	1.0000
	Y	Y	zero	0.00000	0.05000		
2	3	4	pos	0.00000	0.05000	1.00000	1.0000
	Y	D	zero	0.00000	0.05000		

Generator data

No.	Bus	Srated	Ra	X_d''	Xo	Rn	Xn	Con
1	1	1.0000	0.0000	0.200	0.0500	0.0000	0.0400	Y
2	4	1.0000	0.0000	0.200	0.0500	0.0000	0.0400	Y

Zero sequence {Z} matrix

$0.0 + j(0.1144)$	$0.0 + j(0.0981)$	$0.0 + j(0.0163)$	$0.0 + j(0.0000)$
$0.0 + j(0.0981)$	$0.0 + j(0.1269)$	$0.0 + j(0.0212)$	$0.0 + j(0.0000)$
$0.0 + j(0.0163)$	$0.0 + j(0.0212)$	$0.0 + j(0.0452)$	$0.0 + j(0.0000)$
$0.0 + j(0.0000)$	$0.0 + j(0.0000)$	$0.0 + j(0.0000)$	$0.0 + j(0.1700)$

Positive sequence [Z] matrix

$0.0 + j(0.1310)$	$0.0 + j(0.1138)$	$0.0 + j(0.0862)$	$0.0 + j(0.0690)$
$0.0 + j(0.1138)$	$0.0 + j(0.1422)$	$0.0 + j(0.1078)$	$0.0 + j(0.0862)$
$0.0 + j(0.0862)$	$0.0 + j(0.1078)$	$0.0 + j(0.1422)$	$0.0 + j(0.1138)$
$0.0 + j(0.0690)$	$0.0 + j(0.0862)$	$0.0 + j(0.1138)$	$0.0 + j(0.1310)$

The single-line diagram and sequence networks are presented in Figure 4.7.

Suppose bus 3 in the example system represents the fault location and $\bar{Z}_f = 0$. The positive sequence circuit can be reduced to its Thévenin equivalent at bus 3:

$$E_{T1} = 1.0\angle 0° \quad \bar{Z}_{T1} = j0.1422$$

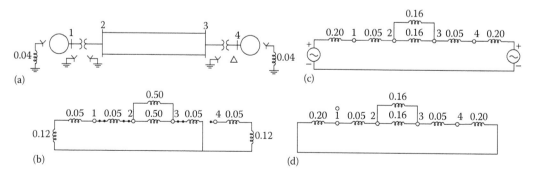

FIGURE 4.7 Example system. (a) Single-line diagram; (b) zero sequence network; (c) positive sequence network; (d) negative sequence network.

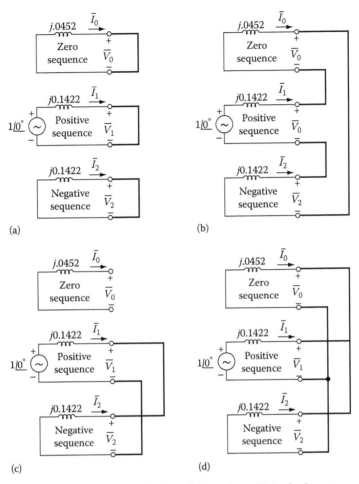

FIGURE 4.8 Example system faults at bus 3. (a) Balanced three-phase; (b) single phase-to-ground; (c) phase-to-phase; (d) double phase-to-ground.

Similarly, the negative and zero sequence Thévenin elements are

$$\overline{E}_{T2} = 0 \quad \overline{Z}_{T2} = j0.1422$$

$$\overline{E}_{T0} = 0 \quad \overline{Z}_{T0} = j0.0452$$

The network interconnections for the four fault types are shown in Figure 4.8. For each of the fault types, compute the currents and voltages at the faulted bus.

4.3.1 Balanced Three-Phase Fault

The sequence networks are shown in Figure 4.8a. Obviously,

$$\overline{V}_0 = \overline{I}_0 = \overline{V}_2 = \overline{I}_2 = 0$$

$$\overline{I}_1 = \frac{1\angle 0^\circ}{j0.1422} = -j7.032; \quad \text{also } \overline{V}_1 = 0$$

To compute the phase values,

$$\begin{bmatrix} \bar{I}_a \\ \bar{I}_b \\ \bar{I}_c \end{bmatrix} = [T]\begin{bmatrix} \bar{I}_0 \\ \bar{I}_1 \\ \bar{I}_2 \end{bmatrix} = \begin{bmatrix} 1 & 1 & 1 \\ 1 & a^2 & a \\ 1 & a & a^2 \end{bmatrix}\begin{bmatrix} 0 \\ -j7.032 \\ 0 \end{bmatrix} = \begin{bmatrix} 7.032\angle-90° \\ 7.032\angle150° \\ 7.032\angle30° \end{bmatrix}$$

$$\begin{bmatrix} \bar{V}_a \\ \bar{V}_b \\ \bar{V}_c \end{bmatrix} = [T]\begin{bmatrix} 0 \\ 0 \\ 0 \end{bmatrix} = \begin{bmatrix} 0 \\ 0 \\ 0 \end{bmatrix}$$

4.3.2 Single Phase-to-Ground Fault

The sequence networks are interconnected as shown in Figure 4.8b.

$$\bar{I}_0 = \bar{I}_1 = \bar{I}_2 = \frac{1\angle0°}{j0.0452 + j0.1422 + j0.1422} = -j3.034$$

$$\begin{bmatrix} \bar{I}_a \\ \bar{I}_b \\ \bar{I}_c \end{bmatrix} = \begin{bmatrix} 1 & 1 & 1 \\ 1 & a^2 & a \\ 1 & a & a^2 \end{bmatrix}\begin{bmatrix} -j3.034 \\ -j3.034 \\ -j3.034 \end{bmatrix} = \begin{bmatrix} -j9.102 \\ 0 \\ 0 \end{bmatrix}$$

The sequence voltages are

$$\bar{V}_0 = -j0.0452(-j3.034) = -1371$$

$$\bar{V}_1 = 1.0 - j0.1422(-j3.034) = 0.5685$$

$$\bar{V}_2 = -j0.1422(-j3.034) = -0.4314$$

The phase voltages are

$$\begin{bmatrix} \bar{V}_a \\ \bar{V}_b \\ \bar{V}_c \end{bmatrix} = \begin{bmatrix} 1 & 1 & 1 \\ 1 & a^2 & a \\ 1 & a & a^2 \end{bmatrix}\begin{bmatrix} -0.1371 \\ 0.5685 \\ -0.4314 \end{bmatrix} = \begin{bmatrix} 0 \\ 0.8901\angle-103.4° \\ 0.8901\angle-103.4° \end{bmatrix}$$

Phase-to-phase and double phase-to-ground fault values are calculated from the appropriate networks (Figure 4.8c and d). Complete results are provided.

Faulted Bus	Phase *a*	Phase *b*	Phase *c*
3	G	G	G

Sequence voltages

Bus	V_0		V_1		V_2	
1	0.0000/	0.0	0.3939/	0.0	0.0000/	0.0
2	0.0000/	0.0	0.2424/	0.0	0.0000/	0.0
3	0.0000/	0.0	0.0000/	0.0	0.0000/	0.0
4	0.0000/	0.0	0.2000/	−30.0	0.0000/	30.0

Phase voltages

Bus	V_a		V_b		V_c	
1	0.3939/	0.0	0.3939/	−120.0	0.3939/	120.0
2	0.2424/	0.0	0.2424/	−120.0	0.2424/	120.0
3	0.0000/	6.5	0.0000/	−151.2	0.0000/	133.8
4	0.2000/	−30.0	0.2000/	−150.0	0.2000/	90.0

Sequence currents

Bus to Bus		I_0		I_1		I_2	
1	2	0.0000/	167.8	3.0303/	−90.0	0.0000/	90.0
1	0	0.0000/	−12.2	3.0303/	90.0	0.0000/	−90.0
2	3	0.0000/	167.8	1.5152/	−90.0	0.0000/	90.0
2	3	0.0000/	167.8	1.5152/	−90.0	0.0000/	90.0
2	1	0.0000/	−12.2	3.0303/	90.0	0.0000/	−90.0
3	2	0.0000/	−12.2	1.5152/	90.0	0.0000/	−90.0
3	2	0.0000/	−12.2	1.5152/	90.0	0.0000/	−90.0
3	4	0.0000/	−12.2	4.0000/	90.0	0.0000/	−90.0
4	3	0.0000/	0.0	4.0000/	−120.0	0.0000/	120.0
4	0	0.0000/	0.0	4.0000/	60.0	0.0000/	−60.0

Faulted Bus	Phase a	Phase b	Phase c
3	G	G	G

Phase currents

Bus to Bus		I_a		I_b		I_c	
1	2	3.0303/	−90.0	3.0303/	150.0	3.0303/	30.0
1	0	3.0303/	90.0	3.0303/	−30.0	3.0303/	−150.0
2	3	1.5151/	−90.0	1.5151/	150.0	1.5151/	30.0
2	3	1.5151/	−90.0	1.5151/	150.0	1.5151/	30.0
2	1	3.0303/	90.0	3.0303/	−30.0	3.0303/	−150.0
3	2	1.5151/	90.0	1.5151/	−30.0	1.5151/	−150.0
3	2	1.5151/	90.0	1.5151/	−30.0	1.5151/	−150.0
3	4	4.0000/	90.0	4.0000/	−30.0	4.0000/	−150.0
4	3	4.0000/	−120.0	4.0000/	120.0	4.0000/	−0.0
4	0	4.0000/	60.0	4.0000/	−60.0	4.0000/	−180.0

Faulted Bus	Phase a	Phase b	Phase c
3	G	0	0

Sequence voltages

Bus	V_0		V_1		V_2	
1	0.0496/	180.0	0.7385/	0.0	0.2615/	180.0
2	0.0642/	180.0	0.6731/	0.0	0.3269/	180.0
3	0.1371/	180.0	0.5685/	0.0	0.4315/	180.0
4	0.0000/	0.0	0.6548/	−30.0	0.3452/	210.0

Phase voltages

Bus	V_a		V_b		V_c	
1	0.4274/	0.0	0.9127/	−108.4	0.9127/	108.4
2	0.2821/	0.0	0.8979/	−105.3	0.8979/	105.3
3	0.0000/	89.2	0.8901/	−103.4	0.8901/	103.4
4	0.5674/	−61.8	0.5674/	−118.2	1.0000/	90.0

Sequence currents

Bus to Bus		I_0		I_1		I_2	
1	2	0.2917/	−90.0	1.3075/	−90.0	1.3075/	−90.0
1	0	0.2917/	90.0	1.3075/	90.0	1.3075/	90.0
2	3	0.1458/	−90.0	0.6537/	−90.0	0.6537/	−90.0
2	3	0.1458/	−90.0	0.6537/	−90.0	0.6537/	−90.0
2	1	0.2917/	90.0	1.3075/	90.0	1.3075/	90.0
3	2	0.1458/	90.0	0.6537/	90.0	0.6537/	90.0
3	2	0.1458/	90.0	0.6537/	90.0	0.6537/	90.0
3	4	2.7416/	90.0	1.7258/	90.0	1.7258/	90.0
4	3	0.0000/	0.0	1.7258/	−120.0	1.7258/	−60.0
4	0	0.0000/	90.0	1.7258/	60.0	1.7258/	120.0

Faulted Bus	Phase a	Phase b	Phase c
3	G	0	0

Phase currents

Bus to Bus		I_a		I_b		I_c	
1	2	2.9066/	−90.0	1.0158/	90.0	1.0158/	90.0
1	0	2.9066/	90.0	1.0158/	−90.0	1.0158/	−90.0
2	3	1.4533/	−90.0	0.5079/	90.0	0.5079/	90.0
2	3	1.4533/	−90.0	0.5079/	90.0	0.5079/	90.0
2	1	2.9066/	90.0	1.0158/	−90.0	1.0158/	−90.0
3	2	1.4533/	90.0	0.5079/	−90.0	0.5079/	−90.0
3	2	1.4533/	90.0	0.5079/	−90.0	0.5079/	−90.0
3	4	6.1933/	90.0	1.0158/	90.0	1.0158/	90.0
4	3	2.9892/	−90.0	2.9892/	90.0	0.0000/	−90.0
4	0	2.9892/	90.0	2.9892/	−90.0	0.0000/	90.0

Faulted Bus	Phase *a*	Phase *b*	Phase *c*
3	0	C	B

Sequence voltages

Bus	V_0		V_1		V_2	
1	0.0000/	0.0	0.6970/	0.0	0.3030/	0.0
2	0.0000/	0.0	0.6212/	0.0	0.3788/	0.0
3	0.0000/	0.0	0.5000/	0.0	0.5000/	0.0
4	0.0000/	0.0	0.6000/	−30.0	0.4000/	30.0

Phase voltages

Bus	V_a		V_b		V_c	
1	1.0000/	0.0	0.6053/	−145.7	0.6053/	145.7
2	1.0000/	0.0	0.5423/	−157.2	0.5423/	157.2
3	1.0000/	0.0	0.5000/	−180.0	0.5000/	−180.0
4	0.8718/	−6.6	0.8718/	−173.4	0.2000/	90.0

Sequence currents

Bus to Bus		I_0		I_1		I_2	
1	2	0.0000/	−61.0	1.5152/	−90.0	1.5152/	90.0
1	0	0.0000/	119.0	1.5152/	90.0	1.5152/	−90.0
2	3	0.0000/	−61.0	0.7576/	−90.0	0.7576/	90.0
2	3	0.0000/	−61.0	0.7576/	−90.0	0.7576/	90.0
2	1	0.0000/	119.0	1.5152/	90.0	1.5152/	−90.0
3	2	0.0000/	119.0	0.7576/	90.0	0.7576/	−90.0
3	2	0.0000/	119.0	0.7576/	90.0	0.7576/	−90.0
3	4	0.0000/	119.0	2.0000/	90.0	2.0000/	−90.0
4	3	0.0000/	0.0	2.0000/	−120.0	2.0000/	120.0
4	0	0.0000/	90.0	2.0000/	60.0	2.0000/	−60.0

Faulted Bus	Phase *a*	Phase *b*	Phase *c*
3	0	C	B

Phase currents

Bus to Bus		I_a		I_b		I_c	
1	2	0.0000/	180.0	2.6243/	180.0	2.6243/	0.0
1	0	0.0000/	180.0	2.6243/	0.0	2.6243/	180.0
2	3	0.0000/	−180.0	1.3122/	180.0	1.3122/	0.0
2	3	0.0000/	−180.0	1.3122/	180.0	1.3122/	0.0
2	1	0.0000/	180.0	2.6243/	0.0	2.6243/	180.0
3	2	0.0000/	−180.0	1.3122/	0.0	1.3122/	180.0
3	2	0.0000/	−180.0	1.3122/	0.0	1.3122/	180.0
3	4	0.0000/	−180.0	3.4641/	0.0	3.4641/	180.0
4	3	2.0000/	−180.0	2.0000/	180.0	4.0000/	0.0
4	0	2.0000/	0.0	2.0000/	0.0	4.0000/	−180.0

Faulted Bus	Phase *a*	Phase *b*	Phase *c*
3	0	G	G

Sequence voltages

Bus	V_0		V_1		V_2	
1	0.0703/	0.0	0.5117/	0.0	0.1177/	0.0
2	0.0909/	0.0	0.3896/	0.0	0.1472/	0.0
3	0.1943/	−0.0	0.1943/	0.0	0.1943/	0.0
4	0.0000/	0.0	0.3554/	−30.0	0.1554/	30.0

Phase voltages

Bus	V_a		V_b		V_c	
1	0.6997/	0.0	0.4197/	−125.6	0.4197/	125.6
2	0.6277/	0.0	0.2749/	−130.2	0.2749/	130.2
3	0.5828/	0.0	0.0000/	−30.7	0.0000/	−139.6
4	0.4536/	−12.7	0.4536/	−167.3	0.2000/	90.0

Sequence currents

Bus to Bus		I_0		I_1		I_2	
1	2	0.4133/	90.0	2.4416/	−90.0	0.5887/	90.0
1	0	0.4133/	−90.0	2.4416/	90.0	0.5887/	−90.0
2	3	0.2067/	90.0	1.2208/	−90.0	0.2943/	90.0
2	3	0.2067/	90.0	1.2208/	−90.0	0.2943/	90.0
2	1	0.4133/	−90.0	2.4416/	90.0	0.5887/	−90.0
3	2	0.2067/	−90.0	1.2208/	90.0	0.2943/	−90.0
3	2	0.2067/	−90.0	1.2208/	90.0	0.2943/	−90.0
3	4	3.8854/	−90.0	3.2229/	90.0	0.7771/	−90.0
4	3	0.0000/	0.0	3.2229/	−120.0	0.7771/	120.0
4	0	0.0000/	−90.0	3.2229/	60.0	0.7771/	−60.0

Faulted Bus	Phase *a*	Phase *b*	Phase *c*
3	0	G	G

Phase currents

Bus to Bus		I_a		I_b		I_c	
1	2	1.4396/	−90.0	2.9465/	153.0	2.9465/	27.0
1	0	1.4396/	90.0	2.9465/	−27.0	2.9465/	−153.0
2	3	0.7198/	−90.0	1.4733/	153.0	1.4733/	27.0
2	3	0.7198/	−90.0	1.4733/	153.0	1.4733/	27.0
2	1	1.4396/	90.0	2.9465/	−27.0	2.9465/	−153.0
3	2	0.7198/	90.0	1.4733/	−27.0	1.4733/	−153.0
3	2	0.7198/	90.0	1.4733/	−27.0	1.4733/	−153.0
3	4	1.4396/	−90.0	6.1721/	−55.9	6.1721/	−124.1
4	3	2.9132/	−133.4	2.9132/	133.4	4.0000/	−0.0
4	0	2.9132/	46.6	2.9132/	−46.6	4.0000/	−180.0

4.4 Further Considerations

Generators are not the only sources in the system. All rotating machines are capable of contributing to fault current, at least momentarily. Synchronous and induction motors will continue to rotate due to inertia and function as sources of fault current. The impedance used for such machines is usually the transient reactance X'_d or the subtransient X''_d, depending on protective equipment and speed of response. Frequently, motors smaller than 50 hp are neglected. Connecting systems are modeled with their Thévenin equivalents.

Although we have used AC circuit techniques to calculate faults, the problem is fundamentally transient since it involves sudden switching actions. Consider the so-called DC offset current. We model the system by determining its positive sequence Thévenin equivalent circuit, looking back into the positive sequence network at the fault, as shown in Figure 4.9. The transient fault current is

$$i(t) = I_{AC}\sqrt{2}\cos(\omega t - \beta) + I_{DC}e^{-t/\tau}$$

This is a first-order approximation and strictly applies only to the three-phase or phase-to-phase fault. Ground faults would involve the zero sequence network also.

$$I_{AC} = \frac{E}{\sqrt{R^2 + X^2}} = \text{rms AC current}$$

$$I_{DC}(t) = I_{DC}e^{-t/\tau} = \text{DC offset current}$$

The maximum initial DC offset possible would be

$$\text{Max } I_{DC} = I_{max} = \sqrt{2}I_{AC}$$

The DC offset will exponentially decay with time constant τ, where

$$\tau = \frac{L}{R} = \frac{X}{\omega R}$$

The maximum DC offset current would be $I_{DC}(t)$

$$I_{DC}(t) = I_{DC}e^{-t/\tau} = \sqrt{2}I_{AC}e^{-t/\tau}$$

The *transient rms* current $I(t)$, accounting for both the AC and DC terms, would be

$$I(t) = \sqrt{I_{AC}^2 + I_{DC}^2(t)} = I_{AC}\sqrt{1 + 2e^{-2t/\tau}}$$

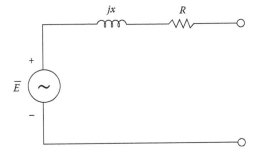

FIGURE 4.9 Positive sequence circuit looking back into faulted bus.

Define a multiplying factor k_i such that I_{AC} is to be multiplied by k_i to estimate the interrupting capacity of a breaker which operates in time T_{op}. Therefore,

$$k_i = \frac{I(T_{op})}{I_{AC}} = \sqrt{1 + 2e^{-2T_{op}/\tau}}$$

Observe that the maximum possible value for k_i is $\sqrt{3}$.

Example 4.1

In the circuit of Figure 4.9, $E = 2400\,V$, $X = 2\,\Omega$, $R = 0.1\,\Omega$, and $f = 60\,Hz$. Compute k_i and determine the interrupting capacity for the circuit breaker if it is designed to operate in two cycles. The fault is applied at $t = 0$.

Solution

$$I_{AC} \cong \frac{2400}{2} = 1200\,A$$

$$T_{op} = \frac{2}{60} = 0.0333s$$

$$\tau = \frac{X}{\omega R} = \frac{2}{37.7} = 0.053$$

$$k_i = \sqrt{1 + 2e^{-2T_{op}/\tau}} = \sqrt{1 + 2e^{-0.0067/0.053}} = 1.252$$

Therefore,

$$I = k_i I_{AC} = 1.252(1200) = 1503\,A$$

The Thévenin equivalent at the fault point is determined by normal sinusoidal steady-state methods, resulting in a first-order circuit as shown in Figure 4.9. While this provides satisfactory results for the steady-state component I_{AC}, the X/R value so obtained can be in serious error when compared with the rate of decay of $I(t)$ as measured by oscillographs on an actual faulted system. The major reasons for the discrepancy are, first of all, that the system, for transient analysis purposes, is actually high-order, and second, the generators do not hold constant impedance as the transient decays.

4.5 Summary

Computation of fault currents in power systems is best done by computer. The major steps are summarized below:

- Collect, read in, and store machine, transformer, and line data in per-unit on common bases.
- Formulate the sequence impedance matrices.
- Define the faulted bus and Z_f. Specify type of fault to be analyzed.
- Compute the sequence voltages.
- Compute the sequence currents.
- Correct for wye-delta connections.
- Transform to phase currents and voltages.

For large systems, computer formulation of the sequence impedance matrices is required. Refer to Further Information for more detail. Zero sequence networks for lines in close proximity to each other (on a common right-of-way) will be mutually coupled. If we are willing to use the same values for positive and negative sequence machine impedances,

$$[Z_1] = [Z_2]$$

Therefore, it is unnecessary to store these values in separate arrays, simplifying the program and reducing the computer storage requirements significantly. The error introduced by this approximation is usually not important. The methods previously discussed neglect the prefault, or load, component of current; that is, the usual assumption is that currents throughout the system were zero prior to the fault. This is almost never strictly true; however, the error produced is small since the fault currents are generally much larger than the load currents. Also, the load currents and fault currents are out of phase with each other, making their sum more nearly equal to the larger components than would have been the case if the currents were in phase. In addition, selection of precise values for prefault currents is somewhat speculative, since there is no way of predicting what the loaded state of the system is when a fault occurs. When it is important to consider load currents, a power flow study is made to calculate currents throughout the system, and these values are superimposed on (added to) results from the fault study.

A term which has wide industrial use and acceptance is the *fault level* or *fault MVA* at a bus. It relates to the amount of current that can be expected to flow out of a bus into a three-phase fault. As such, it is an alternate way of providing positive sequence impedance information. Define

$$\text{Fault level in MVA at bus } i = V_{i_{pu_{nominal}}} I_{i_{pu_{fault}}} S_{3\phi base}$$

$$= (1) \frac{1}{Z_{ii}^1} S_{3\phi base} = \frac{S_{3\phi base}}{Z_{ii}^1}$$

Fault study results may be further refined by approximating the effect of DC offset.

The basic reason for making fault studies is to provide data that can be used to size and set protective devices. The role of such protective devices is to detect and remove faults to prevent or minimize damage to the power system.

4.6 Defining Terms

DC offset—The natural response component of the transient fault current, usually approximated with a first-order exponential expression.

Fault—An unintentional and undesirable conducting path in an electrical power system.

Fault MVA—At a specific location in a system, the initial symmetrical fault current multiplied by the prefault nominal line-to-neutral voltage (×3 for a three-phase system).

Sequence (012) quantities—Symmetrical components computed from phase (*abc*) quantities. Can be voltages, currents, and/or impedances.

References

Anderson P.M., *Analysis of Faulted Power Systems*, Ames, IA: Iowa State Press, 1973.
Elgerd O.I., *Electric Energy Systems Theory: An Introduction*, 2nd edn., New York: McGraw-Hill, 1982.
El-Hawary M.E., *Electric Power Systems: Design and Analysis*, Reston, VA: Reston Publishing, 1983.
El-Hawary M.E., *Electric Power Systems*, New York: IEEE Press, 1995.
General Electric, Short-circuit current calculations for industrial and commercial power systems, Publication GET-3550.

Gross C.A., *Power System Analysis*, 2nd edn., New York: Wiley, 1986.
Horowitz S.H., *Power System Relaying*, 2nd edn., New York: Wiley, 1995.
Lazar I., *Electrical Systems Analysis and Design for Industrial Plants*, New York: McGraw-Hill, 1980.
Mason C.R., *The Art and Science of Protective Relaying*, New York: Wiley, 1956.
Neuenswander J.R., *Modern Power Systems*, Scranton, PA: International Textbook, 1971.
Stagg G. and El-Abiad A.H., *Computer Methods in Power System Analysis*, New York: McGraw-Hill, 1968.
Westinghouse Electric Corporation, *Applied Protective Relaying*, Newark, NJ: Relay-Instrument Division, 1976.
Wood A.J., *Power Generation, Operation, and Control*, New York: Wiley, 1996.

Further Information

For a comprehensive coverage of general fault analysis, see Paul M. Anderson, *Analysis of Faulted Power Systems*, New York: IEEE Press, 1995. Also see Chapters 9 and 10 of *Power System Analysis* by C.A. Gross, New York: Wiley, 1986.

5

Computational Methods for Electric Power Systems

Mariesa L. Crow
Missouri University of
Science and Technology

Electric power systems are some of the largest human-made systems ever built. As with many physical systems, engineers, scientists, economists, and many others strive to understand and predict their complex behavior through mathematical models. The sheer size of the bulk power transmission system forced early power engineers to be among the first to develop computational approaches to solving the equations that describe them. Today's power system planners and operators rely very heavily on the computational tools to assist them in maintaining a reliable and secure operating environment.

The various computational algorithms were developed around the requirements of power system operation. The primary algorithms in use today are the power flow (also known as load flow), optimal power flow (OPF), and state estimation. These are all "steady-state" algorithms and are built up from the same basic approach to solving the nonlinear power balance equations. These particular algorithms do not explicitly consider the effects of time-varying dynamics on the system. The fields of transient and dynamic stability require a large number of algorithms that are significantly different from the powerflow-based algorithms and will therefore not be covered in this chapter.

5.1 Power Flow

The underlying principle of a power flow problem is that given the system loads, generation, and network configuration, the system bus voltages and line flows can be found by solving the nonlinear power flow equations. This is typically accomplished by applying Kirchoff's law at each power system bus throughout the system. In this context, Kirchoff's law can be interpreted as the sum of the powers entering a bus must be zero, or that the power at each bus must be conserved. Since power is comprised of two components, active power and reactive power, each bus gives rise to two equations—one for active power and one for reactive power. These equations are known as the *power flow equations*:

$$0 = \Delta P_i = P_i^{inj} - V_i \sum_{j=1} V_j Y_{ij} \cos(\theta_i - \theta_j - \phi_{ij}) \tag{5.1}$$

$$0 = \Delta Q_i = Q_i^{inj} - V_i \sum_{j=1}^{N} V_j Y_{ij} \sin(\theta_i - \theta_j - \phi_{ij}) \quad i = 1,\ldots,N \tag{5.2}$$

where P_i^{inj} and Q_i^{inj} are the active power and reactive power injected at the bus i, respectively. Loads are modeled by negative power injection. The values V_i and V_j are the voltage magnitudes at bus i and bus j, respectively. The values θ_i and θ_j are the corresponding phase angles. The value $Y_{ij}\angle\phi_{ij}$ is the (ij)th element of the network admittance matrix Y_{bus}. The constant N is the number of buses in the system. The updates ΔP_i and ΔQ_i of Equations 5.1 and 5.2 are called the *mismatch* equations because they give a measure of the power difference, or mismatch, between the calculated power values, as functions of voltage and phase angle, and the actual injected powers.

The formulation in Equations 5.1 and 5.2 is called the *polar* formulation of the power flow equations. If $Y_{ij}\angle\phi_{ij}$ is instead given by the complex sum $g_{ij} + jb_{ij}$, then the power flow equations may be written in *rectangular form* as

$$0 = \Delta P_i = P_i^{inj} - V_i \sum_{j=1}^{N} V_j \left(g_{ij} \cos(\theta_i - \theta_j) + b_{ij} \sin(\theta_i - \theta_j)\right) \tag{5.3}$$

$$0 = \Delta Q_i = Q_i^{inj} - V_i \sum_{j=1}^{N} V_j \left(g_{ij} \sin(\theta_i - \theta_j) - b_{ij} \cos(\theta_i - \theta_j)\right) \quad i = 1, \ldots, N \tag{5.4}$$

There are, at most, $2N$ equations to solve. This number is then further reduced by removing one power flow equation for each known voltage (at voltage controlled buses) and the swing bus angle. This reduction is necessary since the number of equations must equal the number of unknowns in a fully determined system. In either case, the power flow equations are a system of nonlinear equations. They are nonlinear in both the voltage and phase angle. Once the number and form of the nonlinear power flow equations have been determined, the Newton–Raphson method may be applied to numerically solve the power flow equations.

5.1.1 Admittance Matrix

The first step in solving the power flow equations is to obtain the admittance matrix Y for the system. The admittance matrix of a passive network (a network containing only resistors, capacitors, and inductors) may be found by summing the currents at every node in the system. An arbitrary bus i in the transmission network is shown in Figure 5.1. The bus i can be connected to any number of other buses in the system through transmission lines. Each transmission line between buses i and j is represented by a π-circuit with series impedance $R_{ij} + jX_{ij}$ where R_{ij} is the per unit resistance of the transmission line. The per unit line reactance is $X_{ij} = \omega_s L_{ij}$ where ω_s is the system base frequency and L_{ij} is the inductance of the line. In the π-circuit, the line-charging capacitance is represented by two lumped admittances $jB_{ij}/2$ placed at each end of the transmission line. Note that although each transmission-line parameter is in per unit and therefore a unitless quantity, the impedance is given in Ω/Ω whereas the admittance is given as Ω/Ω. The series impedances can also be represented by their equivalent admittance value where

$$y_{ij}\angle\phi_{ij} = \frac{1}{R_{ij} + jX_{ij}}$$

Summing the currents at node i yields

$$I_{i,inj} = I_{i1} + I_{i2} + \cdots + I_{iN} + I_{i10} + I_{i20} + \cdots + I_{iN0} \tag{5.5}$$

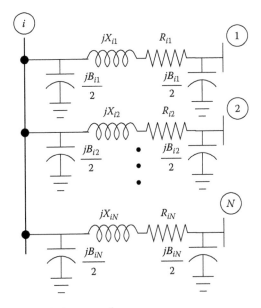

FIGURE 5.1   Bus i in a power transmission network.

$$= \frac{\left(\widehat{V}_i - \widehat{V}_1\right)}{R_{i1} + jX_{i1}} + \frac{\left(\widehat{V}_i - \widehat{V}_2\right)}{R_{i2} + jX_{i2}} + \cdots + \frac{\left(\widehat{V}_i - \widehat{V}_{iN}\right)}{R_{iN} + jX_{iN}}$$

$$+ \widehat{V}\left(j\frac{B_{i1}}{2} + j\frac{B_{i2}}{2} + \cdots + j\frac{B_{iN}}{2} \right) \tag{5.6}$$

$$= \left(\widehat{V}_i - \widehat{V}_1\right) y_{i1}\angle\phi_{i1} + \left(\widehat{V}_i - \widehat{V}_2\right) y_{i2}\angle\phi_{i2} + \cdots + \left(\widehat{V}_i - \widehat{V}_N\right) y_{iN}\angle\phi_{iN}$$

$$+ \widehat{V}_i\left(j\frac{B_{i1}}{2} + j\frac{B_{i2}}{2} + \cdots + j\frac{B_{iN}}{2} \right) \tag{5.7}$$

where $\widehat{V}_i = V_i\angle\theta_i$.

Gathering the voltages yields

$$I_{i,inj} = -\widehat{V}_1 y_{i1}\angle\phi_{i1} - \widehat{V}_2 y_{i2}\angle\phi_{i2} + \cdots + \widehat{V}_i\left(y_{i1}\angle\phi_{i1} + y_{i2}\angle\phi_{i2}Z + \cdots + y_{iN}\angle\phi_{iN}r + j\frac{B_{i1}}{2} + j\frac{B_{i2}}{2} + \cdots + j\frac{B_{iN}}{2} \right)$$

$$+ \cdots - \widehat{V}_N y_{iN}\angle\phi_{iN} \tag{5.8}$$

By noting that there are N buses in the system, there are N equations similar to Equation 5.8. These can be represented in matrix form

$$\begin{bmatrix} I_{1,inj} \\ I_{2,inj} \\ \vdots \\ I_{i,inj} \\ \vdots \\ I_{N,inj} \end{bmatrix} = \begin{bmatrix} Y_{11} & -y_{12}\angle\phi_{12} & \cdots & -y_{1i}\angle\phi_{1i} & \cdots & -y_{1N}\angle\phi_{1N} \\ -y_{21}\angle\phi_{21} & Y_{22} & \cdots & -y_{2i}\angle\phi_{2i} & \cdots & -y_{2N}\angle\phi_{2N} \\ \vdots & \vdots & \ddots & \vdots & \vdots & \vdots \\ -y_{1i}\angle\phi_{i1} & -y_{i2}\angle\phi_{i2} & \cdots & Y_{ii} & \cdots & -y_{iN}\angle\phi_{iN} \\ \vdots & \vdots & \vdots & \vdots & \ddots & \vdots \\ -y_{N1}\angle\phi_{N1} & -y_{N2}\angle\phi_{N2} & \cdots & -y_{Ni}\angle\phi_{Ni} & \cdots & Y_{NN} \end{bmatrix} \begin{bmatrix} \widehat{V}_1 \\ \widehat{V}_2 \\ \vdots \\ \widehat{V}_1 \\ \vdots \\ \widehat{V}_N \end{bmatrix} \tag{5.9}$$

where

$$Y_{ii} = \left(y_{i1} \angle \phi_{i1} + y_{i2} \angle \phi_{i2} + \cdots + y_{iN} \angle \phi_{iN} + j\frac{B_{i1}}{2} + j\frac{B_{i2}}{2} + \cdots + j\frac{B_{iN}}{2} \right) \quad (5.10)$$

The matrix relating the injected currents vector to the bus voltage vector is known as the system *admittance matrix* and is commonly represented as *Y*. A simple procedure for calculating the elements of the admittance matrix is

$Y(i,j)$ negative of the admittance between buses i and j

$Y(i,i)$ sum of all admittances connected to bus i

noting that the line-charging values are shunt admittances and are included in the diagonal elements.

Example 5.1

Find the admittance matrix for the line data given by

i	j	R_{ij}	X_{ij}	B_{ij}
1	2	0.027	0.32	0.15
1	5	0.014	0.18	0.10
2	3	0.012	0.13	0.12
3	4	0.025	0.25	0.00
3	5	0.017	0.20	0.08

Solution

The first step in calculating the admittance matrix is to calculate the off-diagonal elements first. Note that in a passive network, the admittance matrix is symmetric and $Y(i,j) = Y(j,i)$.

The off-diagonal elements are calculated as the negative of the series admittance of each line. Therefore

$$Y(1,2) = Y(2,1) = \frac{-1}{0.027 + j0.32} = 3.1139 \angle 94.82°$$

$$Y(1,5) = Y(5,1) = \frac{-1}{0.014 + j0.18} = 5.5388 \angle 94.45°$$

$$Y(2,3) = Y(3,2) = \frac{-1}{0.012 + j0.13} = 7.6597 \angle 95.27°$$

$$Y(3,4) = Y(4,3) = \frac{-1}{0.025 + j0.25} = 3.9801 \angle 95.71°$$

$$Y(3,5) = Y(5,3) = \frac{-1}{0.017 + j0.20} = 4.9820 \angle 94.86°$$

The diagonal elements are calculated as the sum of all admittances connected to each bus.

Therefore

$$Y(1,1) = \frac{1}{0.027 + j0.32} + \frac{1}{0.014 + j0.18} + j\frac{0.15}{2} + j\frac{0.10}{2} = 8.5281\angle-85.35°$$

$$Y(2,2) = \frac{1}{0.027 + j0.32} + \frac{1}{0.012 + j0.13} + j\frac{0.15}{2} + j\frac{0.12}{2} = 10.6392\angle-84.79°$$

$$Y(3,3) = \frac{1}{0.012 + j0.13} + \frac{1}{0.025 + j0.25} + \frac{1}{0.017 + j0.20}$$

$$+ j\frac{0.12}{2} + j\frac{0.08}{2} = 16.5221\angle-84.71°$$

$$Y(4,4) = \frac{1}{0.025 + j0.25} = 3.9801\angle-84.29°$$

$$Y(5,5) = \frac{1}{0.017 + j0.20} + j\frac{0.08}{2} = 10.4311\angle-85.32°$$

5.1.2 Newton–Raphson Method

The most common approach to solving the power flow equations is to use the iterative Newton–Raphson method. The Newton–Raphson method is an iterative approach to solving continuous nonlinear equations in the form

$$F(x) = \begin{bmatrix} f_1(x_1, x_2,\ldots, x_n) \\ f_2(x_1, x_2,\ldots, x_n) \\ \vdots \\ f_n(x_1, x_2,\ldots, x_n) \end{bmatrix} = 0 \tag{5.11}$$

An iterative approach is one in which an initial guess (x^0) to the solution is used to create a sequence $x^0, x^1, x^2,\ldots$ that (hopefully) converges arbitrarily close to the desired solution vector x^* where $F(x^*) = 0$.

The Newton–Raphson method for n-dimensional systems is given as

$$x^{k+1} = x^k - [J(x^k)]^{-1} F(x^k) \tag{5.12}$$

where

$$x = \begin{bmatrix} x_1 \\ x_2 \\ x_3 \\ \vdots \\ x_n \end{bmatrix}$$

$$F(x^k) = \begin{bmatrix} f_1(x^k) \\ f_2(x^k) \\ f_3(x^k) \\ \vdots \\ f_n(x^k) \end{bmatrix}$$

and the Jacobian matrix $[J(x^k)]$ is given by

$$[J(x^k)] = \begin{bmatrix} \dfrac{\partial f_1}{\partial x_1} & \dfrac{\partial f_1}{\partial x_2} & \dfrac{\partial f_1}{\partial x_3} & \cdots & \dfrac{\partial f_2}{\partial x_n} \\[2ex] \dfrac{\partial f_2}{\partial x_1} & \dfrac{\partial f_2}{\partial x_2} & \dfrac{\partial f_2}{\partial x_3} & \cdots & \dfrac{\partial f_2}{\partial x_n} \\[2ex] \dfrac{\partial f_3}{\partial x_1} & \dfrac{\partial f_3}{\partial x_2} & \dfrac{\partial f_3}{\partial x_3} & \cdots & \dfrac{\partial f_3}{\partial x_n} \\[2ex] \vdots & \vdots & \vdots & \vdots & \vdots \\[2ex] \dfrac{\partial f_n}{\partial x_1} & \dfrac{\partial f_n}{\partial x_2} & \dfrac{\partial f_n}{\partial x_3} & \cdots & \dfrac{\partial f_n}{\partial x_n} \end{bmatrix}$$

Typically, the inverse of the Jacobian matrix $[J(x^k)]$ is not found directly, but rather through LU factorization. Convergence is typically evaluated by considering the norm of the function

$$\left\| F(x^k) \right\| < \varepsilon \tag{5.13}$$

Note that the Jacobian is a function of x^k and is therefore updated every iteration along with $F(x^k)$.

In this formulation, the vector $F(x)$ is the set of power flow equations and the unknown x is the vector of voltage magnitudes and angles. It is common to arrange the Newton–Raphson equations by phase angle followed by the voltage magnitudes as

$$\begin{bmatrix} J_1 & J_2 \\ J_3 & J_4 \end{bmatrix} \begin{bmatrix} \Delta\delta_1 \\ \Delta\delta_2 \\ \Delta\delta_3 \\ \vdots \\ \Delta\delta_N \\ \Delta V_1 \\ \Delta V_2 \\ \Delta V_3 \\ \vdots \\ \Delta Q_N \end{bmatrix} = - \begin{bmatrix} \Delta P_1 \\ \Delta P_2 \\ \Delta P_3 \\ \vdots \\ \Delta P_N \\ \Delta Q_1 \\ \Delta Q_2 \\ \Delta Q_3 \\ \vdots \\ \Delta Q_N \end{bmatrix} \tag{5.14}$$

where

$$\Delta\delta_i = \delta_i^{k+1} - \delta_i^k$$

$$\Delta V_i = V_i^{k+1} - V_i^k$$

and ΔP_i and ΔQ_i are as given in Equations 5.1 and 5.2 and are evaluated at δ^k and V^k. The Jacobian is typically divided into four submatrices, where

$$\begin{bmatrix} J_1 & J_2 \\ J_3 & J_4 \end{bmatrix} = \begin{bmatrix} \dfrac{\partial \Delta P}{\partial \delta} & \dfrac{\partial \Delta P}{\partial V} \\[2ex] \dfrac{\partial \Delta Q}{\partial \delta} & \dfrac{\partial \Delta Q}{\partial V} \end{bmatrix} \tag{5.15}$$

Each submatrix represents the partial derivatives of each of the mismatch equations with respect to each of the unknowns. These partial derivatives yield eight types—two for each mismatch equation, where one is for the diagonal element and the other is for off-diagonal elements. The derivatives are summarized as

$$\frac{\partial \Delta P_i}{\partial \delta_i} = V_i \sum_{j=1}^{N} V_j Y_{ij} \sin(\delta_i - \delta_j - \phi_{ij}) + V_i^2 Y_{ii} \sin \phi_{ii} \tag{5.16}$$

$$\frac{\partial \Delta P_i}{\partial \delta_j} = -V_i V_j Y_{ij} \sin(\delta_i - \delta_j - \phi_{ij}) \tag{5.17}$$

$$\frac{\partial \Delta P_i}{\partial V_i} = -\sum_{i=1}^{N} V_j Y_{ij} \cos(\delta_i - \delta_j - \phi_{ij}) - V_i V_{ii} \cos \phi_{ii} \tag{5.18}$$

$$\frac{\partial \Delta P_i}{\partial V_j} = -V_i Y_{ij} \cos(\delta_i - \delta_j - \phi_{ij}) \tag{5.19}$$

$$\frac{\partial \Delta Q_i}{\partial \delta_i} = -V_i \sum_{j=1}^{N} V_j Y_{ij} \cos(\delta_i - \delta_j - \phi_{ij}) + V_i^2 Y_{ii} \cos \phi_{ii} \tag{5.20}$$

$$\frac{\partial \Delta Q_i}{\partial \delta_j} = V_i V_j Y_{ij} \cos(\delta_i - \delta_j - \phi_{ij}) \tag{5.21}$$

$$\frac{\partial \Delta Q_i}{\partial V_i} = -\sum_{j=1}^{N} V_j Y_{ij} \sin(\delta_i - \delta_j - \phi_{ij}) + V_i Y_{ii} \sin \phi_{ii} \tag{5.22}$$

$$\frac{\partial \Delta Q_i}{\partial V_j} = -V_i Y_{ij} \sin(\delta_i - \delta_j - \phi_{ij}) \tag{5.23}$$

A common modification to the power flow solution is to replace the unknown update ΔV_i by the normalized value $\Delta V_i / V_i$. This formulation yields a more symmetric Jacobian matrix as the Jacobian submatrices J_2 and J_4 are now multiplied by V_i to compensate for the scaling of ΔV_i by V_i. All partial derivatives of each submatrix then become quadratic in voltage magnitude.

The Newton–Raphson method for the solution of the power flow equations is relatively straightforward to program since both the function evaluations and the partial derivatives use the same expressions. Thus it takes little extra computational effort to compute the Jacobian once the mismatch equations have been calculated.

Example 5.2

Find the voltage magnitudes, phase angles, and line flows for the small power system shown in Figure 5.2 with the following system:

Parameters in Per Unit

Bus	Type	V	P_{gen}	Q_{gen}	P_{load}	Q_{load}
1	Swing	1.02	—	—	0.0	0.0
2	PV	1.00	0.5	—	0.0	0.0
3	PQ	—	0.0	0.0	1.2	0.5

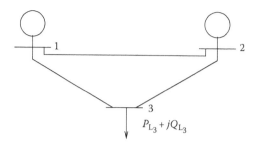

FIGURE 5.2 Example power system.

I	j	R_{ij}	X_{ij}	B_{ij}
1	2	0.02	0.3	0.15
1	3	0.01	0.1	0.1
2	3	0.01	0.1	0.1

Solution

Calculating the admittance matrix for this system yields

$$Y_{\text{bus}} = \begin{bmatrix} 13.1505\angle -84.7148° & 3.3260\angle 93.8141° & 9.9504\angle 95.7106° \\ 3.326\angle 95.7106° & 13.1505\angle -84.7148° & 9.9504\angle 95.7106° \\ 9.9504\angle 95.7106° & 9.9504\angle 95.7106° & 19.8012\angle 84.2606° \end{bmatrix} \quad (5.24)$$

By inspection, this system has three unknowns: δ_2, δ_3, and V_3; thus, three power flow equations are required. These power flow equations are

$$0 = \Delta P_2 = 0.5 - V_2 \sum_{j=1}^{3} V_j Y_{ij} \cos(\delta_2 - \delta_j - \theta_{ij}) \quad (5.25)$$

$$0 = \Delta P_3 = -1.2 - V_3 \sum_{j=1}^{3} V_j Y_{ij} \cos(\delta_3 - \delta_j - \theta_{ij}) \quad (5.26)$$

$$0 = \Delta Q_3 = -0.5 - V_3 \sum_{j=1}^{3} V_j Y_{ij} \sin(\delta_3 - \delta_j - \theta_{ij}) \quad (5.27)$$

Substituting the known quantities for $V_1 = 1.02$, $V_2 = 1.00$, and $\delta_1 = 0$ and the admittance matrix quantities yields

$$\Delta P_2 = 0.5 - (1.00) \begin{bmatrix} (1.02)(3.3260)\cos(\delta_2 - 0 - 93.8141°) \\ +(1.00)(13.1505)\cos(\delta_2 - \delta_2 + 84.7148°) \\ +(V_3)(9.9504)\cos(\delta_2 - \delta_3 - 95.7106°) \end{bmatrix} \quad (5.28)$$

$$\Delta P_3 = -1.2 - (V_3) \begin{bmatrix} (1.02)(9.9504)\cos(\delta_2 - 0 - 95.7106°) \\ +(1.00)(9.9504)\cos(\delta_3 - \delta_2 - 95.7106°) \\ +(V_3)(19.8012)\cos(\delta_3 - \delta_3 - 84.2606°) \end{bmatrix} \quad (5.29)$$

$$\Delta Q_3 = -0.5 - (V_3) \begin{bmatrix} (1.02)(9.9504)\sin(\delta_3 - 0 - 95.7106°) \\ +(1.00)(9.9504)\sin(\delta_3 - \delta_2 - 95.7106°) \\ +(V_3)(19.8012)\sin(\delta_3 - \delta_3 + 84.2606°) \end{bmatrix} \qquad (5.30)$$

The Newton–Raphson iteration for this system is then given by

$$\begin{bmatrix} \dfrac{\partial \Delta P_2}{\partial \delta_2} & \dfrac{\partial \Delta P_2}{\partial \delta_3} & \dfrac{\partial \Delta P_2}{\partial V_3} \\[2mm] \dfrac{\partial \Delta P_3}{\partial \delta_2} & \dfrac{\partial \Delta P_3}{\partial \delta_3} & \dfrac{\partial \Delta P_3}{\partial V_3} \\[2mm] \dfrac{\partial \Delta Q_3}{\partial \delta_2} & \dfrac{\partial \Delta Q_3}{\partial \delta_3} & \dfrac{\partial \Delta Q_3}{\partial V_3} \end{bmatrix} \begin{bmatrix} \Delta \delta_2 \\ \Delta \delta_3 \\ \Delta V_3 \end{bmatrix} = - \begin{bmatrix} \Delta P_2 \\ \Delta P_3 \\ \Delta Q_3 \end{bmatrix} \qquad (5.31)$$

where

$$\frac{\partial \Delta P_2}{\partial \delta_2} = 3.3925 \sin(\delta_2 - 93.8141°)$$
$$+ 9.9504 V_3 \sin(\delta_2 - \delta_3 - 94.7106°)$$

$$\frac{\partial \Delta P_2}{\partial \delta_3} = -9.9504 V_3 \sin(\delta_2 - \delta_3 - 95.7106°)$$

$$\frac{\partial \Delta P_2}{\partial V_3} = -9.9504 \cos(\delta_2 - \delta_3 - 95.7106°)$$

$$\frac{\partial \Delta P_3}{\partial \delta_2} = -9.9504 V_3 \sin(\delta_3 - \delta_2 - 95.7106°)$$

$$\frac{\partial \Delta P_3}{\partial \delta_3} = 10.1494 V_3 \sin(\delta_3 - 95.7106°)$$
$$+ 9.9504 V_3 \sin(\delta_3 - \delta_2 - 95.7106°)$$

$$\frac{\partial \Delta P_3}{\partial V_3} = -10.1494 \cos(\delta_3 - 95.7106°)$$
$$- 9.9504 \cos(\delta_3 - \delta_2 - 95.7106°)$$
$$- 39.6024 V_3 \cos(84.2606°)$$

$$\frac{\partial \Delta Q_3}{\partial \delta_2} = 9.9504 V_3 \cos(\delta_3 - \delta_2 - 95.7106°)$$

$$\frac{\partial \Delta Q_3}{\partial \delta_2} = -10.1494 V_3 \cos(\delta_3 - 95.7106°)$$
$$- 9.9504 V_3 \cos(\delta_3 - \delta_2 - 95.7106°)$$

$$\frac{\partial \Delta Q_3}{\partial V_3} = -10.1494 \sin(\delta_3 - 95.7106°)$$

$$- 9.9504 \sin(\delta_3 - \delta_2 - 95.7106°)$$

$$- 39.6024 V_3 \sin(84.2606°)$$

One of the underlying assumptions of the Newton–Raphson iteration is that the higher order terms of the Taylor series expansion upon which the iteration is based are negligible only if the initial guess is sufficiently close to the actual solution to the nonlinear equations. Under most operating conditions, the voltages throughout the power system are within ±10% of the nominal voltage and therefore fall in the range $0.9 \leq V_i \leq 1.1$ per unit. Similarly, under most operating conditions the phase angle differences between adjacent buses are typically small. Thus if the swing bus angle is taken to be zero, then all phase angles throughout the system will also be close to zero. Therefore in initializing a power flow, it is common to choose a "flat start" initial condition. That is, all voltage magnitudes are set to 1.0 per unit and all angles are set to zero.

Iteration 1

Evaluating the Jacobian and the mismatch equations at the flat start initial conditions yields

$$[J^0] = \begin{bmatrix} -13.2859 & 9.9010 & 0.9901 \\ 9.9010 & -20.000 & -1.9604 \\ -0.9901 & 2.0000 & -19.4040 \end{bmatrix}$$

$$\begin{bmatrix} \Delta P_2^0 \\ \Delta P_3^0 \\ \Delta Q_3^0 \end{bmatrix} = \begin{bmatrix} 0.5044 \\ -1.1802 \\ -0.2020 \end{bmatrix}$$

Solving

$$[J^0] \begin{bmatrix} \Delta \delta_2^1 \\ \Delta \delta_3^1 \\ \Delta V_3^1 \end{bmatrix} = \begin{bmatrix} \Delta P_2^0 \\ \Delta P_3^0 \\ \Delta Q_3^0 \end{bmatrix}$$

by LU factorization yields

$$\begin{bmatrix} \Delta \delta_2^1 \\ \Delta \delta_3^1 \\ \Delta V_3^1 \end{bmatrix} = \begin{bmatrix} -0.0096 \\ -0.0621 \\ -0.0163 \end{bmatrix}$$

Therefore

$$\delta_2^1 = \delta_2^0 + \Delta \delta_2^1 = 0 - 0.0096 = -0.0096$$

$$\delta_3^1 = \delta_3^0 + \Delta \delta_3^1 = 0 - 0.0621 = -0.0621$$

$$V_3^1 = V_3^0 + \Delta V_3^1 = 1 - 0.0163 = 0.9837$$

Note that the angles are given in *radians* and not degrees. The error at the first iteration is the largest absolute value of the mismatch equations, which is

$$\varepsilon^1 = 1.1802$$

One quick check of this process is to note that the voltage update V_3^1 is slightly less than 1.0 per unit, which would be expected given the system configuration. Note also that the diagonals of the Jacobian are all equal or greater in magnitude than the off-diagonal elements. This is because the diagonals are summations of terms, whereas the off-diagonal elements are single terms.

Iteration 2

Evaluating the Jacobian and the mismatch equations at the updated values δ_2^1, δ_3^1, and V_3^1 yields

$$[J^1] = \begin{bmatrix} -13.1597 & 9.7771 & 0.4684 \\ 9.6747 & -19.5280 & -0.7515 \\ -1.4845 & 3.0929 & -18.9086 \end{bmatrix}$$

$$\begin{bmatrix} \Delta P_2^1 \\ \Delta P_3^1 \\ \Delta Q_3^1 \end{bmatrix} = \begin{bmatrix} 0.0074 \\ -0.0232 \\ -0.0359 \end{bmatrix}$$

Solving for the update yields

$$\begin{bmatrix} \Delta\delta_2^2 \\ \Delta\delta_3^2 \\ \Delta V_3^2 \end{bmatrix} = \begin{bmatrix} -0.0005 \\ -0.0014 \\ -0.0021 \end{bmatrix}$$

and

$$\begin{bmatrix} \delta_2^2 \\ \delta_3^2 \\ V_3^2 \end{bmatrix} = \begin{bmatrix} -0.0101 \\ -0.0635 \\ 0.9816 \end{bmatrix}$$

where

$$\varepsilon^2 = 0.0359$$

Iteration 3

Evaluating the Jacobian and the mismatch equations at the updated values δ_2^2, δ_3^2, and V_3^2 yields

$$[J^2] = \begin{bmatrix} -13.1392 & 9.7567 & 0.4600 \\ 9.6530 & -19.4831 & -0.7213 \\ -1.4894 & -3.1079 & -18.8300 \end{bmatrix}$$

$$\begin{bmatrix} \Delta P_2^0 \\ \Delta P_3^0 \\ \Delta Q_3^0 \end{bmatrix} = \begin{bmatrix} 0.1717 \\ -0.5639 \\ -0.9084 \end{bmatrix} \times 10^{-4}$$

Solving for the update yields

$$\begin{bmatrix} \Delta\delta_2^2 \\ \Delta\delta_3^2 \\ \Delta V_3^2 \end{bmatrix} = \begin{bmatrix} -0.1396 \\ -0.3390 \\ -0.5273 \end{bmatrix} \times 10^{-5}$$

and

$$\begin{bmatrix} \delta_2^3 \\ \delta_3^2 \\ V_3^3 \end{bmatrix} = \begin{bmatrix} -0.0101 \\ -0.0635 \\ 0.9816 \end{bmatrix}$$

where

$$\varepsilon^3 = 0.9084 \times 10^{-4}$$

At this point, the iterations have converged since the mismatch is sufficiently small and the values are no longer changing significantly.

The last task in power flow is to calculate the generated reactive powers, the swing bus active power output and the line flows. The generated powers can be calculated directly from the power flow equations:

$$P_i^{inj} = V_i \sum_{j=1}^{N} V_j Y_{ij} \cos(\theta_i - \theta_j - \phi_{ij})$$

$$Q_i^{inj} = V_i \sum_{j=1}^{N} V_j Y_{ij} \sin(\theta_i - \theta_j - \phi_{ij})$$

Therefore

$$P_{gen,1} = P_1^{inj} = 0.7087$$

$$Q_{gen,1} = Q_1^{inj} = 0.2806$$

$$Q_{gen,2} = Q_2^{inj} = -0.0446$$

The total active power losses in the system are the difference between the sum of the generation and the sum of the loads, in this case:

$$P_{loss} = \sum P_{gen} - \sum P_{load} = 0.7087 + 0.5 - 1.2 = 0.0087\,pu \tag{5.32}$$

The line losses for line i–j are calculated at both the sending and receiving ends of the line. The apparent power leaving bus i to bus j on line i–j is

$$S_{ij} = V_i\angle\delta_i\left(I_{ij} + j\frac{B_{ij}}{2}V_i\angle\theta_i\right)^* \tag{5.33}$$

$$= V_i\angle\delta_i\left(\left(\frac{V_i\angle\theta_i - V_j\angle\theta_j}{R_{ij} + jX_{ij}}\right) + j\frac{B_{ij}}{2}V_i\angle\theta_i\right)^* \tag{5.34}$$

and the power received at bus j from bus i on line i–j is

$$S_{ji} = V_j \angle \delta_j \left(I_{ji} - j\frac{B_{ij}}{2} V_j \angle \theta_j \right)^* \tag{5.35}$$

Thus

$$P_{ij} = \mathrm{Re}\{S_{ij}\} = V_i V_j Y_{ij} \cos(\delta_i - \delta_j - \phi_{ij}) - V_i^2 Y_{ij} \cos\phi_{ij} \tag{5.36}$$

$$Q_{ij} = \mathrm{Im}\{S_{ij}\} = V_i V_j Y_{ij} \sin(\delta_i - \delta_j - \phi_{ij}) + V_i^2 \left(Y_{ij} \sin\phi_{ij} - \frac{B_{ij}}{2} \right) \tag{5.37}$$

Similarly, the powers P_{ji} and Q_{ji} can be calculated. The active power loss on any given line is the difference between the active power sent from bus i and the active power received at bus j. Calculating the reactive power losses is more complex since the reactive power generated by the line-charging (shunt capacitances) must also be included.

5.2 Optimal Power Flow

The basic objective of the OPF is to find the values of the system state variables and/or parameters that minimize some cost function of the power system. The types of cost functions are system dependent and can vary widely from application to application and are not necessarily strictly measured in terms of dollars. Examples of engineering optimizations can range from minimizing

- Active power losses
- Particulate output (emissions)
- System energy
- Fuel costs of generation

to name a few possibilities. The basic formulation of the OPF can be represented as minimizing a defined cost function subject to any physical or operational constraints of the system:
minimize

$$\begin{aligned} f(x,u) \quad & x \in R^n \\ & u \in R^n \end{aligned} \tag{5.38}$$

subject to

$$g(x,u) = 0 \quad \text{equality constraints} \tag{5.39}$$

$$h(x,u) = 0 \quad \text{inequality constraints} \tag{5.40}$$

where
  x is the vector of system states
  u is the vector of system parameters

The basic approach is to find the vector of system parameters that when substituted into the system model will result in the state vector x that minimizes the cost function $f(x,u)$.

In an unconstrained system, the usual approach to minimizing the cost function is to set the function derivatives to zero and then solve for the system states from the set of resulting equations. In the majority of applications, however, the system states at the unconstrained minimum will not satisfy the constraint equations. Thus, an alternate approach is required to find the constrained minimum. One approach is to introduce an additional set of parameters λ, frequently known as *Lagrange multipliers*, to impose the constraints on the cost function. The augmented cost function then becomes

$$\text{minimize } f(x,u) - \lambda g(x,u) \tag{5.41}$$

The augmented function in Equation 5.41 can then be minimized by solving for the set of states that result from setting the derivatives of the augmented function to zero. Note that the derivative of Equation 5.41 with respect to λ effectively enforces the equality constraint of Equation 5.39.

Example 5.3

Minimize

$$C : \frac{1}{2}(x^2 + y^2) \tag{5.42}$$

subject to the following constraint:

$$2x - y = 5$$

Solution

Note that the function to be minimized is the equation for a circle. The unconstrained minimum of this function is the point at the origin with $x = 0$ and $y = 0$, which defines a circle with a radius of zero length. However, the circle must also intersect the line defined by the constraint equation; thus, the constrained circle must have a nonzero radius. The augmented cost function becomes

$$C^* : \frac{1}{2}(x^2 + y^2) - \lambda(2x - y - 5) \tag{5.43}$$

where λ represents the Lagrange multiplier. Setting the derivatives of the augmented cost function to zero yields the following set of equations:

$$0 = \frac{\partial C^*}{\partial x} = x - 2\lambda$$

$$0 = \frac{\partial C^*}{\partial y} = y + \lambda$$

$$0 = \frac{\partial C^*}{\partial \lambda} = 2x - y - 5$$

Solving this set of equations yields $[x \quad y \quad \lambda]^T = [2 \quad -1 \quad 1]^T$. The cost function of Equation 5.42 evaluated at the minimum of the augmented cost function is

$$C : \frac{1}{2}[(2)^2 + (-1)^2] = \frac{5}{2}$$

If there is more than one equality constraint (i.e., if $g(x, u)$ of Equation 5.39 is a vector of functions) then λ becomes a vector of multipliers and the augmented cost function becomes

$$C^*: f(x,u) - [\lambda]^T g(x,u) \qquad (5.44)$$

where the derivatives of C^* become

$$\left[\frac{\partial C^*}{\partial \lambda}\right] = 0 = g(x,u) \qquad (5.45)$$

$$\left[\frac{\partial C^*}{\partial x}\right] = 0 = \left[\frac{\partial f}{\partial x}\right] - \left[\frac{\partial g}{\partial x}\right]^T [\lambda] \qquad (5.46)$$

$$\left[\frac{\partial C^*}{\partial u}\right] = 0 = \left[\frac{\partial f}{\partial u}\right] = \left[\frac{\partial g}{\partial u}\right]^T [\lambda] \qquad (5.47)$$

Note that for any *feasible* solution, Equation 5.45 is satisfied, but the feasible solution may not be the optimal solution that minimizes the cost function. In this case, $[\lambda]$ can be obtained from Equation 5.46 and then only

$$\left[\frac{\partial C^*}{\partial u}\right] \neq 0$$

This vector can be used as a gradient vector $[\nabla C]$, which is orthogonal to the contour of constant values of the cost function C. Thus,

$$[\lambda] = \left[\left[\frac{\partial g}{\partial x}\right]^T\right]^{-1}\left[\frac{\partial f}{\partial x}\right] \qquad (5.48)$$

which gives

$$\nabla C = \left[\frac{\partial C^*}{\partial u}\right] = \left[\frac{\partial f}{\partial u}\right] - \left[\frac{\partial g}{\partial u}\right]^T [\lambda] \qquad (5.49)$$

$$= \left[\frac{\partial f}{\partial u}\right] - \left[\frac{\partial g}{\partial u}\right]^T \left[\left[\frac{\partial g}{\partial x}\right]^T\right]^{-1}\left[\frac{\partial f}{\partial x}\right] \qquad (5.50)$$

This relationship provides the foundation of the optimization method known as the *steepest descent* algorithm.

5.2.1 Steepest Descent Algorithm

1. Let $k = 0$. Guess an initial vector $u^k = u^0$.
2. Solve the (possibly nonlinear) system of Equation 5.45 for a feasible solution x.
3. Calculate C^{k+1} and ∇C^{k+1} from Equation 5.50. If $\|C^{k+1} - C^k\|$ is less than some predefined tolerance, stop.
4. Calculate the new vector $u^{k+1} = u^k - \gamma \nabla C$, where γ is a positive number, which is the user-defined "stepsize" of the algorithm.
5. $k = k+1$. Go to step 2.

In the steepest descent method, the u vector update direction is determined at each step of the algorithm by choosing the direction of the greatest change of the augmented cost function C^*. The direction of steepest descent is perpendicular to the tangent of the curve of constant cost. The distance between adjustments is analogous to the stepsize γ of the algorithm. Thus the critical part of the steepest descent algorithm is the choice of γ. If γ is chosen small, then convergence to minimum value is more likely, but may require many iterations, whereas a large value of γ may result in oscillations about the minimum.

Example 5.4

Minimize

$$C : x_1^2 + 2x_2^2 + u^2 = f(x_1, x_2, u) \tag{5.51}$$

subject to the following constraints:

$$0 = x_1^2 - 3x_2 + u - 3 \tag{5.52}$$

$$0 = x_1 + x_2 - 4u - 2 \tag{5.53}$$

Solution

To find ∇C of Equation 5.50, the following partial derivatives are required:

$$\left[\frac{\partial f}{\partial u}\right] = 2u$$

$$\left[\frac{\partial f}{\partial x}\right] = \begin{bmatrix} 2x_1 \\ 4x_2 \end{bmatrix}$$

$$\left[\frac{\partial g}{\partial u}\right]^T = [1 - 4]$$

$$\left[\frac{\partial g}{\partial x}\right] = \begin{bmatrix} 2x_1 & -3 \\ 1 & 1 \end{bmatrix}$$

yielding

$$\nabla C = \left[\frac{\partial f}{\partial u}\right] - \left[\frac{\partial g}{\partial u}\right]^T \left[\left[\frac{\partial g}{\partial x}\right]^T\right]^{-1} \left[\frac{\partial f}{\partial x}\right]$$

$$= 2u - [1 - 4]\left[\begin{bmatrix} 2x_1 & -3 \\ 1 & 1 \end{bmatrix}^T\right]^{-1} \begin{bmatrix} 2x_1 \\ 4x_2 \end{bmatrix}$$

Iteration 1

Let $u = 1$, $\gamma = 0.05$, and choose a stopping criterion of $\varepsilon = 0.0001$. Solving f or x_1 and x_2 yields two values for each with a corresponding cost function:

$$x_1 = 1.7016 \quad x_2 = 0.2984 \quad f = 4.0734$$

$$x_1 = -4.7016 \quad x_2 = 6.7016 \quad f = 23.2828$$

The first set of values leads to the minimum cost function, so they are selected as the operating solution. Substituting $x_1 = 1.7016$ and $x_2 = 0.2984$ into the gradient function yields $\nabla C = 10.5705$ and the new value of u becomes

$$u^{(2)} = u^{(1)} - \gamma \Delta C$$

$$= 1 - (0.05)(10.5705)$$

$$= 0.4715$$

Iteration 2

With $u = 0.4715$, solving for x_1 and x_2 again yields two values for each with a corresponding cost function:

$$x_1 = 0.6062 \quad x_2 = -0.7203 \quad f = 1.6276$$

$$x_1 = -3.6062 \quad x_2 = 3.4921 \quad f = 14.2650$$

The first set of values again leads to the minimum cost function, so they are selected as the operating solution. The difference in cost functions is

$$\left| C^{(1)} - C^{(2)} \right| = \left| 4.0734 - 1.6276 \right| = 2.4458$$

which is greater than the stopping criterion. Substituting these values into the gradient function yields $\nabla C = 0.1077$ and the new value of u becomes

$$u^{(3)} = u^{(2)} - \gamma \Delta C$$

$$= 0.4715 - (0.05)(0.1077)$$

$$= 0.4661$$

Iteration 3

With $u = 0.4661$, solving for x_1 and x_2 again yields two values for each with a corresponding cost function:

$$x_1 = 0.5921 \quad x_2 = -0.7278 \quad f = 1.6271$$

$$x_1 = 3.5921 \quad x_2 = 3.4565 \quad f = 14.1799$$

The first set of values again leads to the minimum cost function, so they are selected as the operating solution. The difference in cost functions is

$$\left| C^{(2)} - C^{(3)} \right| = \left| 1.6276 - 1.6271 \right| = 0.005$$

which is greater than the stopping criterion. Substituting these values into the gradient function yields $\nabla C = 0.0541$ and the new value of u becomes

$$u^{(4)} = u^{(3)} - \gamma \Delta C$$

$$= 0.4661 - (0.05)(0.0541)$$

$$= 0.4634$$

Iteration 4

With $u = 0.4634$, solving for x_1 and x_2 again yields two values for each with a corresponding cost function:

$$x_1 = 0.5850 \quad x_2 = -0.7315 \quad f = 1.6270$$

$$x_1 = 3.5850 \quad x_2 = 3.4385 \quad f = 14.1370$$

The first set of values again leads to the minimum cost function, so they are selected as the operating solution. The difference in cost functions is

$$\left| C^{(3)} - C^{(4)} \right| = \left| 1.6271 - 1.6270 \right| = 0.001$$

which satisfies the stopping criterion. Thus, the values $x_1 = 0.5850$, $x_2 = -0.7315$, and $u = 0.4634$ yield the minimum cost function $f = 1.6270$.

Many power system applications, such as the power flow, offer only a snapshot of the system operation. Frequently, the system planner or operator is interested in the effect that making adjustments to the system parameters will have on the power flow through lines or system losses. Rather than making the adjustments in a random fashion, the system planner will attempt to optimize the adjustments according to some objective function. This objective function can be chosen to minimize generating costs, reservoir water levels, or system losses, among others. The OPF problem is to formulate the power flow problem to find system voltages and generated powers within the framework of the objective function. In this application, the inputs to the power flow are systematically adjusted to maximize (or minimize) a scalar function of the power flow state variables. The two most common objective functions are minimization of generating costs and minimization of active power losses.

The time frame of OPF is on the order of minutes to 1 h; therefore it is assumed that the optimization occurs using only those units that are currently on-line. The problem of determining whether or not to engage a unit, at what time, and for how long is part of the *unit commitment* problem and is not covered here. The minimization of active transmission losses saves both generating costs and creates a higher generating reserve margin.

Example 5.5

Consider again the three machine system of Example 5.2 except that bus 3 has been converted to a generator bus with a voltage magnitude of 1.0 pu. The cost functions of the generators are

$$C_1: P_1 + 0.0625P_1^2 \text{ \$/h}$$

$$C_2: P_2 + 0.0125P_2^2 \text{ \$/h}$$

$$C_3: P_3 + 0.0250P_3^2 \text{ \$/h}$$

Find the optimal generation scheduling of this system.

Solution

Following the steepest descent procedure, the first step is to develop an expression for the gradient ∇C, where

$$\nabla C = \left[\frac{\partial f}{\partial u} \right] - \left[\frac{\partial g}{\partial u} \right]^T \left[\left[\frac{\partial g}{\partial x} \right]^T \right]^{-1} \left[\frac{\partial f}{\partial x} \right] \qquad (5.54)$$

where f is the sum of the generator costs:

$$f : C_1 + C_2 + C_3 = P_1 + 0.0625P_1^2 + P_2 + 0.0125P_2^2 + P_3 + 0.0250P_3^2$$

g is the set of load flow equations:

$$g_1 : 0 = P_2 - P_{L2} - V_2 \sum_{i=1}^{3} V_i Y_{2i} \cos(\delta_2 - \delta_i - \phi_{2i})$$

$$g_2 : 0 = P_3 - P_{L3} - V_3 \sum_{i=1}^{3} V_i Y_{3i} \cos(\delta_3 - \delta_i - \phi_{3i})$$

where P_{Li} denotes the active power load at bus i, the set of inputs u is the set of independent generation settings:

$$u = \begin{bmatrix} P_2 \\ P_3 \end{bmatrix}$$

and x is the set of unknown states

$$x = \begin{bmatrix} \delta_2 \\ \delta_3 \end{bmatrix}$$

The generator setting P_1 is not an input because it is the slack bus generation and cannot be independently set. From these designations, the various partial derivatives required for ∇C can be derived:

$$\left[\frac{\partial g}{\partial u} \right] = \begin{bmatrix} 1 \\ 1 \end{bmatrix} \tag{5.55}$$

$$\left[\frac{\partial g}{\partial x} \right] = \begin{bmatrix} \dfrac{\partial g_1}{\partial \delta_2} & \dfrac{\partial g_1}{\partial \delta_3} \\ \dfrac{\partial g_2}{\partial \delta_2} & \dfrac{\partial g_2}{\partial \delta_3} \end{bmatrix} \tag{5.56}$$

where

$$\frac{\partial g_1}{\partial \delta_2} = V_2(V_1 Y_{12} \sin(\delta_2 - \delta_1 - \phi_{21}) + V_3 Y_{13} \sin(\delta_2 - \delta_3 - \phi_{23})) \tag{5.57}$$

$$\frac{\partial g_1}{\partial \delta_3} = V_2 V_3 Y_{32} \sin(\delta_2 - \delta_3 - \phi_{23}) \tag{5.58}$$

$$\frac{\partial g_2}{\partial \delta_2} = V_3 V_2 Y_{23} \sin(\delta_3 - \delta_2 - \phi_{32}) \tag{5.59}$$

$$\frac{\partial g_2}{\partial \delta_3} = V_3(V_1 Y_{13} \sin(\delta_3 - \delta_1 - \phi_{31}) + V_2 Y_{23} \sin(\delta_3 - \delta_2 - \phi_{32})) \tag{5.60}$$

and

$$\left[\frac{\partial f}{\partial u}\right] = \begin{bmatrix} 1+0.025P_2 \\ 1+0.050P_3 \end{bmatrix} \tag{5.61}$$

Finding the partial derivative $[\partial f/\partial x]$ is slightly more difficult since the cost function is not written as a direct function of x. Recall, however, that P_1 is not an input, but is actually a quantity that depends on x, i.e.,

$$P_1 = V_1(V_1 Y_{11} \cos(\delta_1 - \delta_1 - \phi_{11})$$

$$+ V_2 Y_{12} \cos(\delta_1 - \delta_2 - \phi_{12}) + V_3 Y_{13} \cos(\delta_1 - \delta_3 - \phi_{13})) \tag{5.62}$$

Thus, using the chain rule,

$$\left[\frac{\partial f}{\partial x}\right] = \left[\frac{\partial f}{\partial P_1}\right]\left[\frac{\partial P_1}{\partial x}\right] \tag{5.63}$$

$$= (1+0.125P_1)\begin{bmatrix} V_1 V_2 Y_{12} \sin(\delta_1 - \delta_2 - \phi_{12}) \\ V_1 V_3 Y_{13} \sin(\delta_1 - \delta_3 - \phi_{13}) \end{bmatrix} \tag{5.64}$$

If the initial values of $P_2 = 0.56\,\mathrm{pu}$ and $P_3 = 0.28\,\mathrm{pu}$ are used as inputs, then the power flow yields the following states: $[\delta_2\,\delta_3] = [0.0286 - 0.0185]$ in radians and $P_1 = 0.1152$. Converting the generated powers to megawatt and substituting these values into the partial derivatives yields

$$\left[\frac{\partial g}{\partial u}\right] = \begin{bmatrix} 1 & 0 \\ 0 & 1 \end{bmatrix} \tag{5.65}$$

$$\left[\frac{\partial g}{\partial x}\right] = \begin{bmatrix} -13.3267 & 9.9366 \\ 9.8434 & -19.9219 \end{bmatrix} \tag{5.66}$$

$$\left[\frac{\partial f}{\partial u}\right] = \begin{bmatrix} 15.0000 \\ 15.0000 \end{bmatrix} \tag{5.67}$$

$$\left[\frac{\partial f}{\partial x}\right] = 15.4018\begin{bmatrix} -52.0136 \\ 155.8040 \end{bmatrix} \tag{5.68}$$

which yields

$$\nabla C = \begin{bmatrix} -0.3256 \\ -0.4648 \end{bmatrix} \tag{5.69}$$

Thus, the new values for the input generation are

$$\begin{bmatrix} P_2 \\ P_3 \end{bmatrix} = \begin{bmatrix} 560 \\ 280 \end{bmatrix} - \gamma\begin{bmatrix} -0.3256 \\ -0.4648 \end{bmatrix} \tag{5.70}$$

With $\gamma = 1$, the updated generation is $P_2 = 560.3$ and $P_3 = 280.5\,\mathrm{MW}$.

Proceeding with more iterations until the gradient is reduced to less than a user-defined value yields the final generation values for all of the generators:

$$\begin{bmatrix} P_1 \\ P_2 \\ P_3 \end{bmatrix} = \begin{bmatrix} 112.6 \\ 560.0 \\ 282.7 \end{bmatrix} MW$$

which yields a cost of $7664/MWh.

Often the steepest descent method may indicate that either states or inputs lie outside of their physical constraints. For example, the algorithm may result in a power generation value that exceeds the physical maximum output of the generating unit. Similarly, the resulting bus voltages may lie outside of the desired range (usually ±10% of unity). These are violations of the *inequality constraints* of the problem. In these cases, the steepest descent algorithm must be modified to reflect these physical limitations. There are several approaches to account for limitations and these approaches depend on whether or not the limitation is on the input (independent) or on the state (dependent).

5.2.2 Limitations on Independent Variables

If the application of the steepest descent algorithm results in an updated value of input that exceeds the specified limit, then the most straightforward method of handling this violation is simply to set the input state equal to its limit and continue with the algorithm except with one less degree of freedom.

Example 5.6

Repeat Example 5.5 except that the generators must satisfy the following limitations:

$$80 \le P_1 \le 1200 \, MW$$

$$450 \le P_2 \le 750 \, MW$$

$$150 \le P_3 \le 250 \, MW$$

Solution

From the solution of Example 5.5, the output of generator 3 exceeds the maximum limit of 0.25 pu. Therefore after the first iteration in the previous example, P_3 is set to 0.25 pu. The new partial derivatives become

$$\left[\frac{\partial g}{\partial u} \right] = \begin{bmatrix} 0 \\ 1 \end{bmatrix} \tag{5.71}$$

$$\left[\frac{\partial g}{\partial x} \right] = \text{same} \tag{5.72}$$

$$\left[\frac{\partial f}{\partial u} \right] = [1 + 0.025 P_2] \tag{5.73}$$

$$\left[\frac{\partial f}{\partial x} \right] = \text{same} \tag{5.74}$$

From the constrained steepest descent, the new values of generation become

$$\begin{bmatrix} P_1 \\ P_2 \\ P_3 \end{bmatrix} = \begin{bmatrix} 117.1 \\ 588.3 \\ 250.0 \end{bmatrix} MW$$

with a cost of \$7703/MWh, which is higher than the unconstrained cost of generation of \$7664/MWh. As more constraints are added to the system, the system is moved away from the optimal operating point, increasing the cost of generation.

5.2.3 Limitations on Dependent Variables

In many cases, the physical limitations of the system are imposed upon states that are dependent variables in the system description. In this case, the inequality equations are functions of x and must be added to the cost function. Examples of limitations on dependent variables include maximum line flows or bus voltage levels. In these cases, the value of the states cannot be independently set, but must be enforced indirectly. One method of enforcing an inequality constraint is to introduce a *penalty function* into the cost function. A penalty function is a function that is small when the state is far away from its limit, but becomes increasingly larger the closer the state is to its limit. Typical penalty functions include

$$p(h) = e^{kh} \quad k > 0 \tag{5.75}$$

$$p(h) = x^{2n} e^{kh} \quad n, k > 0 \tag{5.76}$$

$$p(h) = a x^{2n} e^{kh} + b e^{kh} \quad n, k, a, b > 0 \tag{5.77}$$

and the cost function becomes

$$C^*: C(u, x) + \lambda^T g(u, x) + p(h(u, x) - h^{\max}) \tag{5.78}$$

This cost equation is then minimized in the usual fashion by setting the appropriate derivatives to zero. This method not only has the advantage of simplicity of implementation, but also has several disadvantages. The first disadvantage is that the choice of penalty function is often a heuristic choice and can vary by application. A second disadvantage is that this method cannot enforce *hard* limitations on states, i.e., the cost function becomes large if the maximum is exceeded, but the state is allowed to exceed its maximum. In many applications this is not a serious disadvantage. If the power flow on a transmission line slightly exceeds its maximum, it is reasonable to assume that the power system will continue to operate, at least for a finite length of time. If, however, the physical limit is the height above ground for an airplane, then even a slightly negative altitude will have dire consequences. Thus the use of penalty functions to enforce limits must be used with caution and is not applicable for all systems.

Example 5.7

Repeat Example 5.5, except use penalty functions to limit the power flow across line 2–3 to 0.4 pu.

Solution

The power flow across line 2–3 in Example 5.5 is given by

$$P_{23} = V_2 V_3 Y_{23} \cos(\delta_2 - \delta_3 - \phi_{23}) - V_2^2 Y_{23} \cos \phi_{23}$$

$$= 0.467 \, \text{pu} \tag{5.79}$$

If P_{23} exceeds 0.4 pu, then the penalty function

$$p(h) = [1000V_2 V_3 Y_{23} \cos(\delta_2 - \delta_3 - \phi_{23}) - 1000V_2^2 Y_{23} \cos\phi_{23} - 400]^2 \qquad (5.80)$$

will be appended to the cost function. The partial derivatives remain the same with the exception of $[\|\, f/\, \|x]$, which becomes

$$\left[\frac{\partial f}{\partial x} \right] = \left[\frac{\partial f}{\partial P_1} \right]\left[\frac{\partial P_1}{\partial x} \right] + \left[\frac{\partial f}{\partial P_{23}} \right]\left[\frac{\partial P_{23}}{\partial x} \right] \qquad (5.81)$$

$$= (1+0.125P_1)\begin{bmatrix} V_1 V_2 Y_{12} \sin(\delta_1 - \delta_2 - \phi_{1,2}) \\ V_1 V_3 Y_{13} \sin(\delta_1 - \delta_3 - \phi_{1,3}) \end{bmatrix} + 2(P_{23} - 400)\begin{bmatrix} -V_2 V_3 Y_{23} \sin(\delta_2 - \delta_3 - \phi_{23}) \\ V_2 V_3 Y_{23} \sin(\delta_2 - \delta_3 - \phi_{23}) \end{bmatrix} \qquad (5.82)$$

Proceeding with the steepest gradient algorithm iterations yields the final constrained optimal generation scheduling:

$$\begin{bmatrix} P_1 \\ P_2 \\ P_3 \end{bmatrix} = \begin{bmatrix} 128.5 \\ 476.2 \\ 349.9 \end{bmatrix} \text{MW}$$

and $P_{23} = 400$ MW. The cost for this constrained scheduling is \$7882/MWh, which is slightly greater than the non constrained cost.

In the case where hard limits must be imposed, an alternate approach to enforcing the inequality constraints must be employed. In this approach, the inequality constraints are added as additional equality constraints with the inequality set equal to the limit (upper or lower) that is violated. This in essence introduces an additional set of Lagrangian multipliers. This is often referred to as the dual-variable approach, because each inequality has the potential of resulting in two equalities: one for the upper limit and one for the lower limit. However, the upper and lower limit cannot be simultaneously violated; thus, out of the possible set of additional Lagrangian multipliers only one of the two will be included at any given operating point and thus the dual limits are mutually exclusive.

Example 5.8

Repeat Example 5.7 using the dual-variable approach.

Solution

By introducing the additional equation

$$P_{23} = V_2 V_3 Y_{23} \cos(\delta_2 - \delta_3 - \phi_{23}) - V_2^2 Y_{23} \cos\phi_{23} = 0.400 \,\text{pu} \qquad (5.83)$$

to the equality constraints, an additional equation gets added to the set of $g(x)$. Therefore an additional unknown must be added to the state vector x to yield a solvable set of equations (three equations in three unknowns). Either P_{G2} or P_{G3} can be chosen as the additional unknown. In this example, P_{G3} will be chosen. The new system Jacobian becomes

$$\left[\frac{\partial g}{\partial x} \right] = \begin{bmatrix} \dfrac{\partial g_1}{\partial x_1} & \dfrac{\partial g_1}{\partial x_2} & \dfrac{\partial g_1}{\partial x_3} \\[2ex] \dfrac{\partial g_2}{\partial x_1} & \dfrac{\partial g_2}{\partial x_2} & \dfrac{\partial g_2}{\partial x_3} \\[2ex] \dfrac{\partial g_3}{\partial x_1} & \dfrac{\partial g_3}{\partial x_2} & \dfrac{\partial g_3}{\partial x_3} \end{bmatrix} \qquad (5.84)$$

where

$$\frac{\partial g_1}{\partial x_1} = V_2(V_1 Y_{12} \sin(\delta_2 - \delta_1 - \phi_{21}) + V_3 Y_{13} \sin(\delta_2 - \delta_3 - \phi_{23}))$$

$$\frac{\partial g_1}{\partial x_2} = -V_2 V_3 Y_{32} \sin(\delta_2 - \delta_3 - \phi_{23})$$

$$\frac{\partial g_1}{\partial x_3} = 0$$

$$\frac{\partial g_2}{\partial x_1} = -V_3 V_2 Y_{23} \sin(\delta_3 - \delta_2 - \phi_{32})$$

$$\frac{\partial g_2}{\partial x_2} = V_3(V_1 Y_{13} \sin(\delta_3 - \delta_1 - \phi_{31}) + V_2 Y_{23} \sin(\delta_3 - \delta_2 - \phi_{32})$$

$$\frac{\partial g_2}{\partial x_3} = 1$$

$$\frac{\partial g_3}{\partial x_1} = -V_2 V_3 Y_{23} \sin(\delta_2 - \delta_3 - \phi_{23})$$

$$\frac{\partial g_3}{\partial x_2} = V_2 V_3 Y_{23} \sin(\delta_2 - \delta_3 - \phi_{23})$$

$$\frac{\partial g_3}{\partial x_3} = 0$$

and

$$\left[\frac{\partial g}{\partial u}\right] = \begin{bmatrix} 1 \\ 0 \\ 0 \end{bmatrix}; \quad \left[\frac{\partial f}{\partial u}\right] = [1 + 0.025 P_{G2}]$$

Similar to Example 5.5, the chain rule is used to obtain $[\partial f/\partial x]$:

$$\left[\frac{\partial f}{\partial x}\right] = \left[\frac{\partial C}{\partial P_{G1}}\right]\left[\frac{\partial P_{G1}}{\partial x}\right] + \left[\frac{\partial C}{\partial P_{G3}}\right]\left[\frac{\partial P_{G3}}{\partial x}\right] \tag{5.85}$$

$$= (1 + 0.125 P_{G1}) \begin{bmatrix} V_1 V_2 Y_{12} \sin(\delta_1 - \delta_2 - \phi_{12}) \\ V_1 V_3 Y_{13} \sin(\delta_1 - \delta_3 - \phi_{13}) \\ 0 \end{bmatrix}$$

$$+ (1 + 0.050 P_{G3}) \begin{bmatrix} V_3 V_2 Y_{32} \sin(\partial_3 - \partial_2 - \phi_{32}) \\ -V_3\left(V_1 Y_{13} \sin(\delta_3 - \delta_1 - \phi_{31}) + V_2 Y_{23} \sin(\delta_3 - \delta_2 - \phi_{32})\right) \\ 0 \end{bmatrix} \tag{5.86}$$

Substituting these partial derivatives into the expression for ∇C of Equation 5.54 yields the same generation scheduling as Example 5.7.

5.3 State Estimation

In many physical systems, the system operating condition cannot be determined directly by an analytical solution of known equations using a given set of known, dependable quantities. More frequently, the system operating condition is determined by the measurement of system states at different points throughout the system. In many systems, more measurements are made than are necessary to uniquely determine the operating point. This redundancy is often purposely designed into the system to counteract the effect of inaccurate or missing data due to instrument failure. Conversely, not all of the states may be available for measurement. High temperatures, moving parts, or inhospitable conditions may make it difficult, dangerous, or expensive to measure certain system states. In this case, the missing states must be estimated from the rest of the measured information of the system. This process is often known as *state estimation* and is the process of estimating unknown states from measured quantities. State estimation gives the "best estimate" of the state of the system in spite of uncertain, redundant, and/or conflicting measurements. A good state estimation will smooth out small random errors in measurements, detect and identify large measurement errors, and compensate for missing data. This process strives to minimize the error between the (unknown) true operating state of the system and the measured states.

The set of measured quantities can be denoted by the vector z, which may include measurements of system states (such as voltage and current) or quantities that are functions of system states (such as power flows). Thus,

$$z^{\text{true}} = Ax \tag{5.87}$$

where
    x is the set of system states
    A is usually not square

The error vector is the difference between the measured quantities z and the true quantities:

$$e = z - z^{\text{true}} = z - Ax \tag{5.88}$$

Typically, the minimum of the square of the error is desired to negate any effects of sign differences between the measured and true values. Thus, a state estimator endeavors to find the minimum of the squared error, or a *least squares minimization*:

$$\text{minimize } \|e\|^2 = e^T \cdot e = \sum_{i=1}^{m} \left[z_i - \sum_{j=1}^{m} a_{ij} x_j \right]^2 \tag{5.89}$$

The squared error function can be denoted by $U(x)$ and is given by

$$U(x) = e^T \cdot e = (z - Ax)^T (z - Ax) \tag{5.90}$$

$$= (z^T - x^T A^T)(z - Ax)p \tag{5.91}$$

$$= (z^T z - z^T Ax - x^T A^T z + x^T A^T Ax) \tag{5.92}$$

Note that the product $z^T Ax$ is a scalar and so it can be equivalently written as

$$z^T Ax = (z^T Ax)^T = x^T A^T zp$$

Therefore the squared error function is given by

$$U(x) = z^T z - 2x^T A^T z + x^T A^T Ax \tag{5.93}$$

The minimum of the squared error function can be found by an unconstrained optimization where the derivative of the function with respect to the states x is set to zero:

$$\frac{\partial U(x)}{\partial x} = 0 - 2A^T z + 2A^T z + 2A^T Ax \tag{5.94}$$

Thus,

$$A^T Ax = A^T z \tag{5.95}$$

Thus, if $b = A^T z$ and $\hat{A} = A^T A$, then

$$\hat{A}x = b \tag{5.96}$$

This state vector x is the best estimate (in the squared error) to the system operating condition from which the measurements z were taken. The measurement error is given by

$$e = z^{\text{meas}} - Ax \tag{5.97}$$

In power system state estimation, the estimated variables are the voltage magnitudes and the voltage phase angles at the system buses. The input to the state estimator is the active and reactive powers of the system, measured either at the injection sites or on the transmission lines. The state estimator is designed to give the best estimates of the voltages and phase angles minimizing the effects of the measurement errors. All instruments add some degree of error to the measured values, but the problem is how to quantify this error and account for it during the estimation process.

If all measurements are treated equally in the least squares solution, then the less accurate measurements will affect the estimation as significantly as the more accurate measurements. As a result, the final estimation may contain large errors due to the influence of inaccurate measurements. By introducing a weighting matrix to emphasize the more accurate measurements more heavily than the less accurate measurements, the estimation procedure can then force the results to coincide more closely with the measurements of greater accuracy. This leads to the weighted least squares estimation:

$$\text{minimize } \|e\|^2 = e^T \cdot e = \sum_{i=1}^{m} w_i \left[z_i - \sum_{j=1}^{m} a_{ij} x_j \right]^2 \tag{5.98}$$

where w_i is a weighting factor reflecting the level of confidence in the measurement z_i.

In general, it can be assumed that the introduced errors have normal (Gaussian) distribution with zero mean and that each measurement is independent of all other measurements. This means that each measurement error is as likely to be greater than the true value as it is to be less than the true value. A zero mean Gaussian distribution has several attributes. The standard deviation of a zero mean Gaussian distribution is denoted by σ. This means that 68% of all measurements will fall within $\pm\sigma$ of the expected value, which is zero in a zero mean distribution. Further, 95% of all measurements will fall within $\pm2\sigma$ and 99% of all measurements will fall within $\pm3\sigma$. The variance of the measurement distribution is given by σ^2. This implies that if the variance of the measurements is relatively small, then the majority of measurements are close to the mean. One interpretation of this is that accurate measurements lead to small variance in the distribution.

This relationship between accuracy and variance leads to a straightforward approach from which a weighting matrix for the estimation can be developed. With measurements taken from a particular meter, the smaller the variance of the measurements (i.e., the more consistent they are), the greater the level of confidence in that set of measurements. A set of measurements that have a high level of confidence should have a higher weighting than a set of measurements that have a larger variance (and therefore less confidence). Therefore, a plausible weighting matrix that reflects the level of confidence in each measurement set is the inverse of the covariance matrix. Thus, measurements that come from instruments with good consistency (small variance) will carry greater weight than measurements that come from less accurate instruments (high variance). Thus, one possible weighting matrix is given by

$$W = R^{-1} = \begin{bmatrix} \dfrac{1}{\sigma_1^2} & 0 & \cdots & 0 \\ 0 & \dfrac{1}{\sigma_2^2} & \cdots & 0 \\ \vdots & \vdots & \vdots & \vdots \\ 0 & 0 & \cdots & \dfrac{1}{\sigma_m^2} \end{bmatrix} \tag{5.99}$$

where R is the covariance matrix for the measurements.

As in the linear least squares estimation, the nonlinear least squares estimation attempts to minimize the square of the errors between a known set of measurements and a set of weighted nonlinear functions:

$$\text{minimize } f = \|e\|^2 = e^T \cdot e = \sum_{i=1}^{m} \frac{1}{\sigma^2} [z_i - h_i(x)]^2 \tag{5.100}$$

where $x \in R^n$ is the vector of unknowns to be estimated, $z \in R^m$ is the vector of measurements, σ_i^2 is the variance of the ith measurement, and $h(x)$ is the nonlinear function vector relating x to z, where the measurement vector z can be a set of geographically distributed measurements, such as voltages and power flows

In state estimation, the unknowns in the nonlinear equations are the state variables of the system. The state values that minimize the error are found by setting the derivatives of the error function to zero:

$$F(x) = H_x^T R^{-1}[z - h(x)] = 0 \tag{5.101}$$

where

$$H_x = \begin{bmatrix} \dfrac{\partial h_1}{\partial x_1} & \dfrac{\partial h_1}{\partial x_2} & \cdots & \dfrac{\partial h_1}{\partial x_n} \\ \dfrac{\partial h_2}{\partial x_1} & \dfrac{\partial h_2}{\partial h_2} & \cdots & \dfrac{\partial h_2}{\partial x_n} \\ \vdots & \vdots & \vdots & \vdots \\ \dfrac{\partial h_m}{\partial x_1} & \dfrac{\partial h_m}{\partial x_2} & \cdots & \dfrac{\partial h_m}{\partial x_n} \end{bmatrix} \tag{5.102}$$

Note that Equation 5.101 is a set of nonlinear equations that must be solved using Newton–Raphson or another iterative numerical solver. In this case, the Jacobian of $F(x)$ is

$$J_F(x) = H_x^T(x) R^{-1} \frac{\partial}{\partial x}[z - h(x)] \tag{5.103}$$

$$= -H_x^T(x) R^{-1} H_x(x) \tag{5.104}$$

and the Newton–Raphson iteration becomes

$$[H_x^T(x^k)R^{-1}H_x(x^k)][x^{k-1}-x^k] = H_x^T(x^k)R^{-1}[z-h(x^k)] \qquad (5.105)$$

At convergence, $xk+1$ is equal to the set of states that minimize the error function f of Equation 5.100.

Example 5.9

The SCADA system for the power network shown in Figure 5.3 reports the following measurements and variances:

z_i	State	Measurement	Variance (s^2)
1	V_3	0.975	0.010
2	P_{13}	0.668	0.050
3	Q_{21}	−0.082	0.075
4	P_3	−1.181	0.050
5	Q_2	−0.086	0.075

Estimate the power system states.

Solution

The first step in the estimation process is to identify and enumerate the unknown states. In this example, the unknowns are $[x_1 \quad x_2 \quad x_3]^T = [\delta_2 \quad \delta_3 \quad V_3]^T$. After the states are identified, the next step in the estimation process is to identify the appropriate functions $h(x)$ that correspond to each of the measurements. The nonlinear function that is being driven to zero to minimize the weighted error is

$$F(x) = H_x^T R^{-1}[z-h(x)] = 0 \qquad (5.106)$$

where the set of $z - h(x)$ is given by

$$z_1 - h_1(x) = V_3 - x_3$$

$$z_2 - h_2(x) = P_{13} - \left(V_1 x_3 Y_{12}\cos(-x_2-\phi_{13}) - V_1^2 Y_{13}\cos\phi_{13}\right)$$

$$z_3 - h_3(x) = Q_{21} - \left(V_2 V_1 Y_{21}\sin(x_1-\phi_{21}) + V_2^2 Y_{21}\sin\phi_{21}\right)$$

$$z_4 - h_4(x) = P_3 - \left(x_3 V_1 Y_{31}\cos(x_2-\phi_{31}) + x_3 V_2 Y_{32}\cos(x_2-x_1-\phi_{32}) + x_3^2 Y_{33}\cos\phi_{33}\right)$$

$$z_5 - h_5(x) = Q_2 - \left[V_2 V_1 Y_{21}\sin(x_1-\phi_{21}) - V_2^2 Y_{22}\sin\phi_{22} + V_2 x_3 Y_{23}\sin(x_1-x_2-\phi_{23})\right]$$

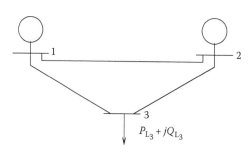

FIGURE 5.3 Example power system.

and the matrix of partial derivatives for the set of functions in Equation 5.106 is

$$
H_x =
\begin{bmatrix}
0 & 0 \\
0 & V_1 x_3 Y_{13} \sin(-x_2 - \phi_{13}) \\
V_1 V_2 Y_{21} \cos(x_1 - \phi_{21}) & 0 \\
x_3 V_2 Y_{21} \sin(x_2 - x\phi_{32}) & -x_3 V_1 Y_{31} \sin(x_2 - \phi_{31}) - x_3 V_2 Y_{32} \sin(x_2 - x_1 - \phi_{32}) \\
V_1 V_2 Y_{21} \cos(x_1 - \phi_{21}) + V_2 x_3 Y_{23} \cos(x_1 - x_2 - \phi_{23}) & -V_2 x_3 Y_{23} \cos(x_1 - x_2 - \phi_{23})
\end{bmatrix}
$$

$$
\times
\begin{bmatrix}
1 \\
V_1 Y_{13} \cos(-x_2 - \phi_{13}) \\
0 \\
V_1 Y_{31} \cos(x_2 - \phi_{31}) + V_2 Y_{32} \cos(x_2 - x_1 - \phi_{32}) + 2 x_3 Y_{33} \cos\phi_{33} \\
V_2 Y_{23} \sin(x_1 - x_2 - \phi_{23})
\end{bmatrix}
\tag{5.107}
$$

The covariance matrix of the measurements is

$$
R =
\begin{bmatrix}
\dfrac{1}{0.010^2} & & & & & \\
& \dfrac{1}{0.050^2} & & & & \\
& & \dfrac{1}{0.075^2} & & & \\
& & & \dfrac{1}{0.050^2} & & \\
& & & & \dfrac{1}{0.075^2} & \\
\end{bmatrix}
\tag{5.108}
$$

The Newton–Raphson iteration to solve for the set of states x that minimize the weighted errors is

$$
[H_x^T(x^k) R^{-1} H_x(x^k)][x^{k-1} - x^k] = H_x^T(x^k) R^{-1}[z - h(x^k)]
\tag{5.109}
$$

Iteration 1

The initial condition for the state estimation solution is the same flat start as for the power flow equations; namely, all angles are set to zero and all unknown voltage magnitudes are set to unity. The measurement functions $h(x)$ evaluated at the initial conditions are

$$
h(x^0) =
\begin{bmatrix}
1.0000 \\
0.0202 \\
-0.0664 \\
-0.0198 \\
-0.1914
\end{bmatrix}
$$

The matrix of partials evaluated at the initial condition yields

$$H_x^0 = \begin{bmatrix} 0 & 0 & 1.0000 \\ 0 & -10.0990 & -1.0099 \\ -0.2257 & 0 & 0 \\ -9.9010 & 20.0000 & 1.9604 \\ -1.2158 & 0.9901 & -9.9010 \end{bmatrix}$$

The nonlinear functions in Equation 5.106 are

$$F(x^0) = \begin{bmatrix} 0.5655 \\ -1.4805 \\ -0.2250 \end{bmatrix}$$

The incremental updates for the states are

$$\Delta x^1 = \begin{bmatrix} -0.0119 \\ -0.0625 \\ -0.0154 \end{bmatrix}$$

leading to the updated states

$$\begin{bmatrix} \delta_2^1 \\ \delta_3^1 \\ V_3^1 \end{bmatrix} = \begin{bmatrix} -0.0119 \\ -0.0625 \\ -0.9846 \end{bmatrix}$$

where δ_2 and δ_3 are in radians. The error at iteration 0 is

$$\varepsilon^0 = 1.4805$$

Iteration 2

The updated values are used to recalculate the Newton–Raphson iterations:

$$h(x^1) = \begin{bmatrix} 0.9846 \\ 0.6585 \\ -0.0634 \\ -1.1599 \\ -0.724 \end{bmatrix}$$

The matrix of partials is

$$H_x^1 = \begin{bmatrix} 0 & 0 & 1.0000 \\ 0 & -9.9858 & -0.3774 \\ -0.2660 & 0 & 0 \\ -9.6864 & 19.5480 & 0.7715 \\ -0.7468 & 0.4809 & -9.9384 \end{bmatrix}$$

The nonlinear function evaluated at the updated values yields

$$F(x^1) = \begin{bmatrix} 0.0113 \\ -0.0258 \\ 0.0091 \end{bmatrix}$$

The incremental updates for the states are

$$\Delta x^2 = \begin{bmatrix} 0.0007 \\ -0.0008 \\ 0.0013 \end{bmatrix}$$

leading to the updated states

$$\begin{bmatrix} \delta_2^2 \\ \delta_3^2 \\ V_3^2 \end{bmatrix} = \begin{bmatrix} -0.0113 \\ -0.0633 \\ 0.9858 \end{bmatrix}$$

The error at Iteration 1 is

$$\varepsilon^1 = 0.0258$$

The iterations are obviously converging. At convergence, the states that minimize the weighted measurement errors are

$$x = \begin{bmatrix} -0.0113 \\ -0.0633 \\ 0.9858 \end{bmatrix}$$

This concludes the discussion of the most commonly used computational methods for power system analysis. This chapter describes only the basic approaches to power flow, OPF, and state estimation. These methods have been utilized for several decades, yet improvements in accuracy and speed are constantly being proposed in the technical literature.

II

Power System Transients

Pritindra Chowdhuri

Pritindra Chowdhuri received his BSc in physics and MSc in applied physics from University of Calcutta, Kolkata, India, MS in electrical engineering from Illinois Institute of Technology, Chicago, Illinois, and DEng in engineering science from Rensselaer Polytechnic Institute, Troy, New York.

He has worked for General Electric Company in Pittsfield, Massachusetts, Schenectady, New York, and Erie, Pennsylvania, as a development engineer and for Los Alamos National Laboratory in Los Alamos, New Mexico, as a staff member. He joined Tennessee Technological University (TTU) in 1986 as a professor of electrical engineering in the Center for Electric Power (later Center for Energy Systems Research) and the Department of Electrical Engineering. He retired from TTU in 2005 as emeritus professor.

6

Characteristics of Lightning Strokes

Francisco De la Rosa
Electric Power Systems

6.1 Introduction

Lightning, one of the most spectacular events of Mother Nature, started to appear significantly demystified after Franklin showed its electric nature with his famous electrical kite experiment in 1752. Although a great deal of research on lightning followed Franklin's observation, lightning continues to be a topic of considerable interest for investigation (Uman, 1969, 1987). This is particularly true for the improved design of electric power systems, since lightning-caused interruptions and equipment damage during thunderstorms stand as the leading causes of failures in the electric utility industry. It is prudent to state that in spite of the impressive amount of lightning research conducted mostly during the last 50 years, the physics of the phenomenon is not yet fully understood. The development of powerful digital recorders with enough bandwidth to capture the microstructure of lightning waveforms and the advent of digital computers and fiber optic communication along with sophisticated direction finding sensors during this period facilitated the extraordinary evolution of lightning monitoring and detection techniques. This allowed for a more accurate description of lightning and the characterization of lightning parameters that are relevant for power system protection.

On a worldwide scale, most lightning currents (over 90%) are of negative polarity. However, it has to be acknowledged that in some parts of the world, mostly over the Northern Hemisphere, the fraction of positive lightning currents can be substantial. A formal assessment of the effects of positive lightning on electric power systems will require using the corresponding parameters, since they may show large deviations from negative flashes in peak current, charge, front duration, and flash duration as it will be described.

6.2 Lightning Generation Mechanism

6.2.1 First Strokes

The wind updrafts and downdrafts that take place in the atmosphere create a charging mechanism that separates electric charges, leaving negative charge at the bottom and positive charge at the top of the cloud. As charge at the bottom of the cloud keeps growing, the potential difference between cloud and ground, which is positively charged, grows as well. This process will continue until air breakdown occurs. See Figure 6.1.

The way in which a cloud-to-ground flash develops involves a stepped leader that starts traveling downward following a preliminary breakdown at the bottom of the cloud. This involves a positive pocket of charge, as illustrated in Figure 6.1. The stepped leader travels downward in steps several tens of meters in length and pulse currents of at least 1 kA in amplitude (Uman, 1969). When this leader is near ground, the potential to ground can reach values as large as 100 MV before the attachment process with one of the upward streamers is completed. Figure 6.2 illustrates the final step when the upward streamer developing from a transmission line tower intercepts the downward leader.

The connection point-to-ground is not decided until the downward leader is some tens of meters above the ground plane. The downward leader will be attached to one of the growing upward streamers developing from elevated objects such as trees, chimneys, power lines, telecommunication towers, etc. It is actually under this principle that lightning protection rods work, that is, they have to be strategically located so that they can trigger upward streamers that can develop into attachment points to downward leaders approaching the protected area. For this to happen, upward streamers developing from

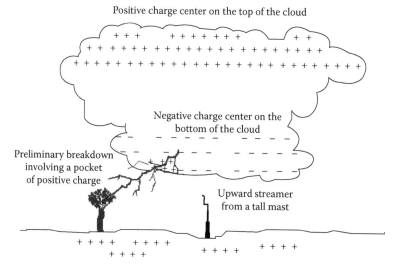

FIGURE 6.1 Separation of electric charge within a thundercloud.

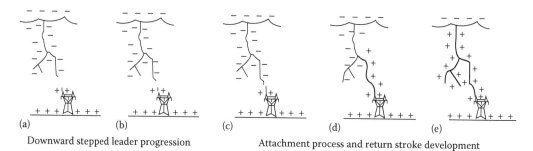

(a)   (b)   (c)   (d)   (e)

Downward stepped leader progression   Attachment process and return stroke development

FIGURE 6.2 Attachment between downward and upward leaders in a cloud-to-ground flash.

TABLE 6.1 Lightning Current Parameters for Negative Flashes

Parameters	Units	Sample Size	Value Exceeding in 50% of the Cases
Peak current (minimum 2 kA)	kA		
First strokes		101	30
Subsequent strokes		135	12
Charge (total charge)	C		
First strokes		93	5.2
Subsequent strokes		122	1.4
Complete flash		94	7.5
Impulse charge (excluding continuing current)	C		
First strokes		90	4.5
Subsequent strokes		117	0.95
Front duration (2 kA to peak)	μs		
First strokes		89	5.5
Subsequent strokes		118	1.1
Maximum di/dt	kA/μs		
First strokes		92	12
Subsequent strokes		122	40
Stroke duration (2 kA to half peak value on the tail)	μs		
First strokes		90	75
Subsequent strokes		115	32
Action integral ($\int i^2 dt$)	A^2		
First strokes	s	91	5.5×10^4
Subsequent strokes		88	6.0×10^3
Time interval between strokes	ms	133	33
Flash duration	ms		
All flashes		94	13
Excluding single-stroke flashes		39	180

Source: Adapted from Berger, K. et al., *Electra*, 41, 23, 1975.

protected objects within the shielded area have to compete unfavorably with those developing from the tip of the lightning rods, which are positioned at a higher elevation.

Just after the attachment process takes place, the charge that is lowered from the cloud base through the leader channel is conducted to ground as a breakdown current pulse, known as the return stroke, travels upward along the channel. The return stroke velocity is around one-third the speed of light. The median peak current value associated to the return stroke is reported to be on the order of 30 kA, with rise time and time to half values around 5 and 75 μs, respectively. See Table 6.1 adapted from Berger et al. (1975).

Associated to this charge-transfer mechanism (an estimated 5 C total charge is lowered to ground through the stepped leader) are the electric and magnetic field changes that can last several milliseconds. These fields can be registered at remote distances from the channel and it is under this principle that lightning sensors work to produce the information necessary to monitor cloud-to-ground lightning.

6.2.2 Subsequent Strokes

After the negative charge from the cloud base has been transferred to ground, additional charge can be made available on the top of the channel when discharges known as J and K processes take place within the cloud (Uman, 1969). This can lead to a number of subsequent strokes of lightning following the first stroke. A so-called dart leader develops from the top of the channel lowering charges, typically of 1 C, until recently believed to follow the same channel of the first stroke. Studies conducted in the past few years, however, suggest that around half of all lightning discharges to earth, both single- and multiple-stroke flashes, may strike ground at more than one point, with separation between channel terminations on ground varying from 0.3 to 7.3 km and a geometric mean of 1.3 km (Thottappillil et al., 1992).

Generally, dart leaders develop no branching and travel downward at velocities of around 3×10^6 m/s. Subsequent return strokes have peak currents usually smaller than first strokes but faster zero-to-peak rise times. The mean interstroke interval is about 60 ms, although intervals as large as a few tenths of a second can be involved when a so-called continuing current flows between strokes (this happens in 25%–50% of all cloud-to-ground flashes). This current, which is on the order of 100 A, is associated to charges of around 10 C and constitutes a direct transfer of charge from cloud to ground (Uman, 1969).

The percentage of single-stroke flashes presently suggested by CIGRE of 45% (Anderson and Eriksson, 1980) are considerably higher than the following figures recently obtained from experimental results: 17% in Florida (Rakov et al., 1994), 14% in New Mexico (Rakov et al., 1994), 21% in Sri Lanka (Cooray and Jayaratne, 1994), and 18% in Sweden (Cooray and Perez, 1994).

6.3 Parameters of Importance for Electric Power Engineering

6.3.1 Ground Flash Density

Ground flash density, frequently referred to as GFD or N_g, is defined as the number of lightning flashes striking ground per unit area and per year. Usually it is a long-term average value and ideally it should take into account the yearly variations that take place within a solar cycle—believed to be the period within which all climatic variations that produce different GFD levels occur.

A 10-year average GFD map of the continental United States (*IEEE Guide*, 2005) obtained by and reproduced here with permission from Vaisala, Inc. of Tucson, Arizona, is presented in Figure 6.3. Note the considerable large GFD levels affecting remarkably the state of Florida, as well as all the southern states along the Gulf of Mexico (Alabama, Mississippi, Louisiana, and Texas). High GFD levels are also

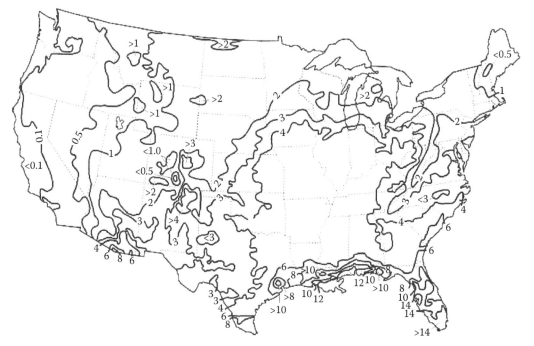

FIGURE 6.3 Ten-year average GFD map of the United States. (Reproduced from Vaisala, Inc. of Tucson, AZ., Standards Information Network, How to protect your house and its contents from lightning, *IEEE Guide for Surge Protection of Equipment Connected to AC Power and Communication Circuits*, IEEE Press, New York, 2005. With permission.)

observed in the southeastern states of Georgia and South Carolina. To the west side, Arizona is the only state with GFD levels as high as 8 flashes/km²/year. The lowest GFD levels (<0.5 flashes/km²/year) are observed in the western states, notably in California, Oregon, and Washington on the Pacific Ocean, in a spot area of Colorado and in the northeastern state of Maine on the Atlantic Ocean.

It is interesting to mention that a previous (a 5-year average) version of this map showed levels of around 6 flashes/km²/year also in some areas of Illinois, Iowa, Missouri, and Indiana, not seen in the present version. This is often the result of short-term observations that do not reflect all climatic variations that take place in a longer time frame.

The low incidence of lightning does not necessarily mean an absence of lightning-related problems. Power lines, for example, are prone to failures even if GFD levels are low when they pass through high-resistivity soils like deserts or when lines span across hills or mountains, where ground wire or lightning arrester earthing becomes difficult. An exception to this may be a procedure being tested by some utilities to protect high GFD (frequently stricken) transmission line structures on elevated spots in mountainous areas by installing surge arresters directly across the insulator strings (Munukutla et al., 2010). This may prove to be a cost-effective method to deal with lightning-related outages in transmission lines where achieving low footing resistance values may become prohibitive.

The GFD level is an important parameter to consider for the design of electric power and telecommunication facilities. This is due to the fact that power line performance and damage to power and telecommunication equipment are considerably affected by lightning. Worldwide, lightning accounts for most of the power supply interruptions in distribution lines and it is a leading cause of failures in transmission systems. In the United States alone, an estimated 30% of all power outages are lightning related on annual average, with total costs approaching $1 billion (Kithil, 1998).

In De la Rosa et al. (1998), it is discussed how to determine GFD as a function of TD (thunder days or keraunic level) or TH (thunder hours). This is important where GFD data from lightning location systems are not available. Basically, any of these parameters can be used to get a *rough* approximation of GFD. Using the expressions described in Anderson et al. (1984) and MacGorman et al. (1984), respectively,

$$N_g = 0.04TD^{1.25} \text{ flashes/km}^2\text{/year} \tag{6.1}$$

$$N_g = 0.054TH^{1.1} \text{ flashes/km}^2\text{/year} \tag{6.2}$$

6.3.2 Current Peak Value

Regarding current peak values, first strokes are associated with peak currents around two to three times larger than subsequent strokes. According to De la Rosa et al. (1998), electric field records, however, suggest that subsequent strokes with higher electric field peak values may be present in one out of three cloud-to-ground flashes. These may be associated with current peak values greater than the first stroke peak.

Tables 6.1 and 6.2 are summarized and adapted from Berger et al. (1975) for negative and positive flashes, respectively. They present statistical data for 127 cloud-to-ground flashes, 26 of them positive, measured in Switzerland. These are the types of lightning flashes known to hit flat terrain and structures of moderate height. This summary, for simplicity, shows only the 50% or statistical value, based on the log-normal approximations to the respective statistical distributions. These data are amply used as primary reference in the literature on both lightning protection and lightning research.

The action integral is an interesting concept, that is, the energy that would be dissipated in a 1 Ω resistor if the lightning current were to flow through it. This is a parameter that can provide some insight into the understanding of forest fires and on damage to power equipment, including surge arresters, in power line installations. All the parameters presented in Tables 6.1 and 6.2 are estimated from current oscillograms with the shortest measurable time being 0.5 μs (Berger and Garbagnati, 1984). It is thought that the distribution of front duration might be biased toward larger values and the distribution of di/dt toward smaller values (De la Rosa et al., 1998).

TABLE 6.2 Lightning Current Parameters for Positive Flashes

Parameters	Units	Sample Size	Value Exceeding in 50% of the Cases
Peak current (minimum 2 kA)	kA	26	35
Charge (total charge)	C	26	80
Impulse charge (excluding continuing current)	C	25	16
Front duration (2 kA to peak)	μs	19	22
Maximum di/dt	kA/μs	21	2.4
Stroke duration (2 kA to half peak value on the tail)	μs	16	230
Action integral ($\int i^2 dt$)	A²s	26	6.5×10^5
Flash duration	ms	24	85

Source: Adapted from Berger, K. et al., *Electra*, 41, 23, 1975.

6.3.3 Correlation between Current and Other Parameters of Lightning

Lightning parameters are sometimes standardized for the purpose of the assessment of lightning performance of specific power line designs (IEEE Std 1410-1997, 1997). Although this can be adequately used to determine effectiveness of different lightning protection methods in a comparative basis, it is important to understand the limitations that this approach may encompass: Lightning parameters may show considerable deviations often caused by spatial and temporal variations (Torres, 1998). On the other hand, gathering data on lightning parameters other than current makes it an impossible task. Among the parameters which can be associated with lightning damage are lightning peak current (i_p), rate of rise (T_{front} and peak di/dt), charge transfer (Q_{imp} and Q_{flash}), and energy (action integral), as described in De la Rosa et al. (2000). Unfortunately, contemporary lightning detection systems (LDS) are not able to provide accurate estimates of many of these parameters, including the current, since they are designed only to sense and record radiated electric and magnetic fields within hundreds of kilometers from the source. Nevertheless, it seems tangible to envision that even with the limited accuracy of lightning current derived from LDS, it will be possible to infer other lightning parameters directly linked to lightning damage. This will allow us to continue assessing the efficacy of lightning mitigation methods based on records of lightning current and other parameters obtained from LDSs.

Table 6.3 shows an interesting correlation study conducted by Dellera (1997) where he found moderate-to-high correlation coefficients between lightning peak current with other parameters measured in a number of research experiments that involved lightning strikes to instrumented towers. This work presented a way to obtain estimates of the total probability of a specific range of simultaneous values (e.g., $i > I_o$ and $Q > Q_o$), thereby refining the probability estimates for the conditions under which a lightning-related failure may occur. Table 6.3 comprises relevant findings for positive flashes, negative first strokes, and negative subsequent strokes. The table entries for parameters in negative discharges provide more accurate results, due to broader bandwidth (higher frequency) recording instruments. Table entries that are not filled in (—) were not analyzed by Dellera.

To illustrate the interpretation of the table, we observe that a moderate-to-high correlation is found between lightning current and all but peak rate of rise in positive flashes. Therefore, extreme heating should be expected in arcing or transient currents conducted through protective devices following insulation flashover produced by positive lightning flashes. Note that lightning parameters associated with heat are charge and action integral and that rate-of-change of lightning current is connected with inductive effects, which do not show strong from the correlation table.

Similarly, overvoltages from negative lightning strokes should be associated with peak current producing large inductive overvoltages especially due to subsequent strokes because of their larger di/dt (by a factor of 3 or more) relative to first strokes. Heating effects, however, are loosely correlated with peak current, since correlation coefficient for the total charge (Q_{flash}) is poor. It is possible that consideration

TABLE 6.3 Correlation between Lightning Parameters

Lightning Parameter	Correlation (Correlation Coefficient)	Data Source
Positive flashes		
Front time (T_{front})	Low (0.18)	Berger and Garbagnati (1984)
Peak rate of rise (di/dt)	Moderate (0.55)	Berger and Garbagnati (1984)
Impulse charge (Q_{imp})	High (0.77)	Berger and Garbagnati (1984)
Flash charge (Q_{flash})	Moderate (0.59)	Berger and Garbagnati (1984)
Impulse action integral (W_{imp})	—	
Flash action integral (W_{flash})	High (0.76)	Berger and Garbagnati (1984)
Negative first strokes		
Front time (T_{front})	Low (—)	Weidman and Krider (1984)[a]
Peak rate of rise (di/dt)	Moderate/high (—)	Weidman and Krider (1984)[a]
Impulse charge (Q_{imp})	High (0.75)	Berger and Garbagnati (1984)
Flash charge (Q_{flash})	Low (0.29)	Berger and Garbagnati (1984)
Impulse action integral (W_{imp})	High (0.86)	Berger and Garbagnati (1984)
Flash action integral (W_{flash})	—	
Negative subsequent strokes		
Front time (T_{front})	Low (0.13)	Fisher et al. (1993)
Peak rate of rise (di/dt)	High (0.7–0.8)	Fisher et al. (1993) and Leteinturier et al. (1990)[b]
Impulse charge (Q_{imp})	—	
Impulse action integral (W_{imp})	—	

Source: Adapted from Dellera, L., Lightning parameters for protection: And updated approach, *CIGRE 97 SC33*, WG33.01, 17 IWD, August 1997.

[a] Inferred from measurements of electric fields propagated over salt water.

[b] 30%–90% slope, which corresponds to an "average" di/dt (triggered lightning studies).

of all stroke peak currents and interstroke intervals in negative flashes, which are available in some modern LDS, will prove a more deterministic means to infer heating due to negative flashes.

6.4 Incidence of Lightning to Power Lines

One of the most accepted expressions to determine the number of direct strikes to an overhead line in an open ground with no nearby trees or buildings is described by Eriksson (1987):

$$N = N_g \left(\frac{28h^{0.6} + b}{10} \right) \tag{6.3}$$

where
    h is the pole or tower height (m)—negligible for distribution lines
    b is the structure width (m)
    N_g is the GFD (flashes/km²/year)
    N is the number of flashes striking the line/100 km/year

For unshielded distribution lines, this is comparable to the fault index due to direct lightning hits. For transmission lines, this is an indicator of the exposure of the line to direct strikes (the response of the line being a function of overhead ground wire shielding angle on one hand and on conductor-tower surge impedance and footing resistance on the other hand).

Note the dependence of the incidence of strikes to the line with height of the structure. This is important since transmission lines are several times taller than distribution lines, depending on their operating voltage level.

Also important is that in the real world, power lines are to different extents shielded by nearby trees or other objects along their corridors. This will decrease the number of direct strikes estimated by Equation 6.3 to a degree determined by the distance and height of the objects. In IEEE Std. 1410-1997, a shielding factor is proposed to estimate the shielding effect of nearby objects to the line. An important aspect of this reference work is that objects within 40 m from the line, particularly if they are equal or higher than around 20 m, can attract most of the lightning strikes that would otherwise hit the line. Likewise, the same objects would produce insignificant shielding effects if located beyond 100 m from the line. On the other hand, sectors of lines extending over hills or mountain ridges may increase the number of strikes to the line.

The aforementioned effects may in some cases cancel one another so that the estimation obtained form Equation 6.3 can still be valid. However, it is recommended that any assessment of the incidence of lightning strikes to a power line be performed by taking into account natural shielding and orographic conditions (terrain undulations) along the line route. This also applies when identifying troubled sectors of the line for the installation of metal oxide surge arresters to improve its lightning performance. For example, those segments of a distribution feeder crossing over hills with little natural shielding would greatly benefit from surge arrester protection.

Finally, the inducing effects of lightning, also described in Anderson et al. (1984) and De la Rosa et al. (1998), have to be considered to properly estimate distribution line lightning performance or when estimating the outage rate improvement after application of any mitigation action. Lightning strokes terminating on ground close to distribution lines have the potential to develop overvoltages large enough to cause insulation flashover. Under certain conditions, like in circuits without grounded neutral, with low critical flashover voltages, high GFD levels, or located on high-resistivity terrain, the number of outages produced by close lightning can considerably surpass those due to direct strikes to the line.

6.5 Conclusions

Parameters that are important for the assessment of lightning performance of power transmission and distribution lines or for evaluation of different protection methods are lightning current and GFD. The former provides a means to appraise the impact of direct hits on power lines or substations and the latter provides an indication of how often this phenomenon may occur. There are, however, other lightning parameters that can be related to the probability of insulation flashover and heating effects in surge protective devices, which are difficult to obtain from conventional lightning detection equipment. Some correlation coefficients observed between lightning peak current and other relevant parameters obtained from significant experiments in instrumented towers are portrayed in this review.

Aspects like different methods available to calculate shielding failures and back flashovers in transmission lines or the efficacy of remedial measures to improve lightning performance of electrical networks are not covered in this summarizing review. Among these, overhead ground wires, metal oxide surge arresters, increased insulation, or use of wood as an arc-quenching device can only be mentioned. The reader is encouraged to further look at the suggested references or to get experienced advice for a more comprehensive understanding on the subject.

References

Anderson, R.B. and Eriksson, A.J., Lightning parameters for engineering applications, *Electra*, 69, 65–102, March 1980.
Anderson, R.B., Eriksson, A.J., Kroninger, H., Meal, D.V., and Smith, M.A., Lightning and thunderstorm parameters, *IEE Lightning and Power Systems Conference Publication No. 236*, London, U.K., pp. 57–61, 1984.

Berger, K., Anderson, R.B., and Kroninger, H., Parameters of lightning flashes, *Electra*, 41, 23–37, July 1975.

Berger, K. and Garbagnati, E., Lightning current parameters, results obtained in Switzerland and in Italy, in *Proceedings of URSI Conference*, Florence, Italy, p. 13, 1984.

Cooray, V. and Jayaratne, K.P.S., Characteristics of lightning flashes observed in Sri Lanka in the tropics, *J. Geophys. Res.*, 99, 21,051–21,056, 1994.

Cooray, V. and Perez, H., Some features of lightning flashes observed in Sweden, *J. Geophys. Res.*, 99, 10,683–10,688, 1994.

De la Rosa, F., Cummins, K., Dellera, L., Diendorfer, G., Galvan, A., Huse, J., Larsen, V., Nucci, C.A., Rachidi, F., Rakov, V., Torres, H. et al., Characterization of lightning for applications in electric power systems, *CIGRE Report#172*, TF33.01.02, December 2000.

De la Rosa, F., Nucci, C.A., and Rakov, V.A., Lightning and its impact on power systems, in *Proceedings of International Conference on Insulation Coordination for Electricity Development in Central European Countries*, Zagreb, Croatia, p. 44, 1998.

Dellera, L., Lightning parameters for protection: An updated approach, *CIGRE 97 SC33*, WG33.01, 17 IWD, August 1997.

Eriksson, A.J., The incidence of lightning strikes to power lines, *IEEE Trans. Power Delivery*, 2(2), 859–870, July 1987.

Fisher, R.J., Schnetzer, G.H., Thottappillil, R., Rakov, V.A., Uman, M.A., and Goldberg, J.D., Parameters of triggered-lightning flashes in Florida and Alabama, *J. Geophys. Res.*, 98(D12), 22,887–22, 902, 1993.

IEEE Guide for Surge Protection of Equipment Connected to AC Power and Communications Circuits, IEEE Press, New York, 2005, Standards Information Network, IEEE Press, p. 3.

IEEE Std 1410–1997, *IEEE Guide for Improving the Lightning Performance of Electric Power Distribution Lines*, IEEE PES, December 1997, Section 5.

Kithil, R., Lightning protection codes: Confusion and costs in the USA, in *Proceedings of the 24th International Lightning Protection Conference*, Birmingham, U.K., September 16, 1998.

Leteinturier, C., Weidman, C., and Hamelin, J., Current and electric field derivatives in triggered lightning return strokes, *J. Geophys. Res.*, 95(D1), 811–828, 1990.

MacGorman, D.R., Maier, M.W., and Rust, W.D., Lightning strike density for the contiguous United States from thunderstorm duration records, NUREG/CR-3759, Office of Nuclear Regulatory Research, U.S. Nuclear Regulatory Commission, Washington, DC, 44 pp., 1984.

Munukutla, K., Vittal, V., Heydt, G., Chipman, D., and Brian Keel, D., A practical evaluation for surge arrester placement for transmission line lightning protection, *IEEE Trans. Power Delivery*, 25(3), 1742–1748, July 2010.

Rakov, M.A., Uman, M.A., and Thottappillil, R., Review of lightning properties from electric field and TV observations, *J. Geophys. Res.*, 99, 10,745–10,750, 1994.

Thottappillil, R., Rakov, V.A., Uman, M.A., Beasley, W.H., Master, M.J., and Shelukhin, D.V., Lightning subsequent stroke electric field peak greater than the first stroke and multiple ground terminations, *J. Geophys. Res.*, 97, 7503–7509, 1992.

Torres, H., Variations of lightning parameter magnitudes within space and time, in *24th International Conference on Lightning Protection*, Birmingham, U.K., September 1998.

Uman, M.A. *Lightning*, Dover, New York, 1969, Appendix E.

Uman, M.A., *The Lightning Discharge, International Geophysics Series*, Vol. 39, Academic Press, Orlando, FL, Chapter 1, 1987.

Weidman, C.D. and Krider, E.P., Variations à l'échelle submicroseconde des champs électromagnetiqués rayonnes par la foudre, *Ann. Telecommun.*, 39, 165–174, 1984.

7

Overvoltages Caused by Direct Lightning Strokes

Pritindra
Chowdhuri
*Tennessee Technological
University*

A lightning stroke is defined as a direct stroke if it hits either the tower or the shield wire or the phase conductor. This is illustrated in Figure 7.1. When the insulator string at a tower flashes over by direct hit either to the tower or to the shield wire along the span, it is called a backflash; if the insulator string flashes over by a strike to the phase conductor, it is called a shielding failure for a line shielded by shield wires. Of course, for an unshielded line, insulator flashover is caused by backflash when the stroke hits the tower or by direct contact with the phase conductor. In the analysis of performance and protection of power systems, the most important parameter that must be known is the insulation strength of the system. It is not a unique number. It varies according to the type of the applied voltage, for example, DC, AC, lightning, or switching surges. For the purpose of lightning performance, the insulation strength has been defined in two ways: basic impulse insulation level (BIL) and critical flashover voltage (CFO or V_{50}). BIL has been defined in two ways. The statistical BIL is the crest value of a standard (1.2/50 μs) lightning impulse voltage, which the insulation will withstand with a probability of 90% under specified conditions. The conventional BIL is the crest value of a standard lightning impulse voltage, which the insulation will withstand for a specific number of applications under specified conditions. CFO or V_{50} is the crest value of a standard lightning impulse voltage, which the insulation will withstand during 50% of the applications. In this chapter, we will use the conventional BIL as the insulation strength under lightning impulse voltages. Analysis of direct strokes to overhead lines can be divided into two classes: unshielded lines and shielded lines. The first discussion involves the unshielded lines.

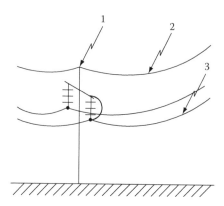

FIGURE 7.1 Illustration of direct lightning strokes to line: (1) backflash caused by direct stroke to tower, (2) backflash caused by direct stroke to shield wire, and (3) insulator flashover by direct stroke to phase conductor (shielding failure).

7.1 Direct Strokes to Unshielded Lines

If lightning hits one of the phase conductors, the return-stroke current splits into two equal halves, each half traveling in either direction of the line. The traveling current waves produce traveling voltage waves, which are given by

$$V = \frac{Z_o I}{2} \tag{7.1}$$

where

I is the return-stroke current

Z_o is the surge impedance of the line given by $Z_o = (L/C)^{1/2}$, and L and C are the series inductance and capacitance to ground per meter length of the line

These traveling voltage waves stress the insulator strings from which the line is suspended as these voltages arrive at the succeeding towers. The traveling voltages are attenuated as they travel along the line by ground resistance and mostly by the ensuing corona enveloping the struck line. Therefore, the insulators of the towers adjacent to the struck point are most vulnerable. If the peak value of the voltage, given by Equation 7.1, exceeds the BIL of the insulator, then it might flash over, causing an outage. The minimum return-stroke current that causes an insulator flashover is called the critical current, I_c, of the line for the specified BIL. Thus, following Equation 7.1,

$$I_c = \frac{2\text{BIL}}{Z_o} \tag{7.2}$$

Lightning may hit one of the towers. The return-stroke current then flows along the struck tower and over the tower-footing resistance before being dissipated in the earth. The estimation of the insulator voltage in that case is not simple, especially because there has been no consensus about the modeling of the tower in estimating the insulator voltage. In the simplest assumption, the tower is neglected. Then, the tower voltage, including the voltage of the cross arm from which the insulator is suspended, is the voltage drop across the tower-footing resistance given by $V_{tf} = IR_{tf}$, where R_{tf} is the tower-footing resistance. Neglecting the power-frequency voltage of the phase conductor, this is then the voltage across the insulator.

It should be noted that this voltage will be of opposite polarity to that for stroke to the phase conductor for the same polarity of the return-stroke current.

Neglecting the tower may be justified for short towers. The effect of the tower for transmission lines must be included in the estimation of the insulator voltage. For these cases, the tower has also been represented as an inductance. Then the insulator voltage is given by $V_{ins} = V_{tf} + L(dI/dt)$, where L is the inductance of the tower.

However, it is known that voltages and currents do travel along the tower. Therefore, the tower should be modeled as a vertical transmission line with a surge impedance, Z_t, where the voltage and current waves travel with a velocity, v_t. The tower is terminated at the lower end by the tower-footing resistance, R_{tf}, and at the upper end by the lightning channel that can be assumed to be another transmission line of surge impedance, Z_{ch}. Therefore, the traveling voltage and current waves will be repeatedly reflected at either end of the tower while producing voltage at the cross arm, V_{ca}. The insulator from which the phase conductor is suspended will then be stressed at one end by V_{ca} (to ground) and at the other end by the power-frequency phase-to-ground voltage of the phase conductor. Neglecting the power-frequency voltage, the insulator voltage, V_{ins}, will be equal to the cross-arm voltage, V_{ca}. This is schematically shown in Figure 7.2a. The initial voltage traveling down the tower, V_{to}, is $V_{to}(t) = Z_t I(t)$, where $I(t)$ is the initial tower current that is a function of time, t. The voltage reflection coefficients at the two ends of the tower are given by

$$a_{r1} = \frac{R_{tf} - Z_t}{R_{tf} + Z_t} \quad \text{and} \quad a_{r2} = \frac{Z_{ch} - Z_t}{Z_{ch} + Z_t} \tag{7.3}$$

Figure 7.2b shows the lattice diagram of the progress of the multiply reflected voltage waves along the tower. The lattice diagram, first proposed by Bewley (1951), is the space–time diagram that shows the position and direction of motion of every incident and reflected and transmitted wave on the system at every instant of time. In Figure 7.2, if the heights of the tower and the cross arm are h_t and h_{ca}, respectively, and the velocity of the traveling wave along the tower is v_t, then the time of travel from the tower top to its foot is $\tau_t = h_t/v_t$, and the time of travel from the cross arm to the tower foot is $\tau_{ca} = h_{ca}/v_t$.

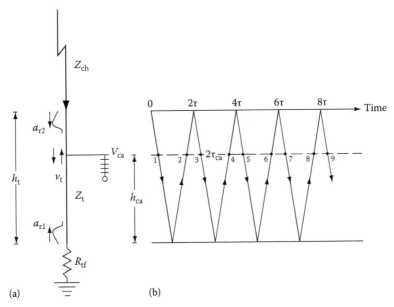

(a)   (b)

FIGURE 7.2 Lightning channel striking tower top: (a) schematic of struck tower; (b) voltage lattice diagram.

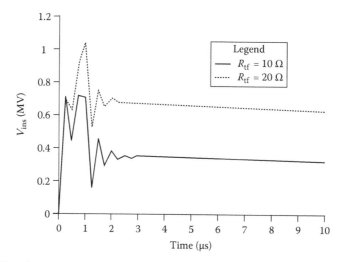

FIGURE 7.3 Profiles of insulator voltage for an unshielded line for a lightning stroke to tower. Tower height = 30 m; cross-arm height = 27 m; phase-conductor height = 25 m; cross-arm width = 2 m; return-stroke current = 1/50 ms 30 kA; $Z_t = 100\ \Omega$; $Z_{ch} = 500\ \Omega$.

In Figure 7.2b, the two solid horizontal lines represent the positions of the tower top and the tower foot, respectively. The broken horizontal line represents the cross-arm position. It takes $(\tau_t - \tau_{ca})$ seconds for the traveling wave to reach the cross arm after lightning hits the tower top at $t = 0$. This is shown by point 1 in Figure 7.2b. Similarly, the first reflected wave from the tower foot (point 2 in Figure 7.2b) reaches the cross arm at $t = (\tau_t + \tau_{ca})$. The first reflected wave from the tower top (point 3 in Figure 7.2b) reaches the cross arm at $t = (3\tau_t - \tau_{ca})$. The downward-moving voltage waves will reach the cross arm at $t = (2n - 1)$ $\tau_t - \tau_{ca}$, and the upward-moving voltage waves will reach the cross arm at $t = (2n - 1)\ \tau_t + \tau_{ca}$, where $n = 1, 2, \ldots, n$. The cross-arm voltage, $V_{ca}(t)$, is then given by

$$V_{ca}(t) = \sum_{n=1}^{n} (a_{r1}a_{r2})^{n-1} V_{to}(t - (2n-1)\tau_t + \tau_{ca})u(t - (2n-1)\tau_t + \tau_{ca})$$

$$+ a_{r1} \sum_{n=1}^{n} (a_{r1}a_{r2})^{n-1} V_{to}(t - (2n-1)\tau_t - \tau_{ca})u(t - (2n-1)\tau_t - \tau_{ca}) \qquad (7.4)$$

The voltage profiles of the insulator voltage, $V_{ins}(=V_{ca})$, for two values of tower-footing resistances, R_{tf}, are shown in Figure 7.3. It should be noticed that the V_{ins} is higher for higher R_{tf} and that it approaches the voltage drop across the tower-footing resistance (IR_{tf}) with time. However, the peak of V_{ins} is significantly higher than the voltage drop across R_{tf}. Higher peak of V_{ins} will occur for (1) taller tower and (2) shorter front time of the stroke current (Chowdhuri, 2004).

7.2 Direct Strokes to Shielded Lines

One or more conductors are strung above and parallel to the phase conductors of single- and double-circuit overhead power lines to shield the phase conductors from direct lightning strikes. These shield wires are generally directly attached to the towers so that the return-stroke currents are safely led to ground through the tower-footing resistances. Sometimes, the shield wires are insulated from the towers by short insulators to prevent power-frequency circulating currents from flowing in the closed-circuit loop formed by the shield wires, towers, and the earth return. When lightning strikes the shield wire, the short insulator flashes over, connecting the shield wire directly to the grounded towers.

For a shielded line, lightning may strike either a phase conductor or the shield wire or the tower. If it strikes a phase conductor but the magnitude of the current is below the critical current level, then no outage occurs. However, if the lightning current is higher than the critical current of the line, then it will precipitate an outage, which is called the shielding failure. In fact, sometimes, shielding is so designed that a few outages are allowed, the objective being to reduce excessive cost of shielding. However, the critical current for a shielded line is higher than that for an unshielded line because the presence of the grounded shield wire reduces the effective surge impedance of the line. The effective surge impedance of a line shielded by one shield wire is given by (Chowdhuri, 2004)

$$Z_{eq} = Z_{11} - \frac{Z_{12}^2}{Z_{22}} \tag{7.5}$$

$$Z_{11} = 60\ell n \frac{2h_p}{r_p}; \quad Z_{22} = 60\ell n \frac{2h_s}{r_s}; \quad Z_{12} = 60\ell n \frac{d_{p's}}{d_{ps}} \tag{7.6}$$

where
  h_p and r_p are the height and the radius of the phase conductor
  h_s and r_s are the height and the radius of the shield wire
  $d_{p's}$ is the distance from the shield wire to the image of the phase conductor in the ground
  d_{ps} is the distance from the shield wire to the phase conductor

Z_{11} is the surge impedance of the phase conductor in the absence of the shield wire, Z_{22} is the surge impedance of the shield wire, and Z_{12} is the mutual surge impedance between the phase conductor and the shield wire.

It can be shown that either for strokes to tower or for strokes to shield wire, the insulator voltage will be the same if the attenuation caused by impulse corona on the shield wire is neglected (Chowdhuri, 2004). For a stroke to tower, the return-stroke current will be divided into three parts; two parts going to the shield wire in either direction from the tower and the third part to the tower. Thus, lower voltage will be developed along the tower of a shielded line than that for an unshielded line for the same return-stroke current, because lower current will penetrate the tower. This is another advantage of a shield wire. The computation of the cross-arm voltage, V_{ca}, is similar to that for the unshielded line, except for the following modifications in Equations 7.3 and 7.7:

1. The initial tower voltage is equal to IZ_{eq}, instead of IZ_t as for the unshielded line, where Z_{eq} is the impedance as seen from the striking point, that is,

$$Z_{eq} = \frac{0.5 Z_s Z_t}{0.5 Z_s + Z_t} \tag{7.7}$$

   where $Z_s = 60\ell n(2h_s/r_s)$ is the surge impedance of the shield wire.
2. The traveling voltage wave moving upward along the tower, after being reflected at the tower foot, encounters three parallel branches of impedances, the lightning-channel surge impedance, and the surge impedances of the two halves of the shield wire on either side of the struck tower. Therefore, Z_{ch} in Equation 7.3 should be replaced by $0.5 Z_s Z_{ch}/(0.5 Z_s + Z_{ch})$.

The insulator voltage, V_{ins}, for a shielded line is not equal to V_{ca}, as for the unshielded line. The shield-wire voltage, which is the same as the tower-top voltage, V_{tt}, induces a voltage on the phase conductor by electromagnetic coupling. The insulator voltage is, then, the difference between V_{ca} and this coupled voltage:

$$V_{ins} = V_{ca} - k_{sp} V_{tt} \tag{7.8}$$

where $k_{sp} = Z_{12}/Z_{22}$. It can be seen that the electromagnetic coupling with the shield wire reduces the insulator voltage. This is another advantage of the shield wire. To compute V_{tt}, we go back to Figure 7.2. As the cross arm is moved toward the tower top, τ_{ca} approaches τ_t, and naturally, at tower top $\tau_{ca} = \tau_t$. Then, except the wave 1, the pairs of upward-moving and downward-moving voltages (e.g., 2 and 3, and 4 and 5) arrive at the tower top at the same time. Substituting $\tau_{ca} = \tau_t$ in Equation 7.4, and writing $a_{t2} = 1 + a_{r2}$, we get V_{tt}:

$$V_{tt}(t) = V_{to}u(t) + a_{t2}a_{r1}\sum_{n=1}^{n}(a_{r1}a_{r2})^{n-1}V_{to}(t - 2n\tau_t)u(t - 2n\tau_t) \tag{7.9}$$

From Equation 7.3

$$a_{t2} = 1 + a_{r2} = \frac{2Z_{ch}}{Z_{ch} + Z_t} \tag{7.10}$$

The coefficient, a_{t2}, is called the coefficient of voltage transmission.

When lightning strikes the tower, equal voltages (IZ_{eq}) travel along the tower as well as along the shield wire in both directions. The voltages on the shield wire are reflected at the subsequent towers and arrive back at the struck tower at different intervals as voltages of opposite polarity (Chowdhuri, 2004). Generally, the reflections from the nearest towers are of any consequence. These reflected voltage waves lower the tower-top voltage. The tower-top voltage remains unaltered until the first reflected waves arrive from the nearest towers. The profiles of the insulator voltage for the same line as in Figure 7.3 but with a shield wire are shown in Figure 7.4. Comparing Figures 7.3 and 7.4, it should be noticed that the insulator voltage is significantly reduced for a shielded line for a stroke to tower. This reduction is possible because (1) a part of the stroke current is diverted to the shield wire, thus reducing the initial tower-top voltage ($V_{to} = I_t Z_t$, $I_t < I$), and (2) the electromagnetic coupling between the shield wire and the phase conductor induces a voltage on the phase conductor, thus lowering the voltage difference across the insulator ($V_{ins} = V_{ca} - k_{sp}V_{tt}$).

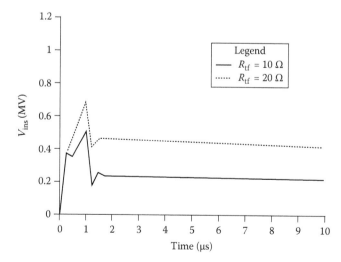

FIGURE 7.4 Profiles of insulator voltage for a shielded line for lightning stroke to tower. Tower height = 30 m; cross-arm height = 27 m; phase-conductor height = 25 m; cross-arm width = 2 m; return-stroke current = 1/50 ms 30 kA; Z_t = 100 V; Z_{ch} = 500 V.

7.2.1 Shielding Design

Striking distance of the lightning stroke plays a crucial role in the design of shielding. Striking distance is defined as the distance through which a descending stepped leader will strike a grounded object. Armstrong and Whitehead (1968) and Brown and Whitehead (1969) proposed a simple relation between the striking distance, r_s, and the return-stroke current, I (in kA), of the form

$$r_s = aI^b \ (\text{m}) \tag{7.11}$$

where a and b are constants. The most frequently used value of a is 8 or 10, and that of b is 0.65. Let us suppose that a stepped leader with prospective return-stroke current of I_s is descending near a horizontal conductor, P (Figure 7.5a). Its striking distance, r_s, will be given by Equation 7.11. It will hit the surface of the earth when it penetrates a plane which is r_s meters above the earth. The horizontal conductor will be struck if the leader touches the surface of an imaginary cylinder of radius, r_s, with its center at the center of the conductor. The attractive width of the horizontal conductor will be ab in Figure 7.5a. It is given by

$$ab = 2\omega_p = 2\sqrt{r_s^2 - (r_s - h_p)^2} = 2\sqrt{h_p(2r_s - h_p)} \quad \text{for } r_s > h_p \tag{7.12a}$$

and

$$ab = 2\omega_p = 2r_s \quad \text{for } r_s \le h_p \tag{7.12b}$$

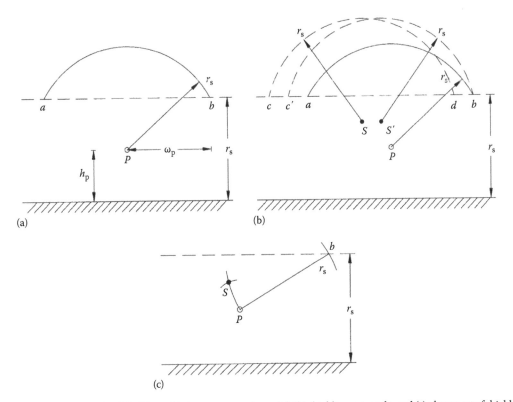

(a)

(b)

(c)

FIGURE 7.5 Principle of shielding: (a) electrogeometric model, (b) shielding principle, and (c) placement of shield wire for perfect shielding.

where h_p is the height of the conductor. For a multiconductor line with a separation distance, d_p, between the outermost conductors, the attractive width will be $2\omega_p + d_p$.

Now, if a second horizontal conductor, S, is placed near P, the attractive width of S will be cd (Figure 7.5b). If S is intended to completely shield P, then the cylinder around S and the r_s-plane above the earth's surface must completely surround the attractive cylinder around P. However, as Figure 7.5b shows, an unprotected width, db, remains. Stepped leaders falling through db will strike P. If S is repositioned to S' so that the point d coincides with b, then P is completely shielded by S.

The procedure to place the conductor, S, for perfect shielding of P is shown in Figure 7.5c. Knowing the critical current, I_c, from Equation 7.2, the corresponding striking distance, r_s, is computed from Equation 7.11. A horizontal straight line is drawn at a distance, r_s, above the earth's surface. An arc of radius, r_s, is drawn with P as center, which intersects the r_s-line above earth at b. Then, an arc of radius, r_s, is drawn with b as center. This arc will go through P. Now, with P as radius another arc is drawn of radius r_{sp}, where r_{sp} is the minimum required distance between the phase conductor and a grounded object. This arc will intersect the first arc at S, which is the position of the shield wire for perfect shielding of P.

Figure 7.6 shows the placement of a single shield wire above a three-phase horizontally configured line for shielding. In Figure 7.6a, the attractive cylinders of all three phase conductors are contained within the attractive cylinder of the shield wire and the r_s-plane above the earth. However, in Figure 7.6b where the critical current is lower, the single shield wire at S cannot perfectly shield the two outer phase conductors. Raising the shield wire helps in reducing the unprotected width; but, in this case, it cannot completely eliminate shielding failure. As the shield wire is raised, its attractive width increases until the shield-wire height reaches the r_s-plane above the earth, where the attractive width is the largest, equal to the diameter of the r_s-cylinder of the shield wire. Raising the shield-wire height further will not help. In this case, either the insulation strength of the line should be increased (i.e., the critical current increased) or two shield wires should be used.

Figure 7.7 shows the use of two shield wires. In Figure 7.7a, all three phase conductors are completely shielded by the two shield wires. However, for smaller I_c (i.e., smaller r_s), part of the attractive cylinder of the middle phase conductor is left exposed (Figure 7.7b). This shows that the middle phase conductor may experience shielding failure even when the outer phase conductors are perfectly shielded. In that case, either the insulation strength of the line should be increased or the height of the shield wires raised or both.

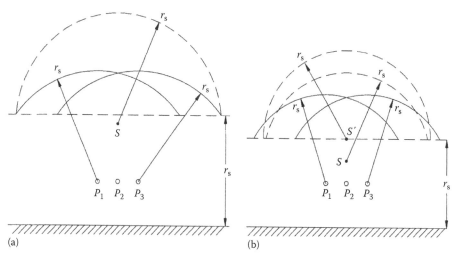

FIGURE 7.6 Shielding of three-phase horizontally configured line by single shield wire: (a) perfect shielding and (b) imperfect shielding.

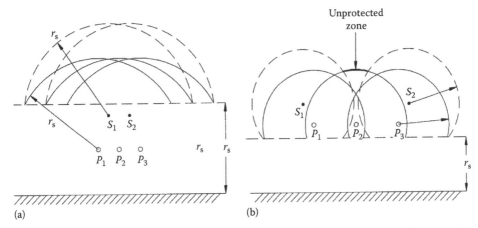

FIGURE 7.7 Shielding of three-phase horizontally configured line by two shield wires: (a) perfect shielding and (b) imperfect shielding.

Successful design of shielding depends on the accurate knowledge of the striking distance. The stepped leader parameters (cloud height, charge distribution along the leader) as well as the heights of the tower and the shield wires play prominent roles in determining the striking distances to the shield (tower and shield wires). The leader parameters are stochastic. However, no statistical data on these parameters, except the polarity, are available now. Extensive discussion on the striking distance can be found in Chowdhuri et al. (2007).

7.3 Significant Parameters

The most significant parameter in estimating the insulator voltage is the return-stroke current, that is, its peak value, waveshape, and statistical distributions of the amplitude and waveshape. The waveshape of the return-stroke current is generally assumed to be double exponential where the current rapidly rises to its peak exponentially and subsequently decays exponentially:

$$I(t) = I_o(e^{-a_1 t} - e^{-a_2 t}) \tag{7.13}$$

The parameters, I_o, a_1, and a_2, are determined from the given peak, I_p, the front time, t_f, and the time to half value, t_h, during its subsequent decay. However, the return-stroke current can also be simulated as a linearly rising and linearly falling wave:

$$I(t) = \alpha_1 t u(t) - \alpha_2 (t - t_f) u(t - t_f) \tag{7.14}$$

where

$$\alpha_1 = \frac{I_p}{t_f} \quad \text{and} \quad \alpha_2 = \frac{2t_h - t_f}{2t_f(t_h - t_f)} I_p \tag{7.15}$$

I_o, a_1, and a_2 of the double exponential function in Equation 7.13 are not very easy to evaluate. In contrast, α_1 and α_2 of the linear function in Equation 7.14 are easy to evaluate as given in Equation 7.15. The results from the two waveshapes are not significantly different, particularly for lightning currents where t_f is on the order of a few microseconds and t_h is several tens of microseconds. As t_h is very long compared to t_f, the influence of t_h on the insulator voltage is not significant. Therefore, any convenient number can be assumed for t_h (e.g., 50 ms) without the loss of accuracy.

The statistical variations of the peak return-stroke current, I_p, fit the log-normal distribution (Popolansky, 1972). The probability density function of I_p, $p(I_p)$, can then be expressed as

$$p(I_p) = \frac{1}{\sqrt{2\pi} I_p \sigma_{\ell n I_p}} e^{-0.5\left(\frac{\ell n I_p - \ell n I_{pm}}{\sigma_{\ell n I_p}}\right)^2} \tag{7.16}$$

where

$\sigma_{\ell n I p}$ is the standard deviation of $\ell n I_p$
I_{pm} is the median value of the return-stroke current, I_p

The cumulative probability, P_c, that the peak current in any lightning flash will exceed I_p kA can be determined by integrating Equation 7.16 as follows:

Substituting,

$$u = \frac{\ell n I_p - \ell n I_{pm}}{\sqrt{2}\sigma_{\ell n I_p}} \tag{7.17}$$

$$P_c(I_p) = \frac{1}{\sqrt{\pi}} \int_u^\infty e^{-u^2} du = 0.5 \,\mathrm{erfc}(u) \tag{7.18}$$

The probability density function, $p(t_f)$, of the front time, t_f, can be similarly determined by replacing I_{pm} and $\sigma_{\ell n I p}$ by the corresponding t_{fm} and $\sigma_{\ell n t_f}$ in Equations 7.17 and 7.18. Assuming no correlation between I_p and t_f, the joint probability density function of I_p and t_f is $p(I_p, t_f) = p(I_p)p(t_f)$. The equation for $p(I_p, t_f)$ becomes more complex if there is a correlation between I_p and t_f (Chowdhuri, 2004). The joint probability density function is then given by

$$p(I_p, t_f) = \frac{(0.5/e^{1-\rho^2})\left(f_1 - 2\rho\sqrt{f_1 \cdot f_2} + f_2\right)}{2\pi(I_p \cdot t_f)\left(\sigma_{\ell n I_p} \cdot \sigma_{\ell n t_f}\right)\sqrt{1-\rho^2}} \tag{7.19}$$

where

$$f_1 = \left(\frac{\ell n I_p - \ell n I_{pm}}{\sigma_{\ell n I_p}}\right)^2; \quad f_2 = \left(\frac{\ell n t_f - \ell n t_{fm}}{\sigma_{\ell n t_f}}\right)^2$$

and ρ = coefficient of correlation. The statistical parameters (I_{pm}, $\sigma_{\ell n I p}$, t_{fm}, and $\sigma_{\ell n t_f}$) have been analyzed in Anderson and Eriksson (1980) and Eriksson (2005) and are given in Chowdhuri (2004) and IEEE = PES Task Force 15.09 (2005):

$$t_{fm} = 3.83\,\mu s; \quad \sigma_{\ell n t_f} = 0.553$$

$$\text{For } I_p < 20\,\text{kA}: I_{pm} = 61.1\,\text{kA}; \quad \sigma_{\ell n I_p} = 1.33$$

$$\text{For } I_p > 20\,\text{kA}: I_{pm} = 33.3\,\text{kA}; \quad \sigma_{\ell n I_p} = 0.605$$

Besides I_p and t_f, the ground flash density, n_g, is the third significant parameter in estimating the lightning performance of power systems. The ground flash density is defined as the average number of lightning flashes per square kilometer per year in a geographic region. It should be borne in mind that the lightning activity in a particular geographic region varies by a large margin from year to year. Generally, the ground flash density is averaged over 10 years. In the past, the index of lightning severity was the keraunic level, that is, the number of thunder days in a region, because that was the only parameter available. Several empirical equations have been used to relate keraunic level with n_g. However, there has been a concerted effort in many parts of the world to measure n_g directly, and the measurement accuracy has also been improved in recent years.

7.4 Outage Rates by Direct Strokes

The outage rate is the ultimate gauge of lightning performance of a transmission line. It is defined as the number of outages caused by lightning per 100 km of line length per year. One needs to know the attractive area of the line in order to estimate the outage rate. The line is assumed to be struck by lightning if the stroke falls within the attractive area. The electrical shadow method has been used to estimate the attractive area. According to the electrical shadow method, a line of height, h_p meters, will attract lightning from a distance of $2\,h_p$ m from either side. Therefore, for a 100 km length, the attractive area will be $0.4\,h_p$ km². This area is then a constant for a specific overhead line of given height and is independent of the severity of the lightning stroke (i.e., I_p). The electrical shadow method has been found to be unsatisfactory in estimating the lightning performance of an overhead power line. Now, the electrogeometric model is used in estimating the attractive area of an overhead line. The attractive area is estimated from the striking distance, which is a function of the return-stroke current, I_p, as given by Equation 7.11. Although it has been suggested that the striking distance should also be a function of other variables (Chowdhuri and Kotapalli, 1989), the striking distance as given by Equation 7.11 is being universally used.

The first step in the estimation of outage rate is the determination of the critical current. If the return-stroke current is less than the critical current, then the insulator will not flash over if the line is hit by the stepped leader. If one of the phase conductors is struck, such as for an unshielded line, then the critical current is given by Equation 7.2. However, for strikes either to the tower or to the shield wire of a shielded line, the critical current is not that simple to compute if the multiple reflections along the tower are considered as in Equation 7.4 or 7.9. For these cases, it is best first to compute the insulator voltage by Equation 7.4 or 7.9 for a return-stroke current of 1 kA, and then estimate the critical current by taking the ratio between the insulation strength and the insulator voltage caused by 1 kA of return-stroke current of the specified front time, t_f, bearing in mind that the insulator voltage is a function of t_f.

Methods of estimation of the outage rate for unshielded and shielded lines will be somewhat different. Therefore, they are discussed separately.

7.4.1 Unshielded Lines

The vertical towers and the horizontal phase conductors coexist for an overhead power line. In that case, there is a race between the towers and the phase conductors to catch the lightning stroke. Some lightning strokes will hit the towers and some will hit the phase conductors. Figure 7.8 illustrates how to estimate the attractive areas of the towers and the phase conductors.

The tower and the two outermost phase conductors are shown in Figure 7.8. In the cross-sectional view, a horizontal line is drawn at a height r_s from the earth's surface, where r_s is the striking distance corresponding to the return-stroke current, I_s. A circle (cross-sectional view of a sphere) is drawn with radius, r_s, and center at the tip of the tower, cutting the line above the earth at a and b. Two circles (representing cylinders) are drawn with radius, r_s, and centers at the outermost phase conductors, cutting the line above the earth again at a and b. The horizontal distance between the tower tip and either a or b is ω_t.

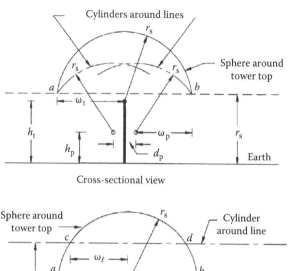

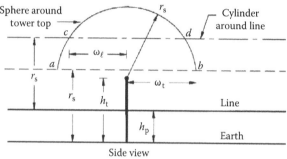

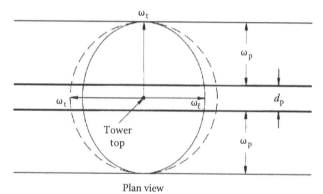

FIGURE 7.8   Attractive areas of tower and horizontal conductors.

The side view of Figure 7.8 shows where the sphere around the tower top penetrates both the r_s-plane (a and b) above ground and the cylinders around the outermost phase conductors (c and d). The projection of the sphere around the tower top on the r_s-plane is a circle of radius, ω_t, given by

$$\omega_t = \sqrt{r_s^2 - (r_s - h_t)^2} = \sqrt{h_t(2r_s - h_t)} \tag{7.20}$$

The projection of the sphere on the upper surface of the two cylinders around the outer phase conductors will be an ellipse with its minor axis, $2\omega_\ell$, along a line midway between the two outer phase conductors and parallel to their axes; the major axis of the ellipse will be $2\omega_t$, as shown in the plan view of Figure 7.8. ω_ℓ is given by

$$\omega_\ell = \sqrt{r_s^2 - (r_s - h_t + h_p)^2} \tag{7.21}$$

If a lightning stroke with return-stroke current, I_s or greater, falls within the ellipse, then it will hit the tower. It will hit one of the phase conductors if it falls outside the ellipse but within the width $(2\omega_p + d_p)$; it will hit the ground if it falls outside the width $(2\omega_p + d_p)$. Therefore, for each span length, ℓ_s, the attractive areas for the tower (A_t) and for the phase conductors (A_p) will be

$$A_t = \pi\omega_t\omega_\ell \qquad (7.22a)$$

and

$$A_p = (2\omega_p + d_p)\ell_s - A_t \qquad (7.22b)$$

The previous analysis was performed for the shielding current of the overhead line when the sphere around the tower top and the cylinders around the outer phase conductors intersect the r_s-plane above ground at the same points (points a and b in Figure 7.8). In this case, $2\omega_t = 2\omega_p + d_p$. The sphere and the cylinders will intersect the r_s-plane at different points for different return-stroke currents; their horizontal segments (widths) can be similarly computed. The equation for ω_t is given earlier. The equation for ω_p is given in Equation 7.12. Due to conductor sag, the effective height of a conductor is lower than that at the tower. The effective height is generally assumed as

$$h_p = h_{pt} - \frac{2}{3}(\text{midspan sag}) \qquad (7.23)$$

where h_{pt} is the height of the conductor at tower.

The critical current, i_{cp}, for stroke to a phase conductor is computed from Equation 7.2. It should be noted that i_{cp} is independent of the front time, t_f, of the return-stroke current. The critical current, i_{ct}, for stroke to tower is a function of t_f. Therefore, starting with a short t_f, such as 0.5 ms, the insulator voltage is determined with 1 kA of tower injected current; then, the critical tower current for the selected t_f is determined by the ratio of the insulation strength (e.g., BIL) to the insulator voltage determined with 1 kA of tower injected current. The procedure for estimating the outage rate is started with the lower of the two critical currents (i_{cp} or i_{ct}). If i_{cp} is the lower one, which is usually the case, the attractive areas, A_p and A_t, are computed for that current. If $i_{cp} < i_{ct}$, then this will not cause any flashover if it falls within A_t. In other words, the towers act like partial shields to the phase conductors. However, all strokes with i_{cp} and higher currents falling within A_p will cause flashover. The cumulative probability, $P_c(i_{cp})$, for strokes with currents i_{cp} and higher is given by Equation 7.18. If there are n_{sp} spans per 100 km of the line, then the number of outages for lightning strokes falling within A_p along the 100 km stretch of the line will be

$$nfpo = n_g P_c(i_{cp})p(t_f)\Delta t_f n_{sp} A_p \qquad (7.24)$$

where
$p(t_f)$ is the probability density function of t_f
Δt_f is the front step size

The stroke current is increased by a small step (e.g., 500 A), Δi ($i = i_{cp} + \Delta i$), and the enlarged attractive area, A_{p1}, is calculated. All strokes with currents i and higher falling within A_{p1} will cause outages. However, outage rate for strokes falling within A_p for strokes i_{cp} and greater has already been computed in Equation 7.24. Therefore, only the additional outage rate, Δnfp, should be added to Equation 7.24:

$$\Delta nfp = n_g P_c(i)p(t_f)\Delta t_f n_{sp}\Delta A_p \qquad (7.25)$$

where $\Delta A_p = A_{p1} - A_p$. The stroke current is increased in steps of Δi and the incremental outages are added until the stroke current is very high (e.g., 200 kA) when the probability of occurrence becomes acceptably low. Then, the front time, t_f, is increased by a small step, Δt_f, and the computations are repeated until the probability of occurrence of higher t_f is low (e.g., $t_f = 10.5\,\mu s$). In the mean time, if the stroke current becomes equal to i_{ct}, then the outages due to strokes to the tower should be added to the outages caused by strokes to the phase conductors. The total outage rate is then given by (Chowdhuri et al., 2002)

$$nfo = nfp + nft \tag{7.26a}$$

$$nfp = nfpo + n_g n_{sp} \sum_{t_f} \sum_i P_c(i) p(t_f) \Delta t_f \Delta A_p \tag{7.26b}$$

and

$$nft = nfto + n_g n_t \sum_{t_f} \sum_i P_c(i) p(t_f) \Delta t_f \Delta A_t \tag{7.26c}$$

With digital computers, the total outage rates can be computed within a few seconds.

7.4.2 Shielded Lines

For strokes to the shield wire, the voltage at the adjacent towers will be the same as that for stroke to the tower for the same stroke current. Therefore, there will be only one critical current for strokes to shielded lines, unlike the unshielded lines. The critical current for shielded lines can be computed similar to that for the unshielded lines, except that Equation 7.8 is now used instead of Equation 7.4.

Otherwise, the computation for shielded lines is similar to that for unshielded lines. h_p and d_p for the phase conductors are replaced by h_s and d_s, which are the shield-wire height and the separation distance between the shield wires, respectively. For a line with a single shield wire, $d_s = 0$. Generally, shield wires are attached to the tower at its top. However, the effective height of the shield wire is lower than that of the tower due to sag. The effective height of the shield wire, h_s, can be computed from Equation 7.23 by replacing h_{pt} by h_{st}, the shield-wire height at tower.

The backflashover rates of a horizontally configured three-phase overhead line with two shield wires are computed. The dimensions of the line are shown in Table 7.1. The backflashover rates with and without shield wires are plotted in Figure 7.9 as function of the insulation withstand voltage. It should be noted that the outage rate of the unshielded line decreases much slower than that for the shielded line. This happens because the phase conductor is struck by lightning more often than the towers, and the critical current for strikes to the phase conductor is significantly smaller than that for strike to a tower.

Tower surge impedance = 100 Ω; tower-footing resistance = 10 Ω; return-stroke surge impedance = 500 Ω; ground flash density = 10/km²/year; correlation coefficient = 0.47.

TABLE 7.1   Dimensions of a Horizontally Configured Distribution Line

Cross-Arm Height (m)	Phase-Conductor Height (m)	Phase-Conductor Separation (m)	Shield-Wire Height (m)	Shield-Wire Separation (m)	Span Length (m)
9.72	10.00	3.66	13.05	3.66	100.0

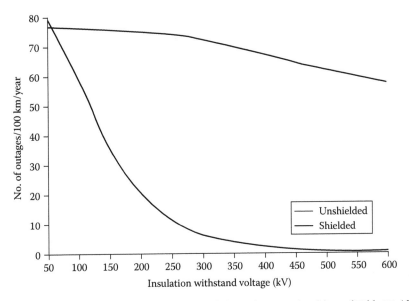

FIGURE 7.9　Outage rates of the horizontally configured three-phase overhead line of Table 7.1 (shielded and unshielded) caused by direct strokes to the line.

7.5　Effects of Induction for Direct Strokes

In the analysis of the insulator-string voltage, V_{ins}, it was assumed that V_{ins} has two components: (1) one due to the cross-arm voltage, V_{ca}, caused by the multiply reflected traveling voltage waves along the struck tower and (2) the other due to electromagnetic coupling of the phase conductor to the shield wire for a shielded line. However, two other component voltages contribute to the total voltage across the insulator string (Chowdhuri et al., 2002; Chowdhuri, 2004). The third voltage component is induced by the electromagnetic fields of the charge and current of the return-stroke channel (Figure 7.10). The stroke channel being only a few meters from the phase conductors, the induction effect should not be ignored. The fourth voltage component is induced across the insulator string by the multiply reflected traveling current waves along the struck tower (Figure 7.10). The voltage induced

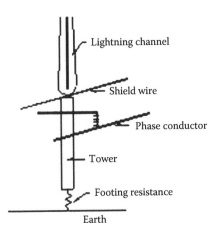

FIGURE 7.10　Direct stroke to a tower.

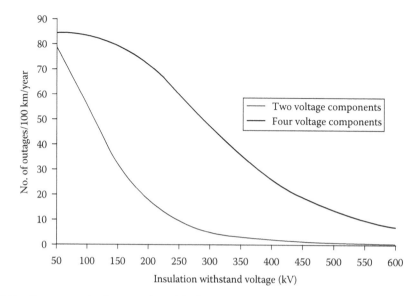

FIGURE 7.11 Outage rates by direct strokes of the horizontally configured three-phase overhead line of Table 7.1 considering two and four voltage components across insulator string. Tower surge impedance = 100 Ω; tower-footing resistance = 10 Ω; velocity of traveling-wave current in tower = 0.8 p.u.; cloud height = 3 km; return-stroke surge impedance = 500 Ω; ground flash density = 10/km²/year; correlation coefficient = 0.47.

across the insulator string by the rapid change of the tower current can be significant because of the proximity of the tower to the insulators.

The third voltage component can be computed by the analysis shown in Chapter 8, with the difference that, in the present case, the stroke hits the tower top instead of the ground. This difference is manifested in the inducing voltage, V_i, which is given in Equations 8.2 and 8.3. The scalar potential, φ, and the vector potential, A, in Equation 8.3 can be computed from Chapter 8 of Chowdhuri (2004) with the difference that the lower and the upper limits of z' (height of a specified point along the stepped leader) are the tower height, h_t, and the cloud height, h_c, for strike to the tower top. Therefore, the voltage induced on the phase conductor for a tower strike is computed for two cloud heights (h_c and h_t), and then the second induced voltage (for h_t) is subtracted from the first induced voltage (due to h_c).

The fourth voltage component, that is, the induced voltage due to the traveling current along the tower, can be computed similar to the normal case for stroke to ground, with the following modifications:

1. In the computation of the inducing voltage for a strike to ground, both the electrostatic component of the upper charge column and the magnetic component of the lower current column are considered. For the traveling current waves along the tower, only magnetic component of the electromagnetic field is considered.
2. After each reflection at either end of the struck tower, both the magnitude and the direction of travel change for the traveling current waves. Therefore, new computations must be superimposed on the initial computation after each reflection.

The effects of the third and fourth voltage components on the estimation of outage rates are shown in Figure 7.11 for a horizontally configured three-phase distribution line with two shield wires.

7.6 Ground Impedance and Corona under Lightning

The lightning current flowing through the overhead lines eventually finds its way to ground through the ground electrodes to the soil. The impedance of the ground electrodes can be computed as discussed in Chisholm (2007). The electrical characteristics of the soil in which the ground electrodes are buried play a significant role in the overall grounding impedance. The electrical resistivity of the soil bed varies from day to day, season to season, and year to year, depending upon the amount of rainfall and temperature. The soil is ionized under large lightning current. The ionization becomes more intense with larger currents. The critical electric field for soil ionization varies for the same soil, depending on its compactness, moisture content and temperature, and varying soil electrical conductivity and permittivity (Manna and Chowdhuri, 2007). All these need to be known to estimate the grounding impedance under lightning currents. Many of these parameters are not known. However, research is going on.

References

Anderson, R.B. and Eriksson, A.J., Lightning parameters for engineering applications, *Electra*, 69, 65–102, 1980.

Armstrong, H.R. and Whitehead, E.R., Field and analytical studies of transmission line shielding, *IEEE Trans. Power Appar. Syst.*, PAS-87, 270–281, 1968.

Bewley, L.V., *Traveling Waves on Transmission Systems*, 2nd edn., John Wiley, New York, 1951.

Brown, G.W. and Whitehead, E.R., Field and analytical studies of transmission line shielding: Part II, *IEEE Trans. Power Appar. Syst.*, PAS-88, 617–626, 1969.

Chisholm, W.A., Transmission system transients—Grounding, power systems, *Electric Power Engineering Handbook*, 2nd edn., L.L. Grigsby, (ed.), CRC Press, Boca Raton, FL, 2007.

Chowdhuri, P., *Electromagnetic Transients in Power Systems*, 2nd edn., Research Studies Press, Baldock, Hertfordshire, U.K., 2004.

Chowdhuri, P. and Kotapalli, A.K., Significant parameters in estimating the striking distance of lightning strokes to overhead lines, *IEEE Trans. Power Delivery*, 4, 1970–1981, 1989.

Chowdhuri, P., Li, S., and Yan, P., Rigorous analysis of back-flashover outages caused by direct lightning strokes to overhead power lines, *IEE Proc.—Gener. Transm. Distrib.*, 149, 58–65, 2002.

Chowdhuri, P., Tajali, G.R., and Yuan, X., Analysis of striking distances of lightning strokes to vertical towers, *IET Gener. Transm. Distrib.*, 1, 879–886, 2007.

Ericsson, A.J., Notes on lightning parameters, CIGRE Note 33–86 (WG33-01) IWD, 15 July 1986.

IEEE = PES Task Force 15.09, Parameters of lightning strokes: A review, *IEEE Trans. Power Delivery*, 20, 346–358, 2005.

Manna, T. K. and Chowdhuri, P., Generalised equation of soil critical electric field E_c based on impulse tests and measured soil electrical parameters, *IET Gener. Transm. Distrib.*, 1, 811–817, 2007.

Popolansky, F., Frequency distribution of amplitudes of lightning currents, *Electra*, 22, 139–147, 1972.

8

Overvoltages Caused by Indirect Lightning Strokes

Pritindra
Chowdhuri
*Tennessee Technological
University*

A direct stroke is defined as a lightning stroke when it hits a shield wire, a tower, or a phase conductor. An insulator string is stressed by very high voltages caused by a direct stroke. An insulator string can also be stressed by high transient voltages when a lightning stroke hits the nearby ground. An indirect stroke is illustrated in Figure 8.1.

During the early years of power transmission, protection against a direct lightning stroke was thought to be impossible. Therefore, power lines were designed with overhead ground wires to protect against transient voltages induced from nearby lightning strokes to ground. As the transmission voltages became higher and higher with the consequent increase in the insulation levels of overhead lines, it was realized that the transient voltages induced from nearby lightning strokes were not the chief source of disturbance, and that protection against direct lightning strokes is indeed technically feasible and economically justifiable (Fortescue, 1930). Now emphasis is put on protection against both direct and indirect lightning strokes. As the reliability of distribution power lines has become increasingly important, many researchers have been studying the effects of lightning-induced voltages on power distribution lines. These various researches have been reviewed by Chowdhuri et al. (2001).

The voltage induced on a line by an indirect lightning stroke has four components:

1. The charged cloud above the line induces bound charges on the line while the line itself is held electrostatically at ground potential by the neutrals of connected transformers and by leakage over the insulators. When the cloud is partially or fully discharged, these bound charges are released and travel in both directions on the line giving rise to the traveling voltage and current waves.

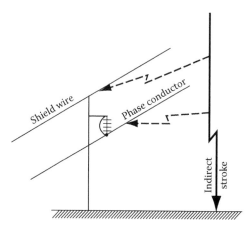

FIGURE 8.1   Illustration of direct and indirect lightning strokes.

2. The charges lowered by the stepped leader further induce charges on the line. When the stepped leader is neutralized by the return stroke, the bound charges on the line are released and thus produce traveling waves similar to that caused by the cloud discharge.
3. The residual charges on the upper part of the return stroke induce an electrostatic field in the vicinity of the line and hence an induced voltage on it.
4. The rate of change of current in the return stroke produces a magnetically induced voltage on the line.

If the lightning has subsequent strokes, then the subsequent components of the induced voltage will be similar to one or the other of the four components discussed earlier.

The magnitudes of the voltages induced by the release of the charges bound either by the cloud or by the stepped leader are small compared with the voltages induced by the return stroke. Therefore, only the electrostatic and the magnetic components induced by the return stroke are considered in the following analysis. The initial computations are performed with the assumption that the charge distribution along the leader stroke is uniform, and that the return-stroke current is rectangular. However, the result with the rectangular current wave can be transformed to that with currents of any other waveshape by the convolution integral (Duhamel's theorem). It was also assumed that the stroke is vertical and that the overhead line is loss free and the earth is perfectly conducting. The vertical channel of the return stroke is shown in Figure 8.2, where the upper part consists of a column of residual charge that is neutralized by the rapid upward movement of the return-stroke current in the lower part of the channel.

Figure 8.3 shows a rectangular system of coordinates where the origin of the system is the point where lightning strikes the surface of the earth. The line conductor is located at a distance y_o meters from the origin, having a mean height of h_p meters above ground and running along the x-direction. The origin of time ($t = 0$) is assumed to be the instant when the return stroke starts at the earth level.

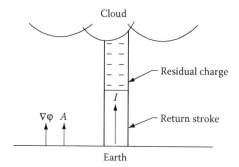

FIGURE 8.2   Return stroke with residual charge column.

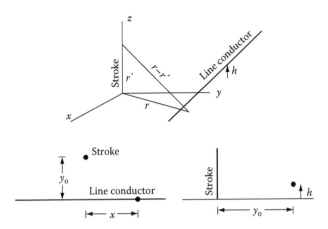

FIGURE 8.3 Coordinate system of line conductor and lightning stroke.

8.1 Inducing Voltage

The total electric field created by the charge and the current in the lightning stroke at any point in space is

$$E_i = E_{ei} + E_{mi} = -\nabla\varphi - \frac{\partial A}{\partial t} \tag{8.1}$$

where φ is the inducing scalar potential created by the residual charge at the upper part of the return stroke and A is the inducing vector potential created by the upward-moving return-stroke current (Figure 8.2). φ and A are called the retarded potentials, because these potentials at a given point in space and time are determined by the charge and current at the source (i.e., the lightning channel) at an earlier time; the difference in time (i.e., the retardation) is the time required to travel the distance between the source and the field point in space with a finite velocity, which in air is $c = 3 \times 10^8$ m/s. These electromagnetic potentials can be deduced from the distribution of the charge and the current in the return-stroke channel. The next step is to find the inducing electric field (Equation 8.1). The inducing voltage, V_i, is the line integral of E_i:

$$V_i = -\int_0^{h_p} E_i \cdot dz = -\int_0^{h_p} E_{ei} \cdot dz - \int_0^{h_p} E_{mi} \cdot dz = V_{ei} + V_{mi} \tag{8.2}$$

As the height, h_p, of the line conductor is small compared with the length of the lightning channel, the inducing electric field below the line conductor can be assumed to be constant and equal to that on the ground surface:

$$V_i = \left(\nabla\varphi + \frac{\partial A}{\partial t}\right) \cdot h_p \tag{8.3}$$

The inducing voltage will act on each point along the length of the overhead line. However, because of the retardation effect, the earliest time, to, the disturbance from the lightning channel will reach a point on the line conductor would be

$$t_o = \frac{\sqrt{x^2 + y_o^2}}{c} \tag{8.4}$$

Therefore, the inducing voltage at a point on the line remains zero until $t = t_o$. Hence,

$$V_i = \psi(x,t)u(t - t_o) \tag{8.5}$$

where $u(t - t_o)$ is the shifted unit step function. The continuous function, $\psi(x,t)$, is the same as Equation 8.3, and is given, for a negative stroke with uniform charge density along its length, by (Rusck, 1958)

$$\psi(x,t) = -\frac{60I_o h_p}{\beta}\left[\frac{1 - \beta^2}{\sqrt{\beta^2 c^2(t - t_o)^2 + (1 - \beta^2)r^2}} - \frac{1}{\sqrt{h_c^2 + r^2}}\right] \tag{8.6}$$

where

 I_o is the step-function return-stroke current (A)
 h_p is the height of line above ground (m)
 β is the v/c, v is the velocity of return stroke (m/s)
 r is the distance of point x on line from point of strike (m)
 h_c is the height of cloud charge center above ground (m)

The inducing voltage is the voltage at a field point in space with the same coordinates as a corresponding point on the line conductor, but without the presence of the line conductor. The inducing voltage at different points along the length of the line conductor will be different. In the overhead line, these differences will tend to be equalized by the flow of current, as it is a good conductor of electricity. Therefore, the actual voltage between a point on the line and the ground below it will be different from the inducing voltage at that point. This voltage, which can actually be measured on the line conductor, is defined as the induced voltage. The calculation of the induced voltage is the primary objective.

8.2 Induced Voltage

Neglecting losses, an overhead line may be represented as consisting of distributed series inductance L (H/m) and distributed shunt capacitance C (F/m). The effect of the inducing voltage will then be equivalent to connecting a voltage source along each point of the line (Figure 8.4). The partial differential equation for such a configuration will be

$$-\frac{\partial V}{\partial x}\Delta x = L\Delta x\frac{\partial I}{\partial t} \tag{8.7}$$

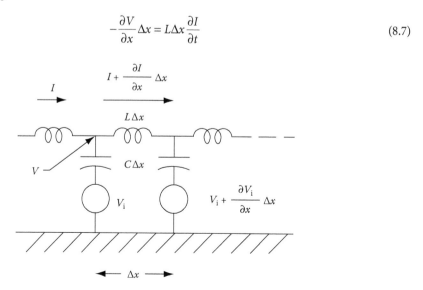

FIGURE 8.4   Equivalent circuit of transmission line with inducing voltage.

and

$$-\frac{\partial I}{\partial x}\Delta x = C\Delta x\frac{\partial}{\partial t}(V - V_i) \tag{8.8}$$

Differentiating Equation 8.7 with respect to x, and eliminating I, the equation for the induced voltage can be written as

$$\frac{\partial^2 V}{\partial x^2} - \frac{1}{c^2}\frac{\partial^2 V}{\partial t^2} = -\frac{1}{c^2}\frac{\partial^2 V_i}{\partial t^2} = F(x,t) \tag{8.9}$$

where

$$c = \frac{1}{\sqrt{LC}} = \frac{1}{\sqrt{\mu_o\varepsilon_o}} = 3\times10^8\,\text{m/s} \tag{8.10}$$

In Laplace transform,

$$\frac{\partial^2 V(x,s)}{\partial x^2} - \frac{s^2}{c^2}V(x,s) = -\frac{s^2}{c^2}V_i(x,s) = F(x,s) \tag{8.11}$$

Equation 8.9 is an inhomogeneous wave equation for the induced voltage along the overhead line. It is valid for any charge distribution along the leader channel and any waveshape of the return-stroke current. Its solution can be obtained by assuming $F(x,t)$ to be the superposition of impulses that involves the definition of Green's function (Morse and Feshbach, 1950).

8.3 Green's Function

To obtain the voltage caused by a distributed source, $F(x)$, the effects of each elementary portion of the source are calculated and then integrated for the whole source. If $G(x;x')$ is the voltage at a point x along the line caused by a unit impulse source at a source point x', the voltage at x caused by a source distribution $F(x')$ is the integral of $G(x;x')F(x')$ over the whole domain (a,b) of x' occupied by the source, provided that $F(x')$ is a piecewise continuous function in the domain $a \le x' \le b$:

$$V(x) = \int_a^b G(x;x')F(x')\,dx' \tag{8.12}$$

The function $G(x;x')$, called Green's function, is, therefore, a solution for a case that is homogeneous everywhere except at one point. Green's function, $G(x;x')$, has the following properties:

1. $G(x;x' + 0) - G(x;x' - 0) = 0$ $\tag{8.13}$

2. $\left(\dfrac{dG}{dx}\right)_{x'+0} - \left(\dfrac{dG}{dx}\right)_{x'-0} = 1$ $\tag{8.14}$

3. $G(x;x')$ satisfies the homogeneous equation everywhere in the domain, except at the point $x = x'$
4. $G(x;x')$ satisfies the prescribed homogeneous boundary conditions.

Green's function can be found by converting Equation 8.11 to a homogeneous equation and replacing $V(x,s)$ by $G(x;x',s)$:

$$\frac{\partial^2 G(x;x',s)}{\partial x^2} - \frac{s^2}{c^2} G(x;x',s) = 0 \tag{8.15}$$

The general solution of Equation 8.15 is given by

$$G(x;x',s) = Ae^{sx/c} + Be^{-sx/c} \tag{8.16}$$

The constants A and B are found from the boundary conditions and from the properties of Green's function.

8.4 Induced Voltage of a Doubly Infinite Single-Conductor Line

The induced voltage at any point, x, on the line can be determined by invoking Equation 8.12, where $G(x;x')\,F(x')$ is the integrand. $F(x')$ is a function of the amplitude and waveshape of the inducing voltage, V_i (Equation 8.5), whereas Green's function, $G(x;x')$, is dependent on the boundary conditions of the line and the properties of Green's function. In other words, it is a function of the line configuration and is independent of the lightning characteristics. Therefore, it is appropriate to determine Green's function first.

8.4.1 Evaluation of Green's Function

As Green's function is finite for $x \to -\infty$ and $x \to +\infty$,

$$G_1 = Ae^{sx/c} \quad \text{for } x < x'; \quad G_2 = Be^{-sx/c} \quad \text{for } x > x'$$

From Equation 8.13,

$$Ae^{sx'/c} = Be^{-(sx'/c)}, \quad \text{i.e., } B = Ae^{2sx'/c}$$

From Equation 8.14,

$$A = \frac{c}{2s}e^{-sx/c}$$

Hence,

$$B = -\frac{c}{2s}e^{sx/c}$$

$$G_1(x;x',s) = -\frac{c}{2s}\exp\left(-\frac{s(x'-x)}{c}\right) \quad \text{for } x < x' \tag{8.17}$$

and

$$G_2(x;x',s) = -\frac{c}{2s}\exp\left(\frac{s(x'-x)}{c}\right) \quad \text{for } x > x'$$ (8.18)

By applying Equation 8.12,

$$V(x,s) = -\frac{c}{2s}\int_{-\infty}^{x} e^{s(x'-x)/c}F(x',s)dx' - \frac{c}{2s}\int_{x}^{\infty} e^{-s(x'-x)/c}F(x',s)dx' = V_1(x,s) + V_2(x,s)$$ (8.19)

8.4.2 Induced Voltage Caused by Return-Stroke Current of Arbitrary Waveshape

The induced voltage caused by return-stroke current, $I(t)$, of arbitrary waveshape can be computed from Equation 8.11 by several methods. In method I, the inducing voltage, V_i, due to $I(t)$ is found by applying Duhamel's integral (Haldar and Liew, 1988):

$$V_i = \frac{d}{dt}\int_{0}^{t} I(t-\tau)V_{istep}(x',\tau)d\tau$$ (8.20)

where V_{istep} is the inducing voltage caused by a unit step-function current. In other words,

$$V_{istep}(x',\tau) = \psi_o(x',\tau)u(\tau-t_o)$$ (8.21)

where

$$\psi_o(x',\tau) = \frac{\psi(x',\tau)}{I_o}$$

and $\psi(x',\tau)$ is given in Equation 8.6. Inserting Equation 8.21 in Equation 8.20 and taking Laplace transform of V_i in Equation 8.20,

$$V_i(x',s) = sI(s)\psi_o(x',s)e^{-st_o}$$ (8.22)

and

$$F(x',s) = -\frac{s^2}{c^2}V_i(x',s) = -\frac{s^2}{c^2}I(s)\psi_o(x',s)e^{-st_o}$$ (8.23)

Replacing $F(x',s)$ in Equation 8.19 by Equation 8.23, the induced voltage, $V(x,s)$, is

$$V(x,s) = \frac{1}{2c}\left[sI(s)\left\{s\int_{-\infty}^{x}\psi_o(x',s)e^{-s\left(t_o-\frac{x'-x}{c}\right)}dx' + s\int_{x}^{\infty}\psi_o(x',s)e^{-s\left(t_o+\frac{x'-x}{c}\right)}dx'\right\}\right]$$ (8.24)

Inverting to time domain by convolution integral,

$$V(x,t) = \frac{1}{2c}\int_0^t \frac{d}{dt}I(t-\tau)\left[\frac{d}{d\tau}\int_{-\infty}^x \psi_0\left(x',\tau+\frac{x'-x}{c}\right)u\left(\tau-t_o+\frac{x'-x}{c}\right)dx'\right]d\tau$$

$$+\frac{1}{2c}\int_0^t \frac{d}{dt}I(t-\tau)\left[\frac{d}{d\tau}\int_x^\infty \psi_0\left(x',\tau-\frac{x'-x}{c}\right)u\left(\tau-t_o-\frac{x'-x}{c}\right)dx'\right]d\tau = V_1(x,t)+V_2(x,t) \quad (8.25)$$

Because of the shifted unit step function in $V_1(x,t)$,

$$\tau \geq t_o - \frac{x'-x}{c}$$

In the limit,

$$\tau = t_o - \frac{x_{o1}-x}{c} = \frac{\sqrt{x_{o1}^2+y_o^2}}{c} - \frac{x_{o1}-x}{c}$$

or

$$x_{o1} = \frac{y_o^2-(c\tau-x)^2}{2(c\tau-x)} \quad (8.26)$$

Similarly,

$$\text{for } V_2(x,t): x_{o2} = \frac{(c\tau+x)^2-y_o^2}{2(c\tau+x)} \quad (8.27)$$

Replacing $-\infty$ by x_{o1} in $V_1(x,t)$ and $+\infty$ by x_{o2} in $V_2(x,t)$ in Equation 8.25,

$$V_1(x,t) = \frac{1}{2c}\int_0^t \frac{d}{dt}I(t-\tau)\frac{d}{d\tau}\left\{\int_{x_{o1}}^x \psi_0\left(x,\tau+\frac{x'-x}{c}\right)dx'\right\}u(\tau-t_o)d\tau \quad (8.28)$$

and

$$V_2(x,t) = \frac{1}{2c}\int_0^t \frac{d}{dt}I(t-\tau)\frac{d}{d\tau}\left\{\int_x^{x_{o2}} \psi_0\left(x,\tau-\frac{x'-x}{c}\right)dx'\right\}u(\tau-t_o)d\tau \quad (8.29)$$

A lightning return-stroke current can be represented by a linearly rising and linearly falling wave with sufficient accuracy (Figure 8.5) (Chowdhuri, 2004):

$$I(t) = \alpha_1 tu(t) - \alpha_2(t-t_f)u(t-t_f) \quad (8.30)$$

where

$$\alpha_1 = \frac{I_p}{t_f} \quad \text{and} \quad \alpha_2 = \frac{2t_H-t_f}{2t_f(t_H-t_f)}I_p \quad (8.31)$$

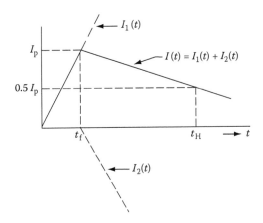

FIGURE 8.5   A linearly rising and falling lightning return-stroke current.

It will be evident from Equation 8.30 that $V_1(x,t)$ in Equation 8.28 will have two components: one component, $V_{11}(x,t)$, will be a function of $I_1(t)$, and the other component, $V_{21}(x,t)$, will be a function of $I_2(t)$, that is, $V_1(x,t) = V_{11}(x,t) + V_{21}(x,t)$. Similarly, $V_2(x,t) = V_{12}(x,t) + V_{22}(x,t)$. After integration and simplifying Equation 8.28, $V_{11}(x,t)$ can be written as

$$V_{11}(x,t) = -\frac{\alpha_1 h_p}{\beta} \times 10^{-7} u(t - t_o) \left[(1 - \beta^2) \ell n \frac{f_{11}(\tau = t) f_{21}(\tau = t_o)}{f_{11}(\tau = t_o) f_{21}(\tau = t)} + \ell n \frac{f_{31}(\tau = t)}{f_{31}(\tau = t_o)} \right] \qquad (8.32)$$

where

$$f_{11}(\tau) = m_{11} + \sqrt{m_{11}^2 + a_{11}^2}; \quad f_{21}(\tau) = m_{21} + \sqrt{m_{21}^2 + a_{11}^2}; \quad f_{31} = x_{o1} + \sqrt{x_{o1}^2 + y_o^2 + h_c^2};$$

$$m_{11} = x + \beta^2(c\tau - x); \quad m_{21} = x_{o1} + \beta^2(c\tau - x); \quad a_{11}^2 = (1 - \beta^2)[y_o^2 + \beta^2(c\tau - x)^2]$$

The expression for $V_{21}(x,t)$ is similar to Equation 8.32, except that α_1 is replaced by $(-\alpha_2)$, and t is replaced by $(t - t_f)$. The computation of $V_2(x,t)$ is similar; namely,

$$V_{12}(x,t) = -\frac{\alpha_1 h_p}{\beta} \times 10^{-7} u(t - t_o) \left[(1 - \beta^2) \ell n \frac{f_{12}(\tau = t) f_{22}(\tau = t_o)}{f_{12}(\tau = t_o) f_{22}(\tau = t)} - \ell n \frac{f_{32}(\tau = t)}{f_{32}(\tau = t_o)} \right] \qquad (8.33)$$

where

$$f_{12}(\tau) = m_{12} + \sqrt{m_{12}^2 + a_{12}^2}; \quad f_{22} = m_{22} + \sqrt{m_{22}^2 + a_{12}^2}; \quad f_{32} = x_{o2} + \sqrt{x_{o2}^2 + y_o^2 + h_c^2};$$

$$m_{12} = x_{o2} - \beta^2(c\tau + x); \quad m_{22} = x - \beta^2(c\tau + x); \quad a_{12}^2 = (1 - \beta^2)[y_o^2 + \beta^2(c\tau + x)^2]$$

$V_{22}(x,t)$ can be similarly determined by replacing α_1 in Equation 8.33 by $(-\alpha_2)$ and replacing t by $(t - t_f)$. The second method of determining the induced voltage, $V(x,t)$, is to solve Equation 8.19, for a unit step-function return-stroke current, and then find the induced voltage for the given return-stroke current waveshape by applying Duhamel's integral (Chowdhuri and Gross, 1967; Chowdhuri, 1989a). The solution of Equation 8.19 for a unit step-function return-stroke current is given by (Chowdhuri, 1989a):

$$V_{step}(x,t) = (V_{11} + V_{12} + V_{21} + V_{22}) u(t - t_o) \qquad (8.34)$$

where

$$V_{11} = \frac{30h_{\mathrm{p}}(1-\beta^2)}{\beta^2(ct-x)^2+y_{\mathrm{o}}^2}\left[\beta(ct-x)+\frac{(ct-x)x-y_{\mathrm{o}}^2}{\sqrt{c^2t^2+((1-\beta^2)/\beta^2)(x^2+y_{\mathrm{o}}^2)}}\right] \tag{8.35}$$

$$V_{12} = -\frac{30h_{\mathrm{p}}}{\beta}\left[1-\frac{1}{\sqrt{k_1^2+1}}-\beta^2\right]\frac{1}{ct-x} \tag{8.36}$$

$$V_{21} = \frac{30h_{\mathrm{p}}(1-\beta^2)}{\beta^2(ct+x)^2+y_{\mathrm{o}}^2}\left[\beta(ct+x)-\frac{(ct+x)x+y_{\mathrm{o}}^2}{\sqrt{c^2t^2+((1-\beta^2)/\beta^2)(x^2+y_{\mathrm{o}}^2)}}\right] \tag{8.37}$$

$$V_{22} = -\frac{30h_{\mathrm{p}}}{\beta}\left[1-\frac{1}{\sqrt{k_2^2+1}}-\beta^2\right]\frac{1}{ct+x} \tag{8.38}$$

$$k_1 = \frac{2h_{\mathrm{c}}(ct-x)}{y_{\mathrm{o}}^2+(ct-x)^2} \tag{8.39}$$

and

$$k_2 = \frac{2h_{\mathrm{c}}(ct+x)}{y_{\mathrm{o}}^2+(ct+x)^2} \tag{8.40}$$

The expressions for the induced voltage, caused by a linearly rising and falling return-stroke current, are given in Appendix 8.A.

The advantage of method II is that once the induced voltage caused by a step-function return-stroke current is computed, it can be used as a reference in computing the induced voltage caused by currents of any given waveshape by applying Duhamel's integral, thus avoiding the mathematical manipulations for every given waveshape. However, the mathematical procedures are simpler for method I than that for method II.

A third method to solve Equation 8.9 is to apply numerical method, which bypasses all mathematical complexities (Agrawal et al., 1980). However, the accuracy of the numerical method strongly depends upon the step size of computation. Therefore, the computation of the induced voltage of long lines, greater than 1 km, becomes impractical.

8.5 Induced Voltages on Multiconductor Lines

Overhead power lines are usually three-phase lines. Sometimes several three-phase circuits are strung from the same tower. Shield wires and neutral conductors are part of the multiconductor system. The various conductors in a multiconductor system interact with each other in the induction process for lightning strokes to nearby ground. The equivalent circuit of a two-conductor system is shown in Figure 8.6. Extending to an n-conductor system, the partial differential equation for the induced voltage, in matrix form, is (Chowdhuri and Gross, 1969; Cinieri and Fumi, 1979; Chowdhuri, 1990, 2004):

$$\frac{\partial^2[V]}{\partial x^2}-\frac{1}{c^2}\frac{\partial^2[V]}{\partial t^2} = -[L][C_{\mathrm{g}}]\frac{\partial^2[V_{\mathrm{i}}]}{\partial t^2} = -[M]\frac{\partial^2[V_{\mathrm{i}}]}{\partial t^2} \tag{8.41}$$

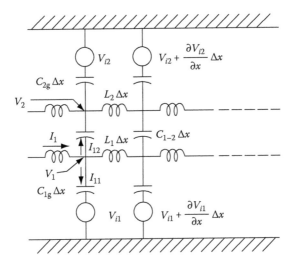

FIGURE 8.6 Equivalent circuit of a two-conductor system.

where $[L]$ is an $n \times n$ matrix whose elements are

$$L_{rr} = 2 \times 10^{-7} \ell n \frac{2h_r}{r_r}; \quad L_{rs} = 2 \times 10^{-7} \ell n \frac{d_{r's}}{d_{rs}}$$

$[C_g]$ is an $n \times n$ diagonal matrix whose elements are $C_{jg} = C_{j1} + C_{j2} + + C_{jn}$, where C_{jr} is an element of an $n \times n$ matrix, $[C] = [p]^{-1}$ and

$$p_{rr} = 18 \times 10^9 \ell n \frac{2h_r}{r_r}; \quad p_{rs} = 18 \times 10^9 \ell n \frac{d_{r's}}{d_{rs}}$$

where

h_r and r_r are the height above ground and the radius of the rth conductor
$d_{r's}$ is the distance between the image of the rth conductor below earth and the sth conductor
d_{rs} is the distance between the rth and sth conductors

From Equation 8.41, for the jth conductor,

$$\frac{\partial^2 V_j}{\partial x^2} - \frac{1}{c^2} \frac{\partial^2 V_j}{\partial t^2} = -\left(M_{j1} \frac{\partial^2 V_{j1}}{\partial t^2} + \cdots + M_{jj} \frac{\partial^2 V_{jj}}{\partial t^2} + \cdots + M_{jn} \frac{\partial^2 V_{jn}}{\partial t^2} \right) \tag{8.42}$$

If the ratio of the inducing voltage of the mth conductor to that of the jth conductor is k_{mj} ($m = 1, 2, \ldots, n$), then

$$\frac{\partial^2 V_j}{\partial x^2} - \frac{1}{c^2} \frac{\partial^2 V_j}{\partial t^2} - c^2 (M_{j1}k_{1j} + \cdots + M_{jj} + \cdots + M_{jn}k_{nj}) \frac{1}{c^2} \frac{\partial^2 V_{ij}}{\partial t^2} = (c^2 M_j) F_j(x,t) \tag{8.43}$$

where

$$M_j = M_{j1}k_{1j} + \cdots + M_{jj} + \cdots + M_{jn}k_{nj} \quad \text{and} \quad F_j(x,t) = \frac{1}{-c^2} \frac{\partial^2 V_{ij}}{\partial t^2} \tag{8.44}$$

If the *j*th conductor in its present position existed alone, the partial differential equation of its induced voltage, V_{js}, would be the same as Equation 8.9, that is,

$$\frac{\partial^2 V_{js}}{\partial x^2} - \frac{1}{c^2}\frac{\partial^2 V_{js}}{\partial t^2} = F_j(x,t) \tag{8.45}$$

Therefore, the ratio of the induced voltage of the *j*th conductor in an *n*-conductor system to that of a single conductor at the same position would be

$$\frac{V_j}{V_{js}} = M_j c^2 \tag{8.46}$$

The inducing voltage being nearly proportional to the conductor height and the lateral distance of the stroke point being significantly larger than the separation distance between phase conductors, the presence of other conductors in a horizontally configured line will be minimal. On the other hand, for a vertically configured line, the induced voltage of the highest conductor will be lower than that for the same conductor without any neighboring conductors. Similarly, the lowest conductor voltage will be pulled up by the presence of the neighboring conductors of higher elevation, and the middle conductor will be the least affected by the presence of the other conductors (Chowdhuri, 2004).

8.6 Effects of Shield Wires on Induced Voltages

If there are $(n + r)$ conductors, of which *r* conductors are grounded (*r* shield wires), then the partial differential equation for the induced voltages of the *n* number of phase conductors is given by Chowdhuri and Gross (1969), Cinieri and Fumi (1979), and Chowdhuri (1990, 2004):

$$\frac{\partial^2 [V_n]}{\partial x^2} - \frac{1}{c^2}\frac{\partial^2 [V_n]}{\partial t^2} = -[L'][C_{gn}]\frac{\partial^2 [V_{in}]}{\partial t^2} = -[M_g]\frac{\partial^2 [V_{in}]}{\partial t^2} \tag{8.47}$$

The matrix $[L']$ is obtained by partitioning the $(n + r) \times (n + r)$ inductance matrix of the $(n + r)$ conductors, and putting $[L'] = [L_{nn}] - [L_{nr}][L_{rr}]^{-1}[L_{rn}]$, where

$$[L]_{(n+r),(n+r)} = \begin{bmatrix} L_{nn} & L_{nr} \\ L_{rn} & L_{rr} \end{bmatrix} \tag{8.48}$$

$[C_{gn}]$ is an $n \times n$ diagonal matrix, each element of which is the sum of the elements of the corresponding row, up to the *n*th row, of the original $(n + r) \times (n + r)$ capacitance matrix of the $(n + r)$ conductors, $[C] = [p]^{-1}$, where $[p]$ is the matrix of the potential coefficients of the $(n + r)$ conductors. The *j*th element of $[C_{gn}]$ is given by

$$C_{jgn}(j \leq n) = \sum_{k=1}^{n+r} C_{jk} \tag{8.49}$$

From Equation 8.47, the induced voltage of the *j*-th conductor is

$$\frac{\partial^2 V_j}{\partial x^2} - \frac{1}{c^2}\frac{\partial^2 V_j}{\partial t^2} = -c^2(M_{gi1}k_{1j} + \cdots + M_{gij} + \cdots + M_{gin}k_{nj})\frac{1}{c^2}\frac{\partial^2 V_{ij}}{\partial t^2} = -(c^2 M_{gi})F_j(x,t) \tag{8.50}$$

Defining the protective ratio as the ratio of the induced voltages on the jth conductor with and without the shield wires in place,

$$\text{Protective ratio} = \frac{M_{gj}}{M_j} \tag{8.51}$$

where M_j is given by Equation 8.44.

8.7 Stochastic Characteristics of Lightning Strokes

The voltage induced on an overhead line is caused by the interaction between various lightning return-stroke parameters and the parameters of the line. The most important return-stroke parameters are (a) peak current, I_p, (b) current front time, t_f, (c) return-stroke velocity, v, and (d) ground flash density, n_g. These parameters are stochastic in nature.

Analysis of field data shows that the statistical variation of the peak, I_p, and the time to crest, t_f, of the return-stroke current fit lognormal distribution (Anderson and Eriksson, 1980). The probability density function, $p(I_p)$, of I_p then can be expressed as

$$p(I_p) = \frac{e^{-0.5 f_1}}{I_p \cdot \sigma_{\ell n I_p} \cdot \sqrt{2\pi}} \tag{8.52}$$

where

$$f_1 = \left(\frac{\ell n I_p - \ell n I_{pm}}{\sigma_{\ell n I_p}} \right)^2 \tag{8.53}$$

and $\sigma_{\ell n I_p}$ is the standard deviation of $\ell n I_p$, and I_{pm} is the median value of I_p. Similarly, the probability density function of t_f can be expressed as

$$p(t_f) = \frac{e^{-0.5 f_2}}{t_f \cdot \sigma_{\ell n t_f} \cdot \sqrt{2\pi}} \tag{8.54}$$

where

$$f_2 = \left(\frac{\ell n t_f - \ell n t_{fm}}{\partial_{\ell n t_f}} \right)^2 \tag{8.55}$$

The joint probability density function, $p(I_p, t_f)$, is given by

$$p(I_p, t_f) = \frac{e(0.5/(1-\rho^2))(f_1 - 2\rho\sqrt{f_1 \cdot f_2} + f_2)}{2\pi(I_p \cdot t_f)(\sigma_{\ell n I_p} \cdot \sigma_{\ell n t_f})\sqrt{1-\rho^2}} \tag{8.56}$$

where ρ is the coefficient of correlation. The statistical parameters of return-stroke current are as follows (Anderson and Eriksson, 1980; Eriksson, 1986; IEEE = PES Task Force 15.09, 2005):

Median time to crest, $t_{fm} = 3.83\,\mu s$; Log (to base e) of standard deviation, $\sigma(\ell n t_f) = 0.553$.
For $I_p < 20\,kA$: Median peak current, $I_{pm1} = 61.1\,kA$; Log (to base e) of standard deviation, $\sigma(\ell n I_{p1}) = 1.33$.
For $I_p > 20\,kA$: Median peak current, $I_{pm2} = 33.3\,kA$; Log (to base e) of standard deviation, $\sigma(\ell n I_{p2}) = 0.605$.

Correlation coefficient, $\rho = 0.47$.

Field data on the return-stroke velocity are limited. Lundholm (1957) and Rusck (1958) proposed the following empirical relationship between the return-stroke peak current and its velocity from the available field data:

$$v = \frac{c}{\sqrt{1 + (500/I_\mathrm{p})}} \quad \text{(m/s)} \tag{8.57}$$

where

c is the velocity of electromagnetic fields in free space = 3×10^8 m/s
I_p is the return-stroke peak current, kA

The ground flash density, n_g (number of flashes to ground per km^2 per year), varies regionally and seasonally. However, the average ground flash density maps around the globe are available to estimate lightning-caused outages.

8.8 Estimation of Outage Rates Caused by Nearby Lightning Strokes

The knowledge of the basic impulse insulation level (BIL) is essential for estimating the outage rate of an overhead power line. With this knowledge, the electrogeometric model is constructed to estimate the attractive area (Figure 8.7). According to the electrogeometric model, the striking distance of a lightning stroke is proportional to the return-stroke current. The following relation is used to estimate this striking distance, r_s:

$$r_\mathrm{s} = 8I_\mathrm{p}^{0.65} \quad \text{(m)} \tag{8.58}$$

where I_p in kA is the peak of the return-stroke current. In the cross-sectional view of Figure 8.7, a horizontal line (representing a plane) is drawn at a distance of r_s meters from the ground plane corresponding to the return-stroke current, I_p. A circular arc is drawn with its center on the conductor, P, and r_s as radius. This represents a cylinder of attraction above the line conductor. The circular arc and the horizontal line intersect at points A and B. The strokes with $I = I_\mathrm{p}$ falling between A and B will strike the conductor, resulting in direct strokes; those falling outside AB will hit the ground, inducing voltages on the line. The horizontal projection of A or B is $y_{\mathrm{o}1}$, which is given by

$$y_{\mathrm{o}1} = r_\mathrm{s} \quad \text{for } r_\mathrm{s} \le h_\mathrm{p} \tag{8.59a}$$

$$y_{\mathrm{o}1} = \sqrt{r_\mathrm{s}^2 - (r_\mathrm{s} - h_\mathrm{p})^2} \quad \text{for } r_\mathrm{s} > h_\mathrm{p} \tag{8.59b}$$

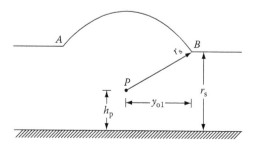

FIGURE 8.7 Electrogeometric model for estimating the least distance of ground strike.

and y_{o1} is the shortest distance of a lightning stroke of the given return-stroke current from the overhead line, which will result in a flash to ground.

To compute the outage rate, the return-stroke current, I_p, is varied from 1 to 200 kA in steps of 0.5 kA (Chowdhuri, 1989c, 2004). The current front time, tf, is varied from 0.5 to 10.5 μs in steps of 0.5 μs. At each current level, the shortest possible distance of the stroke, y_{o1}, is computed from Equation 8.59. Starting at $t_f = 0.5$ μs, the induced voltage is calculated as a function of time and compared with the given BIL of the line. If the BIL is not exceeded, then the next higher level of current is chosen. If the BIL is exceeded, then the lateral distance of the stroke from the line, y, is increased by Δy (e.g., 1 m), and the induced voltage is recalculated and compared with the BIL of the line. The lateral distance, y, is progressively increased until the induced voltage does not exceed BIL. This distance is called y_{o2}. For the selected I_p and t_f, the induced voltage will then exceed the BIL of the line and cause line flashover, if the lightning stroke hits the ground between y_{o1} and y_{o2} along the length of the line. For a 100 km sector of the line, the attractive area, A, will be (Figure 8.8)

$$A = 0.2(y_{o2} - y_{o1})\ \text{km}^2 \tag{8.60}$$

The joint probability density function, $p(I_p, t_f)$, is then computed from Equation 8.56 for the selected $I_p - t_f$ combination. If n_g is the ground flash density of the region, the expected number of flashovers per 100 km per year for that particular $I_p - t_f$ combination will be

$$\text{nfo} = p(I_p, t_f) \cdot \Delta I_p \cdot \Delta t_f \cdot n_g \cdot A \tag{8.61}$$

where
ΔI_p is the current step
Δt_f is the front-time step

The front time, t_f, is then increased by $\Delta t_f = 0.5$ μs to the next step, and nfo for the same current but with the new t_f is computed and added to the previous nfo. Once $t_f = 10.5$ μs is reached, the return-stroke current is increased by $\Delta I_p = 0.5$ kA, and the whole procedure is repeated until the limits $I_p = 200$ kA and $t_f = 10.5$ μs are reached. The cumulative nfo will then give the total number of expected line flashovers per 100 km per year for the selected BIL.

The lightning-induced outage rates of the horizontally configured three-phase overhead line of Chapter 7 are plotted in Figure 8.9. The effectiveness of the shield wire, as shown in the figure, is optimistic, bearing in mind that the shield wire was assumed to be held at ground potential. The shield wire

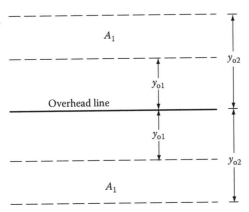

FIGURE 8.8 Attractive area of lightning ground flash to cause line flashover. $A = 2A_1 = 0.2(y_{o2} - y_{o1})\ \text{km}^2$.

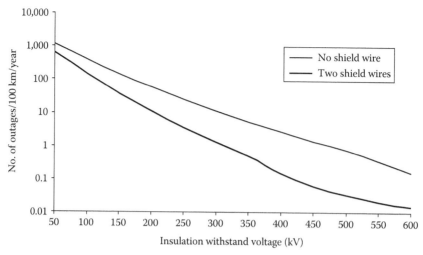

FIGURE 8.9   Outage rates of overhead line by indirect strokes vs. BIL.

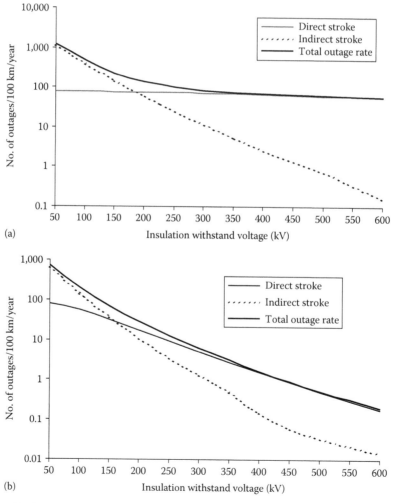

FIGURE 8.10   Total outage rates of overhead line vs. BIL. (a) Unshielded line, and (b) shielded line.

will not be held at ground potential under transient conditions. Therefore, the effectiveness of the shield wire will be less than the idealized case shown in Figure 8.9.

8.9 Estimation of Total Outage Rates

Outages of an overhead power line may be caused by both direct and indirect lightning strokes, especially if the voltage withstand level of the line is low, for example, power distribution lines. Outages caused by direct strokes are caused by either backflash (i.e., lightning hitting the tower or the shield wire) or by shielding failure if lightning hits one of the phase conductors. Outages caused by direct strokes were discussed in Chapter 7. Outages caused by indirect strokes are caused by lightning strokes hitting the nearby ground, which is discussed in this chapter. Therefore, information from both these chapters will be required to estimate the total outage rates of overhead lines caused by lightning. The outage rates of the three-phase horizontally configured overhead line of Chapter 7 are shown in Figure 8.10a for the unshielded line and in Figure 8.10b for the shielded line.

The effects of the variation of different parameters of the power line and also of the lightning stroke should also be considered in estimating the total outages of the line (Chowdhuri, 1989b). The power lines will be of finite length, terminated by transformers, underground cables, and lightning arresters. These terminations will generate multiple reflections. Therefore, power lines of finite lengths should also be considered for specific applications (Chowdhuri, 1991).

8.A Appendix A: Voltage Induced by Linearly Rising and Falling Return-Stroke Current

$$V(x,t) = V_1(x,t)u(t - t_o) + V_2(x,t)u(t - t_{of})$$

where

$$V_1(x,t) = \frac{30\alpha_1 h_p}{\beta c}\left[b_o \cdot \ell n \frac{f_{12}}{f_{11}} + 0.5\ell n(f_{13})\right]; \quad V_2(x,t) = -\frac{30\alpha_2 h_p}{\beta c}\left[b_o \cdot \ell n \frac{f_{12a}}{f_{11a}} + 0.5\ell n(f_{13a})\right]$$

$$b_o = 1 - \beta^2; \quad t_{of} = t_o + t_f; \quad t_{tf} = t - t_f$$

$$f_1 = m_1 + (ct - x)^2 - y_o^2; \quad f_2 = m_1 - (ct - x)^2 + y_o^2$$

$$f_3 = m_0 + (ct - x)^2 + y_o^2; \quad f_4 = m_0 + (ct_o - x)^2 - y_o^2$$

$$f_5 = n_1 + (ct + x)^2 - y_o^2; \quad f_6 = n_1 - (ct + x)^2 + y_o^2$$

$$f_7 = n_o - (ct_o + x)^2 + y_o^2; \quad f_8 = n_o + (ct_o + x)^2 - y_o^2$$

$$f_9 = b_o(\beta^2 x^2 + y_o^2) + \beta^2 c^2 t^2(t + \beta^2); \quad f_{10} = 2\beta^2 ct\sqrt{\beta^2 c^2 t^2 + b_o(x^2 + y_o^2)}$$

$$f_{11} = \frac{c^2 t^2 - x^2}{y_o^2}; \quad f_{12} = \frac{f_9 - f_{10}}{b_o^2 y_o^2}; \quad f_{13} = \frac{f_1 \cdot f_3 \cdot f_5 \cdot f_7}{f_2 \cdot f_4 \cdot f_6 \cdot f_8}$$

$$f_{1a} = m_{1a} + (ct_{tf} - x)^2 - y_o^2; \quad f_{2a} = m_{1a} - (ct_{tf} - x)^2 + y_o^2$$

$$f_{3a} = f_3; \quad f_{4a} = f_4; \quad f_{7a} = f_7; \quad f_{8a} = f_8$$

$$f_{5a} = n_{1a} + (ct_{tf} + x)^2 - y_o^2; \quad f_{6a} = n_{1a} - (ct_{tf} + x)^2 + y_o^2$$

$$f_{9a} = b_o(\beta^2 x^2 + y_o^2) + \beta^2 c^2 t_{tf}^2 (1 + \beta^2); \quad f_{10a} = 2\beta^2 ct_{tf}\sqrt{\beta^2 c^2 t_{tf}^2 + b_o(x^2 + y_o^2)}$$

$$f_{11a} = \frac{c^2 t_{tf}^2 - x^2}{y_o^2}; \quad f_{12a} = \frac{f_{9a} - f_{10a}}{b_o^2 y_o^2}; \quad f_{13a} = \frac{f_{1a} \cdot f_{3a} \cdot f_{5a} \cdot f_{7a}}{f_{2a} \cdot f_{4a} \cdot f_{6a} \cdot f_{8a}}$$

$$m_o = \sqrt{[(ct_o - x)^2 + y_o^2]^2 + 4h_c^2(ct_o - x)^2}; \quad m_1 = \sqrt{[(ct - x)^2 + y_o^2]^2 + 4h_c^2(ct - x)^2}$$

$$n_o = \sqrt{[(ct_o + x)^2 + y_o^2]^2 + 4h_c^2(ct_o + x)^2}; \quad n_1 = \sqrt{[(ct + x)^2 + y_o^2]^2 + 4h_c^2(ct + x)^2}$$

$$m_{1a} = \sqrt{[(ct_{tf} - x)^2 + y_o^2]^2 + 4h_c^2(ct_{tf} - x)^2}; \quad n_{1a} = \sqrt{[(ct_{tf} + x)^2 + y_o^2]^2 + 4h_c^2(ct_{tf} + x)^2}$$

References

Agrawal, A.K., Price, H.J., and Gurbaxani, S.H., Transient response of multiconductor transmission lines excited by a nonuniform electromagnetic field, *IEEE Trans. Electromagn. Compat.*, EMC-22, 119–129, 1980.

Anderson, R.B. and Eriksson, A.J., Lightning parameters for engineering applications, *Electra*, 69, 65–102, 1980.

Chowdhuri, P., Analysis of lightning-induced voltages on overhead lines, *IEEE Trans. Power Delivery*, 4, 479–492, 1989a.

Chowdhuri, P., Parametric effects on the induced voltages on overhead lines by lightning strokes to nearby ground, *IEEE Trans. Power Delivery*, 4, 1185–1194, 1989b.

Chowdhuri, P., Estimation of flashover rates of overhead power distribution lines by lightning strokes to nearby ground, *IEEE Trans. Power Delivery*, 4, 1982–1989, 1989c.

Chowdhuri, P., Lightning-induced voltages on multiconductor overhead lines, *IEEE Trans. Power Delivery*, 5, 658–667, 1990.

Chowdhuri, P., Response of overhead lines of finite length to nearby lightning strokes, *IEEE Trans. Power Delivery*, 6, 343–351, 1991.

Chowdhuri, P., *Electromagnetic Transients in Power Systems*, 2nd edn., Research Studies Press, Baldock, Hertfordshire, U.K., 2004.

Chowdhuri, P. and Gross, E.T.B., Voltage surges induced on overhead lines by lightning strokes, *Proc. IEE (U.K.)*, 114, 1899–1907, 1967.

Chowdhuri, P. and Gross, E.T.B., Voltages induced on overhead multiconductor lines by lightning strokes, *Proc. IEE (U.K.)*, 116, 561–565, 1969.

Chowdhuri, P., Li, S., and Yan, P., Review of research on lightning-induced voltages on an overhead line, *IEE Proc.—Gener. Transm. Distrib.(U.K.)*, 148, 91–95, 2001.

Cinieri, E. and Fumi, A., The effect of the presence of multiconductors and ground wires on the atmospheric high voltages induced on electrical lines (in Italian), *L'Energia Elettrica*, 56, 595–601, 1979.

Eriksson, A.J., Notes on lightning parameters for system performance estimation, CIGRE Note 33–86 (WG33-01) IWD, 15 July 1986.

Fortescue, C.L., Direct strokes—Not induced surges—Chief cause of high-voltage line flashover, *The Electric Journal*, 27, 459–462, 1930.

Haldar, M.K. and Liew, A.C., Alternative solution for the Chowdhuri-Gross model of lightning-induced voltages on power lines, *Proc. IEE (U.K.)*, 135, 324–329, 1988.

IEEE = PES Task Force 15.09, Parameters of lightning strokes: A review, *IEEE Trans. Power Delivery*, 20, 346–358, 2005.

Lundholm, R., Induced overvoltage-surges on transmission lines and their bearing on the lightning performance at medium voltage networks, *Trans. Chalmers Univ. Technol., Gothenburg, Sweden*, 188, 1–117, 1957.

Morse, P.M. and Feshbach, H., *Methods of Theoretical Physics*, Vol. 1, McGraw-Hill, New York, 1950, Chap. 7.

Rusck, S., Induced lightning over-voltages on power-transmission lines with special reference to the overvoltage protection of low-voltage networks, *Trans. R. Inst. Tech., Stockholm, Sweden*, 120, 1–118, 1958.

9

Switching Surges

Stephen R. Lambert
Shawnee Power
Consulting, LLC

Switching surges occur on power systems as a result of instantaneous changes in the electrical configuration of the system, and such changes are mainly associated with switching operations and fault events. These overvoltages generally have crest magnitudes which range from about 1 to 3 pu for phase-to-ground surges and from about 2 to 4 pu for phase-to-phase surges (in pu on the phase to ground crest voltage base) with higher values sometimes encountered as a result of a system resonant condition. Waveshapes vary considerably with rise times ranging from 50 μs to thousands of μs and times to half-value in the range of hundreds of μs to thousands of μs. For insulation testing purposes, a waveshape having a time to crest of 250 μs with a time to half-value of 2000 μs is often used.

The following addresses the overvoltages associated with switching various power system devices. Possible switching surge magnitudes are indicated, and operations and areas of interest that might warrant investigation when applying such equipment are discussed.

9.1 Transmission Line Switching Operations

Surges associated with switching transmission lines (overhead, SF_6, or cable) include those that are generated by line energizing, reclosing (three phase and single phase operations), fault initiation, line dropping (deenergizing), fault clearing, etc. During an energizing operation, for example, closing a circuit breaker at the instant of crest system voltage results in a 1 pu surge traveling down the transmission line and being reflected at the remote, open terminal. The reflection interacts with the incoming wave on the phase under consideration as well as with the traveling waves on adjacent phases. At the same time, the waves are being attenuated and modified by losses. Consequently, it is difficult to accurately predict the resultant waveshapes without employing sophisticated simulation tools such as a transient network analyzer (TNA) or digital programs such as the Electromagnetic Transients Program (EMTP).

Transmission line overvoltages can also be influenced by the presence of other equipment connected to the transmission line—shunt reactors, series or shunt capacitors, static var systems, surge arresters, etc. These devices interact with the traveling waves on the line in ways that can either reduce or increase the severity of the overvoltages being generated.

When considering transmission line switching operations, it can be important to distinguish between "energizing" and "reclosing" operations, and the distinction is made on the basis of whether the line's inherent capacitance retains a trapped charge at the time of line closing (reclosing operation) or whether no trapped charge exists (an energizing operation). The distinction is important as the magnitude of the switching surge overvoltage can be considerably higher when a trapped charge is present; with higher magnitudes, insulation is exposed to increased stress, and devices such as surge arresters will, by necessity, absorb more energy when limiting the higher magnitudes. Two forms of trapped charges can exist—DC and oscillating. A trapped charge on a line with no other equipment attached to the line exists as a DC trapped charge, and the charge can persist for some minutes before dissipating (Beehler, 1964). However, if a transformer (power or wound potential transformer) is connected to the line, the charge will decay rapidly (usually in less than 0.5 s) by discharging through the saturating branch of the transformer (Marks, 1969). If a shunt reactor is connected to the line, the trapped charge takes on an oscillatory waveshape due to the interaction between the line capacitance and the reactor inductance. This form of trapped charge decays relatively rapidly depending on the Q of the reactor, with the charge being reduced by as much as 50% within 0.5 s.

Figures 9.1 and 9.2 show the switching surges associated with reclosing a transmission line. In Figure 9.1 note the DC trapped charge (approximately 1.0 pu) that exists prior to the reclosing operation (at 20 μs). Figure 9.2 shows the same case with an oscillating trapped charge (a shunt reactor was present on the line) prior to reclosing. Maximum surges were 3.0 for the DC trapped charge case and 2.75 pu for the oscillating trapped charge case (both occurred on phase c).

The power system configuration behind the switch or circuit breaker used to energize or reclose the transmission line also affects the overvoltage characteristics (shape and magnitude) as the traveling wave interactions occurring at the junction of the transmission line and the system (i.e., at the circuit breaker) as well as reflections and interactions with equipment out in the system are important. In general, a stronger system (higher short circuit level) results in somewhat lower surge magnitudes than a weaker system, although there are exceptions. Consequently, when performing simulations to predict overvoltages, it is usually important to examine a variety of system configurations (e.g., a line out of service or contingencies) that might be possible and credible.

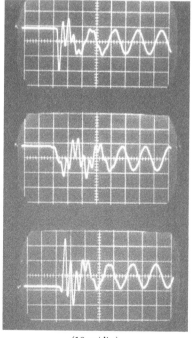

(10 μs/div.)

FIGURE 9.1   DC trapped charge.

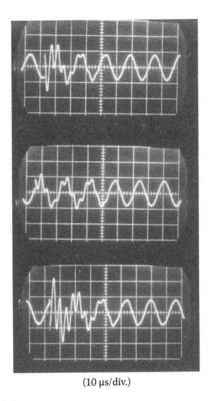

(10 μs/div.)

FIGURE 9.2   Oscillating trapped charge.

Single phase switching as well as three phase switching operations may also need to be considered. On EHV transmission lines, for example, most faults (approximately 90%) are single phase in nature, and opening and reclosing only the faulted phase rather than all three phases, reduces system stresses. Typically, the overvoltages associated with single phase switching have a lower magnitude than those that occur with three phase switching (Koschik et al., 1978).

Switching surge overvoltages produced by line switching are statistical in nature—that is, due to the way that circuit breaker poles randomly close (excluding specially modified switchgear designed to close on or near voltage zero), the instant of electrical closing may occur at the crest of the system voltage, at voltage zero, or somewhere in between. Consequently, the magnitude of the switching surge varies with each switching event. For a given system configuration and switching operation, the surge voltage magnitude at the open end of the transmission line might be 1.2 pu for one closing event and 2.8 pu for the next (Hedman et al., 1964; Johnson et al., 1964), and this statistical variation can have a significantly impact on insulation design (see Chapter 14 on insulation coordination).

Typical switching surge overvoltage statistical distributions (160 km line, 100 random closings) are shown in Figures 9.3 and 9.4 for phase-to-ground and phase-to-phase voltages (Lambert, 1988), and the surge magnitudes indicated are for the highest that occurred on any phase during each closing. With no surge limiting action (by arresters or circuit breaker preinsertion resistors), phase-to-ground surges varied from 1.7 to 2.15 pu with phase-to-phase surges ranging from 2.2 to 3.7 pu. Phase-to-phase surges can be important to line-connected transformers and reactors as well as to transmission line phase-to-phase conductor separation distances when line uprating or compact line designs are being considered.

Figure 9.3 also demonstrates the effect of the application of surge arresters on phase-to-ground surges, and shows the application of resistors preinserted in the closing sequence of the circuit breaker (400 Ω for 5.56 ms) is even more effective than arresters in reducing surge magnitude. The results shown on

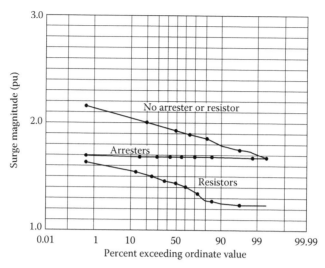

FIGURE 9.3 Phase-to-ground overvoltage distribution.

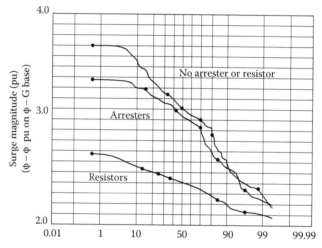

FIGURE 9.4 Phase-to-phase overvoltage distribution.

Figure 9.4, however, indicate that while resistors are effective in limiting phase-to-phase surges, arresters applied line to ground are generally not very effective at limiting phase-to-phase overvoltages.

Line dropping (deenergizing) and fault clearing operations also generate surges on the system, although these typically result in phase-to-ground overvoltages having a maximum value of 2–2.2 pu. Usually the concern with these operations is not with the phase-to-ground or phase-to-phase system voltages, but rather with the recovery voltage experienced by the switching device. The recovery voltage is the voltage which appears across the interrupting contacts of the switching device (a circuit breaker for example) following current extinction, and if this voltage has too high a magnitude, or in some instances rises to its maximum too quickly, the switching device may not be capable of successfully interrupting.

The occurrence of a fault on a transmission line also can result in switching surge type overvoltages, especially on parallel lines. These voltages usually have magnitudes on the order of 1.8–2.2 pu and are usually not a problem (Kimbark and Legate, 1968; Madzarevic et al., 1977).

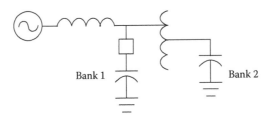

FIGURE 9.5   Voltage magnification circuit.

9.2 Series Capacitor Bank Applications

Installation of a series capacitor bank in a transmission line (standard or thyristor controlled) has the potential for increasing the magnitude of phase-to-ground and phase-to-phase switching surge overvoltages due to the trapped charges that can be present on the bank at the instant of line reclosing. In general, surge arresters limit the phase-to-ground and phase-to-phase overvoltages to acceptable levels; however, one problem that can be serious is the recovery voltage experienced by circuit breakers when clearing faults on a series compensated line. Depending the bank's characteristics and on fault location with respect to the bank's location, a charge can be trapped on the bank, and this trapped charge can add to the surges already being generated during the fault clearing operation (Wilson, 1972). The first circuit breaker to clear is sometimes exposed to excessive recovery voltages under such conditions.

9.3 Shunt Capacitor Bank Applications

Energizing a shunt capacitor bank typically results in maximum overvoltages of about 2 pu or less. However, there are two conditions where significant overvoltages can be generated. One involves a configuration (shown on Figure 9.5) where two banks are separated by a significant inductance (e.g., a transformer) (Schultz et al., 1959). When one bank is switched, if the system inductance and bank 1 capacitance has the same natural frequency as that of the transformer leakage inductance and the bank 2 capacitance, then a voltage magnification can take place.

Another configuration that can result in damaging overvoltages involves energizing a capacitor bank with a transformer terminated transmission line radially fed from the substation at which the capacitor bank is located (Jones and Fortson, 1985). During bank switching, phase-to-phase surges are imposed on the transformer, and because these are not very well suppressed by the usual phase-to-ground application of surge arresters, transformer failures have been known to result. Various methods to reduce the surge magnitude have included the application of controlled circuit breaker closing techniques (closing near voltage zero), and resistors or reactors preinserted in the closing sequence of the switching devices.

Restriking of the switching device during bank deenergizing can result in severe line-to-ground overvoltages of 3–5 pu or more (rarely) (Johnson et al., 1955; Greenwood, 1971). Surge arresters are used to limit the voltages to acceptable levels, but at higher system voltages, the energy discharged from the bank into the arrester can exceed the arrester's capability.

9.4 Shunt Reactor Applications

Switching of shunt reactors (and other devices characterized as having small inductive currents such as transformer magnetizing currents, motor starting currents, etc.) can generate high phase-to-ground overvoltages as well as severe recovery voltages (Greenwood, 1971), especially on lower voltage equipment such as reactors applied on the tertiary of transformers. Energizing the devices seldom generates high overvoltages, but overvoltages generated during deenergizing, as a result of current chopping by the switching device when interrupting the small inductive currents, can be significant. Neglecting damping, the phase-to-ground overvoltage magnitude can be estimated by

$$V = i\sqrt{\frac{L}{C}} \tag{9.1}$$

where

i is the magnitude of the chopped current (0 to perhaps as high as 10 A or more)
L is the reactor's inductance
C is the capacitance of the reactor (on the order of a few thousand picofarads)

When C is small, especially likely with dry-type reactors often used on transformer tertiaries, the surge impedance term can be large, and hence the overvoltage can be excessive.

To mitigate the overvoltages, surge arresters are sometimes useful, but the application of a capacitor on the terminals of the reactor (or other equipment) have a capacitance on the order of 0.25–0.5 µF is very helpful. In the equation above, note that if C is increased from pF to µF, the surge impedance term is dramatically reduced, and hence the voltage is reduced.

References

Beehler, J.E., Weather, corona, and the decay of trapped energy on transmission lines, *IEEE Trans. Power Appar. Syst.* 83, 512, 1964.

Greenwood, A., *Electrical Transients in Power Systems*, John Wiley & Sons, New York, 1971.

Hedman, D.E., Johnson, I.B., Titus, C.H., and Wilson, D.D., Switching of extra-high-voltage circuits, II—Surge reduction with circuit breaker resistors, *IEEE Trans. Power Appar. Syst.* 83, 1196, 1964.

Johnson, I.B., Phillips, V.E., and Simmons, Jr., H.O., Switching of extra-high-voltage circuits, I—system requirements for circuit breakers, *IEEE Trans. Power Appar. Syst.* 83, 1187, 1964.

Johnson, I.B., Schultz, A.J., Schultz, N.R., and Shores, R.R., Some fundamentals on capacitance switching, *AIEE Trans. Power Appar. Syst.* PAS-74, 727, 1955.

Jones, R.A. and Fortson, Jr., H.S., Considerations of phase-to-phase surges in the application of capacitor banks, *IEEE PES Summer Meeting*, Vancouver, Canada, 1985, 85 SM 400–7.

Kimbark, E.W. and Legate, A.C., Fault surge versus switching surge: A study of transient overvoltages caused by line-to-ground faults, *IEEE Trans. Power Appar. Syst.* PAS-87, 1762, 1968.

Koschik, V., Lambert, S.R., Rocamora, R.G., Wood, C.E., and Worner, G., Long line single-phase switching transients and their effect on station equipment, *IEEE Trans. Power Appar. Syst.* PAS-97, 857, 1978.

Lambert, S.R., Effectiveness of zinc oxide surge arresters on substation equipment probabilities of flashover, *IEEE Trans. Power Delivery* 3(4), 1928, 1988.

Madzarevic, V., Tseng, F.K., Woo, D.H., Niebuhr, W.D., and Rocamora, R.G., Overvoltages on EHV transmission lines due to fault and subsequent bypassing of series capacitors, *IEEE PES Winter Meeting*, New York, January 1977, F77 237–1.

Marks, L.W., Line discharge by potential transformers, *IEEE Trans. Power Appar. Syst.* PAS-88, 293, 1969.

Schultz, A.J., Johnson, J.B., and Schultz, N.R., Magnification of switching surges, *AIEE Trans. Power Appar. Syst.* 77, 1418, 1959.

Wilson, D.D., Series compensated lines—voltages across circuit breakers and terminals caused by switching, *IEEE PES Summer Meeting*, San Francisco, CA, 1972, T72 565–0.

10

Very Fast Transients

Juan A.
Martinez-Velasco
*Universitat Politecnica
de Catalunya*

Electromagnetic transient phenomena in power systems are caused by switching operations, faults, and other disturbances, such as lightning strokes. They may appear with a wide range of frequencies that vary from DC to several MHz and occur on a time scale that goes from microseconds to several cycles. Due to frequency-dependent behavior of power components and difficulties for developing models accurate enough for a wide frequency range, the frequency ranges are classified into groups, with overlapping between them. An accurate mathematical representation of each power component can generally be developed for a specific frequency range (CIGRE, 1990; IEC, 2006).

Very fast transients (VFT), also known as very fast front transients, belong to the highest frequency range of transients in power systems. According to the classification proposed by the CIGRE Working Group 33-02, VFT may vary from 100 kHz up to 50 MHz (1990). According to IEC 60071-1, the shape of a very fast front overvoltage is "usually unidirectional with time to peak <0.1 μs, total duration <3 ms, and with superimposed oscillations at frequency 30 kHz < f < 100 MHz" (1993). In practice, the term VFT is restricted to transients with frequencies above 1 MHz.

Several causes can originate these transients in power systems: Disconnector operations and faults within gas-insulated substations (GIS), switching of motors and transformers with short connections to the switchgear, and certain lightning conditions (IEC 60071-2, 1996).

This chapter is exclusively dedicated to explaining the origin, and to analyze the propagation and the effects of VFT in GIS. These transients have a rise time in the range of 4–100 ns and are normally followed by oscillations ranging from 1 to 50 MHz. Their magnitude is usually below 2 per unit (pu) of the line-to-neutral voltage crest, although they can also reach values higher than 2.5 pu. These values are generally below the BIL of the GIS and connected equipment of lower voltage classes. VFT in GIS are of greater concern at the highest voltages, for which the ratio of the BIL to the system voltage is lower (Yamagata et al., 1996). External VFT can be dangerous for secondary and adjacent equipment.

A description of GIS layouts and main components is out of the scope of this chapter. Readers are referred to the specialized literature (IEEE WG, 1982, 1989, 1993; IEEE Std, 2011; CIGRE, 2000; Ryan, 2001; Arora et al., 2005; Bolin, 2007; CIGRE, 2009).

10.1 Origin of VFT in GIS

VFT within a GIS are usually generated by disconnect switch operations, although other events, such as the closing of a grounding switch or a fault, can also cause VFT.

A large number of pre- or restrikes can occur during a disconnector operation due to the relatively slow speed of the moving contact (Ecklin et al., 1980). Figure 10.1 shows a very simple configuration used to explain the general switching behavior and the pattern of voltages on opening and closing of a disconnector at a capacitive load (Boggs et al., 1982). During an opening operation, sparking occurs as soon as the voltage between the source and the load exceeds the dielectric strength across contacts. After a restrike, a high-frequency current will flow through the spark and equalize the capacitive load voltage to the source voltage. The potential difference across the contacts will fall and the spark will extinguish. The subsequent restrike occurs when the voltage between contacts reaches the new dielectric strength level that is determined by the speed of the moving contact and other disconnector characteristics. The behavior during a closing operation is very similar, and the load-side voltage will follow the supply voltage until the contact is made. For a discussion of the physics involved in the restrikes and prestrikes of a disconnect switch operation; see Boggs et al., 1982.

The scheme shown in Figure 10.2 will be very useful to illustrate the generation of VFT due to a disconnector operation. The breakdown of a disconnector when it is closing originates two surges V_L and V_S

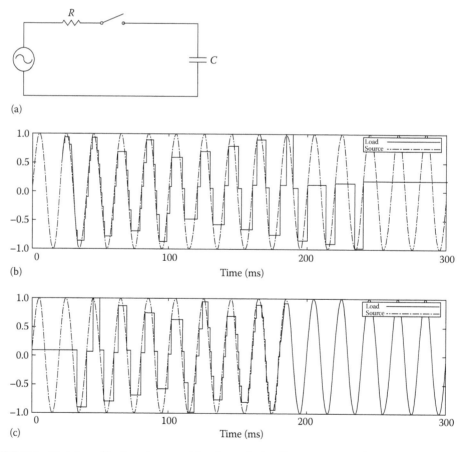

FIGURE 10.1 Variation of load- and source-side voltages during disconnector switching: (a) scheme of the circuit, (b) opening operation, and (c) closing operation.

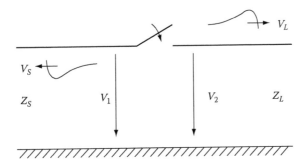

FIGURE 10.2   Generation of VFT.

which travel outward in the bus duct and back into the source side, respectively. The magnitude of both traveling surges is given by

$$V_L = \frac{Z_L}{Z_S + Z_L}(V_1 - V_2) \quad V_S = -V_L \tag{10.1}$$

where Z_S and Z_L are the surge impedances on the source and on the load side, respectively. V_1 is the intercontact spark voltage, while V_2 is the trapped charge voltage at the load side.

Steep-fronted traveling surges can also be generated in case of a line-to-ground fault, as the voltage collapse at the fault location occurs in a similar way as in the disconnector gap during striking.

10.2 Propagation of VFT in GIS

VFT in GIS can be divided into internal and external. Internal transients can produce overvoltages between the inner conductor and the enclosure, while external transients can cause stress on secondary and adjacent equipment. A summary about the propagation and main characteristics of both types of phenomena follows.

10.2.1 Internal Transients

Breakdown phenomena across the contacts of a disconnector during a switch operation or a line-to-ground fault generate very short rise time traveling waves which propagate in either direction from the breakdown location. As a result of the fast rise time of the wave front, the propagation throughout a substation must be analyzed by representing GIS sections as low-loss distributed parameter transmission lines, each section being characterized by a surge impedance and a transit time. Traveling waves are reflected and refracted at every point where they encounter a change in the surge impedance. The generated transients depend on the GIS configuration and on the superposition of the surges reflected and refracted on line discontinuities such as breakers, "T" junctions, or bushings. As a consequence of multiple reflections and refractions, traveling voltages can increase above the original values and very high frequency oscillations occur.

The internal damping of the VFT influencing the highest frequency components is determined by the spark resistance. Skin effects due to the aluminum enclosure can be generally neglected. The main portion of the damping of the VFT occurs by outcoupling at the transition to the overhead line. Due to the traveling wave behavior of the VFT, the overvoltages caused by disconnector switches show a spatial distribution. Normally, the highest overvoltage stress is reached at the open end of the load side.

Overvoltages are dependent on the voltage drop at the disconnector just before striking, and on the trapped charge that remains on the load side of the disconnector. For a normal disconnector with a slow speed, the maximum trapped charge reaches 0.5 pu resulting in a most unfavorable voltage collapse of 1.5 pu. For these cases, the resulting overvoltages are in the range of 1.7 pu and reach 2 pu for very specific cases. For a high-speed disconnector, the maximum trapped charge could be 1 pu and the highest overvoltages reach values up to 2.5 pu. Although values larger than 3 pu have been reported, they have been derived by calculation using unrealistic simplified simulation models. The main frequencies depend on the length of the GIS sections affected by the disconnector operation and are usually in the range of 1–50 MHz, although much higher frequencies have been reported (Rao et al., 2005).

The following examples will be useful to illustrate the influence of some parameters on the frequency and magnitude of VFT in GIS. Figure 10.3 shows two very simple cases: A single bus duct and a "T" junction in which GIS components are modeled as lossless distributed parameter transmission lines. The source side is represented as a step-shaped source in series with a resistance, which is a simplified modeling of an infinite length bus duct. The surge impedance of all bus sections is 50 Ω. For the simplest configuration, the reflections of the traveling waves at both terminals of the duct will produce, when the source resistance is neglected, a pulse-shaped transient of constant magnitude—2 pu—and constant frequency at the open terminal. The frequency of this pulse can be calculated from the following expression:

$$f = \frac{1}{4\tau} \tag{10.2}$$

where τ is the transit time of the line (i.e., the bus duct). If the propagation velocity is close to that of light, the frequency, in MHz, of the voltage generated at the open terminal will be

$$f \approx \frac{75}{d} \tag{10.3}$$

where d is the duct length, in meters. When a more realistic representation of the source is used (e.g., $R = 40\ \Omega$), the maximum overvoltage at the open terminal will depend on the voltage at the disconnector just before striking, and on the trapped charge which remains on the load side.

Overvoltages can reach higher values in more complex GIS configurations. The simulations performed for the "T" configuration shown in Figure 10.3 gave in all cases higher values than in the previous case, being node 4 the location where the highest overvoltages were originated.

10.2.2 External Transients

Internally generated VFT propagate throughout the GIS and reach bushings where they cause transient enclosure voltages (TEV) and traveling waves that propagate along the overhead transmission line. An explanation about the generation of external transients and some comments on their main characteristics follow.

10.2.2.1 Transient Enclosure Voltages

TEV, also known as transient ground potential rises (TGPR), are short-duration high-voltage transients that appear on the enclosure of the GIS through the coupling of internal transients to enclosure at enclosure discontinuities. The simplified circuit shown in Figure 10.4 is used to explain the generation of TEV (Meppelink et al., 1989). At the GIS—air interface, three transmission lines can be distinguished: The coaxial GIS transmission line, the transmission line formed by the bushing conductor and the overhead line, and the GIS enclosure-to-ground transmission line. When an internal wave propagates to the gas-to-air bushing, a portion of the transient is coupled onto the overhead transmission line, and a portion is coupled onto the GIS enclosure-to-ground transmission line. The wave that propagates along the enclosure-to-ground transmission line is the TEV. The usual location for these voltages is the transition GIS-overhead line at an air bushing, although they can also emerge at visual inspection ports, insulated spacers for CTs, or insulated flanges at GIS/cables interfaces.

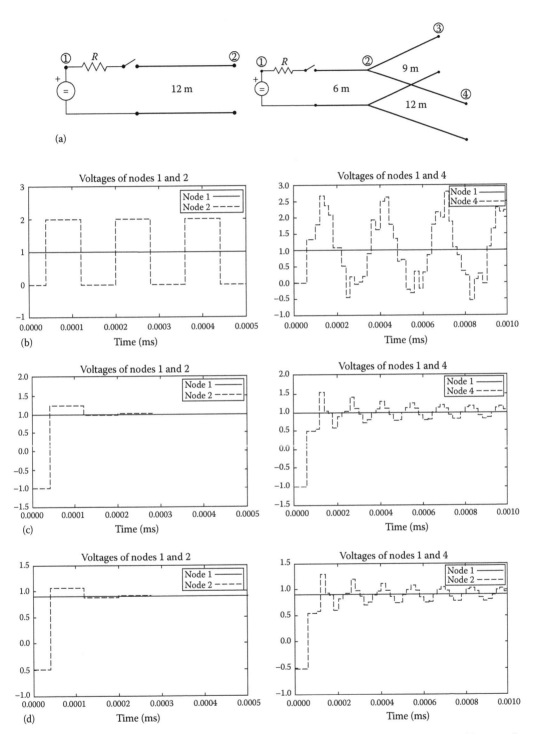

FIGURE 10.3 VFT overvoltages in GIS: (a) scheme of the network; (b) $R = 0$, $V_1 = 1$ pu, $V_2 = 0$; (c) $R = 40\,\Omega$, $V_1 = 1$ pu, $V_2 = -1$ pu; and (d) $R = 40\,\Omega$, $V_1 = 0.9$ pu, $V_2 = -0.5$ pu.

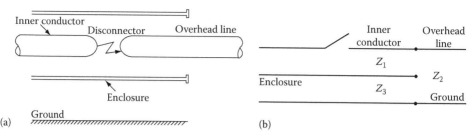

FIGURE 10.4 Generation of TEV: (a) GIS–air transition and (b) single-line diagram.

TEV waveforms have at least two components; the first one has a short initial rise time and is followed by high-frequency oscillations, in the range of 5–10 MHz, determined by the lengths of various sections of the GIS. The second component is of lower frequency, hundreds of kHz, and is often associated with the discharge of capacitive devices with the earthing system. Both components are damped quickly as a result of the lossy nature of the enclosure-to-ground plane transmission mode. TEV generally persists for a few microseconds. The magnitude varies along the enclosure; it can be in the range of 0.1–0.3 pu of the system voltage, and reaches the highest magnitude near the GIS–air interface. Mitigation methods include short length leads, low impedance grounding, and the installation of metal–oxide arresters across insulating spacers.

10.2.2.2 Transients on Overhead Connections

A portion of the VFT traveling wave incident at a gas–air transition is coupled onto the overhead connection and propagates to other components. This propagation is lossy and results in some increase of the waveform rise time. In general, external waveforms have two different characteristics: The overall waveshape, which is dictated by lumped circuit parameters such as the capacitance of voltage transformers or line and earthing inductance, with a rise time in the range of a few hundred nanoseconds, and a fast front portion, which is dictated by transmission line effects, with a rise time in the range of 20 ns. A fast rise time of the initial portion is possible as capacitive components, such as bushings, can be seen as physically long and cannot be treated as lumped elements; the magnitude is generally lower than that of internal VFT and it is usually reduced by discontinuities in the transmission path, with a voltage rate-of-rise in the range of 10–30 kV/μs.

10.2.2.3 Transient Electromagnetic Fields

VFT caused by switching operations and faults can produce high-frequency electromagnetic interference (EMI), which can couple into lower voltage control circuits and electronic equipment, and cause some stress. Transient electromagnetic fields (EMF) leak out into the environment through discontinuities such as gas-to-air bushing, gas-to-cable termination, nonmetallic viewing ports, or insulated flanges, and get coupled to the control equipment or data cables present in the GIS. They produce transient currents and voltages on the shield of cables and equipment. In addition, there can be conducted mechanisms that couple VFT currents to the control wiring, so a portion of the bus transient current may appear at the terminals of the relay or data-acquisition systems connected to them (Rao et al., 2005). All of these coupling modes add different waveshapes, frequency content, and relative phase shift, and result in waveshapes different from those due to any one of the coupling mechanisms acting alone.

EMI issues in GIS have been investigated in a number of studies. Measurements and model predictions for transient cable and wire currents and voltages caused by switching operations in substations were presented by Thomas et al. (1994) and Wiggins et al. (1994). More recently, Rao et al. (2005) estimated VFT currents at various locations in a GIS for different switching operations, considering the configuration of the substation and the characteristics of the components. In further studies, they analyzed the induced currents and voltages in control cables (Rao et al., 2007a), and the shielding effectiveness of bushing and bus section by computing the transient EM field emission levels from the gas-to-air bushing through the composite insulator housing and from the gas-insulated bus duct through the non-metallic flange of the support insulator (Rao et al., 2007b, 2008). The EMF radiated as a consequence of switching operations in a GIS were also studied and measured by Nishiwaki et al. (1995) to analyze the malfunction of power electronic equipment, and by Hoshino et al. (2001) to discriminate the radiation pattern created from partial discharges from that produced when switching GIS equipment.

10.3 Modeling Guidelines and Simulation

Due to the origin and the traveling nature of VFT, modeling of GIS components makes use of electrical equivalent circuits composed of lumped elements and distributed parameter lines. At very high frequencies, the skin losses can produce an important attenuation; however, these losses are usually neglected, which produces conservative results. Only the dielectric losses in some components (e.g., capacitively graded bushing) need be taken into account. The calculation of internal transients may be performed using distributed parameter models for which only an internal mode (conductor-enclosure) is taken into account, and assuming that the external enclosure is perfectly grounded. If TEV is a concern, then a second mode (enclosure-ground) is to be considered.

The next two sections present modeling guidelines to represent GIS equipment in computation of internal and external transients (Matsumura and Nitta, 1981; Dick et al., 1982; Fujimoto et al., 1982, 1986; Ogawa et al., 1986; Witzmann, 1987; CIGRE WG, 1988a; IEEE TF, 1996). They make use of single-phase models and very simple representations. Depending on the substation layout and the study to be performed, three-phase models for inner conductors (Miri and Binder, 1995; Miri and Stojkovic, 2001) or the outer enclosures (Dick et al., 1982) should be considered. More advanced guidelines have been analyzed and proposed by Haznadar et al. (1992).

10.3.1 Computation of Internal Transients

A short explanation about the representation of the most important GIS components follows.

10.3.1.1 Bus Ducts

A bus duct can be represented as a lossless transmission line. The surge impedance and the travel time are calculated from the physical dimensions of the duct. The inductance and the capacitance per unit length of a horizontal single-phase coaxial cylinder configuration, as that shown in Figure 10.5, are given by the following expressions:

$$L_1' = \frac{\mu_0}{2\pi} \ln \frac{R}{r} \tag{10.4}$$

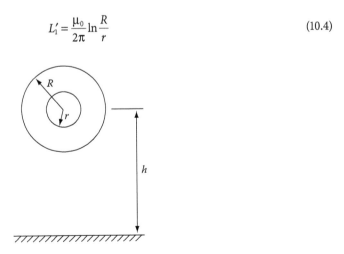

FIGURE 10.5 Coaxial bus duct cross section.

$$C_1' = \frac{2\pi\varepsilon}{\ln R/r} \quad (\varepsilon \approx \varepsilon_0) \tag{10.5}$$

from where the following form for the surge impedance is derived:

$$Z_1 = \sqrt{\frac{L_1'}{C_1'}} = \frac{1}{2\pi}\sqrt{\frac{\mu_0}{\varepsilon}}\ln\frac{R}{r} \approx 60\ln\frac{R}{r} \tag{10.6}$$

A different approach should be used for vertically oriented bus sections (Miri and Binder, 1995). As for the propagation velocity, empirical corrections are usually needed to adjust its value. Experimental results show that the propagation velocity in GIS ducts is close to 0.95–0.96 of the speed of light (Fujimoto et al., 1984).

Other equipment, such as elbows, can also be modeled as lossless transmission lines.

10.3.1.2 Surge Arresters

Experimental results have shown that if switching operations in GIS do not produce voltages high enough to cause metal–oxide surge arrester to conduct, then the arrester can be modeled as a capacitance-to-ground. However, when the arrester conducts, the model should take into account the steep front wave effect, since the voltage developed across the arrester for a given discharge current increases as the time to crest of the current increases, but reaches crest prior to the crest of the current (CIGRE, 1990). A detailed model must represent each internal shield and block individually, and include the travel times along shield sections, as well as capacitances between these sections, capacitances between blocks and shields, and the blocks themselves.

10.3.1.3 Circuit Breakers

A closed breaker can be represented as a lossless transmission line whose electrical length is equal to the physical length, with the propagation velocity reduced to 0.95–0.96 of the speed of light. The representation of an open circuit breaker is more complicated due to internal irregularities. In addition, circuit breakers with several chambers contain grading capacitors, which are not arranged symmetrically. The electrical length must be increased above the physical length due to the effect of a longer path through the grading capacitors, while the speed of propagation must be decreased due to the effects of the higher dielectric constant of these capacitors.

10.3.1.4 Spark Dynamics

The behavior of the spark in disconnector operations can be represented by a dynamically variable resistance with a controllable collapse time. Several models have been proposed to reproduce the arc behavior, being the Toepler's spark law the most popular one. According to Toepler, the instantaneous value of the spark resistance is inversely proportional to the charge conducted through the spark channel

$$R = \ell \cdot \frac{k}{\int_0^t i\,dt} \tag{10.7}$$

where
  ℓ is the length of the discharge gap
  i is the discharge current
  k is the Toepler's constant

For air insulation, the constant k is assumed to be between 0.004 and 0.005 $V \cdot s^{-1} \cdot m^{-1}$. Some works have shown that this constant depends on several factors (e.g., the SF_6 concentration, the pressure, or the inhomogeneity of the electrical field) whose influence can be significant (Osmokrovic et al., 1992; Singha and Joy Thomas, 2003). The variation with respect to the gas pressure and the SF_6 concentration shows an increasing trend, being the breakdown voltage a major controlling parameter.

Other models used for representing the spark gap may be based on the classic dynamic arc representation (CIGRE, 1993; Miri and Stojkovic, 2001; Martinez-Velasco and Popov, 2009).

In general, this representation does not affect the magnitude of the maximum VFT overvoltages, but it can introduce a significant damping on internal transients (Yanabu et al., 1990).

10.3.1.5 Gas-to-Air Bushings

A bushing gradually changes the surge impedance from that of the GIS to that of the line. A simplified model may consist of several transmission lines in series with a lumped resistor representing losses; the surge impedance of each line section increases as the location goes up the bushing. If the bushing is distant from the point of interest, the resistor can be neglected and a single line section can be used. A detailed model must consider the coupling between the conductor and shielding electrodes, and include the representation of the grounding system connected to the bushing (Fujimoto and Boggs, 1988; Ardito et al., 1992).

10.3.1.6 Power Transformers

At very high frequencies, a winding of a transformer may be represented as a capacitive network consisting of series capacitances between turns and coils, and shunt capacitances between turns and coils to the grounded core and transformer tank; the saturation of the magnetic core can be neglected, as well as leakage impedances (Chowdhuri, 2005). Such representation can reproduce the highly nonuniform voltage distribution developed along a transformer winding under steeped front voltages.

When analyzing the interaction of a transformer with the GIS and voltage transfer is not a concern, an accurate representation can be obtained by developing a circuit that matches the frequency response of the transformer at its terminals (Gustavsen, 2004). When voltage transfer has to be calculated, a black box model can be also derived from the frequency response of the transformer at its terminals (Morched et al., 1993; Gustavsen, 2004).

Several approaches have been developed to estimate the voltage distribution along transformer windings under VFT; they are based on the use of detailed lumped- or distributed-parameter models (Shibuya et al., 2001; Popov et al., 2003, 2007; Liang et al., 2006; Hosseini et al., 2008; Shintemirov et al., 2009). Fujita et al. (2007) have found that a shell-type transformer can be simply represented by a capacitance even when resonances occur in the winding. Details on the computation of transformer parameters for application of the various models have been presented by Degeneff (2007), Kulkarni and Khaparde (2004), and de León et al. (2009).

10.3.1.7 Three-Phase Models

The aforementioned models are based on a single-phase representation for all GIS components. However, in some GIS, usually at subtransmission levels, the three phases can be enclosed in a single encapsulation. In that case, the representation of some GIS components (e.g., ducts, elbows, T-junctions) has to include the three phases (Miri and Stojkovic, 2001). Models for three-phase ducts can be derived by using the supporting routine CABLE CONSTANT available in some transients tools (Dommel, 1996).

Tests performed with three-phase GIS switchgear have shown that the amplitude of phase-to-phase voltage transients may be higher than that between phase-to-ground, and the phase-to-phase capacitive coupling of the busbar system can raise the trapped charge left at the load side of the installation after disconnection to values above 1 pu (Smeets et al., 2000).

Table 10.1 shows a summary of modeling guidelines for representing GIS equipment in the calculation of internal VFT when single-phase models can be applied.

TABLE 10.1 Modeling Guidelines for Simulation of VFT in GIS

Component	Representation for VFT
Bus duct	Distributed-parameter transmission line
Spacer	Lumped capacitance to ground
Elbow	Distributed-parameter transmission line with capacitance to ground at the terminal
Surge arrester	A model including the protective characteristics in parallel with arrester capacitance and in series with distributed parameters lines that represent arrester leads
	A capacitance to ground, when arrester does not conduct
Circuit breaker	Capacitive ladder network, when opened
	Lumped capacitance, when closed
	Nonlinear resistance, when sparking
Disconnector	Distributed-parameter transmission line, when opened
	Lumped capacitance, when closed
	Nonlinear resistance, when sparking
Bushing	Gas filled type: Distributed-parameter transmission line with capacitance to ground at the terminals
	Capacitive type: Several transmission lines in series with a lumped resistor representing losses
Power transformer	A circuit synthesized from its frequency response, when seen from its terminals (it can be simplified as a capacitance to ground)
	Black box model derived from its frequency response, when voltage transfer is of concern
	Complex lumped/distributed-parameter circuit, including turn-to-turn parameters, when voltage distribution in windings is of concern
Current transformer	Distributed-parameter transmission line with capacitance to ground at the terminal
Line/cable termination	Resistance matching the surge impedance of the line/cable

10.3.2 Computation of TEV

At very high frequencies, currents are constrained to flow along the surface of the conductors and do not penetrate through them. The inside and the outside of a GIS enclosure are distinct, so that transients generated within the substation do not appear on the outside surface of the enclosure until discontinuities in the sheath are encountered. These discontinuities occur at gas-to-air terminations, GIS–cable transitions, or external core current transformers. The modeling of the GIS for computation of TEV must include the effects of the enclosure, the representation of ground straps, and the earthing grid.

10.3.2.1 Enclosures

A GIS–air termination can be modeled as a junction of three transmission lines, each with its own surge impedance (Figure 10.4). This equivalent network can be analyzed using lossless transmission line models to determine reflected and transmitted waves. The surge impedance of the enclosure-to-ground transmission line (Figure 10.4) is derived from the following forms:

$$L_3' \approx \frac{\mu_0}{2\pi} \ln \frac{2h}{R} \tag{10.8}$$

$$C_3' \approx \frac{2\pi\varepsilon_0}{\ln 2h/R} \tag{10.9}$$

$$Z_3 \approx \sqrt{\frac{L_3'}{C_3'}} = \frac{1}{2\pi} \sqrt{\frac{\mu_0}{\varepsilon_0}} \ln \frac{2h}{R} = 60 \ln \frac{2h}{R} \tag{10.10}$$

The basic mechanism of TEV is defined by the refraction of waves from the internal coaxial bus duct to the enclosure sheath-to-ground system. Figure 10.6 shows the scattering coefficients involved in an

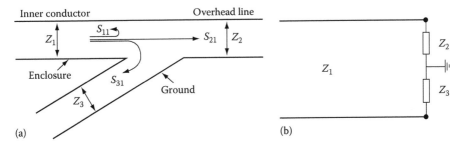

FIGURE 10.6 GIS–air transition. Scattering coefficients: (a) schematic diagram and (b) equivalent circuit.

air–SF$_6$ transition, and the equivalent circuit to be used for calculating these coefficients. The coefficients S_{ji} represent the refraction of waves from line i into line j.

The coefficient S_{11}, which is also the reflection coefficient at the transition, is given by

$$S_{11} = \frac{(Z_2 + Z_3) - Z_1}{Z_1 + Z_2 + Z_3} \tag{10.11}$$

The refraction coefficient at the transition is then

$$r_t = 1 + S_{11} = \frac{2(Z_2 + Z_3)}{Z_1 + Z_2 + Z_3} \tag{10.12}$$

The magnitude of the transmitted wave onto the outside of the enclosure sheath is given by following scattering coefficient:

$$S_{31} = r_t \frac{-Z_3}{Z_2 + Z_3} = -\frac{2Z_3}{Z_1 + Z_2 + Z_3} \tag{10.13}$$

The negative sign means that there is an inversion of the waveform with respect to the internal transient.

10.3.2.2 Ground Straps

TEV propagates back from the gas-to-air termination into the substation on the transmission line defined by the enclosure and the ground plane. The first discontinuity in the propagation is generally a ground strap. For TEV rise times, most ground straps are too long and too inductive for effective grounding. Ground leads may have a significant effect on the magnitude and waveshape of TEV. This effect can be explained by considering two mechanisms (Fujimoto et al., 1982).

First, the ground lead may be seen as a vertical transmission line whose surge impedance varies with height; when the transient reaches the ground strap, a reflected wave is originated that reduces the magnitude of the transmitted wave, with the reduction expressed by the coefficient

$$\frac{2Z_g}{2Z_g + Z_3} \tag{10.14}$$

where Z_g is the surge impedance of the ground strap. As Z_g is usually much larger than Z_3, the attenuation produced by the ground strap will usually be small.

Second, when the portion of the wave that propagates down the ground strap meets the low impedance of the ground grid, a reflected wave is produced that propagates back to the enclosure where it will tend to reduce the original wave.

The representation of a ground lead as a constant surge impedance is not strictly correct. It has a continuously varying surge impedance, so that a continuous reflection occurs as a wave propagates down the lead. The ground strap can be divided into sections, each one represented by a surge impedance calculated from the following expression:

$$Z_s = 60 \ln \frac{2\sqrt{2}\,h}{r} \tag{10.15}$$

where
r is the lead radius
h is the average height of the section (Fujimoto et al., 1982)

A constant inductor model may be adequate for straps with travel time less than the surge rise time, while a nonuniform impedance model may be necessary for very long straps.

10.3.2.3 Earthing Grid

The representation of the earthing grid at TEV frequencies is a very complex task. Furthermore, this grid may not be designed to carry very high frequency currents, as no standards for very high frequency earthing systems are currently available. A simplified modeling may be used by representing the earthing grid as a low-value constant resistance.

Advanced models for GIS components in computation of TEV might consider a frequency-dependent impedance for ground straps, a frequency-dependent model for the enclosure-to-ground line (which could take into account earth losses), and the propagation of phase-to-phase modes on the three enclosures (Fujimoto et al., 1982).

10.3.3 Testing and Simulation

The development of GIS started in the 1960s and was basically made during the 1970s. A complete bibliography that covered early works on gaseous insulation for high-voltage equipment was presented in 1982 (IEEE Subcommittee, 1982). This list of references was later expanded (IEEE Subcommittee, 1989, 1993); see also (IEEE Std, 1993).

A significant number of works published during the 1980s were aimed at analyzing and understanding the origin and propagation of VFT in GIS, both internally and externally (Matsumura and Nitta, 1981; Narimatsu et al., 1981; Boggs et al., 1982; Dick et al., 1982; Ford and Geddes, 1982; Fujimoto et al., 1982; Blahous and Gysel, 1983; Nishiwaki et al., 1983; Gorablenkow et al., 1984; Kynast and Luehrmann; 1984; Murase et al., 1985; Fujimoto et al., 1986; Ogawa et al., 1986; Yoshida et al., 1986; Boersma, 1987; CIGRE, 1988b, 1998; Fujimoto and Boggs, 1988; Meppelink et al., 1989), as well as to develop techniques for mitigating effects caused by VFT (Ishikawa et al., 1981; Harrington and El-Faham, 1985; Lalot et al., 1986; Ozawa et al., 1986; Fujimoto et al., 1988b). Many of those works included both testing and simulations, so the approaches proposed for representing the GIS equipment could be validated. Guidelines for simulation of VFT in GIS were proposed in different papers (Witzmann, 1987; CIGRE, 1988a).

Some subsequent works were motivated by the development of GIS equipment operating at higher voltages (Toda et al., 1992; Yamagata et al., 1996, 1999), although testing and simulation works were continuously performed (Okabe et al., 1991; Lui and Hiley, 1994; Vinod Kumar et al., 2001), and more advanced and sophisticated models were proposed (Haznadar et al., 1992; Ardito et al., 1992; Osmokrovic et al., 1992; IEEE TF, 1996; Singha and Joy Thomas, 2003), covering both single- and three-phase models (Miri and Binder, 1995; Miri and Stojkovic, 2001).

The impact of EMF currents caused by switching operations within a GIS has been another field of research. Since the transient voltages in control circuits depend on the nature of the radiated EM fields and protecting these circuits against coupled transients is important for a reliable GIS operation, the transient EMI environment needs to be fully characterized and compared with equipment susceptibility levels for damage assessment. Rao et al. (2005) estimated the VFT currents that appear at various locations in a GIS for different switching operations, and considering different configurations (i.e., different duct lengths, terminal component capacitances, and duct section branches). The goal was to obtain the most relevant characteristics of the VFT currents (i.e., amplitude, attenuation of the amplitude with distance and time, dominant frequency components, and variation in the frequency content with distance) to assess the impact that the radiated EM fields caused by these currents could have on control circuits.

Some of the main characteristics of VFT caused by switching operations, as well as their propagation within and outside the GIS, have been discussed in the previous sections. They can be summarized as follows:

- The amplitude of the voltages depends on the voltage difference between the contacts prior to breakdown and the surge impedances of the bus ducts.
- Waveforms at a particular GIS location have identical waveshapes but their magnitude depend on the trapped charge at the load side, being the voltages at a particular location related by the following relationship (Lui and Hiley, 1994):

$$v(q_2) = \frac{(1+q_2)v(q_1)-(q_2-q_1)}{1+q_1} \tag{10.16}$$

  where $v(q_i)$ is the voltage with trapped charge q_i.
- The resulting surges propagate with very little attenuation, and overvoltages may occur as a result of reflection and refraction at discontinuities in surge impedance within the substation.
- VFT overvoltages are characterized by very short rise times (5–20 ns), short durations (a few microseconds), high rates of change of voltage (as high as several megavolts per microsecond), relatively low magnitudes (1.5–2.5 pu), and a high frequency of occurrence (some tens to hundreds of individual transients of varying amplitude for each operation).
- VFT overvoltages are a function of time and of location, and depend on the operational configuration of the GIS.
- The pattern of variation of VFT overvoltage peak along the nodes of the GIS caused by a disconnector switch operation is different from that caused by a circuit breaker operation (Vinod Kumar et al., 2001): (a) The circuit breaker operation results in the highest VFT overvoltage level; (b) peak magnitudes of VFT are higher close to the switching point in the case of disconnector operations, but they are higher at the junction of the GIS with overhead line in the case of circuit breaker operation. In any case, the steepest front time and the highest frequency of VFT overvoltages occur close to the switching point.

Lee et al. (2011) analyzed the transfer of VFT overvoltages from the EHV side of a GIS to the MV switchgear. The study is based on a real substation and was motivated by the insulation breakdown caused to the MV switchgear by some switch operation at the EHV side. The insulation failure provoked the blackout of a nuclear power plant. The simulations proved that VFT overvoltages higher than seven times the rated line-to-ground peak voltage can be transferred into the MV system. They also concluded that phase coupling in three-phase gas-insulated line and common bus duct cannot be ignored as they lead to much more dispersed spread of peak voltages under a different operation angle, and single-phase simulation could result in underestimation of the VFT overvoltage in the MV system.

Several gas mixtures with different compositions of SF_6 have been considered as an alternative to pure SF_6 in GIS equipment. Singha and Joy Thomas (2001) carried out simulations and experimental measurements of VFT overvoltage characteristics with SF_6-N_2 mixtures at different gap spacings and gas pressures. They reported a time to breakdown between 4 and 20 ns, depending on the field inhomogeneity, while the transient peaks occurred at about 200 ns after the gap had broken down. Results did show that the VFT overvoltage magnitude in N_2, SF_6, and its mixture increase with increasing concentration of SF_6 in N_2. On the other hand, the overvoltage factors increase with pressure, although the way in which factors vary depends on the gap. The variations in the overvoltage factor are directly related to the breakdown process. Changes in the VFT overvoltage rise time were not observed with short gaps for which the formative time lags do not play any important role in the discharge growth. A decrease of 10%–20% of VFT overvoltage magnitudes can be achieved with SF_6–N_2 mixtures, although a compromise has to be made with the operational pressure and the dimensions of the GIS apparatus and in this respect such a mixture ratio may prove insufficient.

10.3.4 Statistical Calculation

The largest VFT stresses under normal operating conditions originate from disconnector operations. The level reached by overvoltages is random by nature. The maximum overvoltage produced by a disconnector breakdown depends on the geometry of the GIS, the measuring point, the voltage prior to the transient at the load side (trapped charge), and the intercontact voltage at the time of the breakdown, as shown in Section 10.2.1.

Several works have been performed to determine the statistical distribution of VFT overvoltages in a GIS (Boggs et al., 1982, 1984; Fujimoto et al., 1988a; Yanabu et al., 1990). A very simple expression can be used to calculate the transient overvoltages as a function of time t and position s (Boggs et al., 1984; Fujimoto et al., 1988a):

$$V(t,s) = V_b \cdot K(t,s) + V_q \qquad (10.17)$$

where
  $K(t,s)$ is the normalized response of the GIS
  V_b is the intercontact spark voltage
  V_q is the voltage prior to the transient at the point of interest

As V_b and V_q are random variables, $V(t,s)$ is also random. This equation can be used to estimate worst-case values (Fujimoto et al., 1988a).

The performance of a disconnector during an opening operation can be characterized by the pattern of arcing on a capacitive load (Figure 10.1). A difference in breakdown voltages for the two polarities indicates a dielectric asymmetry. When the asymmetry is large compared to the statistical variance in breakdown voltage, a systematic pattern is originated near the end of the arcing sequence (Boggs et al., 1984). The final trapped charge voltage has a distribution which is very dependent on the asymmetry in the intercontact breakdown voltage.

The dielectric asymmetry of a disconnector is usually a function of contact separation. A disconnector may show a different performance at different operating voltages. A consequence of this performance is that very different stresses will be originated as a result of different operational characteristics.

From the results presented in the literature, the following conclusions may be derived (Boggs et al., 1982, 1984; Fujimoto et al., 1988a; Yanabu et al., 1990):

- The value of the trapped charge is mainly dependent on the disconnect switch characteristics: The faster the switch, the greater the mean value that the trapped charge voltage can reach.
- For slow switches, the probability of a re-/prestrike with the greatest breakdown voltage, in the range 1.8–2 pu, is very small; however, due to the large number of re-/prestrikes that are produced with one operation, this probability should not always be neglected.
- The asymmetry of the intercontact breakdown voltage can also affect the trapped charge distribution; in general, both the magnitude and the range of values are reduced if there is a difference in the breakdown voltage of the gap for positive and negative values.

10.3.5 Validation

The results presented in Figures 10.7 and 10.8 illustrate the accuracy which can be obtained by means of a digital simulation. Figure 10.7 shows a good match between a direct measurement in an actual GIS and a

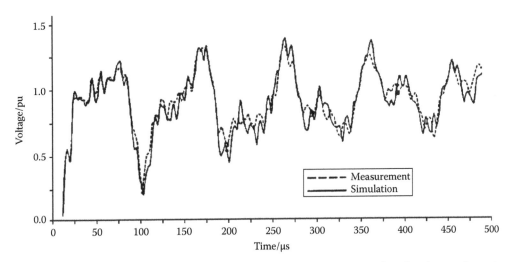

FIGURE 10.7 Comparison of simulation and measurement of disconnect switch–induced overvoltages in a 420 kV GIS. (Copyright 1999 IEEE.)

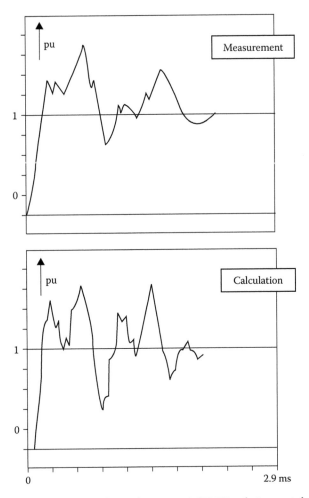

FIGURE 10.8 Measurement and simulation of overvoltages in a 420 kV GIS at closing a switch. (Copyright 1999 IEEE.)

computer result. The simulation was performed including the effects of spacers, flanges, elbows, and other hardware, but neglecting propagation losses (Witzmann, 1987). Figure 10.8 shows that important differences could occur when comparing a measurement and a digital simulation result, although a detailed representation of the GIS was considered. The differences were due to use of low damping equivalent circuits and to limitation of measuring instruments that did not capture very high frequencies (IEEE TF, 1996).

Simulation results were also validated in many other works. Some of these works were aimed at developing models that could be useful for understanding the generation and propagation of VFT waves (Matsumura and Nitta, 1981; Dick et al., 1982; Murase et al., 1985; Ogawa et al., 1986; Fujimoto and Boggs, 1988; Yanabu et al, 1990; Okabe et al., 1991; Haznadar et al., 1992; Ardito et al., 1992; Miri and Stojkovic, 2001). In other works, the comparison of measurements and simulation results was aimed at testing some mitigation solutions (Narimatsu et al., 1981; Ozawa et al., 1986; Yoshida et al., 1986) or checking the features of a new design (Lopez-Roldan et al., 2001). Most of these works were useful for developing and confirming modeling guidelines.

10.4 Effects of VFT on Equipment

The level reached by VFT overvoltages originated by disconnector switching or line-to-ground faults inside a GIS is in general below the BIL of substation and external equipment (Boersma, 1987). However, aging of the insulation of external equipment due to frequent VFT must be considered. TEV is a low-energy phenomenon and it is not considered dangerous to humans; the main concern is in the danger of the surprise–shock effect. External transients can cause interference with or even damage to the substation control, protection, and other secondary equipment (Boersma, 1987; Meppelink et al., 1989). The main effects caused by VFT to equipment and the techniques that can be used to mitigate these effects are summarized next (CIGRE, 1988b).

10.4.1 SF_6 Insulation

Breakdown caused by VFT overvoltages is improbable in a well-designed GIS insulation system during normal operations. The breakdown probability increases with the frequency of the oscillations. In addition, breakdown values can be reduced by insulation irregularities such as edges and fissures. However, at ultrahigh voltage systems (e.g., more than 1000 kV), for which the ratio of BIL to the system voltage is lower, breakdown is more likely to be caused. At these levels, VFT overvoltages can be reduced by using resistor-fitted disconnectors (Yamagata et al., 1996).

Li and Wu (2007) propose the use of ferromagnetic rings to limit the stresses caused by VFT overvoltages. The rings can be mounted on the conductors linked to the disconnector from both sides to effectively limit both the steepness and the amplitudes of overvoltages. Using a frequency-domain approach, the authors analyze the high-frequency suppressing characteristics of different types of ferromagnetic materials and present guidelines for the selection of ferromagnetic rings.

10.4.2 Transformers

Due to steep-fronted wave impulses, direct connected transformers can experience an extremely nonlinear voltage distribution along the high-voltage winding, connected to the oil–SF_6 bushings, and high resonance voltages due to transient oscillations generated within the GIS. Transformers can generally withstand these stresses; however, in critical cases, it may be necessary to install varistors to protect tap changers.

10.4.3 Disconnectors and Breakers

The insulation system of breakers and switches is not endangered by VFT overvoltages generated in adjacent GIS equipment. Ground faults induced by VFT overvoltages have been observed in disconnectors' operations, as residual leader branches can be activated by enhanced field gradient to ground. These faults can be avoided by a proper disconnector design.

10.4.4 Enclosure and Cable Interfaces

TEV can cause sparking across insulated flanges and to insulated busbars of CTs, and can puncture insulation that is intended to limit the spread of circulating currents within the enclosure. TEV can be minimized with a proper design and arrangement of substation masts, keeping ground leads as short and straight as possible in order to minimize the inductance, increasing the number of connections to ground, introducing shielding to prevent internally generated VFT from reaching the outside of the enclosure, and installing voltage limiting varistors where spacers must be employed.

Special measures for the protection of GIS–cable interfaces may be required when VFT originated by line-to-ground faults within the GIS are the triggering mechanism, since the power frequency follow-through current may damage the interface flange. Interface protection should prevent the formation of sparks and control the flow of the fault current, being the reduction of stray inductance the main goal (Fujimoto et al., 1988b). Spark gaps around the insulating flange are not a suitable solution; an increase of the flange insulation level and the use of MOVs are preferable.

10.4.5 Bushings

Very few problems have been reported with capacitively graded bushings. High impedances in the connection of the last graded layer to the enclosure should be avoided.

10.4.6 Secondary Equipment

TEV may interfere with secondary equipment or damage sensitive circuits by raising the housing potential if they are directly connected, or via cable shields to GIS enclosure by emitting free radiation which may induce currents and voltages in adjacent equipment. Correct cable connection procedures may minimize interference. The coupling of radiated energy may be reduced by mounting control cables closely along the enclosure supports and other grounded structures, grounding cable shields at both ends by leads as short as possible, or using optical coupling services. Voltage-limiting devices may have to be installed.

References

Ardito, A., Iorio, R., Santagostino, G., and Porrino, A., Accurate modeling of capacitively graded bushings for calculation of fast transient overvoltages in GIS, *IEEE Trans. Power Delivery*, 7, 1316–1327, 1992.

Arora, A., Becker, G., Boettger, L., Bolin, P., Hopkins, M., and Koch, H., Panel session on gas insulated switchgear and transmission lines, in *Proc. IEEE PES General Meeting*, San Francisco, CA, 2005.

Blahous, T. and Gysel, T., Mathematical investigation of the transient overvoltages during disconnector switching in GIS, *IEEE Trans. Power Appar. Syst.*, 102, 3088–3097, 1983.

Boersma, R., Transient ground potential rises in gas-insulated substations with respect to earthing systems, *Electra*, 110, 47–54, 1987.

Boggs, S.A., Chu, F.Y., Fujimoto, N., Krenicky, A., Plessl, A., and Schlicht, D., Disconnect switch induced transients and trapped charge in gas-insulated substations, *IEEE Trans. Power Appar. Syst.*, 101, 3593–3602, 1982.

Boggs, S.A., Fujimoto, N., Collod, M., and Thuries, E., The modeling of statistical operating parameters and the computation of operation-induced surge waveforms for GIS disconnectors, *CIGRE Session*, Paper No. 13–15, Paris, France, 1984.

Bolin, P., Gas-insulated substations, in *Electric Power Substations Engineering*, McDonald, J.D., Ed., CRC Press, 2nd edn., Boca Raton, FL, 2007, chap. 2.

Chowdhuri, P., *Electromagnetic Transients in Power Systems*, RSP-John Wiley, New York, 2nd edn., 2005, chap. 12.

CIGRE Joint WG 33/23-12, Insulation co-ordination of GIS: Return of experience, on site tests and diagnostic techniques, *Electra*, 176, 66–97, 1998.

CIGRE WG 33/13-09, Very fast transient phenomena associated with gas insulated substations, *CIGRE Session*, Paper 33-13, Paris, France, 1988a.

CIGRE WG 33/13-09, Monograph on GIS Very Fast Transient, *CIGRE*, 1988b.

CIGRE WG 33.02, Guidelines for representation of networks elements when calculating transients, CIGRE Brochure, 1990.

CIGRE WG 13.01, Applications of black box modelling to circuit breakers, *Electra*, 149, 40–71, 1993.

CIGRE WG 23.02, Report on the Second International Survey on High Voltage Gas Insulated Substations (GIS) Service Experience, CIGRE Brochure 150, 2000.

CIGRE WG B3.17, GIS State of the Art 2008, CIGRE Brochure 381, 2009.

Degeneff, R.C., Transient-voltage response, in *Electric Power Transformer Engineering*, Harlow, J.H., Ed., CRC Press, 2nd edn., Boca Raton, FL, 2007, chap. 20.

de León, F., Gómez, P., Martinez-Velasco, J.A., and Rioual, M., Transformers, in *Power System Transients. Parameter Determination*, Martinez-Velasco, J.A., Ed., CRC Press, Boca Raton, FL, 2009, chap. 4.

Dick, E.P., Fujimoto, N., Ford, G.L., and Harvey, S., Transient ground potential rise in gas-insulated substations—Problem identification and mitigation, *IEEE Trans. Power Appar. Syst.*, 101, 3610–3619, 1982.

Dommel, H.W., *EMTP Theory Book*, Bonneville Power Administration, Portland, OR, 1996.

Ecklin, A., Schlicht, D., and Plessl, A., Overvoltages in GIS caused by the operation of isolators, in *Surges in High-Voltage Networks*, Ragaller, K., Ed., Plenum Press, New York, 1980, chap. 6.

Ford, G.L. and Geddes, L.A., Transient ground potential rise in gas insulated substations—Assessment of shock hazard, *IEEE Trans. Power Appar. Syst.*, 101, 3620–3629, 1982.

Fujimoto, N. and Boggs, S.A., Characteristics of GIS disconnector-induced short risetime transients incident on externally connected power system components, *IEEE Trans. Power Delivery*, 3, 961–970, 1988.

Fujimoto, N., Chu, F.Y., Harvey, S.M., Ford, G.L., Boggs, S.A., Tahiliani, V.H., and Collod, M., Developments in improved reliability for gas-insulated substations, *CIGRE Session*, Paper No. 23-11, Paris, France, 1988a.

Fujimoto, N., Croall, S.J., and Foty, S.M., Techniques for the protection of gas-insulated substation to cable interfaces, *IEEE Trans. Power Delivery*, 3, 1650–1655, 1988b.

Fujimoto, N., Dick, E.P., Boggs, S.A., and Ford, G.L., Transient ground potential rise in gas-insulated substations—Experimental studies, *IEEE Trans. Power Appar. Syst.*, 101, 3603–3609, 1982.

Fujimoto, N., Stuckless, H.A., and Boggs, S.A., Calculation of disconnector induced overvoltages in gas-insulated substations, in *Gaseous Dielectrics IV*, Christophorou, L.G. and Pace, M.O., Eds., Pergamon Press, Oxford, U.K., 1984.

Fujita, S., Shibuya, Y., and Ishii, M., Influence of VFT on shell-type transformer, *IEEE Trans. Power Delivery*, 22, 217–222, 2007.

Gorablenkow, J.M., Kynast, E.E., and Luehrmann, H.M., Switching of capacitive currents of disconnectors in gas-insulated substations, *IEEE Trans. Power Appar. Syst.*, 103, 1363–1370, 1984.

Gustavsen, B., Wide band modeling of power transformers, *IEEE Trans. Power Delivery*, 19, 414–422, 2004.

Harrington, R.J. and El-Faham, M.M., Proposed methods to reduce transient sheath voltage rise in gas insulated substations, *IEEE Trans. Power Appar. Syst.*, 104, 1199–1206, 1985.

Haznadar, Z., Carsimamovic, S., and Mahmutcehajic, R., More accurate modeling of gas insulated substation components in digital simulations of very fast electromagnetic transients, *IEEE Trans. Power Delivery*, 7, 434–441, 1992.

Hoshino, T., Kato, K., Hayakawa, N., and Okubo, H., Frequency characteristics of electromagnetic wave radiated from GIS apertures, *IEEE Trans. Power Delivery*, 16, 552–557, 2001.

Hosseini, S.M.H., Vakilian, M., and Gharehpetian, G.B., Comparison of transformer detailed models for fast and very fast transient studies, *IEEE Trans. Power Delivery*, 23, 733–741, 2008.

IEC 60071-1, *Insulation Co-ordination—Part 1: Definitions, Principles and Rules*, Ed. 8.0, 2006.

IEC 60071-2, *Insulation Co-ordination—Part 2: Application Guide*, Ed. 3.0, 1996.

IEEE Std C37.122-2010, IEEE Standard for High Voltage Gas-Insulated Substations Rated Above 52 kV. January 2011.

IEEE TF on Very Fast Transients (Povh, D., Chairman), Modelling and analysis guidelines for very fast transients, *IEEE Trans. Power Delivery*, 11, 2028–2035, 1996.

IEEE WG K9 of the Gas-Insulated Substations Subcommittee (IEEE Substation Committee), Bibliography of gas-insulated substations, *IEEE Trans. Power Appar. Syst.*, 101, 4289–4315, 1982.

IEEE WG K9 of the Gas-Insulated Substations Subcommittee (IEEE Substation Committee), Addendum I to bibliography of gas-insulated substations, *IEEE Trans. Power Delivery*, 4, 1003–1020, 1989.

IEEE WG K9 of the Gas-Insulated Substations Subcommittee (IEEE Substation Committee), Addendum II to bibliography of gas-insulated substations, *IEEE Trans. Power Delivery*, 8, 73–82, 1993.

Ishikawa, M., Oh-hashi, N., Ogawa, Y., Ikeda, M., Miyamoto, H., and Shinagawa, J., An approach to the suppression of sheath surge induced by switching surges in a GIS/power cable connection system, *IEEE Trans. Power Appar. Syst.*, 100, 528–538, 1981.

Kulkarni, S.V. and Khaparde, S.A., *Transformer Engineering. Design and Practice*, Marcel Dekker, New York, 2004, chap. 7. Surge Phenomena in Transformers, pp. 277–325.

Kynast, E.E. and Luehrmann, H.M., Switching of disconnectors in GIS. Laboratory and field tests, *IEEE Trans. Power Appar. Syst.*, 104, 3143–3150, 1984.

Lalot, J., Sabot, A., Kieffer, J., and Rowe, S.W., Preventing earth faulting during switching of disconnectors in GIS including voltage transformers, *IEEE Trans. Power Delivery*, 1, 203–211, 1986.

Lee, C.H., Hsu, S.C., Hsi, P.H., and Chen, S.L., Transferring of VFTO from an EHV to MV system as observed in Taiwan's no. 3 nuclear power plant, *IEEE Trans. Power Delivery*, 26, 1008–1016, 2011.

Li, Q. and Wu, M., Simulation method for the applications of ferromagnetic materials in suppressing high-frequency transients within GIS, *IEEE Trans. Power Delivery*, 22, 1628–1632, 2007.

Liang, G., Sun, H., Zhang, X., and Cui, X., Modeling of transformer windings under very fast transient overvoltages, *IEEE Trans. Electromagn. Compat.*, 48, 621–627, 2006.

Lopez-Roldan, J., Irwin, T., Nurse, S., Ebden, C., and Hansson, J., Design, simulation and testing of an EHV metal-enclosed disconnector, *IEEE Trans. Power Delivery*, 16, 558–563, 2001.

Lui, C.Y. and Hiley, J., Computational study of very fast transients in GIS with special reference to effects of trapped charge and risetime on overvoltage amplitude, *IEE Proc.-Gener. Transm. Distrib.*, 141, 485–490, 1994.

Martinez-Velasco, J.A. and Popov, M., Circuit breakers, in *Power System Transients. Parameter Determination*, Martinez-Velasco, J.A., Ed., CRC Press, Boca Raton, FL, 2009, chap. 7.

Matsumura, S. and Nitta, T., Surge propagation in gas insulated substation, *IEEE Trans. Power Appar. Syst.*, 100, 3047–3054, 1981.

Meppelink, J., Diederich, K., Feser, K., and Pfaff, W., Very fast transients in GIS, *IEEE Trans. Power Delivery*, 4, 223–233, 1989.

Miri, A.M. and Binder, C., Investigation of transient phenomena in inner- and outer systems of GIS due to disconnector operation, in *Proc. Int. Conf. Power Systems Transients*, Lisbon, Portugal, 1995.

Miri, A.M. and Stojkovic, Z., Transient electromagnetic phenomena in the secondary circuits of voltage- and current transformers in GIS (measurements and calculations), *IEEE Trans. Power Delivery*, 16, 571–575, 2001.

Morched, A., Marti, L., and Ottevangers, J., A high frequency transformer model for the EMTP, *IEEE Trans. Power Delivery*, 8, 1615–1626, 1993.

Murase, H., Ohshima, I., Aoyagi, H., and Miwa, I., Measurement of transient voltages induced by disconnect switch operation, *IEEE Trans. Power Appar. Syst.*, 104, 157–165, 1985.

Narimatsu, S., Yamaguchi, K., Nakano, S., and Maruyama, S., Interrupting performance of capacitive current by disconnecting switch for gas insulated switchgear, *IEEE Trans. Power Appar. Syst.*, 100, 2726–2732, 1981.

Nishiwaki, S., Kanno, Y., Sato, S., Haginomori, E., Yamashita, S., and Yanabu, S., Ground fault by restriking surge of SF6 gas-insulated disconnecting switch and its synthetic tests, *IEEE Trans. Power Appar. Syst.*, 102, 219–227, 1983.

Nishiwaki, S., Nojima, K., Tatara, S., Kosakada, M., Tanabe, N., and Yanabu, S., Electromagnetic interference with electronic apparatus by switching surges in GIS - Cable system, *IEEE Trans. Power Delivery*, 10, 739–746, 1995.

Ogawa, S., Haginomori, E., Nishiwaki, S., Yoshida, T., and Terasaka, K., Estimation of restriking transient overvoltage on disconnecting switch for GIS, *IEEE Trans. Power Delivery*, 1, 95–102, 1986.

Okabe, S., Kan, M., and Kouno, T., Analysis of surges measured at 550 kV substations, *IEEE Trans. Power Delivery*, 6, 1462–1468, 1991.

Osmokrovic, P., Krstic, S., Ljevak, M., and Novakovic, D., Influence of GIS parameters on the Toepler constant, *IEEE Trans. Electr. Insul.*, 27, 214–220, 1992.

Ozawa, J., Yamagiwa, T., Hosokawa, M., Takeuchi, S., and Kozawa, H., Suppression of fast transient overvoltage during gas disconnector switching in GIS, *IEEE Trans. Power Delivery*, 1, 194–201, 1986.

Popov, M., van der Sluis, L., Paap, G.C., and De Herdt, H., Computation of very fast transient overvoltages in transformer windings, *IEEE Trans. Power Delivery*, 18, 1268, 2003.

Popov, M., van der Sluis, L., Smeets, R.P.P., and Lopez-Roldan, J., Analysis of very fast transients in layer-type transformer windings, *IEEE Trans. Power Delivery*, 22, 1268–1274, 2007.

Rao, M.M., Thomas, M.J., and Singh, B.P., Frequency characteristics of very fast transient currents in a 245-kV GIS, *IEEE Trans. Power Delivery*, 20, 2450–2457, 2005.

Rao, M.M., Thomas, M.J., and Singh, B.P., Transients induced on control cables and secondary circuit of instrument transformers in a GIS during switching operations, *IEEE Trans. Power Delivery*, 22, 1505–1513, 2007a.

Rao, M.M., Thomas, M.J., and Singh, B.P., Electromagnetic field emission from gas-to-air bushing in a GIS during switching operations, *IEEE Trans. Electromagn. Compat.*, 49, 313–321, 2007b.

Rao, M.M., Thomas, M.J., and Singh, B.P., Shielding effectiveness of the gas-insulated bus duct for transient EM fields generated in a GIS during switching operations, *IEEE Trans. Power Delivery*, 23, 1946–1953, 2008.

Ryan, H.M., Application of gaseous insulants, in *High Voltage Engineering and Testing*, Ryan H.M., Ed., 2nd edn., The Institution of Electrical Engineers, London, U.K., 2001, chap. 3.

Shibuya, Y., Fujita, S., and Tamaki, E., Analysis of very fast transients in transformers, *IEE Proc.-Gener. Transm. Distrib.*, 148, 377–383, 2001.

Shintemirov, A., Tang, W.H., and Wu, Q.H., A hybrid winding model of disc-type power transformers for frequency response analysis, *IEEE Trans. Power Delivery*, 24, 730–739, 2009.

Singha, S. and Joy Thomas, M., Very fast transient overvoltages in GIS with compressed SF_6-N_2 gas mixtures, *IEEE Trans. Dielectr. Electr. Insul.*, 8, 658–664, 2001.

Singha, S. and Joy Thomas, M., Toepler's spark law in a GIS with compressed SF6-N2 mixture, *IEEE Trans. Dielectr. Electr. Insul.*, 10, 498–505, 2003.

Smeets, R.P.P., van der Linden, W.A., Achterkamp, M., Damstra, G.C., and de Meulemeester, E.M., Disconnector switching in GIS: Three-phase testing and phenomena, *IEEE Trans. Power Delivery*, 15, 122–127, 2000.

Thomas, D.E., Wiggins, C.M., Salas, T.M., Nickel, F.S., and Wright, S.E., Induced transients in substation cables: Measurements and models, *IEEE Trans. Power Delivery*, 9, 1861–1868, 1994.

Toda, H., Ozaki, Y., Miwa, I., Nishiwaki, S., Muruyama, Y., and Yanabu, S., Development of 800 kV gas-insulated switchgear, *IEEE Trans. Power Delivery*, 7, 316–323, 1992.

Vinod Kumar, V., Joy Thomas, M., and Naidu, M.S., Influence of switching conditions on the VFTO magnitudes in a GIS, *IEEE Trans. Power Delivery*, 16, 539–544, 2001.

Wiggins, C.M., Thomas, D.E., Nickel, F.S., Wright, S.E., and Salas, T.M., Transient electromagnetic interference in substations, *IEEE Trans. Power Delivery*, 9, 1869–1884, 1994.

Witzmann, R., Fast transients in gas insulated substations. Modelling of different GIS components, in *Proc. 5th Int. Symp. HV Engineering*, Braunschweig, Germany, 1987.

Yamagata, Y., Nakada, Y., Nojirna, K., Kosakada, M., Ozawa, J., and Ishigaki, I., Very fast transients in 1000 kV gas insulated switchgear, in *Proc. IEEE Transm. Distrib. Conf.*, New Orleans, LA, 1999.

Yamagata, Y., Tanaka, K., Nishiwaki, S., Takahashi, N., Kokumai, T., Miwa, I., Komukai, T., and Imai, K., Suppression of VFT in 1100 kV GIS by adopting resistor-fitted disconnector, *IEEE Trans. Power Delivery* 11, 872–880, 1996.

Yanabu, S., Murase, H., Aoyagi, H., Okubo, H., and Kawaguchi, Y., Estimation of fast transient overvoltage in gas-insulated substation, *IEEE Trans. Power Delivery*, 5, 1875–1882, 1990.

Yoshida, T., Sakakibara, T., Terasaka, K., and Miwa, I., Distribution of induced grounding current in large-capacity GIS using multipoint grounding system, *IEEE Trans. Power Delivery*, 1, 120–127, 1986.

11

Transmission System Transients: Grounding

William A.
Chisholm
*Kinectrics/Université du
Quebec à Chicoutimi*

11.1 General Concepts

Electric power systems are often grounded, that is to say "intentionally connected to earth through a ground connection or connections of sufficiently low impedance and having sufficient current-carrying capacity to prevent the buildup of voltages which may result in undue hazard to connected equipment or to persons" (IEEE Std 100, 2000). Grounding affects the dynamic power-frequency voltages of unfaulted phases, and influences the choice of surge protection. Also, the tower footing impedance is an important specification for estimating the transient voltage across insulator strings for a lightning flash to an overhead ground wire, tower or phase conductor with surge arrester.

To mitigate ac fault conditions, systems can be grounded by any of three means (IEEE Std 100, 2000):

- *Inductance grounded*, such that the system zero-sequence reactance is much higher than the positive-sequence reactance, and is also greater than the zero-sequence resistance. The ground-fault current then becomes more than 25% of the three-phase fault current.

- *Resistance grounded*, either directly to ground, or indirectly through a transformer winding. The low-resistance-grounded system permits a higher ground-fault current (on the order of 25 A to several kiloamperes) for selective relay performance.
- *Resonant grounded*, through a reactance with a value of inductive current that balances the power-frequency capacitive component of the ground-fault current during a single line-to-ground fault. With resonant grounding of a system, the net current is limited so that the fault arc will extinguish itself.

Power system transients have a variety of waveshapes, with spectral energy ranging from the power frequency harmonics up to broadband content in the 300-kHz range, associated with 1-μs rise and fall times of lightning currents and insulator breakdown voltages. With the wide frequency content of transient waveshapes, resonant grounding techniques offer no direct benefit in preventing arcing flashover from the grounding system to the insulated phases. Also, resistance grounds that may be effective for power frequency can have an additional inductive voltage rise ($L\, dI/dt$ with $dt = 1\,\mu s$) that dominates the transient response. Both resistive and inductive aspects must be considered in the selection of an appropriate ground electrode for adequate performance during transients such as lightning.

11.2 Material Properties

The earth resistivity (ρ, units of Ωm) dominates the potential rise on ground systems at low frequencies and currents. Near the surface, resistivity changes as a function of moisture, temperature, frequency, and electric field stress. Figure 11.1 shows that this variation can be quite large.

Soil moisture can change over periods of days or even hours, giving significant changes in resistivity especially in surface layers of soil. Reconnaissance of earth resistivity, from traditional four-terminal resistance measurements, is a classic tool in geological prospecting (Keller and Frischknecht, 1982). A current I (A) is injected at the outer two locations in a line of four equally spaced probes. A potential difference U (V) then appears between the inner two probes, which are separated by a distance a (m). The apparent resistivity ρ_a (Ωm) is then defined as

$$\rho_a = 2\pi a \frac{U}{I}$$

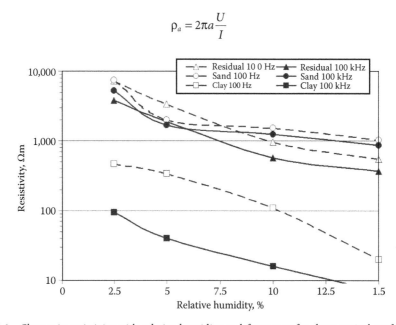

FIGURE 11.1 Change in resistivity with relative humidity and frequency for three typical surface materials. (From Visacro, S. and Portela, C.M., Soil permittivity and conductivity behavior on frequency range of transient phenomena in electric power systems, *Proc. 5th ISH*, Paper 93.06, August 1987.)

At a given location, several measurements of ρ_a are taken at geometrically spaced values of a, such as (a = 1, 3, 9, 27, 81 m or a = 2, 6, 18, 54 m) so that the outer probes from one measurement become the inner probes on the next. When ρ_a is constant with distance, the assumption of a uniform soil model is justified, and the effective resistivity ρ_e for any electrode size is simply ρ_a. However, in many cases, there are two or more layers of contrasting soil. The most difficult case tends to be a thin, conducting top layer (clay, till, sand) over a thick, poorly conducting rock layer ($\rho_1 < \rho_2$). This case will have an increasing value of ρ_a with distance. A simplified interpretation in this case can be given for a practical range of resistivity values: For flat electrodes, the effective resistivity, ρ_e, equals the value of ρ_a observed at a probe spacing of (a = 2s), where s is the maximum extent (e.g., the radius of a ring electrode), or a is the diagonal dimension of the legs of a tower.

11.3 Electrode Dimensions

Five dimensions are relevant for analysis of electrode response under steady-state and transient conditions. In order of decreasing importance, these are as follows:

s: The three-dimensional distance from the center of the electrode to its outermost point. For a spheroid in a conducting half-plane, s = MAX(a,b) where a is the maximum cross-section radius and b is the length in the axis of symmetry. Table 11.1 adapts equations for surface area of a cylinder, spheroid, and box to typical electrodes. Different dimensions dominate the s, g, and A_{Total} terms, depending on the electrode shape. Table 11.1 shows that the three-dimensional extent s of cylinders and prisms are slightly larger than the s for a prolate spheroid of the same depth. The propagation time $\tau = s/c$, calculated from speed-of-light propagation at $c = 3 \times 10^8$ m/s, is used to estimate transient electrode impedance.

g: The geometric radius of the electrode, $g = \left(R_x^2 + R_y^2 + R_z^2\right)^{1/2}$, is used to estimate capacitance. For long, thin, or rectangular shapes, $g = s$; for a disk, $g = \sqrt{2}s$; for a hemisphere, $g = \sqrt{3}s$.

A_{Total}: The surface area in contact with soil. Electrodes with large surface area will have lower resistance, lower impedance, and less susceptibility to unpredictable effects of soil ionization. For objects with concave features, the area of the smallest convex body that can envelop is determined. With this model, a tube, a circular array of wires, or a collinear array of rings have the same area A_{Total} as a solid cylinder of the same dimensions. Buried horizontal wires expose area A_{Total} on both sides of the narrow trench. Table 11.1 provides further interpretations.

L: The total length of wire in a wire frame approximation to a three-dimensional solid, L, is used to correct for the contact resistance, typically with $R_{Contact} \leq \rho_1/L$ as described by Laurent (IEEE Standard 80, 1986). Contact resistance tends to be much smaller than the geometric resistance for most electrodes.

A_{Wire}: The total surface area of the wire of length L to be used to refine the calculation of $R_{Contact}$ for wire frame electrodes (Table 11.1).

The transmission tower foundation tends to have the largest surface area A_{total} and greatest extent g compared to supplemental vertical rods and buried horizontal wires. Thus, a significant fraction of lightning current will flow into the foundation. As described in Grigsby (2011), the foundations may consist of the following:

- One, two, or four reinforced concrete drilled shafts, each with a ratio length/diameter of 2:1 to 10:1
- Directly embedded steel or concrete poles with a length/diameter ratio of 3:1 to 6:1
- Spread foundations with steel lattice (grillage) beneath
- Anchors, providing parallel grounding paths for current flow along tower guy wires, typically screwed into soil or grouted into drilled holes in rock to a length of at least 5 m

Unlike distribution lines, where the pole embedment can be a fixed (10%) fraction of line height, there is no single rule that relates the transmission tower height to the depth of its foundation. Also, there is a tendency to have foundations of reduced dimensions in rock of high electrical resistivity, while

TABLE 11.1 Values of A_{Total}, s, g, L, and A_{Wire} for Typical Ground Electrodes in Half-Plane

Geometry	A_{Total}, Surface Area	s, Maximum Extent	$g = \sqrt{r_x^2 + r_y^2 + r_z^2}$	L	A_{Wire}
Cylindrical Shapes					
Vertical cylindrical rod length l, radius r	$2\pi rl + \pi r^2 \approx 2\pi rl$	$\sqrt{l^2 + r^2}$	$\sqrt{l^2 + r^2 + r^2} \approx l$	l	$= A_{Total}$
Solid cylinder length l, radius r	$2\pi rl + \pi r^2$	$\sqrt{l^2 + r^2}$	$\sqrt{l^2 + 2r^2}$	l	$= A_{Total}$
Buried circular disk at depth h with radius r	$2\pi rh + \pi r^2$	$\sqrt{h^2 + r^2}$	$\sqrt{h^2 + 2r^2}$	h	πr^2
Buried circular ring at depth h with radius r and wire diameter d	"	"	"	$2\pi r$	$2\pi r \cdot \pi d$
Circular disk on surface thickness t, radius r	$2\pi rt + \pi r^2 \approx \pi r^2$	$\sqrt{t^2 + r^2} \approx r$	$\sqrt{t^2 + 2r^2} \approx \sqrt{2}\cdot r$	t	$= A_{Total}$
Spheriodal Shapes					
Oblate (disk-like) half spheroid radius $a >$ thickness b, $\varepsilon = \sqrt{1 - b^2/a^2}$	$\pi a^2 + \dfrac{\pi b^2}{2\varepsilon}\ln\left(\dfrac{1+\varepsilon}{1-\varepsilon}\right)$	a	$\sqrt{2a^2 + b^2}$	b	$= A_{Total}$
Hemisphere of radius r	$2\pi r^2$	r	$\sqrt{3}\cdot r$	r	$= A_{Total}$
Prolate (tube-like) half spheroid radius $a <$ length b, $\varepsilon = \sqrt{1 - a^2/b^2}$	$\pi a^2 + \dfrac{\pi ab}{\varepsilon}\sin^{-1}(\varepsilon)$	b	$\sqrt{2a^2 + b^2}$	a	$= A_{Total}$

Rectangular Shapes

Vertical solid plate, l long, w wide, and t thick	$wt + 2(t+w)l \approx 2lw$	$s = g = \sqrt{t^2/4 + w^2/4 + l^2}$	l	$= A_{Total}$
Two or more vertical rods in straight line of length l, driven to depth h, rod diameter d	$\pi l d + 2\left(1 + \dfrac{\pi d}{2}\right)h$ $\approx 2lh$	"	$nh,\ n = No.\ of\ rods$	$n \cdot \pi d h$
Buried counterpoise of length l at depth h, wire diameter d		$s = g = \sqrt{\dfrac{l^2}{4} + \dfrac{w^2}{4} + h^2}$	$h + l$	$\pi d l$
Conducting box l long by w wide at depth h	$lw + 2(l+w)h$	"	h	$= A_{Total}$
Buried rectangle plate, l long by w wide at depth h, thickness t	"	"	h	$2lw + 2(l+w)t$
Buried rectangular grid l long by w wide at depth h, wire diameter d	"	"	$nl + mw$ where n, $m =$ no. of grids	$(nl + mw) \cdot \pi d$
Buried rectangular loop, l long by w wide at depth h, wire diameter d	"	"	$h + 2(l+w)$	$(h + 2(l+w)) \cdot \pi d$
Four wires from center to corners of Rectangle, l long by w wide at depth h, wire diameter d		"	$h + 2\sqrt{w^2 + l^2}$	$L \cdot \pi d$
Four vertical rods on corners of rectangle, l long by w wide by h deep, rod diameter d	"		$4h$	$4\pi h d$
Surface plate, l long by w wide by t thick	$lw + 2(l+w)t \approx lw$	$s = g = \sqrt{l^2/4 + w^2/4 + t^2}$	t	$= A_{Total}$

foundations need larger dimensions to ensure adequate transfer of mechanical loads in clay materials of low resistivity. Please refer to Grigsby (2011) for additional details.

11.4 Self-Capacitance of Electrodes

Electrode capacitance is easily calculated (Chow and Yovanovich, 1982; Chow and Srivastava, 1988) and offers elegant description of grounding response to transients. Also, the self-capacitance C_{self} to infinity of an arbitrary conducting object in full space has a useful dual relation to its steady-state resistance R in a half-space of conducting medium, given by (Weber, 1950)

$$R = \frac{2\varepsilon_o\rho}{C_{self}} \tag{11.1}$$

where
ε_o is the permittivity constant, 8.854×10^{-12} F/m
ρ is the earth resistivity, in Ωm

The transient impedance of the same arbitrary conducting object can be modeled using the time (τ) it takes to charge up its self-capacitance C_{self}. This time cannot be less than the maximum dimension of the electrode, s, divided by the speed of light. An average surge impedance Z, given by the ratio τ/C, can then be used to relate voltages and currents during any initial surge. The capacitance of an object is approximately (Chow and Yovanovich, 1982)

$$C_{Self} = \varepsilon_o c_f \sqrt{4\pi A} \tag{11.2}$$

where
A is the total surface area of the object, including both sides of disk-like objects
c_f is a correction factor between the capacitance of the object and the capacitance of a sphere with the same surface area

For a wide range of objects, $0.9 < c_f < 1.2$ using $c_f \approx \sqrt{2\gamma}/\ln(4\gamma)$, where γ is the ratio of length to width of the object.

A close estimate of c_f for spheroids is given by the following expression:

$$c_f = \frac{4\pi g}{\sqrt{4\pi A}\ln\left(4\pi e^{\sqrt{3}}g^2/3A\right)} \cong \frac{3.54g}{\sqrt{A}\ln(23.7g^2/A)} \tag{11.3}$$

Again, $g = \sqrt{r_x^2 + r_y^2 + r_z^2}$, the geometric radius. Equation 11.3 is exact for a sphere, and remains valid for a wide range of electrode shapes, from disk to rod.

A surprisingly large amount of the surface area of a solid can be removed without materially affecting its self-capacitance. For example, the capacitance of a solid box of dimensions $1 \times 1 \times 0.4$ m and a surface area of 3.6 m^2 is 56 pF. The capacitance of two parallel loops filling out the same space is 50 pF, even though the loops have only 20% of the surface area of the box (Chow and Srivastava, 1988). The equivalent correction from wire frame approximations to solid objects in ground resistance calculations is known as a "Contact Resistance."

11.5 Initial Transient Response from Capacitance

Once the capacitance of a conducting electrode and its minimum charging time τ have been estimated, its average transient impedance can be computed from the relation $Z = \tau/C_{self}$, obtained from the use of open-circuit stub transmission lines for tuning radio antennas. The charging time τ of the conducting

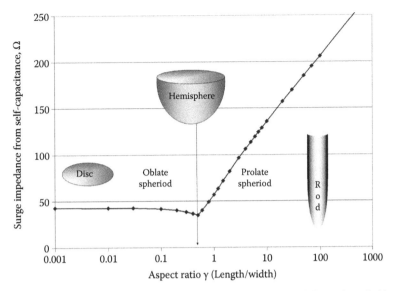

FIGURE 11.2 Relation between transient impedance and aspect ratio: spheroid electrodes in half space.

TABLE 11.2 Transient Impedance of Conducting Electrodes

Shape	Surface Area	3-D Extent s	Capacitance Eq. 11.2	Travel Time	Transient Impedance
Circular disk	$2\pi s^2$	s	$0.9\,\pi\,\varepsilon\,s\,\sqrt{8}$	s/c	47 Ω
Hemisphere	$3\pi s^2$	s	$1.0\,\pi\,\varepsilon\,s\,\sqrt{12}$	s/c	35 Ω
Long cylinder	$2\pi r\,s$	s	$3.3\,\pi\,\varepsilon\,\sqrt{4rs}$	s/c	$s/r = 100$: 210 Ω
			$7.8\,\pi\,\varepsilon\,\sqrt{4rs}$		$s/r = 1{,}000$: 270 Ω
			$20\,\pi\,\varepsilon\,\sqrt{4rs}$		$s/r = 10{,}000$: 340 Ω

electrode in free space is the maximum three-dimensional extent, s, divided by the velocity of light in free space, c. This transient impedance of the ground electrode will be seen only during the charging time from the center of the electrode to its full extent, and the impedance will reduce with increasing time as the electromagnetic fields start to interact with the surroundings, including soil, towers, and overhead ground wires.

The following graph (Figure 11.2) gives the initial transient response of conducting spheroid electrodes (see Table 11.2).

The main observation from Figure 11.2 is that wide, flat electrodes will have inherently better initial transient response than long, thin electrodes. This includes electrodes in both horizontal and vertical planes, giving a physical reason why long leads to remote ground electrodes are not effective under transient conditions.

Chisholm and Janischewskyj (1989) confirmed an apparent initial transient impedance of about 60 Ω for a perfectly conducting ground plane (infinite circular disk) that reduced as a function of time, with a rate of reduction depends on the electromagnetic travel time from tower top to tower base. Baba and Rakov (2007) validated the imperfect reflection from a ground plane using advanced FDTD numerical methods and offered more detailed interpretation. For compact electrodes, the response can be lumped into an inductance element ($L_{Self} = \tau^2/C_{Self}$). The potential rise on the earth electrode can then be estimated from the simple circuit model:

$$U_{electrode} = RI + \left(\frac{\tau^2}{C_{Self}} \right) \frac{dI}{dt} \qquad (11.4)$$

A numerical example for Equation 11.4 is useful. For $\rho = 100\ \Omega m$ and a disk electrode of 5 m radius, the resistance to remote ground will be 3.2 Ω and $\tau^2/C_{self} = 0.8\ \mu H$. For a typical (median) lightning stroke with $I = 30\ kA$ and $dI/dt = 24\ kA/\mu s$ at the peak, the two terms of the peak potential rise V_{Pk} will be

$$V_{Pk} = RI + L\frac{dI}{dt} = (30\ kA)(3.2\ \Omega) + \frac{24\ kA}{\mu s}(0.8\ \mu H) = 96\ kV + 19\ kV$$

The inductive term is desirably low in the example, but it can dominate the response of long, thin electrodes in low-resistivity soil. For distributed electrodes, however, the circuit approximation in Equation 11.4 eventually fails as rate of current rise (dI/dt) increases and surge impedance models must be used as developed next.

11.6 Ground Electrode Impedance: Wire over Perfect Ground

The inductance L per unit length of a distributed grounding connection over a conducting plane can be calculated from the surge impedance Z of the wire and its travel time τ:

$$L = Z\tau = \left(\frac{\tau^2}{C_{Self}}\right) \quad Z = 60\ln\left(\frac{2h}{r}\right) \tag{11.5}$$

where

Z is the surge impedance (Ω) of a wire of radius r (m) over a ground plane at a height h (m)

$\tau = s/c$, where s is the wire length or electrode extent (m) divided by the speed of light c (3.0×10^8 m/s)

11.7 Ground Electrode Impedance: Wire over Imperfect Ground

When the electrode is placed over imperfect ground, the effective return depth of current will increase. Bewley (1963) suggests that the plane for image currents, in his tests of buried horizontal wires, was 61 m (200′) below the earth surface. Some analytical indication of the increase in return depth is given by the normal skin depth, $\delta = 1/\sqrt{\pi f \mu_o/\rho}$, which decreases from 460 m (60 Hz) to 11 m (100 kHz) for a resistivity of $\rho = 50\ \Omega m$. Deri et al (1981) propose a more complete approach, replacing the height h in Equation 11.6 with ($h + p$) where with the depth p to an equivalent perfectly conducting plane is a complex number given by

$$p = \sqrt{\frac{\rho}{j\omega\mu_o}} \tag{11.6}$$

where

ω is 2π times the frequency (Hz)

μ_o is $4\pi \times 10^{-7}$ H/m

ρ is the resistivity in Ωm

For good soil with a resistivity of $\rho = 50\ \Omega m$, the complex depth is 230 $(1 - j)$ m at 60 Hz and 5.6 $(1 - j)$ m at 100 kHz. The complex depth is related to the normal skin depth by the relation $1/p = (1 + j)\ 1/\delta$. The velocity of propagation will also reduce as the wire is brought close to, or buried in, imperfectly conducting ground. Darveniza (2007) simplifies the approach and recommends the following expression for effective height above ground h_{eff}, to be substituted into Equation 11.5:

$$h_{eff} = h + 0.15\sqrt{\rho} \tag{11.7}$$

where

ρ is the soil resistivity (Ωm)

h_{eff} and h are in m

11.8 Analytical Treatment of Complex Electrode Shapes

Simple analytical expressions are documented for a variety of regular electrode shapes (see, e.g., Sunde, 1949; Smythe, 1950; Weber, 1950; Keller and Frischknecht, 1982). However, grounding of electrical systems often consists of several interconnected components, making estimation of footing resistance more difficult. The tower foundation can be a single or (more typically) four concrete cylinders, often reinforced with steel. In the preferred case, the steel is bonded electrically, and a grounding connection is brought out of the form before the concrete is poured. In areas with low soil resistivity, four concrete footings can often provide a low tower resistance without supplemental electrodes.

In some cases on both transmission and distribution systems, a metal grillage (or pole butt-wrap) is installed at the base after excavation. This deep electrode is more effective than a surface electrode of the same area. Also, grillage and pole-wrap electrodes are protected from vandalism and frost damage.

Supplemental grounding electrodes are often installed during line construction or upgrade. The following approaches are used:

- Horizontal conductors are bonded to the tower, and then buried at a practical depth.
- Vertical rods are driven into the soil at some distance from the tower, and then bonded to the tower base, again using bare wires, buried at a practical depth.
- Supplemental guy wires are added to the tower (often for higher mechanical rating) and then grounded using rock or soil anchors at some distance away.

Supplemental grounding should be considered to have a finite lifetime of 5–20 years, especially in areas where the soil freezes in winter. Also, auxiliary electrodes such as rock anchors should be designed to carry their share of impulse current, and to withstand the associated traverse forces.

The resistance of an electrode that envelops all contacts can be used to obtain a good estimate of the combined resistance of a complex, interconnected electrode. From Chisholm and Janischewskyj (1989), the resistance of a solid rectangular electrode is approximately

$$R_{Geometric} = \frac{\rho}{2\pi s} \ln\left(\frac{2\pi e s^2}{A}\right) \cong \frac{\rho}{2\pi s} \ln\left(\frac{17 s^2}{A}\right) \qquad (11.8)$$

where
   s is the three-dimensional distance from the center to the furthest point on the electrode
   A is the convex surface area that would be exposed if the electrode were excavated
   e is the exponential constant, 2.718

The resistance can also be estimated using the geometric radius g rather than the maximum dimension s:

$$R_{Geometric} \cong \frac{\rho}{2\pi g} \ln\left(\frac{11.8 g^2}{A}\right) \qquad (11.9)$$

where
   $g = \sqrt{r_x^2 + r_y^2 + r_z^2}$ is the geometric radius of the electrode, $r_{x,y,z}$ = maximum x, y, and z dimensions
   $11.8 = (2\pi e \sqrt{3})/3$

If the electrode is a wire frame, rather than a solid, then a correction for contact resistance should be added to the geometric resistance:

$$R_{Wire\,Frame} = R_{Geometric} + R_{Contact} = R_{Geometric} + K\frac{\rho_1}{L} \qquad (11.10)$$

where

L is the total length of the wire frame

ρ_1 is the resistivity of the upper layer of soil (the layer next to the wire)

K is a constant that varies from $0.5 \leq K \leq 1.3$ for most electrode shapes depending on fill ratio

Many examples of contact resistance can be found in the literature on grounding (e.g., Sunde 1949; IEEE Standard 80, 1986). Consider for example the resistance of a surface disk of radius s compared to a ring of the same radius, made from wire with a diameter d:

$$R_{Disc} = \frac{\rho}{4s}; \quad R_{Ring} = \frac{\rho}{2\pi^2 s} \ln\left(\frac{4s}{d}\right)$$

$$g = \sqrt{s^2 + s^2 + d^2} \approx \sqrt{2}s; \quad A = \pi s^2; \quad R_{Geometric} = \frac{\rho}{2\pi g} \ln\left(\frac{11.8g^2}{A}\right) = \frac{\rho}{4.4s} \tag{11.11}$$

$$R_{Contact} = R_{Ring} - R_{Disc}$$

The difference between the resistance of a ring of $d = 13\,mm$ wire and a disk of the same radius s can be described using a value of K in Equation 11.10 that ranges from $K = 0.48$ for $s = 2\,m$, to $K = 0.99$ for $s = 10\,m$ ring, to $K = 1.33$ for $s = 30\,m$, depending on the logarithm of the ratio of overall to wire surface areas as follows:

$$K = \frac{1}{2\pi} \ln\left(\frac{A_{Total}}{2A_{Wire}}\right), \quad K > 0 \tag{11.12}$$

Equation 11.12 is valid for most electrode shapes, such as combinations of rings, grids, and rods. When several (N) radial wires meet at a point, the contact resistance coefficient K in Equation 11.10 should use

$$K = \frac{1}{2\pi} \ln\left(\frac{N^2 A_{Total}}{8A_{Wire}}\right), \quad K > 0 \tag{11.13}$$

where

N is the number of wires meeting at a point

A_{Total} is the surface area of the pattern made by the radial array

A_{Wire} is the surface area of the wires themselves

11.9 Numerical Treatment of Complex Electrode Shapes

Solving for the combined resistance of a number of individual electrode elements, such as four foundations of a transmission tower along with some horizontal wires and ground rods, can take several approaches.

The geometric resistance of the entire electrode can be computed, and the contact resistance correction for a wire frame approximation can be added as in Equation 11.11. However, in some cases such as solid cylinders of concrete, the "length" of wire to use is not obvious, and also, the value of K varies depending on whether the electrode is tightly meshed or open.

A second possibility is to develop a matrix of self- and mutual-resistance values of each solid, conducting element. The geometric resistances defined in Equation 11.10 give the self-resistance R_{ii}. The mutual resistance between the centroids of two objects is calculated from their separation d_{ij}:

$$R_{ij} = \frac{\rho}{2\pi d_{ij}} \tag{11.14}$$

A symmetrical matrix of self- and mutual resistances is multiplied by a vector of current values to obtain the potentials at each electrode. The matrix can be built, inverted, and solved in Excel as follows.

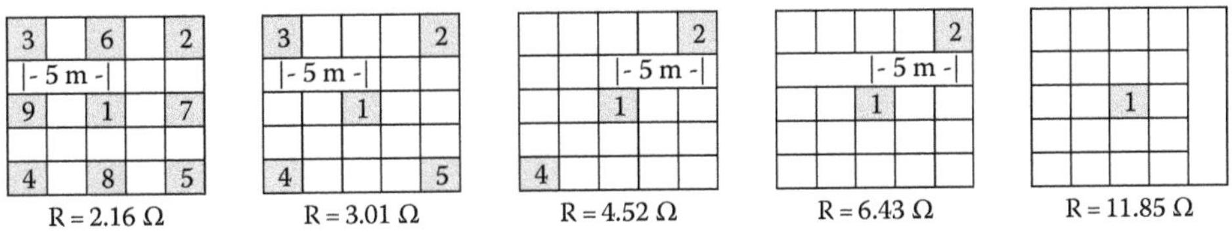

3	6	2
|- 5 m -|	
9	1	7
4	8	5

R = 2.16 Ω

3		2
|- 5 m -|	
	1	
4		5

R = 3.01 Ω

		2
|- 5 m -|	
	1	
4		

R = 4.52 Ω

		2
|- 5 m -|	
	1	

R = 6.43 Ω

		2
|- 5 m -|	
	1	

R = 11.85 Ω

FIGURE 11.3 Numerical solutions of resistance for 10 m rods at 5 m grid spacing (*IEEE Guide* 80/2000).

In our case, we will consider five 3 m long rods of 5 mm radius in ρ = 320 Ωm soil, on a 5 m × 5 m grid as shown in Figure 11.3. The self-resistance is calculated from g = 10 m, $A = 2\pi rl = 0.4\,m^2$ in Equation 11.10 as 119 Ω. The distance between the center rod and all others is d_{1j} = 7.07 m, $d_{23} = d_{34} = d_{45} = d_{25}$ = 10 m, and $d_{24} = d_{35}$ = 14.1 m. The mutual resistances are given by $R_{ij} = \rho/(2\pi d_{ij})$ from Equation 11.14 as follows:

Distances D_{ij} in m					Resistances R_{ij} In Ω (ρ = 320 Ωm)				
0.01	7.07	7.07	7.07	7.07	119.30	7.20	7.20	7.20	7.20
7.07	0.01	10.00	14.14	10.00	7.20	119.30	5.09	3.60	5.09
7.07	10.00	0.01	10.00	14.14	7.20	5.09	119.30	5.09	3.60
7.07	14.14	10.00	0.01	10.00	7.20	3.60	5.09	119.30	5.09
7.07	10.00	14.14	10.00	0.01	7.20	5.09	3.60	5.09	119.30

In Excel, a square matrix can be inverted by highlighting a blank space of the same size (5 × 5 cells in this case), typing in the formula = *MINVERSE(A1:E5)* where the original R_{ij} matrix resides in cells A1:E5, and pressing the *F2* key and then *<Ctrl><Shift><Enter>* at the same time.

The inverted matrix can be multiplied by a vector of unit potentials (1s) to obtain the currents in each electrode. The unit potential divided by the sum of these currents gives the resistance of the combined electrode, including self and mutual effects. The following table completes the numerical example.

Inverse of R_{ij} Matrix (×1000)					Potential (V)		Current = MMULT($[R_{ij}], V_1$)
8.49	−0.46	−0.46	−0.46	−0.46	1		6.65 mA
−0.46	8.44	−0.31	−0.20	−0.31	1		7.15 mA
−0.46	−0.31	8.44	−0.31	−0.20	1		7.15 mA
−0.46	−0.20	−0.31	8.44	−0.31	1		7.15 mA
−0.46	−0.31	−0.20	−0.31	8.44	1		7.15 mA
						Sum	35.27 mA 1/.03527 = 28.4 Ω

To multiply the inverted resistance matrix (say in locations A10:E15) by a vector of unit potentials (say in locations G10:G15), the cells I10:I15 would be highlighted, the formula = *MMULT(A10:E15,G10:G15)* would be entered, and the *F2* key would be pressed and then *<Ctrl><Shift><Enter>* pressed at the same time.

The currents are summed to a total of 35.27 mA, giving a resistance of 28.4 Ω. The five resistances in parallel, ignoring mutual effects, would have a resistance of 23.9 Ω, so the effect at a separation of 5 m is still significant.

A series of numerical calculations are found for a similar set of electrodes in (*IEEE Guide* 80, 2000, p. 183) for 10 m rods on 5 m grid spacing in 100 Ωm soil.

The relative reduction in resistance is largest when the second electrode is added, and additional nearby rods are seen to be less effective.

11.10 Treatment of Multilayer Soil Effects

Generally, the treatment of footing resistance in lightning calculations considers a homogeneous soil with a finite conductivity. This treatment, however, seldom matches field observations, particularly in areas where grounding is difficult. Under these conditions, a thin "overburden" layer of conducting clay,

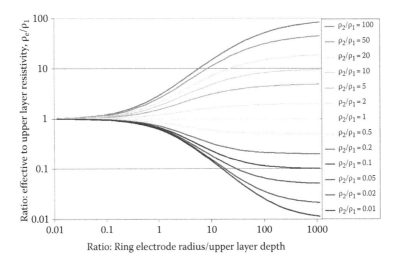

FIGURE 11.4 Relative effective resistivity versus ratio of electrode radius to upper-layer depth, with resistivity ratio as a parameter.

till, or gravel often rests on top of insulating rock. The distribution of resistivity values for a particular overburden material and condition can be narrow, with standard deviations usually less than 10%. However, the variation of overburden depth with distance can be large. Airborne electromagnetic survey techniques at multiple frequencies in the 10–100 kHz range offer a cost-effective method of reconnaissance of the overburden parameters of resistivity and depth.

Once a resistivity survey has established an upper-layer resistivity ρ_1, a layer depth d, and a lower-layer resistivity ρ_2, the equivalent resistivity ρ_e can be computed. For a disk-like electrode buried just below the surface, ρ_e from the elliptic-integral solutions of (Zaborsky, 1955) can approximated with better than 5% accuracy by the following empirical equations:

$$\rho_e = \rho_1 \frac{1 + C(\rho_2/\rho_1)(r/d)}{1 + C(r/d)}$$

$$C = \begin{cases} \rho_1 \geq \rho_2, & \dfrac{1}{1.4 + (\rho_2/\rho_1)^{0.8}} \\[4mm] \rho_1 < \rho_2, & \dfrac{1}{1.4 + (\rho_2/\rho_1)^{0.8} + ((\rho_2/\rho_1)(r/d))^{0.5}} \end{cases} \qquad (11.15)$$

The ratio of effective resistivity to upper-layer resistivity varies with the ratio of electrode radius s to upper-layer depth d, with a small ratio of s/d giving a ratio of unity as shown in Figure 11.4.

Normally, electrode penetration through an upper layer would only be desirable in extreme examples of Case 1 ($\rho_1 \gg \rho_2$). Rather than recomputing the effective resistivity with revised image locations, the effects of the upper layer can be neglected, with the connection through ρ_1 providing only series self-inductance.

11.11 Layer of Finite Thickness over Insulator

A simpler two-layer soil treatment is appropriate for Case 2 when $\rho_2 \gg \rho_1$, or equivalently the reflection coefficient from upper to lower layer Γ_{12} approaches 1. Under these conditions, the following summation in Equation 11.16 describes the resistance of a single hemisphere of radius s in a finitely conducting slab with resistivity ρ_1 and thickness h:

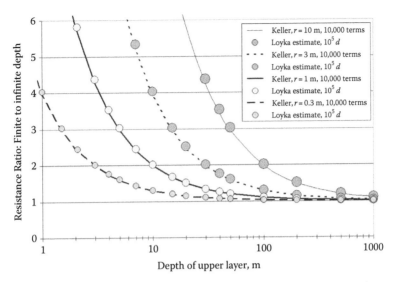

FIGURE 11.5 Asymptotic behavior of hemispherical electrode in conducting layer over insulator.

$$\Gamma_{12} = \frac{\rho_2 - \rho_1}{\rho_2 + \rho_1} \approx 1 \quad \text{for } \rho_2 \gg \rho_1 \quad R_{Keller} = \frac{\rho_1}{2\pi s}\left(1 + 2 \cdot \sum_{n=1}^{\infty} \frac{\Gamma_{12}^n}{\sqrt{1 + (2nh/s)^2}}\right)$$

$$R_{Loyka} \approx \frac{\rho_1}{2\pi s}\left[1 + \frac{s}{h}\ln\left(\frac{1 + \sqrt{1 + (10^5 h/2h)^2}}{1 + \sqrt{1 + (s/2h)^2}}\right)\right] \approx \frac{\rho_1}{2\pi s}\left[1 + \frac{s}{h}\ln\left(\frac{50{,}000}{1 + \sqrt{1 + (s/2h)^2}}\right)\right] \tag{11.16}$$

For a reflection coefficient $\Gamma_{12} = 1$, the Keller series in Equation 11.16 becomes the harmonic series and converges slowly. Loyka (1999) offers a convenient approximation to the sum, developed using the potential from a second hemisphere, located sufficiently far away (e.g., $10^5 d$, where d is the layer thickness) to have no influence. Figure 11.5 shows that the Loyka approximation to the Keller series is good for a wide range of transmission line grounding applications, and that finite upper layer depth has a strong influence on the resistance of the electrodes.

11.12 Treatment of Soil Ionization

Under high electric fields, air will ionize and become effectively a conductor. The transient electric fields needed to ionize air in small volumes of soil, or to flashover across the soil surface, are typically between 100 and 1000 kV/m (Korsuncev, 1958; Liew and Darveniza, 1974; Oettle, 1988). Considering that the potential rise on a small ground electrode can reach 1 MV, the origins of 10 m furrows around small (inadequate) ground electrodes after lightning strikes become clear. Surface arcing activity is unpredictable and may transfer lightning surge currents to unprotected facilities. Thus, power system ground electrodes for lightning protection should have sufficient area and multiplicity to limit ionization.

Korsuncev used similarity analysis to relate dimensionless ratios of s, ρ, resistance R, current I, and critical breakdown gradient E_o as follows:

$$\Pi_1 = \frac{Rs}{\rho} \quad \text{with } \Pi_1^O = \frac{1}{2\pi}\ln\left(\frac{2\pi e s^2}{A}\right) \tag{11.17}$$

$$\Pi_2 = \frac{\rho I}{E_o s^2} \tag{11.18}$$

$$\Pi_1 = \text{MIN}\left(\Pi_1^O, 0.26 \cdot \Pi_2^{-0.31}\right) \tag{11.19}$$

$$R = \frac{\rho \Pi_1}{s} \tag{11.20}$$

where
   s is the three-dimensional distance from the center to the furthest point on the electrode, m
   I is the electrode current, A
   ρ is the resistivity, Ωm
   E_o is the critical breakdown gradient of the soil, usually 300–1000 kV/m
   R is the resistance of the ground electrode under ionized conditions

The calculation of ionized electrode resistance proceeds as follows:

- A value of Π_1^O is calculated from Equation 11.17. This un-ionized value will range from $\Pi_1^O = 0.159$ (for a hemisphere) to 1.07 for a 3 m long, 0.01 m radius cylinder.
- A value of Π_2 is calculated from Equation 11.18. For a 3 m rod at 100 kA in 100 Ωm soil, with E_o of 300 kV/m, the value of $\Pi_2 = 3.7$ is obtained.
- A value of Π_1 is computed from Π_2 using 0.26 $\Pi_2^{-0.31}$. This value, from Equation 11.19, represents the fully ionized sphere with the gradient of E_o at the injected current; for the rod example, $\Pi_1 = 0.173$.
- If the ionized value of Π_1 is not greater than Π_1^O, then there is not enough current to ionize the footing, so the unionized resistance from Π_1^O will be seen for the calculation of resistance in Equation 11.20. For the 3 m rod, ionization reduces the 36 Ω low-current resistance to 6 Ω at 100 kA.

In two-layer soils with sparse electrodes, ionization effects will tend to reduce the contact resistance (from Equation 11.10) without altering the surface area A or characteristic dimension s. This will tend to reduce the influence of ionization, since the geometric resistance of the electrode is unchanged.

11.13 Design Process

The inputs to the design process are the prospective lightning surge current, the local soil resistivity, and the dimensions of the tower base or trial electrode. The low-frequency geometric resistance of the electrode is computed from Equation 11.9 and double-checked with the estimate from Equations 11.15 and 11.20.

   If the electrode is compact (e.g., a driven rod), then the ionization effects should be estimated using Equations 11.18 and 11.19, with the reduced resistance under high-current conditions being given by Equation 11.20.

   If the electrode is distributed (e.g., a buried wire), then the ionized resistance will only reduce the contact resistance term in Equation 11.10, and for practical purposes it will be sufficient to disregard this term and consider only the geometric resistance estimated from Equation 11.8 or Equation 11.9.

   In each case, an apparent transient impedance of the ground electrode should be modeled as an equivalent inductance in series with the resistive rise associated with the geometric resistance.

   The fraction of surge current absorbed by a given electrode will change as its size is adjusted, making the design process iterative.

11.14 Design Recommendations

It is prudent where possible to select an electrode size that does not rely on ionization for adequate transient performance. This can be achieved by inverting the design process as follows.

- Establish Π_1^O, the shape coefficient of the electrode (0.16 for a hemisphere, 0.26 for a cube, and 0.27 for a disk).
- Establish the value of Π_2^O that will just cause ionization (Chisholm and Janischewskyj 1989), inverting Equation 11.19 to obtain

$$\Pi_2^O = 0.0131\left(\Pi_1^O\right)^{-3.24} \tag{11.21}$$

For the example of a disk electrode, the value of $\Pi_2^O = 0.92$ is obtained.
- The extent (in this case radius) of the electrode s needed to prevent ionization in soil of resistivity ρ in Ωm with current I in kA and voltage gradient E_o in kV/m is given by

$$s = \sqrt{\frac{\rho I}{E_o \Pi_2^O}} \tag{11.22}$$

For $\rho = 300\ \Omega$m, $I = 30$ kA, and E_o of 400 kV/m, a disk or ring radius of $s = 9.1$ m will be sufficient to prevent ionization. The geometric resistance of this electrode from Equation 11.20 is 8.9 Ω.

The following advice is especially relevant for transmission towers or other tall structures, where a low-impedance ground is needed to limit lightning-transient overvoltages:

- Choose a wide, flat electrode shape rather than a long, thin shape. Four radial counterpoise wires of 60 m will be two to eight times more effective than a single counterpoise of 240 m under lightning surge conditions.
- Take advantage of natural elements in the structure grounding, such as foundations and guy anchors, by planning for electrical connections and by extending radial wires outward from these points.
- In rocky areas, use modern airborne techniques to survey resistivity and layer depth using several frequencies up to 100 kHz. Place towers where conductive covering material is deep.
- Provide grounding staff with the tools and techniques to preestimate the amount of wire required for target footing impedance values, using simple interpretation of two-layer soil data.
- Near areas where transferred lightning potentials could be dangerous to adjacent objects or systems, use sufficient electrode dimensions to limit ionization, that is to remain on the Π_1^O characteristic.

11.A Appendix A: Relevant IEEE and IEC Standards in Lightning and Grounding

11.A.1 ANSI/IEEE Std 80-1986: *IEEE Guide for Safety in AC Substation Grounding*

Presents essential guidelines for assuring safety through proper grounding at ac substations at all voltage levels. Provides design criteria to establish safe limits for potential differences within a station, under fault conditions, between possible points of contact. Uses a step-by-step format to describe test methods, design and testing of grounding systems. Provides English translations of fundamental papers on grounding by such as Rüdenberg and Laurent that are not widely available.

11.A.2 ANSI/IEEE Std 80-2000: *IEEE Guide for Safety in AC Substation Grounding*

Provides an improved methodology for interpreting two-layer soil resistivity and using the values in the design of ac substations. Provides methods for determining the maximum grid current at substations, some of which also predict the maximum fault currents available on lines close by. Provides a number of new worked examples in appendices.

11.A.3 IEC 62305: *Protection against Lightning*, Edition 2.0, 2010–12

The European Technical Committee 81 of the International Electrotechnical Commission (IEC, www.iec.ch) prepared an authoritative and comprehensive lightning protection standard in five parts as follows:

- Part 1, General principles
- Part 2, Risk management
- Part 3, Physical damage to structures and life hazard
- Part 4, Electrical and electronic systems within structures
- Part 5, Services (Edition 1)

The rationale in these reference documents forms a sound basis for national lightning protection standards of structures and wind turbines.

11.A.4 IEEE Std 81-1983: *IEEE Guide for Measuring Earth Resistivity, Ground Impedance, and Earth Surface Potentials of a Ground System*

The present state of the technique of measuring ground resistance and impedance, earth resistivity, and potential gradients from currents in the earth, and the prediction of the magnitude of ground resistance and potential gradients from scale-model tests are described and discussed. Factors influencing the choice of instruments and the techniques for various types of measurements are covered. These include the purpose of the measurement, the accuracy required, the type of instruments available, possible sources of error, and the nature of the ground or grounding system under test. The intent is to assist the engineer or technician in obtaining and interpreting accurate, reliable data. The test procedures described promote the safety of personnel and property and prevent interference with the operation of neighboring facilities. The standard is under revision as of September 2010.

11.A.5 IEEE Std 81.2-1991: *IEEE Guide for Measurement of Impedance and Safety Characteristics of Large, Extended, or Interconnected Grounding Systems*

Practical instrumentation methods are presented for measuring the ac characteristics of large, extended, or interconnected grounding systems. Measurements of impedance to remote earth, step and touch potentials, and current distributions are covered for grounding systems ranging in complexity from small grids (less than 900 m²) with only a few connected overhead or direct-burial bare concentric neutrals, to large grids (greater than 20,000 m²) with many connected neutrals, overhead ground wires (sky wires), counterpoises, grid tie conductors, cable shields, and metallic pipes. This standard addresses measurement safety; earth-return mutual errors; low-current measurements; power-system staged faults; communication and control cable transfer impedance; current distribution (current splits) in the grounding system; step, touch, mesh, and profile measurements; the foot-equivalent electrode earth resistance; and instrumentation characteristics and limitations.

11.A.6 IEEE Std 367-1996: *IEEE Recommended Practice for Determining the Electric Power Station Ground Potential Rise and Induced Voltage from a Power Fault*

Information for the determination of the appropriate values of fault-produced power station ground potential rise (GPR) and induction for use in the design of protection systems is provided. Included are the determination of the appropriate value of fault current to be used in the GPR calculation; taking into account the waveform, probability, and duration of the fault current; the determination of inducing currents, the mutual impedance between power and telephone facilities, and shield factors; the vectorial summation of GPR and induction; considerations regarding the power station GPR zone of influence; and communications channel time requirements for noninterruptible services. Guidance for the calculation of power station GPR and longitudinal induction (LI) voltages is provided, as well as guidance for their appropriate reduction from worst-case values, for use in metallic telecommunication protection design.

11.A.7 IEEE Std 524a-1993: *IEEE Guide to Grounding during the Installation of Overhead Transmission Line Conductors—Supplement to IEEE Guide to the Installation of Overhead Transmission Line Conductors*

General recommendations for the selection of methods and equipment found to be effective and practical for grounding during the stringing of overhead transmission line conductors and overhead ground wires are provided. The guide is directed to transmission voltages only. The aim is to present in one document sufficient details of present day grounding practices and equipment used in effective grounding and to provide electrical theory and considerations necessary to safeguard personnel during the stringing operations of transmission lines.

11.A.8 IEEE Std 837-2002: *IEEE Standard for Qualifying Permanent Connections Used in Substation Grounding*

Directions and methods for qualifying permanent connections used for substation grounding are provided. Particular attention is given to the connectors used within the grid system, connectors used to join ground leads to the grid system, and connectors used to join the ground leads to equipment and structures. The purpose is to give assurance to the user that connectors meeting the requirements of this standard will perform in a satisfactory manner over the lifetime of the installation provided, that the proper connectors are selected for the application, and that they are installed correctly. Parameters for testing grounding connections on aluminum, copper, steel, copper-clad steel, galvanized steel, stainless steel, and stainless-clad steel are addressed. Performance criteria are established, test procedures are provided, and mechanical, current–temperature cycling, freeze-thaw, corrosion, and fault-current tests are specified.

11.A.9 IEEE Std 1048-2003: *IEEE Guide for Protective Grounding of Power Lines*

Guidelines are provided for safe protective grounding methods for persons engaged in deenergized overhead transmission and distribution line maintenance. They comprise state-of-the-art information on protective grounding as currently practiced by power utilities in North America. The principles of protective grounding are discussed. Grounding practices and equipment, power-line construction, and ground electrodes are covered.

11.A.10 IEEE Std 1050-2004: *IEEE Guide for Instrumentation and Control Equipment Grounding in Generating Stations*

Information about grounding methods for generating station instrumentation and control (I & C) equipment is provided. The identification of I & C equipment grounding methods to achieve both a suitable level of protection for personnel and equipment is included, as well as suitable noise immunity for signal ground references in generating stations. Both ideal theoretical methods and accepted practices in the electric utility industry are presented.

11.A.11 IEEE Std 1243-1997: *IEEE Guide for Improving the Lightning Performance of Transmission Lines*

Procedures for evaluating the lightning outage rate of overhead transmission lines at voltage levels of 69 kV or higher are described. Effects of improved insulation, shielding, coupling and grounding on back-flashover, and shielding failure rates of transmission lines are then discussed. Finally, a description of the IEEE FLASH program to predict lighting outage rates from www.ieee.org/pes-lightning is provided.

11.A.12 IEEE Std 1313.1-1996: *IEEE Standard for Insulation Coordination—Definitions, Principles, and Rules*

The procedure for selection of the withstand voltages for equipment phase-to-ground and phase-to-phase insulation systems is specified. A list of standard insulation levels, based on the voltage stress to which the equipment is being exposed, is also identified. This standard applies to three-phase ac systems above 1 kV.

11.A.13 IEEE Std 1410-2010: *IEEE Guide for Improving the Lightning Performance of Distribution Lines*

Procedures for evaluating the lightning outage rate of overhead distribution lines at voltage levels below 69 kV are described. The guide then identifies factors that contribute to lightning-caused faults on the line insulation of overhead distribution lines and suggested improvements to existing and new constructions, including evaluation of the effect of wood or fiberglass materials in series with standard polymer or ceramic insulators.

References

Baba, Y. and Rakov, V.A., On the Interpretation of ground reflections observed in small-scale experiments simulating lightning strikes to towers, *IEEE Trans. on Electromagnetic Compatibility*, 47(3), August 2005.

Bewley, L.V., *Traveling Waves on Transmission Systems*, New York: Dover, 1963.

Chisholm, W.A. and Janischewskyj, W., Lightning surge response of ground electrodes, *IEEE Trans. Power Delivery*, 4(2), 1329–1337, April 1989.

Chow, Y.L. and Srivastava, K.D., Non-uniform electric field induced voltage calculations, Final Report for Canadian Electrical Association Contract 117 T 317, February 1988.

Chow, Y.L. and Yovanovich, M.M., The shape factor of the capacitance of a conductor, *J. Appl. Phys.*, 53, 8470–8475, December 1982.

Darveniza, M., A practical extension of Rusck's formula for maximum lightning-induced voltages that accounts for ground resistivity, *IEEE Trans. Power Delivery*, 22(1), 605–612, January 2007.

Deri, A., Tevan, G., Semlyen, A., and Castanheira, A., The complex ground return plane – A simplified model for homogeneous and multi-layer earth return, *IEEE Trans. Power Appar. Syst.*, PAS-100(8), 3686–3693, August 1981.

Grigsby, L.L., Transmission line structures (Chapter 9), *Electric Power Generation, Transmission and Distribution, Electric Power Engineering Handbook*, 3rd edn., CRC Press, 2011.

IEEE Standard 100, *The Authoritative Dictionary of IEEE Standard Terms*, 7th edn., Piscataway, NJ: IEEE, 2000.

IEEE Standard 80-2000, *IEEE Guide for Safety in AC Substation Grounding*, Piscataway, NJ: IEEE, 2000.

Keller, G.G. and Frischknecht, F.C., *Electrical Methods in Geophysical Prospecting*, New York: Pergamon, 1982.

Korsuncev, A.V., Application on the theory of similarity to calculation of impulse characteristics of concentrated electrodes, *Elektrichestvo*, 5, 31–35, 1958.

Laurent, P.G., Les bases générales de la technique des mises à la terre dans les installations électriques, *Bulletin de la Société Française des Electriciens*, 1(7), 368–402, July 1951.

Liew, A.C. and Darveniza, M., Dynamic model of impulse characteristics of concentrated earths, *IEE Proc.*, 121(2), 123–135, February 1974.

Loyka, S.L., A simple formula for the ground resistance calculation, *IEEE Trans. EMC*, 41(2), 152–154, May 1999.

Oettle, E.E., A new general estimation curve for predicting the impulse impedance of concentrated earth electrodes, *IEEE Trans. Power Delivery*, 3(4), October 1988.

Smythe, W.R., *Static and Dynamic Electricity*, New York: McGraw-Hill, 1950.

Sunde, E.D., *Earth Conduction Effects in Transmission Systems*, Toronto, Ontario, Canada: Van Nostrand, 1949.

Visacro, S. and Portela, C.M., Soil permittivity and conductivity behavior on frequency range of transient phenomena in electric power systems, *Proc. 5th ISH*, Paper 93.06, Germany: Braunschweig, August 1987.

Weber, E., *Electromagnetic Fields Theory and Applications Volume 1 – Mapping of Fields*, New York: Wiley, 1950.

Zaborsky, J., Efficiency of grounding grids with nonuniform soil, *AIEE Trans.*, 74, 1230–1233, December 1955.

12

Transient Recovery Voltage

Thomas E.
McDermott
Meltran, Inc.

Transient recovery voltage (TRV) refers to the voltage appearing across circuit breaker contacts after it interrupts current. The circuit breaker must withstand TRV in order to complete the current interruption process. For circuit breakers rated above 1 kV, the TRV is a crucial application criterion, along with several other important factors:

1. Voltage rating
2. Interrupting current ratings
3. Capacitive current switching (inrush, outrush, close-and-latch ratings)
4. Out-of-phase switching
5. Generator breakers pose a special case due to asymmetrical current interruption

TRV is not a consideration for low-voltage circuit breakers rated 1 kV or less, but above 1 kV, every circuit breaker application should include TRV analysis. The rest of this section introduces TRV calculations and mitigation, at a level sufficient for educational purposes. For actual design work, the user should have access to one of the standard application guides (IEEE C37.011-2005, 2005; IEC 62271-100, 2008). In some cases, time-domain simulation in an electromagnetic transients (EMT) program is necessary.

12.1 Fault Interruption Process

As a circuit breaker opens to interrupt current, an arc develops between the separating contacts. The current continues to flow through this arc, with a nonlinear voltage drop that produces heat. In AC steady state, the current approaches each zero crossing with approximately linear slope, and when it passes through zero, there is a chance for current interruption. During the first few microseconds after interruption, a post-arc current may flow and produce more heat. At the same time, the arc dissipates heat through radiation, convection, and conduction according to the breaker design. During this initial energy balance period, the arc must lose heat faster than it gains heat. Otherwise, a reignition occurs and interruption cannot occur until the next natural current zero crossing.

Beyond the energy balance period, recovery voltage continues to build up across the still opening breaker contacts. The dielectric strength, which depends on contact separation, must exceed the recovery voltage stress throughout the dielectric stress period. If not, a reignition or restrike occurs and interruption must wait for the next natural current zero crossing.

The circuit breaker must pass both the energy balance and dielectric stress periods, in order to complete a current interruption. Detailed arc physics models have been developed to analyze these processes, but these models are of most use to circuit breaker designers. In applications, the TRV is usually calculated based on the electrical system model only, ignoring interactions between the arc and system models. This calculated TRV stress is then compared to standard TRV test results, which are performed in the lab and already encompass the arc behavior. There may be system TRV studies that require detailed arc models, but they are beyond the scope of this chapter.

There are consequences when the breaker fails to interrupt because of TRV. At a minimum, each reignition or restrike produces a transient overvoltage, until the breaker finally interrupts. If the breaker never succeeds to interrupt, it will require inspection and repair because of thermal damage or dielectric breakdown. TRV failures usually occur when interrupting fault current, in which case a backup circuit breaker will have to interrupt the fault. Backup clearing takes longer, which allows more time for equipment damage to occur from heating or short-circuit forces during the fault. Backup clearing also removes more of the system from service, which may lead to cascading outages. If the TRV failure occurs during a normal (i.e., nonfault) opening, it probably leads to a short circuit in the failed breaker, which also leads to backup clearing and possible cascading effects. TRV analysis is important because it affects protection of the power system.

12.2 Analysis Principles

Hand calculation and computer simulation both play roles in TRV studies. First, a simplified circuit analysis provides guidance on how much of the power system to model, which component parameters are most important, and what results to expect. Second, computer simulation with an EMT model should provide more realistic results and precise evaluation of all important cases. Third, reanalysis of an equivalent circuit can help explain the results and correct modeling errors. IEEE Std. C37.011 is a good source of typical data for equipment capacitance and other important parameters. Circuit breaker modeling guidelines are found in (IEEE PES Task Force, 2005). Guidelines for modeling the balance-of-system are found in (IEC TR 60071, 2004).

TRV phenomena may consist of high-frequency transients in lumped-parameter circuits, along with traveling wave transients in distributed-parameter overhead lines and cables. It is also necessary to consider all three phases, and neutral grounding. These factors make TRV, or any transient analysis, more complicated than steady-state calculations at power frequency.

Figure 12.1 shows a simplified three-phase equivalent circuit on the source side of a circuit breaker, which is interrupting a three-phase grounded fault on its terminals. The three phases do not interrupt simultaneously, and Figure 12.1 shows just the first phase opening. The most severe TRV usually appears on the first phase of the last circuit breaker to clear a fault. The source inductance often has unequal zero sequence and positive sequence values, as in Figure 12.1, which affects the equivalent inductance for TRV analysis:

$$L_{eq} = \frac{3L_0L_1}{L_1 + 2L_0} \tag{12.1}$$

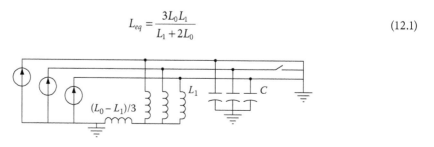

FIGURE 12.1   Three-phase equivalent circuit for first pole to clear.

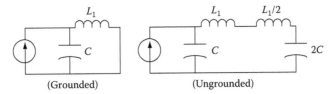

FIGURE 12.2 Single-phase equivalent circuits for the first pole to clear, $L_0 = L_1$.

The equivalent capacitance, C, comes from buswork, power transformers, instrument transformers, bushings, and other equipment. This stray capacitance is usually grounded for the purpose of TRV calculations. The most important capacitance values come from power transformers, coupling capacitor voltage transformers (CCVT), and capacitive voltage transformers (CVT). Because capacitance mitigates TRV, the study should make conservative assumptions about which capacitive equipment will be in service when the breaker opens. In many cases, it is not necessary to explicitly model the buswork between equipment, but Section 12.6 describes one exception to that guideline.

The left-hand side of Figure 12.2 shows the greatest possible simplification of Figure 12.1, with a grounded fault and equal sequence inductances. Each phase operates independently. This lumped circuit has transient response governed by the surge impedance and natural frequency:

$$Z_{lump} = \sqrt{\frac{L}{C}} \tag{12.2}$$

$$f_0 = \frac{1}{2\pi\sqrt{LC}} \tag{12.3}$$

Z_{lump} is the ratio of peak transient voltage to peak transient current in the circuit. With no damping, the transient voltage and current will oscillate around their steady-state values, with 100% overshoot. A parallel resistance in Figure 12.2 will provide damping, and reduce the peak transient voltage and current. If the parallel $R \leq 0.5\ Z_{lump}$, then the circuit response will be critically damped or overdamped. This means the response is exponential rather than oscillatory. If the parallel R is approximately 4 Z_{lump}, the damping factor is 1.7. This means the first transient peak will be only 85% of the undamped peak. The generalized damping curves in Greenwood (1991) provide more detailed information. IEEE Std. C37.011 suggests that practical damping factors range from 1.6 to 1.9, based on measurements. In an EMT model, damping comes primarily from frequency-dependent line and transformer losses, which are sometimes burdensome to model in detail. As an alternative, series or shunt resistors may be added to the EMT model at strategic points, in order to provide high-frequency TRV damping.

The right-hand side of Figure 12.2 illustrates the use of circuit folding and phase symmetry, if the fault is ungrounded, but the sequence inductance values are still equal. The source-side parameters, L_1 and C, do not change. However, the fault current no longer flows to ground, but must return in the two unfaulted phases (top and bottom wires in Figure 12.1). These phases provide parallel paths, with equivalent inductance $L_1/2$ and capacitance $2C$.

Overhead lines and cables connected to a bus would initially appear as shunt resistors in Figures 12.1 and 12.2. Some traveling wave effects are discussed in Section 12.5. For TRV studies, the surge impedances should be calculated at high frequency, or in EMT simulations, a frequency-dependent model can be used. Typical values of the positive sequence overhead line surge impedance are 350 Ω for single conductors and 275 Ω for bundled conductors. During faults, subconductor clashing may increase the effective surge impedance. Given the positive sequence surge impedance, Z_1, and the number of lines connected to a bus, the zero sequence and single-phase equivalent surge impedances are

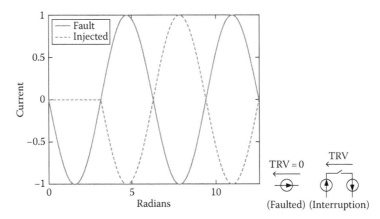

FIGURE 12.3   Current injection to simulate fault interruption.

$$Z_0 \approx 1.6Z_1 \tag{12.4}$$

$$Z_{eq} = \frac{3}{n}\frac{Z_0 Z_1}{Z_1 + 2Z_0} \tag{12.5}$$

A cable's surge impedance, both positive sequence and equivalent, ranges from 20 to 75 Ω.

For hand calculation, the current injection method simulates TRV following fault interruption. Referring to the right-hand side of Figure 12.3, during the fault, current flows through the circuit breaker pole, and the "TRV" across that pole is 0. If the circuit is linear, then we can simulate fault interruption by injecting equal and opposite current through the breaker pole. In the left-hand side of Figure 12.3, interruption is simulated by injecting a current at π radians, and after that the sum of the two currents is 0. In the right-hand side of Figure 12.3, this is done by injecting currents on each side of the open pole. Each injected current source produces a transient voltage at its terminal, and the difference between them is the total TRV. If the fault is grounded, one of the current source terminal voltages may be 0. For hand calculation by Laplace transform, sometimes the injected currents are simplified to linear ramps, valid for a short period of time.

The current injection method is also useful in EMT simulation, because the time of fault interruption is precisely controlled. That helps control numerical oscillations, and also simplifies waveform evaluation when the TRV starts exactly at time zero. The injection current sources should be sinusoids instead of ramps. Of course, the EMT model must be linear in order for superposition to apply. That is not a serious restriction because most TRV studies are done without surge arrester models and without iron core saturation models. Exceptions might apply when studying surge arresters across the breaker terminals (i.e., not to ground), and when studying lower-frequency resonant or dynamic overvoltages, for which iron core saturation may become important.

The most severe TRV comes from a three-phase ungrounded fault, but those faults are quite rare. The standard TRV application for an effectively grounded system is based on a three-phase grounded fault at the breaker terminals.

With EMT simulation, the question will arise about how much of the surrounding power system to model. One of the classic rules of thumb has been "within two buses from the breaker under study." A better rule accounts for the rated TRV time to peak, which increases with the breaker voltage level, and on typical wave travel times for overhead lines. As a guideline, include all buses within 0.48 * (system nominal kV). For example, for a 115 kV system, include all buses within 55 km of the studied bus, along with all transformer and generator sources connected to those buses. This ensures that all traveling wave and remote source effects appear at the station under study, before the breaker's rated TRV reaches its peak value.

12.3 TRV for Transformer-Fed Faults

Transformer-fed faults occur quite often in medium-voltage facilities, and provide a good starting point for understanding TRV. Figure 12.4 shows a fault cleared by a breaker on the secondary of a transformer, with equivalent circuit parameters L_S and C_S. L_S primarily comes from the transformer inductance, while C_S comes primarily from transformer secondary capacitance, and secondary cables between the transformer and breaker. In this example, E is 11.3 kV peak line-to-ground (for a 13.8 kV system), L_S is 0.69 mH, and C_S is 0.11 μF. From Equation 12.3, the natural frequency is 18.3 kHz. The TRV will be a 1-cosine shape defined by Equation 12.6 and plotted in Figure 12.5:

$$\text{TRV}_S = 2E\left[1 - \cos\frac{t}{\sqrt{L_S C_S}}\right] \tag{12.6}$$

In Figure 12.6, the same breaker interrupts a fault on the other side of a load transformer, as might occur during backup clearing. The source-side equivalent circuit does not change, but there is also a load-side equivalent circuit consisting of L_L and C_L, where L_L comes from the load transformer, and C_L comes

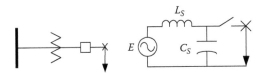

FIGURE 12.4 Single-frequency transformer fed fault TRV.

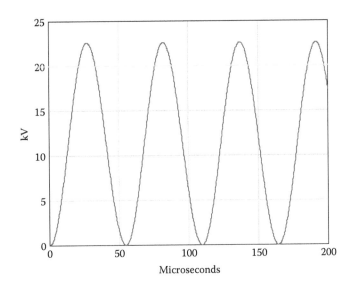

FIGURE 12.5 Single-frequency TRV.

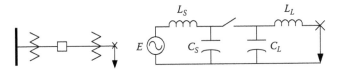

FIGURE 12.6 Double-frequency transformer fed fault TRV.

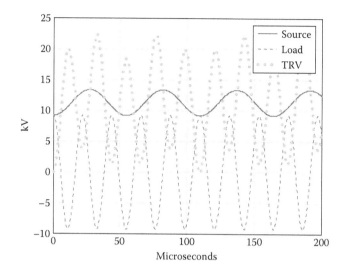

FIGURE 12.7 Double-frequency TRV.

from the load transformer and connected cable. In this example, L_L is 3.03 mH and C_L is 4 nF. A double-frequency TRV will appear across the breaker pole. The source-side component frequency is 18.3 kHz as before, while the load-side frequency is 45.7 kHz from Equation 12.3. The faulted system voltage at the breaker is no longer zero at the time of interruption, because of the load transformer impedance between breaker and fault:

$$E_{bus} = E \frac{L_L}{L_S + L_L} = 9.2 \text{ (kV)} \tag{12.7}$$

The load-side voltage will oscillate around zero, starting from E_{bus}. The source side voltage will oscillate around $E = 11.3$ kV, also starting from E_{bus}. These voltages and the TRV are provided in Equations 12.8 through 12.10 and plotted in Figure 12.7:

$$E_{load} = 9.2 \cos \frac{t}{\sqrt{L_L C_L}} \tag{12.8}$$

$$E_{source} = 9.2 + 2.1 \left[1 - \cos \frac{t}{\sqrt{L_S C_S}} \right] \tag{12.9}$$

$$E_{TRV} = E_{source} - E_{load} \tag{12.10}$$

These examples illustrated single-frequency and double-frequency TRV, and the influence of impedance between breaker and fault. Damping was not considered. System resistance was also ignored, so the fault current and source voltage were 90° out of phase. Considering the actual phase shift would affect E.

12.4 TRV for Capacitor Bank Switch Opening

This example uses the case of capacitor bank de-energization to illustrate the importance of neutral voltage shifts in TRV. These phenomena occur at power frequency, so the voltage across breaker contacts is more properly termed recovery voltage, rather than transient recovery voltage. However, a restrike or reignition of the breaker will produce high-frequency transients. Capacitor banks are usually switched often. Special-purpose breakers or switches may be specified to minimize the possibility of restrikes.

FIGURE 12.8   One phase of grounded capacitor bank opening.

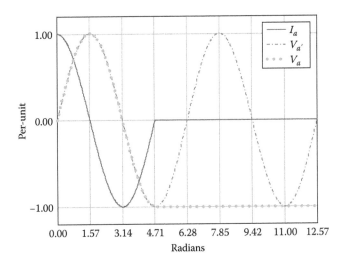

FIGURE 12.9   Grounded capacitor bank recovery voltage.

Figure 12.8 shows one phase of a three-phase grounded capacitor bank, so each phase may be considered independently, and the neutral voltage remains at zero. Figure 12.9 shows that the current leads the voltage by 90°, and when the current passes through zero at 1.5 π radians, the voltage on the bank is at its negative peak. This voltage is trapped on the disconnected capacitance, and will remain at this DC value for typically several minutes. Meanwhile, the source-side voltage continues to oscillate at power frequency. The peak voltage across the switch, $V_{a'} - V_a$, is 2 per unit. The switch should be able to withstand this.

Figure 12.10 shows a three-phase ungrounded capacitor bank. The neutral is actually grounded through stray capacitance, which may be on the order of 1 nF, but the bank neutral voltage is no longer held at zero. The bank neutral can shift and hold a nonzero voltage, supported by the stray capacitance. The actual value of this stray capacitance has little importance to the analysis.

Figure 12.11 shows the three-phase voltages on the bus, and Figure 12.12 shows the capacitor bank currents. Phase A opens first, at which time phases B and C are still energized in series, by the line-to-line voltage between phases B and C. The capacitor bank currents in phases B and C must be equal and opposite.

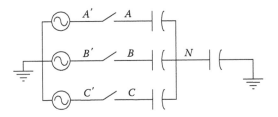

FIGURE 12.10   Ungrounded capacitor bank opening.

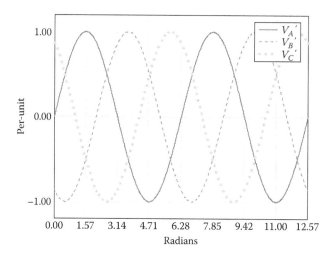

FIGURE 12.11 Ungrounded capacitor bank source voltages.

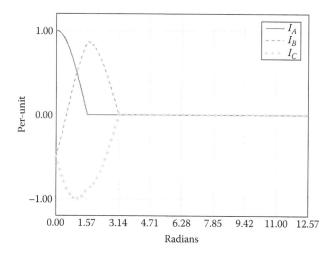

FIGURE 12.12 Ungrounded capacitor bank opening currents.

Figure 12.12 shows these currents have equal and opposite values of 0.866 per unit, and their waveshapes change slope at the point of phase *A* interruption. The phase *B* and *C* currents remain equal and opposite, until they interrupt 1/4 cycle later.

Figure 12.13 shows the voltage on each of the four capacitors shown in Figure 12.10. After phase *A* interruption, 1/4 cycle into the event, the phase *A-N* capacitance has a trapped DC voltage of 1 per unit. This is similar to Figure 12.9 although opposite in polarity. The phase *B-N* and *C-N* capacitances have equal −0.5 per unit voltages at this time, but they continue to be charged or discharged by the phase *B-C* line-to-line voltage for another 1/4 cycle. Based on Figure 12.12, phase *B-N* receives positive current and charge for 1/4 cycle, while phase *C-N* receives negative current and charge over that 1/4 cycle. The change in voltage is 0.866 per unit instead of 1.0 per unit during that time, because each capacitor is now energized by half of the line-to-line voltage, rather than line-to-neutral voltage. When phase *B* and *C* interrupt at 1/2 cycle into the event, they have trapped capacitance voltages 0.366 and 1.366 per unit, respectively.

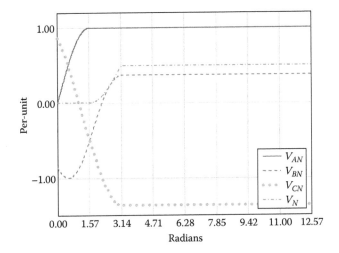

FIGURE 12.13   Ungrounded capacitor bank opening voltages.

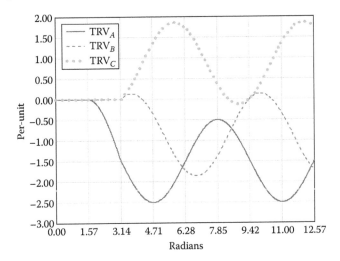

FIGURE 12.14   Ungrounded capacitor bank recovery voltages.

The neutral point voltage has to increase during that second 1/4 cycle of unbalanced charging, and has 0.5 per unit trapped on it when phases B and C interrupt. Figure 12.14 shows the recovery voltage on each switch pole, evaluated from

$$\text{TRV}_A = V_A^{'} - (V_{AN} + V_N)$$

$$\text{TRV}_B = V_B^{'} - (V_{BN} + V_N) \qquad (12.11)$$

$$\text{TRV}_C = V_C^{'} - (V_{CN} + V_N)$$

The peak recovery voltage on phase A is now 2.5 per unit because of the neutral shift, although the recovery voltages on the other two phases are less than 2.0 per unit. Again, the capacitor switching device should be able to withstand this. If phases B and C do not interrupt at their first chance, the neutral voltage may reach a peak of 1.0 per unit and increase the peak TRV to 3.0 per unit. If that causes a restrike, the switch should be designed to minimize the occurrence rate of restrikes.

The textbook (Greenwood, 1991) contains more detail on the effect of source impedance, the transient after a restrike, and the effect of neutral voltage shift on shunt reactor switching.

12.5 TRV for Line-Fed Faults

When interrupting a fault fed at least partially by overhead/underground lines, traveling wave effects and resistive surge impedances play a role in the TRV, which may no longer be oscillatory (Colclaser and Buettner, 1969; Colclaser, 1972). Figure 12.15 shows an example 115 kV substation bus, fed by three transmission lines and a transformer. Two fault scenarios are considered, one at the terminals of a line breaker and another one located a short distance, d, out on the line. These two scenarios are chosen to show the difference between terminal faults and short-line faults on the same breaker. This terminal fault does not produce the highest possible fault current; the highest bus fault is fed by all three lines plus the transformer. If it is possible for a breaker within the station to interrupt the total bus fault current, that case should be included in the study because higher fault current produces higher TRV.

The transformer inductance is 20 mH and the total bus capacitance is 10 nF. From Equation 12.2, the lumped circuit surge impedance is 1414 Ω. The effective surge impedance is 420 Ω for each line, and for two lines in parallel on the source side, the equivalent surge impedance $Z = 210$ Ω. This value is much less than half of the lumped circuit surge impedance, so the TRV will be an exponential waveform. The capacitance is small enough to be ignored for initial analysis.

If the fault current through the line breaker is 20 kA, then the TRV response is given in Equation 12.12, where the angular frequency ω is 377 for a 60 Hz power system (i.e., this case) and 314 for a 50 Hz power system:

$$\tau = \frac{L}{Z} = 95 \ (\mu s)$$

$$E_1 = \sqrt{2} I_{flt} \omega L = 213 \ (kV) \tag{12.12}$$

$$TRV = E_1 \left(1 - \exp\left(\frac{t}{\tau}\right)\right)$$

This TRV is plotted in Figure 12.16. In a real power system, traveling wave reflections from nearby stations would modify this TRV, possibly increasing or decreasing the peak value. It is difficult to analyze these reflections by hand, because they are complicated by multiple line and inductive terminations.

On a 115 kV system, about 9 kA fault current would flow through the transformer, and the remaining 11 kA must come from the two unfaulted lines. If the fault current were higher, then the TRV would increase. On the other hand, higher fault current usually means more parallel source connections, and stronger source equivalents at nearby stations, both of which should help reduce the TRV. This is one of the main reasons for using EMT simulation to perform the study. The EMT model can be set up for

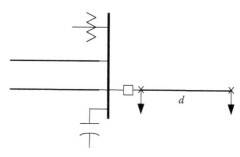

FIGURE 12.15   Bus terminal and short-line fault fed by lines and transformers.

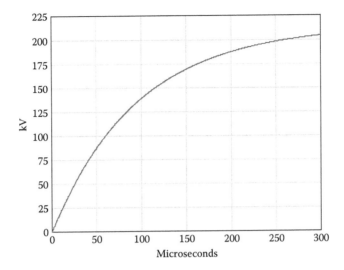

FIGURE 12.16 Bus terminal fault TRV.

transients only, and use current injection sources derived from a separate fault study. Alternatively, the EMT model can be set up and used for both steady-state fault analysis and transient analysis, avoiding the use of current injection. Either way, accurate fault current levels must be used in the study.

If there is no inductive source at the station in Figure 12.15, then the TRV will consist of a linearly rising ramp, until modified by traveling wave reflections from nearby stations. Cables have lower surge impedances than overhead lines, leading to longer time constants for exponential TRV, or lower slopes for linear TRV components.

The second fault location in Figure 12.15 is a distance $d = 3.2$ km from the line breaker terminals. The TRV is lower because the fault current is less, but a saw tooth component on the 3.2 km line segment produces a higher initial rate of rise. Years ago, short-line faults like this led to some breaker failures. Newer versions of the IEEE and IEC standards include short-line fault tests, so that the user does not have to explicitly evaluate them for applications. However, the case provides an instructive example, and the tested short-line fault capability can also be applied to other situations.

Assuming the maximum fault currents were evaluated at 1.05 per unit operating voltage, the first step is to estimate the breaker fault current during a short-line fault. (This could also be done using a separate short-circuit analysis.) Based on typical line reactance of 0.5 Ω/km in positive sequence and 1.2 Ω/km in zero sequence, the total line reactance to the fault is

$$X_L = d\frac{2X_{1L} + X_{0L}}{3} = 2.35\,\Omega \tag{12.13}$$

The source-side equivalent reactance, from two lines and one transformer, comes from the 20 kA terminal fault current:

$$X_S = \frac{1.05 * 115}{\sqrt{3} * 20} = 3.49\,\Omega \tag{12.14}$$

Those two reactances now determine the reduced fault current:

$$I_{SLF} = \frac{1.05 * 115}{\sqrt{3} * (2.35 + 3.49)} = 11.9\text{ kA} \tag{12.15}$$

The bus voltage at the time of interruption is no longer zero, because of the line impedance out to the fault. After interruption, the source-side voltage will approach its 1.05 per unit peak prefault level according to an exponential waveform. Modifying Equation 12.12 accordingly,

$$\tau = \frac{L}{Z} = 95 \ (\mu s)$$

$$E_{1-SLF} = \left[115 * 1.05 * \sqrt{\frac{2}{3}} - X_L I_{SLF} \sqrt{2} \right] = 59 \ (kV) \tag{12.16}$$

$$TRV_{1-SLF} = E_{1-SLF}\left(1 - \exp\left(\frac{t}{\tau}\right) \right)$$

The source-side TRV component, plotted in Figure 12.17, has a lower prospective peak value but the same time constant.

Equation 12.17 defines the rate of rise of the line-side component; $Z_{eff}= 420 \ \Omega$ because it applies to the single faulted line segment

$$R = \sqrt{2}\omega I_{SLF} Z_{eff} * 10^{-6} = 2.69 (kV/\mu s) \tag{12.17}$$

In IEEE standards, a damping factor of 1.6 applies to the line-side component of TRV, meaning that the damped peak reaches 1.6 per unit of the line-side breaker terminal voltage at the instant of interruption. Without damping, the peak would be 2.0 per unit of that line-side breaker terminal voltage. This damping factor is η, and the peak line-side TRV component is

$$U_L = \eta X_L I_{SLF} \sqrt{2} = 63.3 kV \tag{12.18}$$

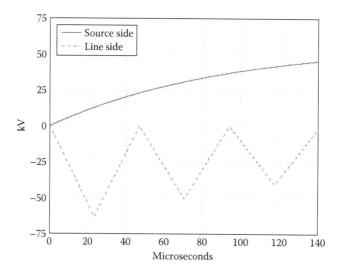

FIGURE 12.17 Short-line fault TRV components.

The time to first peak is $U_L/R = 23.5\,\mu s$. Figure 12.17 includes this damped line-side TRV component. The total TRV is the difference between the two waveforms in Figure 12.17. Compared to Figure 12.16, the peak TRV is lower but the initial rate of rise is more severe (note that for clarity, Figure 12.17 has a shorter timescale than Figure 12.16).

12.6 TRV for Current-Limiting Reactor Faults

Shunt capacitor banks and underground cables often have current-limiting reactors (CLR) installed in series. The CLR may be necessary to limit capacitor bank inrush and outrush currents, especially when multiple capacitors are connected to the same bus. CLR has also been used to reduce fault currents in low-impedance cable systems. The CLR can increase high-frequency TRV stresses on switchgear, and there have been cases of capacitor switch failure when trying to interrupt a CLR-limited fault. When this causes backup clearing of several lines and transformers connected to that bus, more widespread outages and disruptions can occur.

Figure 12.18 shows a simplified view of one station with a faulted capacitor bank at the node CAP, and a breaker at node BRKR tasked to clear that fault. A CLR between the capacitor and breaker has a series inductance, with relatively small shunt and series capacitances associated. The high-frequency lumped-circuit oscillations within the CLR produce high TRV rate of rise on the line side, and possibly lead to failure. Because of the fault location, the capacitor bank's large capacitance provides no TRV mitigation. However, other buswork and equipment capacitance located between CLR and breaker may help in mitigating the TRV.

On the source side of the breaker, a total bus capacitance and external network of lines and transformers determine the source-side TRV, which would be similar to that produced by a bus terminal fault. If the CLR were moved to the source side of the breaker, it would still produce a high-frequency TRV component. That high-frequency TRV, now on the source side instead of load side, would still possibly lead to breaker failure.

Figure 12.19 shows a simplification of the model, by aggregating all capacitance between breaker and CLR into one value, ignoring the inductance of connections between them. The full external system network should still be part of the model, for accurate fault currents and source-side TRV components. The study can focus on changing CLR or breaker parameters to mitigate the TRV.

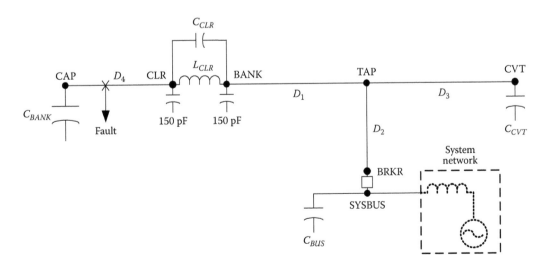

FIGURE 12.18 Model for current-limiting reactor fault TRV.

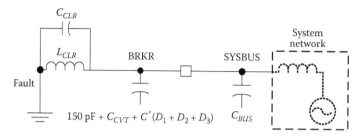

FIGURE 12.19   Reduced model for current-limiting reactor fault TRV.

12.7 Switchgear Tests and Standards

Both IEEE and IEC define circuit breaker test procedures, rating structures, preferred rating values, and TRV evaluation procedures in a set of standards that are updated periodically (IEEE C37, IEC 62271). Most North American utilities use IEEE standards, while other utilities in the world generally use IEC standards. The same choice would apply to medium-voltage and high-voltage industrial or commercial users, who might also need to do TRV studies.

Historically, there have been differences between IEEE and IEC standards, but the IEEE standards have been recently harmonized with IEC. That harmonization process is not yet complete. Even when harmonization is completed, circuit vendors can still offer products that were type-tested to earlier versions of IEEE standards. Circuit breakers can last for many decades in service, and they would have been tested to earlier standards. For both IEEE and IEC standards, there may be changes in procedures and ratings between updates. These are all good reasons to verify, at the beginning of a TRV study, which version of the standards applies to the breakers under evaluation.

At present, the standards mention that TRV evaluation can be based on three-phase grounded terminal faults, and that the vendor is responsible for testing breakers to withstand short-line fault TRV. There may be situations that still call for analysis of three-phase ungrounded faults, short-line faults, out-of-phase switching, or other examples not covered in this chapter.

Figure 12.20 shows the TRV rating for a 123 kV class circuit breaker, interrupting 100%, 60%, 30%, or 10% of its rated fault capability. These ratings apply to older IEEE standards; at 60% and

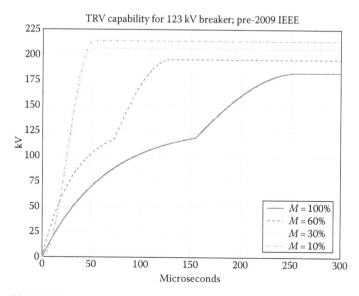

FIGURE 12.20   Old IEEE TRV ratings.

TABLE 12.1 Interpolated TRV Ratings for IEEE and IEC Standards, Breakers Rated 100 kV and Higher

	IEEE			IEC		
M (%)	kE_2	kT_2	R_1 (kV/μs)	K_{af}	ROR (kV/μs)	T_2
10	1.17	0.2	0	1.53	7	0
30	1.13	0.2	0	1.54	5	U_c/ROR
60	1.07	0.5	4	1.50	3	$6T_1$
100	1.00	1.0	2	1.40	2	$4T_1$

higher, the TRV envelope is a composite 1-exponential and 1-cosine shape. At 30% and 10% interrupting rating, and for voltage ratings below 100 kV, only the 1-cosine shape is used. At fault current levels in between the defined four test levels, TRV rating envelopes may be interpolated. The calculated or simulated TRV stress should lie entirely under the applicable envelope in Figure 12.20 for a successful application. The user should also be aware of the TRV "delay line" as described in both IEEE and IEC standards.

Given breaker rating values of E_m = 123 kV and T_2 = 260 μs, the curves in Figure 12.20 may be constructed from Equation 12.19 with interpolation factors from Table 12.1.

$$K_a = 1.4$$

$$K_f = 1.3$$

$$T_2 \leftarrow T_2 kT_2$$

$$U_r = E_m \sqrt{\frac{2}{3}}$$

$$E_2 = K_a K_f U_r kE_2$$

$$\tau = \frac{E_1}{R_1}$$

$$E_{exp} = E_1 \left[1 - \exp\left(-\frac{t}{\tau} \right) \right]$$

$$E_{cos} = 0.5 E_2 \left[1 - \cos\left(\frac{\pi t}{T_2} \right) \right]$$

$$\text{TRV} = \max(E_{exp}, E_{cos})$$

(12.19)

R_1 is a rate of rise, to be interpolated from Table 12.1. If R_1 is 0, the E_{exp} term does not appear in the TRV, only the 1-cosine term, E_{cos}. Table 12.1 also provides multipliers to use with E_2 and T_2, at lower-than-rated fault currents.

Figure 12.21 shows corresponding IEC standard TRV ratings, for the same four test current levels, and the same 123 kV breaker voltage class. At 100% and 60% of rated interrupting current, the TRV envelope is a two-slope characteristic, defined by two points, or four parameters. At 30% and 10% of rated interrupting current, or for any breaker voltage rating below 100 kV, only one point defines the TRV. These are often called "4 parameter" and "2 parameter" TRV ratings. Newer versions of the IEEE

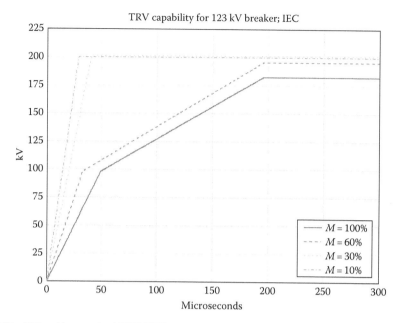

FIGURE 12.21 IEC and harmonized IEEE TRV ratings.

standards will use TRV ratings presented as in Figure 12.21. With reference to interpolated values from Table 12.1, Figure 12.21 is constructed according to

$$K_{pp} = 1.3$$

$$U_r = E_m \sqrt{\frac{2}{3}} \tag{12.20}$$

$$U_c = K_{pp} K_{af} U_r$$

If $M \leq 30\%$ then Equation 12.21 applies

$$T_2 = \frac{U_c}{ROR}$$

$$T_1 = 0.5 T_2 \tag{12.21}$$

$$U_1 = 0.5 U_c$$

If $M > 30\%$ then Equation 12.22 applies, and T_2 is interpolated according to Table 12.1 after T_1 has been determined:

$$U_1 = 0.75 K_{pp} U_r$$

$$T_1 = \frac{U_1}{ROR} \tag{12.22}$$

The TRV is then constructed by linear interpolation between four points: $(0, 0)$, (T_1, U_1), (T_2, U_c), and $(2T_2, U_c)$. Equations 12.20 through 12.22 smoothly transition from two-parameter and four-parameter characteristics, when M increases from 30% to 60%.

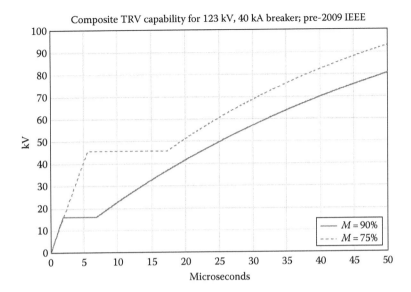

FIGURE 12.22 Composite TRV rating based on short-line fault tests.

Figure 12.22 illustrates the aggregation of a breaker's tested short-line fault TRV capability, with the base TRV rating. By IEEE standards, the short-line fault test may be performed at M values of 70%–95%. The standard test surge impedance $Z = 450\ \Omega$, and this breaker has rated frequency $f = 60\ \text{Hz}$ and rated interrupting current $I_r = 40\ \text{kA}$. The short-line fault test verifies TRV withstand up to a point defined by the linear segment from $(0, 0)$ to (T_{slf}, E_{slf}), as obtained from

$$E_{line} = 1.6(1 - \bar{M})U_r$$

$$R_{slf} = 8.8858^{-6}\, f\bar{M}I_r Z$$

$$T_{slf} = \frac{E_{line}}{R_{slf}} \qquad (12.23)$$

$$E_{srce} = \max\left[0, 2\bar{M}(T_{slf} - 2)\right]$$

$$E_{slf} = E_{line} + E_{srce}$$

After T_{slf}, the TRV rating extends horizontally at voltage E_{slf}, until it encounters the exponential TRV characteristic. The extra short-time TRV capability may be useful in applications that produce a fast-rising TRV.

12.8 TRV Mitigation

If the calculated TRV stress exceeds the breaker ratings, mitigation will be necessary:

1. Consult the circuit breaker vendor, and provide the TRV stress waveform in electronic form. Sometimes the vendor has designed the breaker to withstand higher TRV, and is willing to state that the application is acceptable.
2. Add capacitance to the circuit breaker terminals, or across the terminals. This slows down the TRV by reducing the rate of rise or oscillation frequency. It is one of the most common TRV mitigation techniques. However, the first peak in TRV can increase slightly with more capacitance, so that should be evaluated, too. The capacitance also has to be located close enough to the circuit breaker, and cannot be separated from the breaker through disconnect switch operation or station reconfiguration.

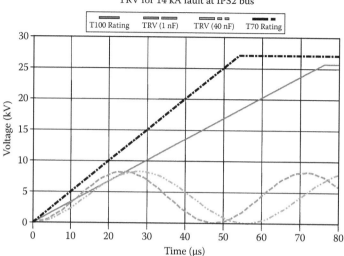

FIGURE 12.23 Mitigation of medium-voltage breaker TRV. (Reproduced with permission from McDermott, T.E. and Dafis, C., Cost-effective ship electrical system simulation, *ISS IX Conference Paper*. Philadelphia, PA, May 25–26, 2011. Copyright 2011 American Society of Naval Engineers.)

3. When a current-limiting reactor is involved, capacitance can be added on the CLR terminals. It may also be possible for the CLR vendor to alter its design to have more internal capacitance.
4. Use a circuit breaker of higher voltage rating. This can be expensive, but it works because the breaker will have significantly higher TRV capability for the same fault current.
5. Use a circuit breaker of higher current rating. This can also be expensive, but it increases the TRV capability because the system fault current will be a lower portion of the breaker rating.
6. Reduce the system fault current, either by splitting a bus or by adding fault current-limiting impedances. The system impedance changes may tend to increase TRV, but that is usually more than offset by the reduction in fault current level.
7. Surge arresters across the breaker contacts have been considered, especially when the TRV comes from dynamic overvoltages. One must also check the arrester energy and temporary overvoltage ratings in this application.
8. Some older circuit breakers have used grading resistors to limit TRV.

Any such measures that change the system response (e.g., items 2, 3, 6, 7, 8) call for a reevaluation of the TRV, with countermeasures in place. Figure 12.23 shows the TRV stress on an example 15 kV, 20 kA circuit breaker interrupting a 14 kA fault (Mc Dermott and Dafis, 2011). With 1 nF total bus capacitance, the TRV exceeds the breaker capability at 100% current interruptions, T100. By increasing the total capacitance to 40 nF, the peak TRV occurs later, and the stress waveform lies completely under the T100 rating. Because the 14 kA fault current is only 70% of the breaker's interrupting capability, the TRV rating may be increased to T70. In that case, the TRV stress waveform lies completely under the T70 rating, even with 1 nF total capacitance.

References

ANSI Std. C37.06-2009, *IEEE Standard for AC High-Voltage Circuit Breakers Rated on a Symmetrical Current Basis–Preferred Ratings and Related Required Capabilities for Voltages above 1000 V*, The Institute of Electrical and Electronic Engineers, New York, 2009.

Colclaser, R. G., The transient recovery voltage application of power circuit breakers, *IEEE Transactions on Power Apparatus and Systems*, 91(5), 1941–1947, May 1972.

Colclaser, R. G. and D. E. Buettner, The traveling-wave approach to transient recovery voltage, *IEEE Transactions on Power Apparatus and Systems*, 88(7), 1028–1035, July 1969.

Greenwood, A. N., *Electrical Transients in Power Systems*, 2nd edn., John Wiley & Sons, New York, 1991.

IEC, Insulation co-ordination–Part 4: Computational guide to insulation co-ordination and modelling of electrical networks, Technical Report TR 60071-4, Edition 1, International Electrotechnical Commission, Geneva, Switzerland, June 2004.

IEC Std. 62271-1, *High-voltage Switchgear and Controlgear—Part 1: Common Specifications*, Edition 1.0, International Electrotechnical Commission, Geneva, Switzerland, October 2007.

IEC Std. 62271-100, *High-voltage Switchgear and Controlgear—Part 100: Alternating-current Circuit-breakers*, Edition 2.0, International Electrotechnical Commission, Geneva, Switzerland, April 2008.

IEEE PES Task Force on Data for Modeling System Transients, Parameter determination for modeling system transients—Part VI: Circuit breakers, *IEEE Transactions on Power Delivery*, 20(3), 2079–2085, July 2005.

IEEE Std. C37.04-1999, *IEEE Standard Rating Structure for AC High-Voltage Circuit Breakers*, The Institute of Electrical and Electronic Engineers, New York, June 1999.

IEEE Std. C37.04-1999/Cor 1-2009, *IEEE Standard for Rating Structure for AC High-Voltage Circuit Breakers Corrigendum 1*, The Institute of Electrical and Electronic Engineers, New York, 2009.

IEEE Std. C37.04a-2003, *IEEE Standard Rating Structure for AC High-Voltage Circuit Breakers Rated on a Symmetrical Current Basis: Amendment 1 Capacitance Current Switching*, The Institute of Electrical and Electronic Engineers, New York, July 2003.

IEEE Std. C37.04b-2008, *IEEE Standard for Rating Structure for AC High-Voltage Circuit Breakers Rated on a Symmetrical Current Basis Amendment 2: To Change the Description of Transient Recovery Voltage for Harmonization with IEC 62271-100*, The Institute of Electrical and Electronic Engineers, New York, 2008.

IEEE Std. C37.09-1999, *IEEE Standard Test Procedure for AC High-Voltage Circuit Breakers Rated on a Symmetrical Current Basis*, The Institute of Electrical and Electronic Engineers, New York, January 2000.

IEEE Std. C37.09-1999/Cor 1-2007, *IEEE Standard Test Procedure for AC High-Voltage Circuit Breakers Rated on a Symmetrical Current Basis–Corrigendum 1*, The Institute of Electrical and Electronic Engineers, New York, September 2007.

IEEE Std. C37.09a-2005, *IEEE Standard Test Procedure for AC High-Voltage Circuit Breakers Rated on a Symmetrical Current Basis: Amendment 1, Capacitance Current Switching*, The Institute of Electrical and Electronic Engineers, New York, March 2005.

IEEE Std. C37.011-2005, *IEEE Application Guide for Transient Recovery Voltage for AC High-Voltage Circuit Breakers*, The Institute of Electrical and Electronic Engineers, New York, September 2005.

13

Surge Arresters

Thomas E.
McDermott
Meltran, Inc.

Surge arresters are shunt, nonlinear, resistive devices that are installed to limit transient overvoltages, thereby protecting equipment insulation. Sometimes they have been called "lightning arresters," and this may be due to the fact that many years ago, surge arresters were tailored to protect against the surge front times typical of lightning. They were much less effective against steeper-fronted overvoltages, or against slow-front switching surges. However, the proper term is "surge arresters," and the modern generation of metal oxide surge arresters is effective against both lightning and switching surges.

The rest of this chapter summarizes metal oxide surge arrester ratings and application principles for system voltage levels above 1 kV. This treatment should suffice for educational purposes. For actual design work, the use of time-domain simulation in an electromagnetic transients program (EMTP) would be highly recommended. The designer should also have access to one of the standard application guides (IEEE C62.22, IEC 60099-5) or the text on insulation coordination by Hileman (1999).

13.1 Arrester Types and Auxiliary Equipment

Nearly 100 years ago, electrode gaps (rod, sphere, or pipe) were used to limit overvoltages on equipment (Sakshaug, 1991). Some of these systems, particularly pipe gaps, may still be in service today. However, the characteristic of gap sparkover voltage vs. surge front time does not match up well with the strength vs. front characteristics of most insulation; that is, it is difficult to coordinate. The next evolutionary step was to add a resistive element in series with the gap, in order to limit the power follow current after an arrester discharge operation. The current limiting would hopefully allow the arrester to clear this power follow current, instead of relying on a nearby breaker or fuse. At the same time, the resistor voltage during a discharge must be low enough that it does not allow an excessive voltage to appear on the protected equipment. These competing requirements led to the use of expensive and complicated nonlinear resistive elements, some involving both solid and liquid materials with high maintenance burdens.

Beginning around 1930, silicon carbide (SiC) was used for the nonlinear resistive elements, leading to much better protective characteristics. Because the SiC would conduct significant current at nominal voltage, it was necessary to provide a sparkover gap that prevents conduction at nominal voltage. After an arrester discharge, these gaps must reseal against the power follow current, otherwise, the arrester would fail thermally. In the mid 1950s, active gaps were developed for SiC arresters. These active gaps contain auxiliary elements that would

1. Preionize the sparkover gap to obtain better surge protective levels
2. Elongate the power follow arc, and move its attachment points, to obtain better interruption performance

SiC arresters were successfully applied on transmission systems up to 345 kV, but some limitations appeared with regard to switching surge protection, energy discharge capability, and pressure relief capability. Having both gaps and SiC blocks, the arrester height increased to the point where it was difficult to vent the pressure buildup during a fault, which limited the arrester's pressure relief rating. Due to their discharge characteristics vs. frequency or front time, the SiC surge arresters were optimized for lightning. They were less effective for steeper-fronted surges and slow-fronted switching surges.

In the mid 1970s, metal oxide surge arresters were developed into commercial products (Sakshaug et al., 1977). The metal oxide blocks are much more nonlinear than silicon carbide, so that they conduct only a few milliamperes at nominal AC voltage. It eventually became possible to dispense with gaps completely, although earlier designs made some use of gaps (see Figure 13.1). Metal oxide surge arresters have several major advantages over the earlier silicon carbide arresters:

1. Active gaps are not necessary, leading to improved reliability.
2. Metal oxide can discharge much more energy per unit volume than silicon carbide.
3. Metal oxide provides better protection across the range of surge wave fronts than silicon carbide, and in fact, protects effectively against switching surges.
4. The decrease in arrester height, caused by eliminating sparkover gaps, leads to higher pressure relief ratings.

Virtually all new applications will use metal oxide surge arresters. Metal oxide has enabled some new applications, like series capacitor protection and overhead line switching surge control that were not possible with silicon carbide. However, many silicon carbide arresters are still in service. Some investigators have noted high silicon carbide arrester failure rates, due to moisture ingress, after several years of service on medium-voltage distribution systems. This experience does not necessarily apply to surge arresters in substations. If such problems arise, it would make sense to systematically replace silicon carbide arresters on a system. Otherwise, assuming that the original application was proper, the older silicon carbide arresters could remain in service.

Figure 13.1 shows the general use of gaps in surge arresters. The gapped design (Figure 13.1a) applies to silicon carbide, whereas the gapless design (Figure 13.1d) applies to the latest generation of metal oxide. One manufacturer used the shunt gap (Figure 13.1b) in early metal oxide arresters. At steady state, both

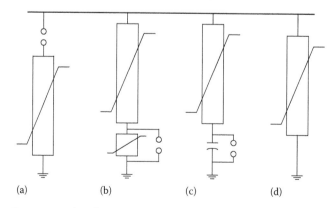

(a)   (b)   (c)   (d)

FIGURE 13.1   Use of gaps in metal oxide surge arresters (a) gapped, (b) shunt gap, (c) graded gap, and (d) gapless.

nonlinear elements would support the nominal voltage, somewhat reducing the current. During a surge discharge, the shunt gap would sparkover to bypass the smaller section of metal oxide, thereby reducing the discharge voltage and providing somewhat better protection. Another manufacturer used the series gap with capacitive grading (Figure 13.1c) in early metal oxide arresters. At steady state, this decreases the voltage on the metal oxide. During a surge discharge, the gap sparks over "immediately" due to the capacitive grading. The latest generations of metal oxide do not need these gaps, although there is some consideration for using gaps to achieve specific goals (e.g., coordinating arresters, withstanding temporary overvoltages).

Recently, surge arresters have been applied to transmission lines for lightning protection of the line insulation. Two varieties have been developed, first a nongapped line arrester (NGLA) as in Figure 13.1d, and then an externally gapped line arrester (EGLA) as in Figure 13.1a.

The surge arrester must be installed on something, such as a transformer tank or a pedestal. It must also be connected to the protected system, typically through a wire or lead. Later, it will be shown that these connections have important effects on the overall protection, especially for steep surges. The pedestal and lead, both length and location, must be considered as part of the overall arrester installation.

On distribution systems, a ground lead disconnector is often used with the surge arrester. If the arrester fails and then conducts current on a steady-state basis, the disconnector will detonate and disconnect the base of the arrester from ground. This should happen in approximately 1 s, or faster. The arrester may then remain connected to the system until maintenance personnel have a chance to replace it. No breaker or fuse need operate to isolate the failed arrester; if the arrester is the only thing that failed, no customers need to lose their electric service. Of course, the arrester is not providing any surge protection during this period with its ground lead disconnected. There would be a clear visual indication that the ground lead has been disconnected; it will be "hanging down" below the arrester. Regular visual inspections are necessary to maintain surge protection whenever ground-lead disconnectors are used.

Many surge arresters in substations or industrial facilities have been installed with "surge counters." These are accessories to be installed in the surge arrester's ground-lead connection. Two functions may be provided:

1. A steady-state current meter, calibrated in mA. If this current increases over time, it may indicate thermal damage to the surge arrester. However, the presence of harmonics or external leakage currents would complicate the assessment.
2. A counter indicates the number of surge current discharges above a certain threshold, which may depend on frequency or front time. Even if the count is accurate, it does not mean that the discharge voltage reached any particular level during those events.

To use surge counters effectively, it is important to track the readings on a regular basis, beginning at the time of commissioning.

13.2 Ratings and Tests

Surge arresters have both a voltage rating and a class, or type. For metal oxide, the important voltage rating is the maximum continuous operating voltage (MCOV), which is the steady-state voltage the arrester could support indefinitely. This rating is most important for metal oxide, because most of these arresters are now gapless and carry a few milliamperes at all times. The MCOV should be at least 1.05 times the system's nominal line-to-ground voltage. There are some cases like distribution feeders with poor voltage regulation, which might require a higher MCOV. As discussed later, short-term temporary overvoltages (TOV) also play an important role in selecting the arrester rating, but still the basic rating is MCOV.

TABLE 13.1 Schedule of IEC
Voltage Rating Steps

Range of Rated Voltage (kV rms)	Step Size (kV rms)
3–30	1
30–54	3
54–96	6
96–288	12
288–396	18
396–756	24

Historically, the SiC arresters had a voltage rating that corresponded to the duty cycle test, with no direct link to MCOV. Given that SiC arresters were always gapped, MCOV was not an important concern. In IEEE standards, the old SiC numerical schedule of voltage ratings has been carried over to metal oxide arresters, and the MCOV is usually about 84% of the arrester voltage rating. For example, a typical 108 kV arrester has an MCOV of 84 kV. Only the MCOV number is important.

The IEC standard defines the voltage rating to be the TOV capability at 10 s, and again the MCOV is about 84% of this value. Table 13.1 shows the schedule of IEC voltage rating steps, and in general these line up with the schedule of IEEE voltage ratings. This table could be used to choose the preliminary arrester rating for some application. However, all surge arrester manufacturers now have their catalog information readily available on the web, and it is better to use their actual data.

The IEEE specifies a set of withstand tests for surge arresters, each of them followed by a period of operation at MCOV. These tests are as follows:

1. High current, short duration: Two $4 \times 10\,\mu s$ current surges are discharged through the arrester. The peak current is 65 kA for station, intermediate, and distribution normal-duty arresters. It is 100 kA for distribution heavy-duty and 40 kA for distribution light-duty arresters. This is followed by 30 min at MCOV.
2. Low current, long duration: For station and intermediate arresters, a transmission pi section is discharged 20 times through the arrester. The switching surge voltage ranges from 2.0 to 2.6 pu and the equivalent line length ranges from 160 to 320 km, depending on the arrester class and voltage rating. For distribution arresters, a 2000 ms square wave of current is discharged 20 times, with a peak current of 250 A for heavy-duty and 75 A for normal-duty or light-duty arresters, followed by 30 min at MCOV.
3. Duty cycle test: While energized at the voltage rating, which is higher than MCOV, discharge an $8 \times 20\,\mu s$ lightning impulse 20 times, with peak ranging from 5 to 20 kA, followed by 2 discharges without power frequency voltage, followed by 30 min at MCOV.
4. Pressure relief test: An AC current of 40–80 kA rms (root mean square) is applied for station class, or 16.1 kA for intermediate class, so that the arrester fails. The pressure must vent well enough that any components lie no farther away than the arrester height. Polymer housings tend to have better pressure relief ratings than porcelain housings.
5. Contamination test: Energized for 1 h at MCOV under contaminated conditions, followed by 30 min at MCOV.
6. Temporary overvoltages: Verifies the TOV capability over a range from 0.02 to 1000 s.
7. Switching surge energy: This is an optional supplement to the low-current, long-duration test, which carries an implied energy discharge duty. The manufacturer typically provides a value for energy capability in kJ per kV of MCOV, which is meant to cover multiple discharges within a period of about 1 min.

The test results for TOV and switching surge energy capability are often used in application studies.

The manufacturer also conducts tests of the arrester protective characteristics. These are always directly useful in modeling and application studies:

1. 8 × 20 μs discharge voltage: Current surges are applied with peak magnitudes of 1.5, 3, 5, 10, and 20 kA, with the resulting peak discharge voltages tabulated. Due to various frequency-dependent effects, the peak voltage occurs before the peak current. For some ratings, the 15 and 40 kA discharge voltages are also tabulated.
2. Front-of-wave (FOW) protective level: This is the voltage having a 0.5 μs time to crest, for a current surge of 5, 10, 15, or 20 kA depending on the rating. To get this value, it is necessary to apply different front times for the current surge and interpolate or extrapolate the results.
3. Switching surge protective level: The peak arrester voltage for a current surge having 45–60 μs front time and a peak of 500–2000 A depending on the rating.

Many vendors provide more detailed information about protective levels as a function of the surge front time, which is helpful in modeling.

Figure 13.2 shows typical maximum 8 × 20 μs discharge voltage and FOW discharge characteristics for station class metal oxide surge arresters, rated 54 kV and above. There is a manufacturing variation within each product line, and the minimum discharge voltage could be a few percent lower than these values. In most cases, the maximum values would be used for application studies. Figure 13.3 shows

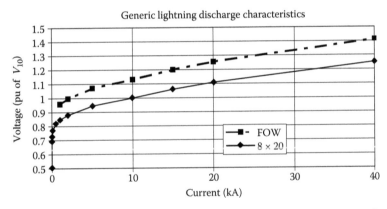

FIGURE 13.2 Metal oxide surge arrester lightning discharge characteristics, for voltage ratings from 54 to 360 kV. V_{10} is the arrester's 8 × 20 μs discharge voltage at 10 kA.

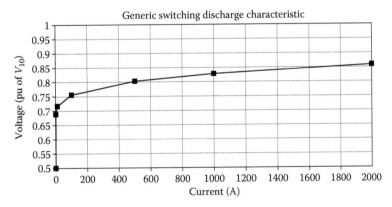

FIGURE 13.3 Metal oxide surge arrester switching discharge characteristic, for voltage ratings from 54 to 360 kV. V_{10} is the arrester's 8 × 20 μs discharge voltage at 10 kA.

a typical maximum switching surge discharge characteristic. All of these are normalized to the published $8 \times 20\,\mu s$ discharge voltage at 10 kA.

The IEEE standards define several classes of surge arresters, based on the withstand test levels discussed earlier. In substations and many transmission line applications, the station class arresters are most common. The intermediate class arresters are used for lightning protection of transmission line towers and for some of the smaller substations. The choice of distribution arrester class depends mainly on how severe the local lightning environment is. Although purchased in significant quantities, the surge arresters are typically not a significant cost item in a single substation.

Among the IEEE standards, C62.11 defines the metal oxide arrester tests and ratings, whereas C62.22 provides the application guidance. The older standards for silicon carbide arresters, C62.1 and C62.2, are still available. These apply to the United States, whereas, in most other parts of the world, the International Electrotechnical Committee (IEC) standards are used. These are denoted IEC 60099 parts 1 through 8, except that there is no part 2 and part 7 is a glossary. For gapless metal oxide arresters, part 4 defines the tests and ratings, while part 5 presents the application guidelines. For gapped arresters, including silicon carbide, part 1 applies. The IEEE standards may be found through www.ieee.org, and the IEC standards through www.iec.ch.

Both IEEE and IEC standards define the discharge characteristics in similar ways. The same applies to MCOV and TOV capabilities. In IEC standards, the rated voltage is actually the TOV capability at 10 s and also the duty cycle test voltage. These differences generally will not concern the user. The main difference is that IEC standards classify arresters by nominal discharge current (1.5, 2.5, 5, 10, or 20 kA) and line discharge class (1–5, or none). An IEEE station class arrester is roughly comparable to an IEC 10 kA arrester, whereas the intermediate and distribution class arresters are roughly comparable to IEC 5 kA arresters. At 500 kV and above, the IEEE classifying current is 15 or 20 kA, roughly corresponding to the IEC nominal discharge current of 20 kA used at these voltage levels. The appropriate standard should be consulted for details, but Osterhout (1992) and Hamel and St. Jean (1992) provide overviews of the differences.

Low-voltage surge arrester applications follow a different set of standards and guides, for system voltage levels below 1 kV, which includes secondary services and customer utilization circuits. The applicable IEEE standards are C62.72 and C62.41 for application guidance, plus others in the C62 family for test requirements. The applicable IEC standards for low-voltage power systems are 61643, parts 1 and 12.

13.3 Selection by TOV

As discussed earlier, the first choice of arrester voltage rating is based on MCOV. That would be the lowest and best choice because it minimizes the arrester discharge voltage for any particular surge, thereby maximizing protective margins. However, it may be necessary to increase the arrester voltage rating to withstand either TOV or surge energy content.

The TOVs come from several sources, most commonly

1. During ground faults, voltage on the unfaulted phases will rise above nominal unless the system is solidly grounded. The duration of this overvoltage depends on protective relaying system response, usually less than 1 s, but up to several hours for delta systems or ungrounded systems.
2. When first energizing a long line or cable, the voltage at the open end will rise above nominal due to the Ferranti effect. The duration depends on how long it takes to switch on shunt compensation or to close the other end of the line. A similar effect occurs after load rejection when only one end of the line opens.
3. Harmonics, ferroresonance, and transformer inrush currents can produce dynamic overvoltages that last for several cycles.

Usually the ground fault TOVs are most important, but there are special cases where simulation of other TOVs should be considered. Given the zero sequence and positive sequence impedances at the point of

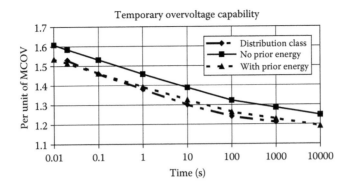

FIGURE 13.4 Temporary overvoltage capability.

a fault, the per-unit unfaulted phase voltage, often called the earth-fault factor (EFF) may be estimated for a single-line-to-ground fault (IEEE C62.22)

$$\mathrm{EFF} = -\frac{\sqrt{3}}{2}\left[\frac{\sqrt{3}K}{2+K} \pm j1\right]$$

$$K = \frac{Z_0 + R_f}{Z_1 + R_f} \tag{13.1}$$

$$Z_0 = R_0 + jX_0$$

$$Z_1 = R_1 + jX_1$$

Note that there are two unfaulted phases to evaluate, corresponding to the $+j1$ and $-j1$ terms. That voltage should be further increased by a factor of 1.05 to account for above-nominal operating voltage. In most cases, the fault resistance, R_f, should be considered zero as it maximizes EFF. For distribution systems, R_f should be equal to the design level of fault resistance that relaying will detect; this is really a matter of policy. For double-line-to-ground faults, the per-unit voltage on the unfaulted phase would be (IEEE C62.22)

$$\mathrm{EFF} = \frac{3K}{2K+1}$$

$$K = \frac{Z_0 + 2R_f}{Z_1 + 2R_f} \tag{13.2}$$

Again, this voltage should be further increased by a factor of 1.05 to account for above-nominal prefault voltage.

The TOV duties from various sources must be compared to the arrester's TOV capability, which may be obtained from the manufacturer's catalog, or taken from Figure 13.4 to use typical data. Some data include TOV capability both with and without earlier surge discharges; the capability with prior discharge, which is always lower, should be used. If the TOV duty exceeds the arrester capability, then a higher voltage rating must be chosen for the arrester, accepting lower protective margins. It is difficult to mitigate TOVs as they typically arise from high-level system design choices.

13.4 Selection by Energy Rating

Surge arresters have a switching surge energy discharge capability higher than implied by the low-current, long-duration discharge test. The vendor may provide this value as kJ per kV of MCOV, or rated voltage with footnotes that specify the number, shape, and interval between surges. For example, if the capability is given as 8 kJ per kV of MCOV for a single discharge, then a 108 kV arrester with an MCOV of 84 kV has a switching surge discharge capability of 672 kJ.

The energy capability may be increased by using two or more columns of metal oxide in parallel. The manufacturer has to match the disks to achieve this and typically two parallel columns provide a bit less than double the energy rating. Note that increasing the arrester voltage rating will not help much, because even though the energy rating increases, the discharge voltage also increases resulting in higher surge energy discharge. Parallel columns will not help with TOV, as TOV is a sustained phenomena supported by the system voltage.

There is some evidence that these energy ratings are actually the point at which failure probability becomes nonzero. Hileman (1999) estimated the probability of failure from several reported tests, assuming a Weibull function to model the failure probability

$$P_F = 1 - 0.5^{((z/4)+1)^5}$$

$$Z = \frac{W_C/W_R - 2.5}{0.375}$$

(13.3)

where
　W_R is the energy rating
　W_C is the calculated energy duty

Under this model, P_F reaches 50% only at 2.5 times the energy rating. Note that standards and vendors do not address this point yet, but consideration of the actual failure probability may avoid overdesign for switching surge energy discharge.

A conservative estimate of the switching surge energy discharge, W_C, is

$$W_C = \frac{1}{2} CE^2$$

(13.4)

For energy in joules, C is the total capacitance in μF and E is the peak voltage in kV. The peak voltage ranges from 2 to 2.5 pu of the peak line-to-neutral voltage, although it may be higher for cases like capacitor bank restrikes and line reclosing. The total capacitance may come from an overhead line, a cable, or a shunt capacitor bank. Better estimates may be obtained through time-domain simulation.

Lightning surge discharge also poses a significant energy duty on surge arresters. However, the switching surge energy rating does not apply directly and the duty might be better described as a charge duty. In either case, the lightning discharge current may be represented as a decaying exponential with total area equal to the stroke charge. Details of the wave front are unimportant for this evaluation. If one arrester discharges all of the lightning stroke current, a conservative estimate of the energy duty would be

$$W_C = QE_d$$

(13.5)

where
　Q is the total stroke charge in coulombs
　E_d is the arrester discharge voltage at the peak lightning stroke current

The actual energy discharge is reduced by nonlinear arrester characteristics and by sharing from nearby arresters. Even though nearby arresters will not have matched characteristics, they still share a significant portion of the energy because of inductances between the arresters and also because the characteristics are more linear at lightning discharge current levels than at switching discharge current levels. Time-domain simulation helps to quantify these effects.

13.5 Arrester Modeling

Surge arresters are both nonlinear and frequency-dependent devices. The only strictly correct model would be a time-domain simulation at the level of material physics. In most practical cases, one must use a nonlinear resistance with a separate linear circuit to represent frequency-dependent effects. Surge arresters also interact with other equipment that has distributed parameters. Usually the other system nonlinearities are not important, but in special cases some of these effects (e.g., corona) may become significant. In general, arrester application studies need to be done with time-domain simulation in the EMTP or one of its variants like the Alternative Transients Program (ATP).

In the absence of EMTP or ATP, the arrester discharge voltage may be approximated for a given surge current peak and waveshape. First, choose the FOW, $8 \times 20\,\mu s$, or switching surge characteristic that best matches the waveshape. If the manufacturer's data are not available, Figures 13.2 and 13.3 may be used with a normalizing value (10 kA discharge) of 2.3 times MCOV. In cases where the discharge current peak is "known," interpolate with this value to estimate the peak discharge voltage.

More often, though, a voltage surge arrives at the arrester location and the discharge current peak is not known in advance. In the simplest case, the arrester is at the end of a line with known surge impedance, Z, incoming surge magnitude, E, and power frequency offset, V_{pf}. The arrester current and voltage may then be estimated according to Hileman (1999)

$$I_A = \frac{2E - E_0 - V_{pf}}{Z + R_A}$$

$$E_d = E_0 + I_A R_A \tag{13.6}$$

$$E_A = E_d + V_{pf}$$

This approximates the arrester discharge characteristic, E_d, with a straight line having slope R_A and intercept E_0, as determined by interpolation on either the manufacturer's data or Figures 13.2 and 13.3. The actual arrester voltage, E_A, is the discharge voltage, E_d, plus the power frequency offset voltage, V_{pf}. To use Equation 13.6, make the first guess at I_A and determine the corresponding E_0 and R_A parameters from the discharge characteristic. Then solve for I_A using the first equation; if it falls outside the linear segment used to estimate R_A and E_0, iterate until convergence on I_A. Then E_d and E_A may be determined directly. Hileman (1999) presents more detailed estimating methods for other situations, but in those cases it may be easier to use EMTP or ATP.

Figure 13.5 shows three frequency-dependent surge arrester models. The Cigre model (Hileman et al., 1990) works well, but it requires the user to program a time-dependent conductance to simulate the arrester turn-on characteristics. It is not directly applicable to EMTP or ATP. The IEEE model (IEEE PES task force) uses high-frequency and low-frequency discharge characteristics represented by A_0 and A_1, respectively, with a two-stage RL filter connecting them. At nominal voltage, the arrester current is primarily capacitive, represented by the shunt C component. Both IEEE and Cigre models require

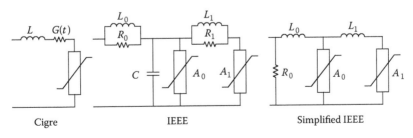

FIGURE 13.5 Frequency-dependent surge arrester models.

iteration of the parameters to represent a particular arrester, before it can be used in application studies. A simplified version of the IEEE model (Magro et al., 2004) has been developed for use "out of the box"; it ignores the capacitive effect.

In the Cigre model, the turn-on conductance begins at zero and increases with time. A reasonable approximation for distribution arresters is

$$\frac{dG}{dt} = \frac{G_{ref}}{T}\left(1 + \frac{G}{G_{ref}}\right)\left(1 + \frac{G}{G_{ref}}\left(\frac{I}{I_{ref}}\right)^2\right)\exp\left(\frac{U}{U_{ref}}\right)$$

$$G_0 = 0$$

$$T = 80 \tag{13.7}$$

$$G_{ref} = \frac{34}{U_{10}}$$

$$I_{ref} = 5.4$$

$$U_{ref} = kU_{10}$$

where
U_{10} is the 10 kA discharge voltage in kV
U is the arrester voltage in kV
I is the arrester current in kA
k is a constant ranging between 0.03 and 0.05, depending on the manufacturer

The series inductance, L, is approximately 1 µH per meter of arrester height for outdoor arresters, and one-third of that for gas-insulated substation (GIS) arresters.

The following are starting values for the IEEE model parameters:

$$L_1 = 15\frac{d}{n}$$

$$R_1 = 65\frac{d}{n}$$

$$L_0 = 0.2\frac{d}{n} \tag{13.8}$$

$$R_0 = 100\frac{d}{n}$$

$$C = 100\frac{n}{d}$$

where
d is the arrester height in meters
n is the number of metal oxide columns in parallel
L is in µH
R is in Ω
C is in pF

Parameters for the simplified model are

$$R_0 = 1e6$$

$$L_0 = \frac{1}{12} \frac{V_{FOW} - V_{10}}{V_{10}} V_n \tag{13.9}$$

$$L_1 = \frac{1}{4} \frac{V_{FOW} - V_{10}}{V_{10}} V_n$$

The inductance values L_0 and L_1 are in μH, V_{FOW} is the FOW discharge voltage in kV, V_{10} is the $8 \times 20\,\mu s$ discharge voltage in kV, and V_n is the arrester voltage rating (not the MCOV), also in kV. R_0 is present for numerical stability and may not be required for some time-domain simulators. R_0 may also be increased by one or two orders of magnitude for extra high voltage (EHV) levels. Whenever V_{FOW} is not available or when the ratio V_{FOW}/V_{10} is more than 1.18, the simplified inductance parameters become

$$L_0 = 0.01V_n$$
$$\tag{13.10}$$
$$L_1 = 0.03V_n$$

In both IEEE models, A_1 closely follows the arrester discharge voltage characteristic for $8 \times 20\,\mu s$ discharge currents. A_0 is 20%–30% above A_1, such that in parallel and considering the RL filter at high frequency, they approximate the arrester's FOW protective level (FOWPL). At low frequency, the filter inductances have little impact and A_1 dominates the model characteristic.

All of the frequency-dependent models include built-in arrester inductance. The arrester lead and pedestal should also be modeled, along with bus, line, or cable lengths to the nearby protected equipment. The lead and pedestal may be represented either as lumped inductance, approximately $1\,\mu H$ per meter, or with traveling wave lines having surge impedance of about $300\,\Omega$.

The separation effect may be approximated for an open-end termination, at which traveling wave reflections will cause doubling before the arrester's limiting effect is seen. The peak voltage at the open end is

$$E_T = E_d + 2S\tau \tag{13.11}$$

where
E_d is the arrester discharge voltage in kV
S is the surge steepness in kV/μs
τ is the travel time in μs to the open line end

For overhead lines or buswork, τ is approximately the distance in meters, divided by 300. For example, if the arrester discharge voltage is 297 kV, the surge steepness is 1000 kV/μs and the distance to the protected transformer is 10 m, then the peak transformer voltage, E_T, is approximately 364 kV. That is an increase of 22% from 297 kV, illustrating the importance of minimizing lead lengths. Note that this estimate ignores the effect of transformer capacitance and power frequency offset, both of which would further increase E_T.

In many cases, the ground resistance does not have to be included. The main exceptions occur when arresters are applied outside of the substation, such as riser poles, distribution transformers, and transmission line towers. In the riser pole application, a ground connection affects the differential-mode surge entering the cable. On distribution lines, the multigrounded neutral conductor allows current to circulate back through nearby arresters. The same thing happens with transmission line arresters installed with overhead shield wires. In all these cases, the ground resistances should be modeled, along with any conductors that are grounded periodically.

If it is necessary to model an older SiC arrester, the $8 \times 20\,\mu s$ discharge characteristic is represented very well by a single exponential power law

$$I_d = kV_d^\alpha \tag{13.12}$$

This must be connected in series with a gap, having a sparkover voltage ranging anywhere from 5% to 20% higher than V_{10}. The exponential power, α, ranges from two to six, with higher values being more typical for SiC surge arresters. If the actual data are no longer available, it may be assumed that the 10 kA discharge voltage is approximately equal to the 10 kA, $8 \times 20\,\mu s$ discharge voltage of a metal oxide arrester having the same voltage rating (Sakshaug, 1991). On that assumption, the value of k becomes, for units of kA and kV,

$$k = \frac{10}{V_{10}^\alpha} \tag{13.13}$$

Typical SiC data may also be found in Greenwood (1991). Both the FOW and switching surge protective levels for SiC arresters would be significantly higher than for metal oxide arresters of the same voltage rating, even when the V_{10} values are approximately equal. Conversely, the α parameter in metal oxide varies with current, so that a single exponential does not provide a good model. Piecewise exponential segments are often used for metal oxide surge arrester models in EMTP and ATP.

13.6 Applications

Perhaps the most important surge arrester application is to protect the equipment in a substation. After selecting the lowest possible arrester voltage rating, one typically applies an arrester as close as possible to each transformer terminal. Then given the transformer BIL, a protective margin is calculated using the $8 \times 20\,\mu s$ discharge voltage at the "coordination current," which is typically 10 kA:

$$M_{pct} = 100\left(\frac{BIL}{V_{10}} - 1\right) \tag{13.14}$$

where
   M_{pct} is the protective margin, in percent
   BIL is the transformer basic insulation level, in kV
   V_{10} is the arrester's 10 kA discharge voltage, in kV

The margin could also be calculated at a different coordination current, such as 5 or 20 kA, in which case V_5 or V_{20} would be used instead of V_{10}. One could also calculate a chopped wave protective margin, using the chopped wave insulation level and FOWPL in place of BIL and V_{10}. Finally, one could calculate a margin for switching surges using the BSL and the arrester's switching surge protective level. In all cases the principle applied is the same and 10%–20% should be considered the minimum acceptable margin. These margins are usually easy to obtain. For example, on a 138 kV system, the transformer BIL is at least 450 kV. If the system is effectively grounded, a 108 kV arrester might be used with a 10 kA discharge voltage of 263 kV, providing a margin of 71%. Using a 120 kV arrester with 297 kV discharge voltage, the margin is still 52%.

As discussed in Chapter 14, many uncertainties and deviations complicate the use of margins. After a long time in service, the insulation strength may not really be at the original BIL and BSL levels. The actual surges also differ from laboratory test waves, so that the BIL and BSL do not describe the insulation strength perfectly. Furthermore, several important electrical circuit phenomena are not considered in the simple use of margins and Figure 13.6 illustrates some of them.

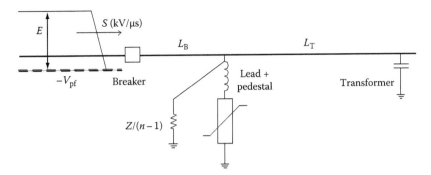

FIGURE 13.6 Station protection by surge arresters.

The surge arrester's lead and pedestal connections should be added to the basic arrester model, either as lumped inductances or as very short distributed-parameter lines. As a result, the voltage to ground at the point of arrester connection will be somewhat higher than the pure arrester discharge voltage. The distances to protected equipment, L_B and L_T in Figure 13.6, should also be represented with distributed-parameter lines. The arrester itself should be represented with a frequency-dependent model, as discussed earlier. At these high frequencies, the transformer should be represented as a capacitance ranging from 1 to 10 nF, with some sensitivity analysis of this parameter. The incoming surge itself is primarily described with a steepness, S, ranging from 1000 to 2000 kV/µs, depending on the transmission line characteristics, lightning environment, and design level of mean time between failures. The surge magnitude, E, is typically based on some percentage of the line insulation critical flashover (CFO) voltage. It arrives on top of an opposite-polarity power frequency offset, so the arriving surge voltage to ground is actually $E - V_{pf}$. All of these factors tend to increase the peak voltage appearing at the transformer, which decreases the protective margin. Note: when using EMTP, the margin is calculated from the simulated peak voltage at the transformer, instead of V_{10}.

If two or more lines enter the station, which is typically the case, the additional lines tend to increase protective margins. As an approximation, these lines may be represented with the line surge impedances to ground, connected in parallel at the point of arrester connection. This is shown in Figure 13.6, where n is the total number of lines entering the station. A better representation would be to represent the bus sections with distributed-parameter lines, and connect the line entrances at points where they actually enter the station.

A further refinement is to represent the surge with a Thevenin equivalent. The simplest such equivalent is a surge voltage of $2E$ on top of a DC voltage equal to $-V_{pf}$, both behind a lumped resistance equal to the line surge impedance, Z. The next refinement would be to represent the line with distributed parameters back to the stroke point, with a tower footing resistance value in place of Z. Even with these refinements, a single-phase model is adequate for the simulation.

Sometimes, the switching surge energy discharge duty on a station arrester will exceed its rating. Typically this happens when large shunt capacitor banks, or long cables, are nearby. The arrester can be ordered with parallel columns, which are matched at the time of manufacturing to achieve effective sharing of the energy duty. If not matched, a small difference in the discharge characteristic could result in one column discharging virtually all of the energy. This would happen even with typical small manufacturing variations of 3% or less.

Once the model in Figure 13.6 has been set up, the arrester rating and location will be finalized to adequately protect each transformer. Protective margins for other equipment will also be available, such as the circuit breaker in Figure 13.6. For many stations, arresters at the transformer terminals will adequately protect all equipment in the station. If not, arresters can be added to protect equipment that is far away from the transformers. This process applies to each nominal voltage level in the substation.

As a further constraint, the arresters on two different voltage levels should be coordinated so that the high-side arrester always operates first for a surge impinging on the high side of the transformer. If the

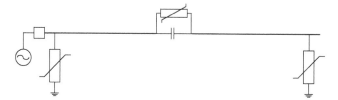

FIGURE 13.7   Transmission line protection by surge arresters.

low-side arresters operate first, they might discharge all of the energy in the high-side surge, leading to possible failure. This coordination is achieved by ensuring 4% margin between the transformer turns ratio and the ratio of arrester protective levels. For example, a 345/138 transformer has a turns ratio of 2.5. The switching surge discharge voltage of the high-side arrester should be no more than 2.5/1.04, or 2.4 times the switching surge discharge voltage of the low-side arrester.

Figure 13.7 shows how surge arresters may be used to protect overhead transmission lines. Arresters on the line side of a circuit breaker, shown to the left in Figure 13.7, can protect the breaker from lightning surges traveling in from the line, while the breaker is open. If the breaker is open for maintenance, a disconnect switch would be opened to isolate it from the line. The need for protection would arise when the breaker opens to clear a lightning-induced fault during a storm, and then a subsequent lightning stroke to the line causes another surge while the breaker is still open. As this condition occurs rarely, some utilities may not use arresters to provide this protection.

On long EHV transmission lines, surge arresters may be applied at each end to mitigate switching surges during line energization and reclosing. A statistical switching study using either EMTP or a transient network analyzer (TNA) is required for this application. The overvoltages will be higher at intermediate points along the line, due to separation effects. The study results should include a profile of probabilistic switching overvoltage parameters vs. line length, to be used with a probabilistic evaluation of the switching surge flashover rate. Commonly used alternatives to these line-end arresters include preinsertion resistors in the circuit breakers or controlled-closing circuit breakers. While often more expensive and more complicated, these alternatives typically produce a flatter overvoltage profile along the line length.

Some EHV transmission lines use series capacitors to reduce their electrical length, and during faults on the line, very high voltages would result on the capacitors. Metal oxide surge arresters typically protect these capacitor banks, as shown in Figure 13.7. The energy dissipation requirements are severe and the manufacturer takes extra care to ensure that the parallel metal oxide columns are matched so that each column shares the energy and current duty. The blocks may also be larger than in typical station-class surge arresters. Although the transient phenomena is at power frequency and subsynchronous frequencies, these energy calculations are normally done using EMTP.

Recently, arresters have been applied to transmission line towers to provide lightning protection of the line insulation. These may be used in place of, or in addition to, overhead shield wires. Some manufacturers have special-purpose line arresters, which are comparable to intermediate or distribution heavy-duty class. Arresters may be used in addition to a shield wire in areas where low footing resistance is difficult to achieve or in high-exposure areas like river crossings. In that case, the arrester energy duty is lessened because the shield wire and footing discharge most of the lightning stroke current. Without a shield wire, the energy duties will increase but are still mitigated by sharing from the arresters on nearby towers.

Arresters on transmission towers can have high protective levels, well above the anticipated switching surge and TOV levels, but still low enough to protect the line insulation during lightning stroke discharges. As a result, these line arresters have a smaller block diameter and lower energy discharge capability than station class arresters. The protective level of an EGLA is approximately equal to the gap sparkover voltage, or the 1 kA discharge voltage of the metal oxide. However, success or failure of this application depends more on the ability to withstand vibration and other mechanical stresses, rather than electrical performance.

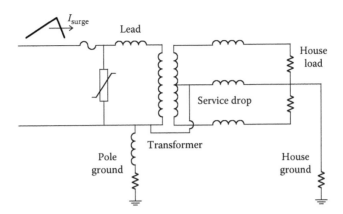

FIGURE 13.8 Distribution transformer protection by surge arresters.

Surge arresters protect virtually all distribution transformers connected to overhead lines, as illustrated in Figure 13.8. It is essential to keep the lead length short. The historical rules of thumb like "2 kV per foot" are grossly understated, because they were based on laboratory test waveshapes that do not represent natural lightning surges. If the arrester is installed between a small fuse and the transformer, then the arrester discharge current must also flow through the fuse, which may cause the fuse to melt unnecessarily. If this becomes a problem, larger fuses or completely self-protected transformers (CSPs), which have internal fuses, could help mitigate it.

Because the arrester is connected between the transformer tank and terminal, pole ground resistance and pole download inductance do not have a direct impact on the transformer's protective margin. However, if the pole ground resistance is too high with respect to the house or customer ground resistance, higher surge currents will circulate in the secondary service drop and the customer load equipment. This may lead to failures due to surges in the transformer secondary, or damage to customer equipment, even when the transformer primary enjoys a comfortable protective margin. It has been suggested that transformers with interlaced secondary windings are less susceptible to these failures. In any case, the secondary surges should be limited by using triplex service drops rather than open-wire drops, and also by keeping the utility pole ground resistance to a reasonably low value.

Given the dispersed installation of transformers on a distribution feeder, surge arresters will be installed at relatively close spacings on the distribution line. Typical average spacings might be from two to four poles in built-up areas, although not necessarily on all phases. These arresters may be used to provide some protection from lightning flashovers of the line insulation. Because of the low insulation levels, typically 100–300 kV CFO, this protection will generally be more effective for induced voltages from nearby lightning strokes, which do not actually hit the line. For direct strokes to the line, energy duty on the arresters may be too high, although there is little field experience to indicate that arresters are failing at high rates due to lightning. If the arresters are not placed at every pole and on every phase, separation effects will be more significant than in a typical substation. Some trials have been done with arresters on just the topmost phase, reasoning that the arrester will convert that phase conductor into a shield wire during a lightning discharge. However, the topmost phase is typically not high enough to serve as an effective shield wire for the two outside phase conductors. Many distribution engineers feel that a direct stroke to the line will result in either a line insulation flashover, or less often, an arrester failure.

Figure 13.9 shows the application of surge arresters at the riser pole, where an underground cable segment connects to an overhead primary distribution feeder. The riser pole arrester limits a surge entering the cable, but surge doubling may occur at the cable end and tap points. If the lightning surge is "bipolar," the surge may even quadruple at the open cable ends (Barker, 1990). If the nominal voltage and cable lengths are low enough and setting aside the small risk of voltage quadrupling, the riser pole arrester may be sufficient. For longer cables, it may be necessary to add arresters at some of the open cable ends

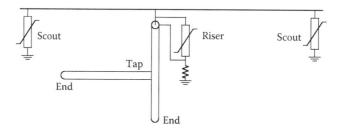

FIGURE 13.9 Riser pole and scout arresters on a distribution system.

or tap points. Before taking that step, the effect of "scout arresters," shown at either end of the overhead feeder section in Figure 13.9, should be included in the study. These arresters will mitigate most surges arriving at the riser pole.

References

Barker, P.P., Voltage quadrupling on a UD cable, *IEEE Trans. Power Delivery*, 5(1), 498–501, January 1990.

Greenwood, A.N., *Electrical Transients in Power Systems,* chap. 16, Protection of systems and equipment against transient overvoltages. 2nd edn., John Wiley & Sons, New York, 1991, pp. 533–534.

Hamel, A. and St. Jean, G., Comparison of ANSI, IEC, and CSA standards' durability requirements on station-type metal oxide surge arresters for EHV power systems, *IEEE Trans. Power Delivery*, 7(3), 1283–1298, July 1992.

Hileman, A.R., *Insulation Coordination for Power Systems*, Marcel Dekker, Inc., New York, 1999, pp. 497–675.

Hileman, A.R., Roguin, J., and Weck, K.H., Metal oxide surge arresters in AC systems—Part v: Protection performance of metal oxide surge arresters, *Electra*, 133, 133–144, December 1990.

IEC Standard 60099-1, Edition 3.1, *Surge Arresters—Part 1: Nonlinear Resistor Type Gapped Surge Arresters for A.C. Systems*, International Electrotechnical Commission, December 1999.

IEC Standard 60099-3, Edition 1.0, *Surge Arresters—Part 3: Artificial Pollution Testing of Surge Arresters*, International Electrotechnical Commission, September 1990.

IEC Standard 60099-4, Edition 2.2, *Surge Arresters—Part 4: Metal Oxide Surge Arresters without Gaps for AC Systems*, International Electrotechnical Commission, May 2009.

IEC Standard 60099-5, Edition 1.1, *Surge Arresters—Part 5: Selection and Application Recommendations*, International Electrotechnical Commission, March 2000.

IEC Standard 60099-6, Edition 1.0, *Surge Arresters—Part 6: Surge Arresters Containing Both Series and Parallel Gapped Structures—Rated 52 kV and Less*, International Electrotechnical Commission, August 2002.

IEC Standard 60099-7, Edition 1.0, *Surge Arresters—Part 7: Glossary of Terms and Definitions*, International Electrotechnical Commission, April 2004.

IEC Standard 60099-8, Edition 1.0, *Surge Arresters—Part 8: Metal-Oxide Surge Arresters with External Series Gap (EGLA) for Overhead Transmission and Distribution Lines of A.C. Systems above 1 kV*, International Electrotechnical Commission, January 2011.

IEC Standard 61643-1, Edition 2.0, *Low-Voltage Surge Protective Devices—Part 1: Surge Protective Devices Connected to Low-Voltage Power Distribution Systems—Requirements and Tests*, International Electrotechnical Commission, May 2003.

IEC Standard 61643-12, Edition 2.0, *Low-Voltage Surge Protective Devices—Part 12: Surge Protective Devices Connected to Low-Voltage Power Distribution Systems—Selection and Application Principles*, International Electrotechnical Commission, November 2008.

IEEE PES Task Force on Data for Modeling System Transients, Parameter determination for modeling system transients—Part v: Surge arresters, *IEEE Trans. Power Delivery*, 20(3), 2073–2078, July 2005.

IEEE Std. C62.11-2005, *IEEE Standard for Metal-Oxide Surge Arresters for AC Power Circuits (>1 kV)*, Institute of Electrical and Electronic Engineers, March 2006.

IEEE Std. C62.11a-2008, *IEEE Standard for Metal-Oxide Surge Arresters for AC Power Circuits (>1 kV) Amendment 1: Short-Circuit Tests for Station Intermediate and Distribution Arresters*, Institute of Electrical and Electronic Engineers, July 2008.

IEEE Std. C62.22-2009, *IEEE Guide for the Application of Metal-Oxide Surge Arresters for Alternating-Current Systems*, Institute of Electrical and Electronic Engineers, July 2009.

IEEE Std. C62.41.1-2002, *IEEE Guide on the Surge Environment in Low-Voltage (1000 V and Less) AC Power Circuits*, Institute of Electrical and Electronic Engineers, November 2002.

IEEE Std. C62.41.2-2002, *IEEE Recommended Practice on Characterization of Surges in Low-Voltage (1000 V and Less) AC Power Circuits*, Institute of Electrical and Electronic Engineers, November 2002.

IEEE Std. C62.72-2007, *IEEE Guide for the Application of Surge-Protective Devices for Low-Voltage (1000 V or Less) AC Power Circuits*, Institute of Electrical and Electronic Engineers, August 2007.

IEEE Std. C62.1-1989, *IEEE Standard for Gapped Silicon Carbide Surge Arresters for AC Power Circuits*, Institute of Electrical and Electronic Engineers, April 1990.

IEEE Std. C62.2-1987, *IEEE Guide for the Application of Gapped Silicon-Carbide Surge Arresters for Alternating-Current Systems*, Institute of Electrical and Electronic Engineers, April 1989.

Magro, M.C., Giannettoni, M., and Pinceti, P., Validation of ZnO surge arrester model for overvoltage studies, *IEEE Trans. Power Delivery*, 19(4), 1692–1695, October 2004.

Osterhout, J.C., Comparison of IEC and U.S. standards for metal oxide surge arresters, *IEEE Trans. Power Delivery*, 7(4), 2002–2011, October 1992.

Sakshaug, E.C., A brief history of AC surge arresters, *IEEE Power Eng. Rev.*, 40, 11–13, August 1991.

Sakshaug, E.C., Kresge, J.S., and Miske, S.A., A new concept in arrester design, *IEEE Trans. Power Appar. Syst.*, 96(2), 647–656, March/April 1977.

14

Insulation Coordination

Stephen R. Lambert
Shawnee Power
Consulting, LLC

14.1 Insulation Coordination

The art of correlating equipment electrical insulation strengths with expected overvoltage stresses so as to result in an acceptable risk of failure while considering economics and operating criteria (McNutt and Lambert, 1992).

Insulation properties can be characterized as self-restoring and nonself-restoring. Self-restoring insulation has the ability to "heal" itself following a flashover, and such insulation media is usually associated with a gas—air, SF_6, etc. Examples include overhead line insulators, station buswork, external bushing surfaces, SF_6 buswork, and even switchgear insulation. With self-restoring insulation, some flashovers are often acceptable while in operation. An EHV transmission line, for example, is allowed to experience occasional line insulator flashovers during switching operations such as energizing or reclosing, or as a result of a lightning flash striking the tower, shield wires, or phase conductors.

Nonself-restoring insulation is assumed to have permanently failed following a flashover, and repairs must be effected before the equipment can be put back into service. Insulation such as oil, oil/paper, and solid dielectrics such as pressboard, cross-link polyethylene, butyl rubbers, etc., are included in this insulation class. Any flashover of nonself-restoring insulation, say within a transformer or a cable, is unacceptable as such events usually result in lengthy outages and costly repairs.

The performance level of self-restoring insulation is usually addressed and defined in terms of the probability of a flashover. Thus, for a specific voltage stress, a given piece of insulation has an expected probability of flashover (pfo), e.g., a 1-m conductor-to-conductor gap exposed to a 490-kV switching surge would be expected to have a 50% chance of flashover; with a 453-kV surge, the gap would be expected to have a 10% chance of flashover, etc. Consequently, when self-restoring insulation is applied, the procedure is to select a gap length that will give the overall desired performance (pfo) as a function of the stress (overvoltages) being applied.

For nonself-restoring insulation, however, any flashover is undesirable and unacceptable, and consequently for application of nonself-restoring insulation, a capability is selected such that the "100%" withstand level (effectively a 0% chance of flashover) of the insulation exceeds the highest expected stress by a suitable margin.

14.2 Insulation Characteristics

Self-restoring (as well as nonself-restoring) insulation has, when exposed to a voltage, a pfo which is dependent on

- Dielectric material (air, SF_6, oil...)
- Waveshape of the stress (voltage)
- Electrode or gap configuration (rod–rod, conductor to structure...)
- Gap spacing
- Atmospheric conditions (for gases)

14.3 Probability of Flashover

Assuming the flashover characteristics of insulation follow a Gaussian distribution, and this is a good assumption for most insulation media (air, SF_6, oil, oil/paper), the statistical flashover characteristics of insulation can be described by the V_{50} or mean value of flashover, and a standard deviation. The V_{50} is a function of the rise time of the applied voltage, and when at a minimum, it is usually known as the CFO or critical flashover voltage.

Consequently, for a given surge level and insulation characteristic, the pfo of a single gap can be described by p, and can be determined by first calculating the number of standard deviations the stress level is above or below the mean:

$$\#\delta = \frac{V_{\text{stress}} - V_{50}}{1 \text{ standard deviation}} \tag{14.1}$$

For air insulation, 1 standard deviation is either 3% of the V_{50} for fundamental frequency (50–60 Hz) voltages and for lightning impulses or 6% of the V_{50} for switching surge impulses. That the standard deviation is a fixed percentage of the V_{50} and is not a function of gap length is very fortuitous and simplifies the calculations. Once the number of standard deviations away from the mean has been found, then by calculation or by entering a table, the probability of occurrence associated with that number of standard deviations is found.

Example: Assume that an insulator has a V_{50} of 1100 kV with a standard deviation of 6%, and a switching overvoltage of 980 kV is applied to the insulation. The stress is 1.82 standard deviations below the mean:

$$\#\delta = \frac{980 - 1100}{0.06 \times 1100}$$

$$= -1.82 \text{ standard deviations below the mean} \tag{14.2}$$

By calculation or table, the probability associated with −1.82 standard deviations below the mean (for a normal distribution) is 3.4%. Thus, there is a 3.4% chance of insulation flashover every time the insulation is exposed to a 980-kV surge.

The physics of the flashover mechanism precludes a breakdown or flashover below some stress level, and this is generally assumed to occur at 3.5–4 standard deviations below the mean.

14.3.1 Multiple Gaps per Phase

The pfo, P_n, for n gaps in parallel (assuming the gaps have the same characteristics and are exposed to the same voltage) can be described by the following equation where p is the pfo of one gap. This mathematical expression defines the probability of one or more gaps flashing over, but practically only one gap of the group will flashover as the first gap to flashover reduces the voltage stress on the other gaps:

$$P_n = 1 - (1-p)^n \qquad (14.3)$$

14.3.2 Multiple Gaps and Multiple Phases

Analysis of some applications may not only require consideration of multiple gaps in a given phase but also of multiple phases. Consider the pfo analysis of a transmission line; during a switching operation for example, multiple towers are exposed to surges and at each tower, each of the three phases is stressed (typically by different surge magnitudes). Thus it is important to consider not only the multiple gaps associated with the multiple towers, but also all three phases often need to be considered to determine the overall line pfo. The overall pfo for a given surge, PFO, can be expressed as

$$PFO = 1 - \left(1 - pfo_{n,a}\right)^g \left(1 - pfo_{n,b}\right)^g \left(1 - pfo_{n,c}\right)^g \qquad (14.4)$$

where
    $pfo_{n,x}$ is the pfo of the x phase for the given surge
    n, g is the number of towers (gaps in parallel)

The simultaneous analysis of all three phases can be important especially when various techniques are used to substantially suppress the surges (Lambert, 1988).

14.4 Flashover Characteristics of Air Insulation

14.4.1 Voltage Waveshape

Waveshapes used for testing and for determining the flashover response of insulation have been standardized by various groups and while there is not 100% agreement, the waveshapes used generally conform to the following:

Fundamental frequency	50 or 60 Hz sine wave (8000 μs rise time)
Switching impulse	200–250 μs by 2000 μs
Lightning impulse	1.2 μs by 50 μs

The impulse waveshapes are usually formed by a double exponential having the time to crest indicated by the first number and the time to 50% of the crest on the tail of the wave indicated by the second number. Thus, a lightning impulse would crest at 1.2 μs and following the crest would fall off to 50% of the crest at 50 μs.

Fundamental frequency characteristics have been published, and typical values are indicated on Figure 14.1 (Aleksandrov et al., 1962; EPRI, 1982).

Equations have also been published or can be developed which define the typical responses to positive polarity switching and lightning impulses (see Figure 14.2). Insulation usually has a lower withstand capability when exposed to positive polarity impulses than when exposed to negative impulses; thus, designs are usually based on positive magnitude impulses.

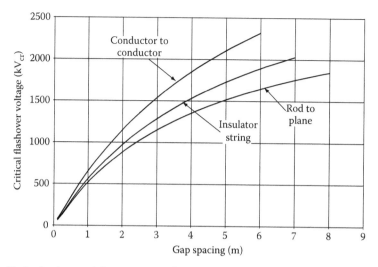

FIGURE 14.1 V_{50} for fundamental frequency waveshapes. (From *Transmission Line Reference Book, 345 kV and Above*, 2nd edn., Electric Power Research Institute, Palo Alto, CA, 1975.)

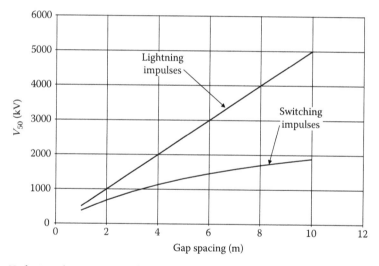

FIGURE 14.2 V_{50} for impulses—positive polarity, rod-plane gap. (From *EHV Transmission Line Reference Book*, Edison Electric Institute, New York, 1968; Gallet, G. et al., *IEEE Trans. Power Appar. Syst.*, PAS-95, 580, 1976.)

For switching surge impulses (gaps ≤ 15 m) (Gallet et al., 1976):

$$V_{50} = k\frac{3400}{1+(8/d)}\,\text{kV} \tag{14.5}$$

For lightning impulses, the following equation can be developed from EEI (1968):

$$V_{50} = k*500*d\,\text{kV} \tag{14.6}$$

where
　　k is an electrode factor reflecting the shape of the electrodes (Paris, 1967)
　　d is the electrode gap spacing in meters

14.4.2 Electrode Configuration

Electrode configuration has a pronounced effect on the V_{50} characteristics, and this is reflected as a gap or electrode factor, k (Paris, 1967). Examples of k are

Rod–plane	1.00
Conductor–structure	1.30
Rod–rod	1.30
Conductor–rope	1.40
Conductor–rod	1.65

14.4.3 Effect of Insulator

The presence of an insulator in a gap tends to reduce the gap factor from those given above, mainly due to the terminal electrode configuration (and intermediate flanges for multiunit column bus support insulators). The reduction increases with increased gap factor and typical correction values may be found on Figure 14.3. Note that these corrections are subject to variations (Thione, 1984).

Rain has little effect on a gap without an insulator; however, rain does reduce the gap factor when an insulator is present. Reductions as high as 20% have been noted; but, in general a reduction of 4%–5% is typical (Thione, 1984).

14.4.4 Effect of Atmospheric Conditions on Air Insulation

V_{50} for gases is affected by temperature, atmospheric pressure, and humidity, and for air the correction can be expressed as

$$V_{50,\text{ambient}} = V_{50,\text{NTP}} \left(\frac{\delta}{H_0} \right)^n \tag{14.7}$$

where
   NTP is the normal temperature and pressure (20°C, 101.3 kPa)
   H_0 is the humidity correction factor
   n is a gap length correction factor
   δ is the relative air density correction factor

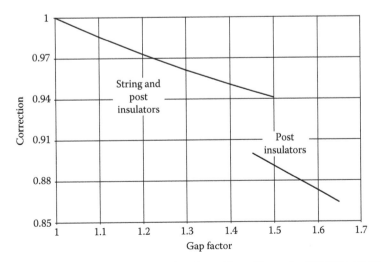

FIGURE 14.3 Gap factor correction for presence of insulator. (From Thione, L., *ELECTRA*, 94, 77, 1984.)

The correction for temperature and pressure, "δ," is known as the RAD (relative air density) correction factor and is expressed by

$$\delta = \frac{0.386 H_{\text{mm of Hg}}}{273 + T} \tag{14.8}$$

where

$H_{\text{mm of Hg}}$ is the atmospheric pressure in mm of Hg
T is the temperature in °C

The humidity correction factor, H_o, is given in IEEE 4 (1978) and can be expressed approximately by

$$H_o \cong 1.1 - 0.00820 * H_{AB}$$

$$\cong 1.1 - 0.008071 * \text{VP} \tag{14.9}$$

where

H_{AB} is the absolute humidity in g/m³
VP is the vapor pressure in mm of Hg

For switching impulses (and fundamental frequency) the effect of the RAD and humidity on V_{50} is, however, a function of the gap length and has less effect on longer gap than on shorter gap lengths. For lengths of 0–1 m the n correction factor is 1.0; from 1 to 6 m, the correction decreases linearly from 1.0 to 0.4; and for lengths greater than 6 m, the factor is 0.4. There is no gap length correction for positive lightning impulses (EEI, 1968; EPRI, 1975, 1982). Other approaches for humidity corrections can be found in Menemenlis et al. (1988), Thione (1984), Feser and Pigini (1987).

14.4.5 Altitude

Corrections for altitude are also important as the insulation capability drops off about 10% per 1000 m as shown in Figure 14.4. There are various equations for the altitude correction factor (ACF) and the following expression is representative of most in use (IEEE 1312, 1999):

$$\text{ACF} = \left(e^{-H_t/8600} \right)^n \tag{14.10}$$

where

H_t is the altitude in meters
n is a gap length correction factor

14.4.6 Insulator Contamination

Insulator contamination is an important issue for fundamental frequency voltage considerations, and the equivalent salt density, ESDD, approach is extensively used as a design tool. The contamination severity is defined by the ESDD in mg/cm², and an insulator creepage distance, in terms of mm/kV$_{\text{rms, phase to phase}}$, can then be selected (IEC 815, 1986). Note that insulator/bushing shed/skirt design has a significant impact on the performance, and some past designs performed poorly due to skirt configuration even though they had large creepage distances. With the ESDD approach insulators are tested to define their expected performance. Table 14.1 shows the relationship between contamination level, ESDD, and recommended creepage distances.

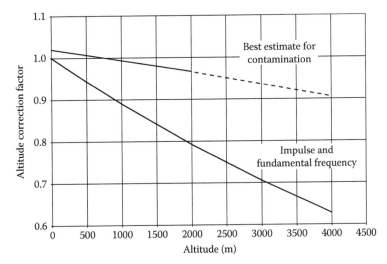

FIGURE 14.4 Altitude correction factors. (From Mizuno, Y. et al., *IEEE Trans. Dielectr. Electr. Insul.*, 4, 286, 1997; *IEEE Standard for Insulation Coordination—Part 2, Application Guide*, Institute of Electrical and Electronic Engineers (IEEE), 1312, 1999.)

TABLE 14.1 Recommended Creepage Distances

Contamination Level	Example	ESDD (mg/cm²)	Minimum Recommended Creepage Distance (mm/kV$_{\text{rms, phase to phase}}$)
Light	Low industrial activity	0.03–0.06	16
Medium	Industrial activity—some exposure to wind from the sea	0.1–0.2	20
Heavy	Industrial area and areas close to the sea	0.3–0.6	25
Very heavy	Heavy industrial or sea coast area	>0.6	31

Altitude also has an effect on the performance of contaminated insulation, and the degradation of capability as a function of altitude may be found in Figure 14.4 (Mizuno et al., 1997).

Example 14.1

Assume 10 identical substation bus support insulators in parallel located in a 500-kV substation located at sea level; this configuration can be described as an air gap, conductor to rod configuration at standard atmospheric conditions. Assume that an overall pfo for the 10 gaps of 0.5% is desired when the configuration is exposed to a switching surge of 939 kV (2.3 pu on a 500-kV system). What is the required gap clearance in meters?

Solution

The desired pfo of one gap, p, then should be

$$\text{pfo}_{10\text{gaps}} = 0.005 = 1 - (1-p)^{10}$$

and

$$p = 1 - (1-0.005)^{1/10}$$

$$= 0.0005011$$

From tables or calculations for a normal or Gaussian distribution, a probability of 0.0005011 corresponds to 3.29 standard deviations below the mean (V_{50}). Therefore, the desired V_{50} value is

$$939\,kV = V_{50}\left(1 - 3.29 * 0.06\right)$$

and

$$V_{50} = 1170 \text{ kV}$$

A standard deviation of 6% is often used for the air gap for switching surge stresses.

With the V_{50} of 1170 kV and noting that a conductor to rod gap has a k factor of 1.65 assumed to be reduced to 1.42 due to the presence of the insulator, the desired gap spacing value can be calculated by

$$1170\,kV = 1.42\frac{3400}{1 + (8/d)}$$

and

$$d = 2.56\,m$$

14.4.7 Application of Surge Arresters

Surge arresters are used to limit overvoltages and as a result, allow reductions in the clearances required for self-restoring gaps (e.g., transmission line towers) as well as the capability required for nonself-restoring insulation such as transformer windings. In most applications the proper approach is to determine the minimum arrester rating, which can be applied without resulting in damage to the arrester and then to define the insulation level required so as to result in an acceptable pfo or risk of failure.

For a transmission line application, for example, although the arrester reduces higher magnitude surges to lower levels, the line is still allowed to have a finite, albeit low, pfo for a specific switching operation. Thus, the arrester, by limiting the higher magnitude surges, allows smaller conductor to tower clearances.

However, when arresters are used to protect a transformer for example, an insulation level, which has a significantly higher capability than the maximum surge allowed by the arrester, is selected. This margin between the arrester protective levels (lightning or switching surge) is a function of various considerations as well as the conservatism of the person applying the arrester/insulation system.

Today, for new applications, only metal oxide (ZnO) arresters are being applied. Although there are certainly many of the gapped, silicon carbide type arresters still in service and which still perform effectively, in what follows, only metal oxide arresters will be considered to protect insulation. Successful application requires that the arrester survives the electrical environment in which it is placed, and the following arrester capabilities must be carefully considered:

MCOV—maximum fundamental frequency continuous operating voltage applied to the arrester
TOV—temporary fundamental frequency overvoltages to which the arrester may be exposed
Energy—the energy which must be absorbed by the arrester when limiting switching surges

14.4.7.1 MCOV

The highest system voltage, which can be continuously applied to the arrester, needs to be determined and the arrester capability, its MCOV rating, should at least be equal to and should usually exceed the highest continuous system voltage by some small margin. For example, if a nominal 345-kV system is never operated above 352 kV, then the maximum continuous voltage, which would be expected to be applied to a line to ground arrester, would be 352 kV/$\sqrt{3}$ = 203.2 kV. With today's typical arresters, the next highest available MCOV capability would be 209 kV and is associated with an arrester rated 258 kV.

14.4.7.2 TOV

On occasion, the fundamental-frequency voltage applied to an arrester will exceed the expected MCOV. Examples include fault conditions during which line to ground voltages on unfaulted phases can rise significantly (as high as phase-to-phase voltage for ungrounded systems); rise in line voltage when energizing a transmission line (Ferranti effect) and voltages which occur during load rejection events—these are usually associated with voltages experienced on a radial transmission line emanating from a generating plant when the load terminal of the line opens unexpectedly.

14.4.7.3 Energy

When an arrester limits switching surges on a transmission line, it can absorb a significant amount of energy, and it can be important to examine events and determine the energy which could be absorbed. Exceeding the arrester's capability could result in immediate damage to the arrester and failure. It is also important to the arrester's TOV capability as absorbing energy heats the arrester material, and application of a significant temporary overvoltage immediately following absorption of a significant amount of energy could result in thermal runaway and arrester failure.

Following selection of an arrester which would be expected to survive the electrical environment (i.e., the minimum rated arrester), the protective levels of the arrester must be correlated with the insulation capability and acceptable margins between the protective levels and the insulation capability achieved.

The protective level or discharge voltage of an arrester is the voltage magnitude to which the arrester will limit the voltage while discharging a surge, and these levels are a function of the waveshape and rise time of the surge as well as the current magnitude of the discharge. In general, the discharge or protective levels considered for coordination with insulation capability are

- A 10-kA, 8 × 20-μs discharge for coordination with the insulation full wave or lightning impulse (BIL) capability and
- A 0.5–2.0-kA, 36 × 90-μs discharge for coordination with the switching impulse capability

There should always be margin between the protective level of the arrester and the insulation capability to allow for uncertainties in arrester protective levels due to surge rise times, discharge currents, and arrester separation distance (faster rise times, higher currents, and longer separation distance or lead lengths generate higher protective levels). Uncertainties in insulation capability include reduced insulation strength due to aging (especially for paper insulation in transformers for example) and limitations of the ability of laboratory dielectric testing to accurately relate to field conditions.

In the author's opinion, a margin of at least 40% is appropriate unless all the uncertainties and the risks are carefully evaluated.

14.4.8 Examples of Surge Arrester Application (Nonself-Restoring Insulation)

14.4.8.1 34.5-kV System Application

Surge arresters are to be applied line to ground at the terminals of a circuit breaker (38-kV rating, 150-kV BIL) used on a solidly grounded 34.5-kV system. The highest expected continuous system voltage is 37 kV, and during fault conditions, the phase to ground voltage can rise to 1.4 pu or 27.9 kV. Faults can persist for 20 cycles.

The maximum line to ground voltage is 37/$\sqrt{3}$ = 21.4 kV and the MCOV of the arrester must meet or exceed this value. An arrester rated 27 kV would be acceptable as it has an MCOV of 22.0 kV. The 1 s TOV capability of the arrester is 31.7 kV, and as this exceeds the 27.9 kV phase to ground voltage expected during faults, the 27-kV arrester meets the TOV criteria as well.

A 27-kV arrester has a 10-kA discharge level of 67.7 kV, and thus the margin between the discharge or protective level and the insulation BIL is (150/67.7 × 100−100) or 121%. This margin is obviously more than adequate, and selection of an arrester rated 27 kV would be appropriate.

14.4.8.2 500-kV System Application

A 500-kV shunt reactor (solidly grounded neutral) is being applied at the end of a 300 km, 500-kV transmission line, and arresters are to be applied line to ground on the terminals of the reactor to limit surges to reasonable levels. The reactor is solidly connected to the line and is switched with the line, and the substation at which the reactor resides is at an altitude of 1800 m. The highest expected continuous system voltage is 550 kV. During line switching operations, the circuit breaker at the reactor terminal may not be closed for some period following energizing of the line/reactor from the other terminal, and the phase to ground voltage at the reactor can be as high as 1.15 pu for as long as 5 min. Arrester energy requirements were determined (by EMTP or TNA simulations of switching operations) to be well within the capability of an arrester rated 396 kV.

The minimum required MCOV is $550/\sqrt{3} = 317.5$ kV. The minimum required TOV is $1.15 \times 500/\sqrt{3} = 332$ kV for 300 s, and for most arresters, such a requirement would correlate with a 1 s TOV rating of 451 kV. An arrester rated 396 kV has a 318-kV MCOV and a 1 s TOV rating of 451 kV; thus, a 396-kV arrester would be the minimum rating that could be used. Of course any arrester rated higher than 396 kV could also be used. The 10-kA lightning (8 × 20 μs waveform) and switching surge (2 kA, 36 × 90 μs) discharge levels for a 396-kV and a 420-kV arrester are

	Discharge Levels	
Rating (kV)	10 kA (kV)	Switching Surge (kV)
396	872	758
420	924	830

BIL values of 1300 and 1425 kV for the reactor's *internal insulation* (i.e., insulation not affected by altitude) could be considered as reasonable candidates for a specification. The corresponding switching impulse levels (SIL) would be 1080 and 1180 kV, respectively, and the following table indicates the margin between the arrester protective levels and the insulation level.

	1300-kV BIL		1425-kV BIL	
Arrester	396 kV (%)	420 kV (%)	396 kV (%)	420 kV (%)
SIL	42	30	56	42
BIL	49	41	63	54

Application of a 420-kV arrester for a 1300-kV BIL insulation level results in margins below 40%, and unless the application is very carefully considered from the point of view of arrester separation distance and lead length, expected maximum discharge current level, wave rise time, etc., a 396-kV arrester would be a better choice. For a 1425-kV BIL, either the 396-kV or the 420-kV arrester would result in sufficient margins.

For *external insulation*, i.e., the reactor bushings, the effect of altitude on the insulation capability needs to be considered. At 1800 m, the insulation has only 81% of the withstand capability demonstrated at sea level or 0 m. For example, the SIL of a 1425-kV bushing (1180 kV at sea level) would be reduced to 956 kV at 1800 m (1180 × 0.81 = 956 kV), and application of even a 396-kV arrester would result in a margin of 26%—hardly acceptable.

Assume that a 420-kV arrester was selected to protect the reactor (the arrester itself is rated for application to 3000 m). The switching surge and 10 kA protective levels are 830 and 924 kV, respectively. With a desired minimum margin of 40%, and correcting for altitude, the minimum SIL and BIL at sea level (0 m) should be

$$\text{Minimum SIL} = \frac{830 \times 1.4}{0.81} = 1435 \text{ kV}$$

$$\text{Minimum BIL} = \frac{924 \times 1.4}{0.81} = 1597 \text{ kV}$$

A 1550-kV BIL bushing would have a 1290-kV SIL, and even if one would accept the slightly less than a 36% margin for the BIL, the SIL margin would only be 26%. A 1675-kV BIL bushing would be expected to have a 1390-kV SIL capability, and so the SIL margin would be 36% with a BIL margin of 47%. The next higher rated bushing (1800-kV BIL) would mean applying 800-kV system class bushings, and their increased size and cost would likely not make for a reasonable design. Consequently, specifying a 1675-kV BIL bushing and accepting the slightly reduced SIL margin would be a reasonable compromise.

14.4.8.3 Effect of Surge Reduction Techniques on Overall PFO

Application of surge arresters to significantly reduce switching surge levels on transmission line and substation insulators can be effective, however, the designer should be aware that the overall PFO of all three phases needs to be considered as it will usually be higher than that found for a single phase by a factor often approaching three. Also for long transmission lines, application of arresters at the line terminals will certainly limit the surges at the terminals but will not limit the surges at other points on the line to the same level. Consequently, the surge distribution along the line may need to be considered (Lambert, 1988; Ribiero et al., 1991).

References

Aleksandrov, G.N., Kizvetter, V.Y., Rudakova, V.M., and Tushnov, A.N., The AC flashover voltages of long air gaps and strings of insulators, *Elektrichestvo*, 6, 27–32, 1962.

EHV Transmission Line Reference Book, Edison Electric Institute, New York, 1968.

Feser, K. and Pigini, A., Influence of atmospheric conditions on the dielectric strength of external insulation, *ELECTRA*, 112, 83–93, 1987.

Gallet, G., Bettler, M., and Leroy, G., Switching impulse results obtained on the outdoor testing area at Renardieres, *IEEE Transactions on Power Apparatus and Systems*, PAS-95(2), 580–585, 1976.

Guide for the Selection of Insulators in Respect of Polluted Conditions, The International Electrotechnical Commission Publication 815, 1986.

IEEE Standard for Insulation Coordination—Part 2, Application Guide, Institute of Electrical and Electronic Engineers (IEEE) 1312, 1999.

IEEE Standard Techniques for High-Voltage Testing, Institute of Electrical and Electronic Engineers (IEEE) 4-1978.

Lambert, S.R., Effectiveness of zinc oxide surge arresters on substation equipment probabilities of flashover, *IEEE Transactions on Power Delivery*, 3(4), 1928–1934, 1988.

McNutt, W.J. and Lambert, S.R., *Transformer Concepts and Applications Course*, Power Technologies, Inc., Schenectady, NY, 1992.

Menemenlis, C., Carrara, G., and Lambeth, P.J., Application of insulators to withstand switching surges in substations, part I: Switching impulse insulation strength, 88 WM 077-0, *IEEE/PES Winter Meeting*, New York, January 31–February 5, 1988.

Mizuno, Y., Kusada, H., and Naito, K., Effect of climatic conditions on contamination flashover voltage of insulators, *IEEE Transactions on Dielectrics and Electrical Insulation*, 4(3), 286–289, 1997.

Paris, L., Influence of air gap characteristics on line-to-ground switching surge strength, *IEEE Transactions on Power Apparatus and Systems*, PAS-86(8), 936–947, 1967.

Ribeiro, J.R., Lambert, S.R., and Wilson, D.D., Protection of compact transmission lines with metal oxide arresters, *CIGRE Leningrad Symposium*, 400-6, S33–S91, 1991.

Thione, L., Evaluation of the switching impulse strength of external insulation, *ELECTRA*, 94, 77–95, 1984.

Transmission Line Reference Book, 345 kV and Above, 1st edn., Electric Power Research Institute, Palo Alto, CA, 1975.

Transmission Line Reference Book, 345 kV and Above, 2nd edn., Electric Power Research Institute, Palo Alto, CA, 1982.

III

Power System Planning (Reliability)

Gerald B. Sheblé

Gerald B. Sheblé, executive advisor and senior director of U.S. Research and Development, has more than 38 years of experience in industry, consulting, expert witness, and academic environments. He has led research, development, and implementation for power system operation and planning using classical analytics and computational adaptive agents. Dr. Sheblé is an internationally recognized expert in the area of electric energy auction markets, capital budgeting, power system analysis, and real option valuation. He is nationally recognized as one of the nation's leading business experts on business trends within the utility industry and on trends forced by external economic drivers. His expertise extends from smart grid, data mining, power system optimization, electric energy market modeling, energy and emission management, control automation, technology innovation, bulk power operations, generation and transmission planning, operator training, energy risk management, distribution operations, energy management systems, to portfolio decision analysis of interchange contracts. His academic work has concentrated on developing concepts, methods, and products for reregulated operation, market analysis and design, equipment scheduling (especially unit commitment), transaction evaluation, reliability analysis by decision analysis, Monte Carlo simulation, failure tree analysis, market simulation and energy economics, decision analysis, and real option valuation of renewable energy equipment.

He has pioneered structure and market definition predicting successful industry reregulation. He has also authored over 100 papers and 50 research documents. He is an associate business scene editor of *IEEE Power Magazine*. His columns are frequently highlighted in respected trade journals for industry innovations. He has been a premier guest or contributor at over 50 specialized workshops or courses throughout Europe and North America in over 24 countries. He has recently been an expert witness on electrical accidents, deficient project management, grounding, and intellectual property rights. He has also been an expert witness on software engineering, software business planning, and industrial trends.

Before joining Quanta Technology, he has worked for Commonwealth Edison, Systems Control Inc., Control Data Corporation, Energy and Control Consultants, Auburn University, Iowa State University, Portland Stat University, University of Porto (Portugal), and INESC Porto.

Dr. Sheblé received his MBA from The University of Iowa (Henry B. Tippie College of Business), his PhD from Virginia Technological and State University, and his MSEE and BSEE from Purdue University. Dr. Sheblé is an IEEE Fellow for his contributions to the development of auction methods as an alternative to power system optimization methods, addressing the deregulation of the electric utility business.

Dr. Gerald Sheblé can be reached at gsheble@quanta-technology.com.

15

Planning Environments

Gerald B. Sheblé
Quanta Technology, LLC

15.1 Introduction

Capacity expansion decisions are made daily by government agencies, private corporations, partnerships, and individuals. Most decisions are small relative to the profit and loss sheet of most companies. (Aggarwal, 1993; Bussey, 1981; Daellenbach, 1984). However, many decisions are sufficiently large to determine the future financial health of the nation, company, partnership, or individual. Capacity expansion of hydroelectric facilities may require the commitment of financial capital exceeding the income of most small countries. Capacity expansion of thermal fossil fuel plants is not as severe, but does require a large number of financial resources including bank loans, bonds for long-term debt, stock issues for more working capital, and even joint-venture agreements with other suppliers or customers to share the cost and the risk of the expansion. This section proposes several mathematical optimization techniques to assist in this planning process. These models and methods are tools for making better decisions based on the uncertainty of future demand, project costs, loan costs, technology change, etc. (Binger, 1998). Although the material presented in this section is only a simple model of the process, it does capture the essence of real capacity expansion problems.

This section relies on a definition of electric power industry restructuring presented in Sheblé (1999). The new environment within this work assumes that the vertically integrated utility has been segmented into a horizontally integrated system (Ilic, 1998). Specifically, generation companies (GENCOs), distribution companies (DISTCOs), and transmission companies (TRANSCOs) exist in place of the old. This work does not assume that separate companies have been formed. It is only necessary that comparable services are available for anyone connected to the transmission grid.

As can be concluded, this description of a deregulated marketplace is an general version of the commodity markets. It needs polishing and expanding. The change in the electric utility business environment is depicted generically below. The functions shown are the emerging paradigm. This work outlines the market organization for this new paradigm.

Attitudes toward restructuring still vary from state to state and from country to country. Many electric utilities in the United States have been reluctant to change the status quo. Electric utilities with high rates are very reluctant to restructure since the customer is expected to leave for the lower prices.

Electric utility companies in regions with low prices are more receptive to change since they expect to pick up more customers. In 1998, California became the first state in the United States to adopt a competitive structure, and other states are observing the outcome. Some offer customer selection of supplier. Some offer markets similar to those established in the United Kingdom, Norway, and Sweden, but not Spain. Several countries have gone to the extreme competitive position of treating electricity as a commodity as seen in New Zealand and Australia (Outhred, 1993). As these markets continue to evolve, governments in all areas of the world will continue to form opinions on what market, operational, and planning structures will suit them best.

15.2 Defining a Competitive Framework

There are many market frameworks that can be used to introduce competition between electric utilities. Almost every country embracing competitive markets for its electric system has done so in a different manner. The methods described here assume an electric marketplace derived from commodities exchanges like the Chicago Mercantile Exchange (CME), Chicago Board of Trade (CBOT), and New York Mercantile Exchange (NYMEX) where commodities (other than electricity) have been traded for many years. NYMEX added electricity futures to their offerings in 1996, supporting this author's previous predictions (Sheblé, 1991, 1992, 1993, 1994) regarding the framework of the coming competitive environment. The framework proposed has similarities to the Norwegian-Sweden electric systems. The proposed structure is partially implemented in New Zealand, Australia, and Spain. The framework is being adapted since similar structures are already implemented in other industries. Thus, it would be extremely expensive to ignore the treatment of other industries and commodities. The details of this framework and some of its major differences from the emerging power markets/pools are described in Sheblé (1999).

These methods imply that the ultimate competitive electric industry environment is one in which retail consumers have the ability to choose their own electric supplier. Often referred to as retail access, this is quite a contrast to the vertically integrated monopolies of the past. Telemarketers are contacting consumers, asking to speak to the person in charge of making decisions about electric service. Depending on consumer preference and the installed technology, it may be possible to do this on an almost real-time basis as one might use a debit card at the local grocery store or gas station. Real-time pricing, where electricity is priced as it is used, is getting closer to becoming a reality as information technology advances. Presently, however, customers in most regions lack the sophisticated metering equipment necessary to implement retail access at this level.

Charging rates that were deemed fair by the government agency, the average monopolistic electric utility of the old environment met all consumer demand while attempting to minimize their costs. During natural or man-made disasters, neighboring utilities cooperated without competitively charging for their assistance. The costs were always passed on to the rate payers. The electric companies in a country or continent were all members of one big happy family. The new companies of the future competitive environment will also be happy to help out in times of disaster, but each offer of assistance will be priced recognizing that the competitor's loss is gain for everyone else. No longer guaranteed a rate of return, the entities participating in the competitive electric utility industry of tomorrow will be profit driven.

15.2.1 Preparing for Competition

Electric energy prices recently rose to more than $7500/MWh in the Midwest (1998) (Midwest ISO, website) due to a combination of high demand and the forced outage of several units. Many midwestern electric utilities bought energy at that high price, and then sold it to consumers for the normal rate. Unless these companies thought they were going to be heavily fined, or lose all customers for a very long time, it may have been more fiscally responsible to terminate services.

Under highly competitive scenarios, the successful supplier will recover its incremental costs as well as its fixed costs through the prices it charges. For a short time, producers may sell below their costs, but will need to make up the losses during another time period. Economic theory shows that eventually, under perfect competition, all companies will arrive at a point where their profit is zero. This is the point at which the company can break even, assuming the average cost is greater than the incremental cost. At this ideal point, the best any producer can do in a competitive framework, ignoring fixed costs, is to bid at the incremental cost. Perfect competition is not often found in the real world for many reasons. The prevalent reason is *technology change*. Fortunately, there are things that the competitive producer can do to increase the odds of surviving and remaining profitable.

The operational tools used and decisions made by companies operating in a competitive environment are dependent on the structure and rules of the power system operation. In each of the various market structures, the company goal is to maximize profit. Entities such as commodity exchanges are responsible for ensuring that the industry operates in a secure manner. The rules of operation should be designed by regulators prior to implementation to be complete and "fair." *Fairness* in this work is defined to include noncollusion, open market information, open transmission and distribution access, and proper price signals. It could call for maximization of social welfare (i.e., maximize everyone's happiness) or perhaps maximization of consumer surplus (i.e., make customers happy).

Changing regulations are affecting each company's way of doing business and to remain profitable, new tools are needed to help companies make the transition from the old environment to the competitive world of the future. This work describes and develops methods and tools that are designed for the competitive component of the electric industry. Some of these tools include software to generate bidding strategies, software to incorporate the bidding strategies of other competitors, and updated common tools like economic dispatch and unit commitment to maximize profit.

15.2.2 Present View of Overall Problem

This work is motivated by the recent changes in regulatory policies of inter-utility power interchange practices. Economists believe that electric pricing must be regulated by free market forces rather than by public utilities commissions. A major focus of the changing policies is "competition" as a replacement for "regulation" to achieve economic efficiency. A number of changes will be needed as competition replaces regulation. The coordination arrangements presently existing among the different players in the electric market would change operational, planning, and organizational behaviors.

Government agencies are entrusted to encourage an open market system to create a competitive environment where generation and supportive services are bought and sold under demand and supply market conditions. The open market system will consist of GENCOs, DISTCOs, TRANSCOs, a central coordinator to provide independent system operation (ISO), and brokers to match buyers and sellers (BROCOs). The long-term planning has been separated under the organization known as regional transmission organization (RTO). The energy market exchange has been absorbed into the ISO instead of being a separate energy mercantile association (EMA). The energy contracts segmented to the distribution companies are often partially segmented to energy service suppliers (ESS), which may include energy service companies (ESCOs) or energy management companies (EMCOs). The services offered to the customers are not different from the traditional average pricing as of this writing. However, the impact of the smart grid is expected to change this dramatically. The interconnection between these groups is shown in Figure 15.1.

The ISO is independent and a dissociated agent for market participants. The roles and responsibilities of the ISO in the new marketplace are yet not clear. This work assumes that the ISO is responsible for coordinating the market players (GENCOs, DISTCOs, and TRANSCOs) to provide a reliable power system functions. Under this assumption, the ISO would require a new class of optimization algorithms to perform price-based operation. Efficient tools are needed to verify that the system remains in operation with all contracts in place. This work proposes an energy brokerage model for all services as a novel framework for price-based optimization. The proposed foundation is used to develop analysis and simulation tools to study the implementation aspects of various contracts in a deregulated environment.

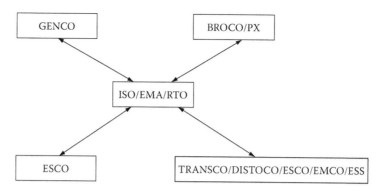

FIGURE 15.1 New organizational structure.

Although it is conceptually clean to have separate functions for the GENCOs, DISTCOs, TRANSCOs, and the ISO, the overall mode of real-time operation is still evolving. Presently, two possible versions of market operations are debated in the industry. One version is based on the traditional power pool concept (POOLCO). The other is based on transactions and bilateral transactions as presently handled by commodity exchanges in other industries. Both versions are based on the premise of price-based operation and market-driven demand. This work presents analytical tools to compare the two approaches. Especially with the developed auction market simulator, POOLCO, multilateral, and bilateral agreements can be studied.

Working toward the goal of economic efficiency, one should not forget that the reliability of the electric services is of the utmost importance to the electric utility industry in North America. In the words of the North American Electric Reliability Council (NERC), reliability in a bulk electric system indicates "the degree to which the performance of the elements of that system results in electricity being delivered to customers within accepted standards and in the amount desired. The degree of reliability may be measured by the frequency, duration, and magnitude of adverse effects on the electric supply." The council also suggests that reliability can be addressed by considering the two basic and functional aspects of the bulk electric system—adequacy and security. In this work, the discussion is focused on the adequacy aspect of power system reliability, which is defined as the static evaluation of the system's ability to satisfy the system load requirements. In the context of the new business environment, market demand is interpreted as the system load. However, a secure implementation of electric power transactions concerns power system operation and stability issues:

1. *Stability issue*: The electric power system is a nonlinear dynamic system comprised of numerous machines synchronized with each other. Stable operation of these machines following disturbances or major changes in the network often requires limitations on various operating conditions, such as generation levels, load levels, and power transmission changes. Due to various inertial forces, these machines, together with other system components, require extra energy (reserve margins and load following capability) to safely and continuously actuate electric power transfer.
2. *Thermal overload issue*: Electrical network capacity and losses limit electric power transmission. Capacity may include real-time weather conditions as well as congestion management. The impact of transmission losses on market power is yet to be understood.
3. *Operating voltage issues*: Enough reactive power support must accompany the real power transfer to maintain the transfer capacity at the specified levels of open access.

In the new organizational structure, the services used for supporting a reliable delivery of electric energy (e.g., various reserve margins, load following capability, congestion management, transmission losses, reactive power support, etc.) are termed supportive services. These have been called "ancillary services" in the past. In this context, the term "ancillary services" is misleading since the services in question are not ancillary but *closely bundled* with the electric power transfer as described earlier. The open market system should consider all of these supportive services as an integral part of power transaction.

This work proposes that supportive services become a competitive component in the energy market. It is embedded so that no matter what reasonable conditions occur, the (operationally) centralized service will have the obligation and the authority to deliver and keep the system responding according to adopted operating constraints. As such, although competitive, it is burdened by additional goals of ensuring reliability rather than open access only. The proposed pricing framework attempts to become economically efficient by moving from cost-based to price-based operation and introduces a mathematical framework to enable all players to be sufficiently informed in decision-making when serving other competitive energy market players, including customers.

15.2.3 Economic Evolution

Some economists speculate that regional commodity exchanges within the United States would be oligopolistic in nature (having a limited numbers of sellers) due to the configuration of the transmission system. Some postulate that the number of sellers will be sufficient to achieve near-perfect competition. Other countries have established exchanges with as few as three players. However, such experiments have reinforced the notion that collusion is all too tempting, and that market power is the key to price determination, as it is in any other market. Regardless of the actual level of competition, companies that wish to survive in the deregulated marketplace must change the way they do business. They will need to develop bidding strategies for trading electricity via an exchange.

Economists have developed theoretical results of how variably competitive markets are supposed to behave under varying numbers of sellers or buyers. The economic results are often valid only when aggregated across an entire industry and frequently require unrealistic assumptions. While considered sound in a macroscopic sense, these results may be less than helpful to a particular company (not fitting the industry profile) that is trying to develop a strategy that will allow it to remain competitive.

GENCOs, energy service suppliers (ESSs), and DISTCOs that participate in an energy commodity exchange must learn to place effective bids in order to win energy contracts. Microeconomic theory states that in the long term, a hypothetical firm selling in a competitive market should price its product at its marginal cost of production. The theory is based on several assumptions (e.g., all market players will behave rationally, all market players have perfect information) that may tend to be true industry-wide, but might not be true for a particular region or a particular firm. As shown in this work, the normal price offerings are based on average prices. Markets are very seldom perfect or in equilibrium.

There is no doubt that deregulation in the power industry will have many far-reaching effects on the strategic planning of firms within the industry. One of the most interesting effects will be the optimal pricing and output strategies generator companies (GENCOs) will employ in order to be competitive while maximizing profits. This case study presents two very basic, yet effective means for a single generator company (GENCO) to determine the optimal output and price of their electrical power output for maximum profits.

The first assumption made is that switching from a government regulated, monopolistic industry to a deregulated competitive industry will result in numerous geographic regions of oligopolies. The market will behave more like an oligopoly than a purely competitive market due to the increasing physical restrictions of transferring power over distances. This makes it practical for only a small number of GENCOs to service a given geographic region.

The strongest force in the economic evolution is the demand for electricity. Customer\demand was assumed to be inelastic under many of the early market analysis studies. This ignored the success of demand side management in the regulated environment (Le et al., 1983; Cohen et al., 1987; Mortensen and Haggerty, 1990; Lee and Chen, 1992; Chu et al., 1993; Rupanagunta et al., 1995; Wei and Chen, 1995; Kurucz et al., 1996) and the cost reduction noticed by industrial and commercial customers (Bentley and Evelyn, 1987; Chen and Sheen 1993). Such asset flexibility is a large leverage in the management of the risk with portfolio management tools used in the competitive market. The drive to install smart grid components provide the infrastructure for demand side management as well as for distributed generation.

Level 1	FERC	SPUC	SPUC	....	SPUC
2			NERC		
3			ICA/ISO/RTO		
4	GENCO	ESS/ESCO/EMCO	TRANSCO	DISTCO	EMA
5	MARKCO				BROCO

FIGURE 15.2   Business environmental model.

15.2.4 Market Structure

Although nobody knows the exact structure of the emerging deregulated industry, this research predicts that regional exchanges (i.e., electricity mercantile associations [EMAs]) will eventually play an important role. Electricity trading of the future will be accomplished through bilateral contracts and EMAs where traders bid for contracts via a double auction. The electric marketplace used in this section has been refined and described by various authors. Fahd and Sheblé (1992) demonstrated an auction mechanism. Sheblé (1994b) described the different types of commodity markets and their operation, outlining how each could be applied in the evolved electric energy marketplace. Sheblé and McCalley (1994) outlined how spot, forward, future, planning, and swap markets can handle real-time control of the system (e.g., automatic generation control) and risk management. Work by Kumar and Sheblé (1996b) brought the above ideas together and demonstrated a power system auction game designed to be a training tool. That game used the double auction mechanism in combination with classical optimization techniques.

In several references (Kumar, 1996a, 1996b; Sheblé 1996; Richter 1997a), a framework is described in which electric energy is only sold to DISTCOs, and electricity is generated by GENCOs (see Figure 15.2). The NERC sets the reliability standards. Along with DISTCOs and GENCOs, ESCOs, ancillary services companies (ANCILCOs), and TRANSCOs interact via contracts. The contract prices are determined through a double auction. Buyers and sellers of electricity make bids and offers that are matched subject to approval of the independent contract administrator (ICA), who ensures that the contracts will result in a system operating safely within limits. The ICA submits information to an ISO for implementation. The ISO is responsible for physically controlling the system to maintain its security and reliability. As of this writing, the ISO has integrated the functions of EMA, MARKCO, and BROCO within itself.

The ESS provide basic or extended services to the customers. ESCOs have provided minimal contracts simply to provide energy. EMCOs provide contracts and services to reduce the costs of energy. The advent of the smart grid enables customers to participate on the hourly transactions and pay by time of use instead of average price. Market companies (MARKCO) enable anyone to invest in the energy markets, especially speculators. Speculators already exist at most trading companies but the general public is not able to participate as in the other commodity markets. Until the general public is enabled to enter the markets, brokerage companies (BROCOs) are not needed.

15.2.5 Fully Evolved Marketplace

The following sections outline the role of a horizontally integrated industry. Many curious acronyms have described generation companies (IPP, QF, Cogen, etc.), transmission companies (IOUTS, NUTS, etc.), and distribution companies (IOUDC, COOPS, MUNIES, etc.). The acronyms used in this work are described in the following sections.

15.2.5.1 Horizontally Integrated

The restructuring of the electric power industry is most easily visualized as a horizontally integrated marketplace. This implies that interrelationships exist between generation (GENCO), transmission (TRANSCO), and distribution (DISTCO) companies as separate entities. Note that independent power producers (IPP), qualifying facilities (QF), etc. may be considered as equivalent generation companies. Nonutility transmission systems (NUTS) may be considered as equivalent transmission companies. Cooperatives and municipal utilities may be considered as equivalent distribution companies. All companies are assumed to be coordinated through a regional Transmission Corporation (or regional transmission group).

15.2.5.2 Federal Energy Regulatory Commission (FERC)

FERC is concerned with the overall operation and planning of the national grid, consistent with the various energy acts and public utility laws passed by Congress. Similar federal commissions exist in other government structures. The goal is to provide a workable business environment while protecting the economy, the customers, and the companies from unfair business practices and from criminal behavior. GENCOs, ESCOs, and TRANSCOs would be under the jurisdiction of FERC for all contracts impacting interstate trade.

15.2.5.3 State Public Utility Commission (SPUC)

SPUCs protect the individual state economies and customers from unfair business practices and from criminal behavior. It is assumed that most DISTCOs would still be regulated by SPUCs under performance-based regulation and not by FERC. GENCOs, ESCOs, and TRANSCOs would be under the jurisdiction of SPUCs for all contracts impacting intrastate trade.

15.2.5.4 Generation Company (GENCO)

The goal for a generation company, which has to fill contracts for the cash and futures markets, is to package production at an attractive price and time schedule. One proposed method is similar to the classic decentralization techniques used by a vertically integrated company. The traditional power system approach is to use Dantzig–Wolfe decomposition. Such a proposed method may be compared with traditional operational research methods used by commercial market companies for a "make or buy" decision.

15.2.5.5 Transmission Company (TRANSCO)

The goal for transmission companies, which have to provide services by contracts, is to package the availability and the cost of the integrated transportation network to facilitate transportation from suppliers GENCOs to buyer ESCOs. One proposed method is similar to oil pipeline networks and energy modeling. Such a proposed method can be compared to traditional network approaches using optimal power flow programs.

15.2.5.6 Distribution Company (DISTCO)

The goal for distribution companies, which have to provide services by contracts, is to package the availability and the cost of the radial transportation network to facilitate transportation from suppliers GENCOs to buyers ESCOs. One proposed method is similar to distribution outlets. Such proposed methods can be compared to traditional network approaches using optimal power flow programs. The disaggregation of the transmission and the distribution system may not be necessary, as both are expected to be regulated as monopolies at the present time.

15.2.5.7 Energy Service Supplier (ESS), Energy Service Company (ESCO), Energy Management Company (EMCO)

A primary tool of the ESS company is to implement demand side management (DSM) to control demand not only to reduce the peak demand but also to provide flexibility for risk management, for alternative ancillary service sources, and for other more complex risk and service enterprises (Daryanian et al., 1989; Ng and Sheblé, 1998; Lee and Chen, 1992, 1993, 1994). A secondary tool is the availability of distributed generation with natural gas micro turbines, wind generators, and solar cells. The advent of more biofuels will provide more flexibility of hybrid diesel automobiles as distributed generation. The eventual implementation of the hydrogen economy as methane gas (hydrogen combined with CO_2) will extend the use of distributed generation with cost-effective storage of CH_4 for micro turbines and hybrid automobiles (Sheblé, 1999a,b).

The goal for energy service companies, which may be large industrial customers or customer pools, is to purchase power at the least cost when needed by consumers. One proposed method is similar to the decision of a retailer to select the brand names for products being offered to the public. Such a proposed method may be compared to other retail outlet shops.

15.2.5.8 Independent System Operator (ISO)

The primary concern is the management of operations. Real-time control (or nearly real-time) must be completely secure if any amount of scheduling is to be implemented by markets. The present business environment uses a fixed combination of units for a given load level, and then performs extensive analysis of the operation of the system. If markets determine schedules, then the unit schedules may not be fixed sufficiently ahead of realtime for all of the proper analysis to be completed by the ISO.

15.2.5.9 Regional Transmission Organization (RTO)

The goal for a regional transmission group, which must coordinate all contracts and bids among the three major types of players, is to facilitate transactions while maintaining system planning. One proposed method is based on discrete analysis of a Dutch auction. Other auction mechanisms may be suggested. Such proposed methods are similar to a warehousing decision on how much to inventory for a future period. As shown later in this work, the functions of the RTG and the ISO could be merged. Indeed, this should be the case based on organizational behavior.

15.2.5.10 Independent Contract Administrator (ICA)

The goal for an ICA is a combination of the goals for an ISO and an RTG. Northern States Power Company originally proposed this term. This term will be used in place of ISO and RTG in the following to differentiate the combined responsibility from the existing ISO companies.

15.2.5.11 Energy Mercantile Association (EMA), Market Company (MARKCO), Brokerage Company (BROCO)

Competition may be enhanced through the various markets: cash, futures, planning, and swap. The cash market facilitates trading in spot and forward contracts. This work assumes that such trading would be on an hourly basis. Functionally, this is equivalent to the interchange brokerage systems implemented in several states. The distinction is that future time period interchange (forward contracts) are also traded.

The futures market facilitates trading of futures and options. These are financially derived contracts used to spread risk. The planning market facilitates trading of contracts for system expansion. Such a market has been proposed by a west coast electric utility. The swap market facilitates trading between all markets when conversion from one type of contract to another is desired. It should be noted that multiple markets are required to enable competition between markets.

The structure of any spot market auction must include the ability to schedule as far into the future as the industrial practice did before deregulation. This would require extending the spot into the future for at least 6 months, as proposed by this author (Sheblé, 1994). Future month production should be traded for actual delivery in forward markets. Future contracts should be implemented at least 18 months into the future if not 3 years. Planning contracts must be implemented for at least 20 years into the future, as recently offered by TVA, to provide an orderly, predictable expansion of the generation and transmission systems. Only then can timely addition of generation and transmission be assured. Finally, a swap market must be established to enable the transfer of contracts from one period (market) to another.

To minimize risk, the use of option contracts for each market should be implemented. Essentially, all of the players share the risk. This is why all markets should be open to the public for general trading and subject to all rules and regulations of a commodity exchange. Private exchanges, not subject to such regulations, do not encourage competition and open price discovery.

Time horizon (Months)						
0 1	2	....	12	....	18	....	360
Spot market	Forward market					
Swap market (Market to market contracts)						
		Futures market			Planning market	

FIGURE 15.3   Interconnection between markets.

The described framework (Sheblé, 1996) allows for cash (spot and forward), futures, and planning markets as shown in Figure 15.3. The *spot market* is most familiar within the electric industry (Schweppe et al. 1988). A seller and a buyer agree (either bilaterally or through an exchange) upon a price for a certain amount of power (MW) to be delivered sometime in the near future (e.g., 10 MW from 1:00 p.m. to 4:00 p.m. tomorrow). The buyer needs the electricity, and the seller wants to sell. They arrange for the electrons to flow through the electrical transmission system. A *forward contract* is a binding agreement in which the seller agrees to deliver an amount of a particular product in a specified quality at a specified time to the buyer.

A futures contract is further into the future than is the spot market. In both the forward and spot contracts, the buyer and seller want physical goods (e.g., the electrons). A *futures contract* is primarily a financial instrument that allows traders to lock in a price for a commodity in some future month. This helps traders manage their risk by limiting potential losses or gains. Futures contracts exist for commodities in which there is sufficient interest and in which the goods are generic enough that it is not possible to tell one unit of the good from another (e.g., 1 MW of electricity of a certain quality, voltage level, etc.).

A futures *option contract* is a form of insurance that gives the option purchaser the right, but not the obligation, to buy (sell) a futures contract at a given price. For each options contract, there is someone "writing" the contract who, in return for a premium, is obligated to sell (buy) at the strike price (see Figure 15.3). Both the options and the futures contracts are financial instruments designed to minimize risk. Although provisions for delivery exist, they are not convenient (i.e., the delivery point is not located where you want it to be located). The trader ultimately cancels his position in the futures market, either with a gain or loss. The physicals are then purchased on the spot market to meet demand with the profit or loss having been locked in via the futures contract.

A *swap* is a customized agreement in which one firm agrees to trade its coupon payment for one held by another firm involved in the swap. Finally, a *planning market* is needed to establish a basis for financing long term projects like transmission lines and power plants (Sheblé, 1993).

15.2.6 Computerized Auction Market Structure

Auction market structure is a computerized market, as shown in Figure 15.4. Each of the agents has a terminal (PC, workstation, etc.) connected to an auctioneer (auction mechanism) and a contract evaluator. Players generate bids (buy and sell) and submit the quotation to the auctioneer. A bid is a specified amount of electricity at a given price. The auctioneer binds bids (matching buyers and sellers) subject to approval of the contract evaluation. This is equivalent to the pool operating convention used in the vertically integrated business environment.

ICA			
Auction mechanism		Contract evaluations	
Communication player to auction			
Player	Player	Player	Player

FIGURE 15.4   Computerized markets.

Time line into future (hours)				Information Level
Real time...Planning horizon				
GENCO	ESCO	MARKCO	BROCO	Player
Bids	Bids	....	Bids	Communication
Spot	Forward	Futures	Planning	Markets
Swap				Market
ICA/ISO/RTO				Coordination
GENCO	ESCO	MARKCO	BROCO	Players
...	...	...	...	...

FIGURE 15.5   Electric market.

The contract evaluator verifies that the network can remain in operation with the new bid in place. If the network cannot operate, then the match is denied. The auctioneer processes all bids to determine which matches can be made. However, the primary problem is the complete specification of how the network can operate and how the agents are treated comparably as the network is operated closer to limits. The network model must include all constraints for adequacy and security.

The major trading objectives are hedging, speculation, and arbitrage. Hedging is a defense mechanism against loss and/or supply shortages. Speculation is assuming an investment risk with a chance for profit. Arbitrage is crossing sales (purchases) between markets for riskless profit. This work assumes that there are four markets commonly operated: forward, futures, planning, and swaps (Figure 15.5).

Forward Market: The forward contracts reflect short term future system conditions. In the forward market, prices are determined at the time of the contract but the transactions occur at some future time. Optimization tools for short term scheduling problems can be enhanced to evaluate trading opportunities in the forward market. For example, short term dispatching algorithms, such as economic unit commitment dispatch, can be used to estimate and earn profit in the forward market.

Futures Market: A futures market creates competition because it unifies diverse and scattered local markets and stabilizes prices. The contracts in the futures market are risky because price movements over time can result in large gains or losses. There is a link between forward markets and futures markets that restricts price volatility. *Options* (options contracts) allow the agent to exercise the right to activate a contract or cancel it. Claims to buy are called "call" options. Claims to sell are called "put" options.

A more detailed discussion of an electric futures contract is discussed in Sheblé (1994b). The components include trading unit, trading hours, trading months, price quotation, minimum price fluctuation, maximum daily price fluctuation, last trading day, exercise of options, option strike prices, delivery, delivery period, alternate delivery procedure, exchange of futures for, or in connection with, physicals, quality specifications, and customer margin requirements. The implementation proposals include optimization techniques at the hourly level (or faster) with the unit commitment selection under the GENCO decision domain (Fahd and Sheblé, 1992; Fahd et al., 1992; Smith, 1993; Post et al., 1995) and integrated unit commitment models with optimal power flow dispatches for each hour (Kumar and Sheblé, 1994, 1996a,b; Schweppe et al., 1988; Dekrajangpetch and Sheblé, 1998).

Swap Market: In the swap market, contract position can be closed with an exchange of physical or financial substitutions. The trader can find another trader who will accept (make) delivery and end the trader's delivery obligation. The acceptor of the obligation is compensated through a price discount or a premium relative to the market rate.

The financial drain inflicted on traders when hedging their operations in the futures market is slightly higher than the one inflicted through direct placement in the forward market. An optimal mix of options, forward commitments, futures contracts, and physical inventories is difficult to assess and depends on hedging, constraints imposed by different contracts, and the cost of different contracts. A clearinghouse such as a swap market handles the exchange of various energy instruments.

Planning Market: The growth of transmission grid requires transmission companies to make contracts based on the expected usage to finance projects. The planning market would underwrite equipment usage subject to the long term commitments to which all companies are bound by the rules of network expansion to maintain a fair marketplace. The network expansion would have to be done to maximize the use of transmission grid for all agents. Collaboration would have to be overseen and prohibited with a sufficiently high financial penalty. The growth of the generation supply similarly requires such markets. However, such a market has been started with the use of franchise rights (options) as established in recent Tennessee Valley Authority connection contracts. This author has published several papers outlining the need for such a market. Such efforts are not documented in this work.

15.2.7 Capacity Expansion Problem Definition

The capacity expansion problem is different for an ESCO, GENCO, TRANSCO, DISTCO, and ANSILCO. This section assumes that the ICA will not own equipment but will only administer the contracts between players. The capacity expansion problem is divided into the following areas: generation expansion, transmission expansion, distribution expansion, and market expansion. ESCOs are concerned with market expansion. GENCOs are concerned with generation expansion. TRANSCOs are concerned with transmission expansion. DISTCOs are concerned with distribution expansion. ANSILCOs are concerned with supportive devices expansion. This author views ancillary services as a misnomer. Such services are necessary supportive services. Thus, the term "supportive" will be used instead of ancillary. Also, since supportive devices are inherently part and parcel of the transmission or distribution system, these devices will be assumed into the TRANSCO and DISTCO functions without loss of generality. Thus, ANSILCOs are not treated separately.

Based on the above idealized view of the marketplace, the following generalizations are made. GENCOs are concerned with the addition of capacity to meet market demands while maximizing profit. Market demands include bilateral contracts with the EMA as well as bilateral contracts with ESCOs or with the ICA. ESCOs are concerned with the addition of capacity of supplying customers with the service desired to maintain market share. ESCOs are thus primarily concerned with the processing of information from marketplace to customer. However, ESCOs are also concerned with additional equipment supplied by DISTCOs or TRANSCOs to provide the level of service required by some customers. ESCOs are thus concerned with all aspects of customer contracts and not just the supply of "electrons."

The ICA is concerned with the operation of the overall system subject to the contracts between the buyers and the sellers and between all players with ICA. The overall goal of the ICA is to enable any customer to trade with any other customer with the quick resolution of contract enforcement available through mercantile associations. The ICA maintains the reliability of the network by resolving the unexpected differences between the contracts, real operation, and unplanned events. The ICA has the authority, through contracts, to buy generation services, supportive services, and/or transmission services, or to curtail contracts if the problems cannot be resolved with such purchases as defined in these contracts. Thus, the ICA has the authority to connect or disconnect generation and demand to protect the integrity of the system. The ICA has the authority to order new transmission or distribution expansion to maintain the system reliability and economic efficiency of the overall system. The economic efficiency is determined by the price of electricity in the cash markets on a periodic basis. If the prices are approximately the same at all points in the network, then the network is not preventing customers from getting to the suppliers. Similarly, the suppliers can get to the buyers. Since all buyers and suppliers are protected from each other through the default clauses of the mercantile agreement, it does not matter which company deals with other companies as the quick resolution of disputes is guaranteed.

This strictness of guarantee is the cornerstone of removing the financial uncertainty at the price of a transaction fee to cover the costs of enforcement.

The goal of each company is different but the tools are the same for each. First, the demand must be predicted for future time periods sufficiently into the future to maintain operation financially and physically. Second, the present worth of the expansion projects has to be estimated. Third, the risks associated with each project and the demand-forecast uncertainty must be estimated. Fourth, the acceptable value at risk acceptable for the company has to be defined. Fifth, the value at risk has to be calculated. Sixth, methods of reducing the value at risk have to be identified and evaluated for benefits. Seventh, the overall portfolio of projects, contracts, strategies, and risk has to be assessed. Only then can management decide to select a project for implementation.

The characteristics of expansion problems include

1. The cost of equipment or facilities should exhibit economies of scale for the same risk level baring technology changes.
2. Time is a primary factor since equipment has to be in place and ready to serve the needs as they arise. Premature installation results in idle equipment. Delayed installation results in lost market share.
3. The risk associated with the portfolio of projects should decrease as time advances.
4. The portfolio has to be revalued at each point when new information is available that may change the project selection, change the strategy, or change the mix of contracts.

The capital expansion problem is often referred to as the "capital budgeting under uncertainty" problem (Aggarwal, 1993). Thus, capital expansion is an exercise in estimating the present net value of future cash flows and other benefits as compared to the initial investment required for the project given the risk associated with the project(s). The key concept is the uncertainty and thus the risk of all business ventures. Uncertainties may be due to estimation (forecasting) and measurement errors. Such uncertainties can be reduced by the proper application of better tools. Another approach is to investment in information technology to coordinate the dissemination of information. Indeed, information technology is one key to the appropriate application of capital expansion.

Another uncertainty factor is that the net present value depends on market imperfections. Market imperfections are due to competitor reactions to each other's strategies, technology changes, and market rule changes (regulatory changes). The options offered by new investment are very hard to forecast. Also the variances of the options to reduce the risk of projects are critical to proper selection of the right project. Management has to constantly revalue the project, change the project (including termination), integrate new information, or modify the project to include technology changes.

Estimates have often been biased by management pressure to move ahead, to not investigate all risks, or to maintain strategies that are not working as planned. Uncertainties in regulations and taxes are often critical for the decision to continue.

There are three steps to any investment plan: investment alternative identification, assessment, selection and management of the investment as events warrant.

Capacity expansion is one aspect of capital budgeting. Marketing and financial investments are also capital budgeting problems. Often, the capacity expansion has to be evaluated not only on the projects merits, but also the merits of the financing bundled with the project.

15.3 Regulated Environment

The regulated industry is structured as a vertically integrated business model. This organization is typified by the structure shown in Figure 15.6. This structure should be compared with the previously defined competitive business environmental model shown in Figure 15.2. The entities previously described function in the same manner in this business model. The difference is that the utility company is an umbrella company or an integrated company with divisions segmented by type of expertise

Level 1	FERC	SPUC
2		NERC
3		UTILITY
4	GENCO/ESS/TRANSCO/DISTCO	
5	NETRELCO	

FIGURE 15.6 Regulated business environmental model.

and equipment. All divisions report to the utility and take direction from the utility in all business operations. The utility seeks tariff approval from the state public utility commission and the federal energy regulatory commission. The interchange between utility companies is approved by both FERC and SPUCs. The interchange capabilities and contract approvals are handled by the network reliability management company (NETRELCO), which coordinates network activities between utilities.

Alternative umbrella company structures are in use within the United States. American Electric Power and Southern Company are umbrella companies that own several utilities in several different states within the United States. Due to SPUC regulations and tariff approvals, utilities are located within a state boundary within the United States for ease of accounting and of regulation.

The Mid-America Interpool Network (MAIN) Company was one such network reliability management company that coordinated maintenance and interchange schedules between utilities in the Midwestern part of the United States. MAIN was announced on November 24, 1964, by American Electric Power, Commonwealth Edison, the Illinois-Missouri Pool, the Indiana Power Pool, and the Wisconsin Planning Group. It later became Mid-America Interconnected Network, Inc. and was the first regional electric reliability council. When the North American Electric Reliability Council (NERC) was later formed, MAIN became 1 of the 10 electric reliability councils that comprised NERC. MAIN served the electric utilities in the Midwest for over 41 years.

The need for a NETRELCO was due to the large amount of interchange between utilities. Interchange is beneficial to lower costs each hour (economy a), to lower costs over a number of sequential hours (economy b), and to increase reliability with more inertia and capacity as a response to disturbances (Wood and Wollenberg, 1996). The types of interchange grew quickly justifying more transmission between companies for short- and long-term economic reasons as well as for reliability reasons (Kelley et al., 1987; Winston and Gibson, 1988; Parker et al., 1989; Rau, 1989; Shirmohammadi et al., 1989; Clayton et al., 1990; Happ, 1990; Svoboda and Oren, 1994; Tabors, 1994; Post et al., 1995; Vojdani et al., 1995). Interchange on the east cost of the United States led to pool operation as demonstrated by the formation of the PJM power pool.

An alternative form was the pool company (POOLCO) such as the Pennsylvania-New Jersey-Maryland (PJM) power pool. PJM was formed in 1927 when three utilities, realizing the benefits and efficiencies possible by interconnecting to share their generating resources as if one utility, formed the world's first continuing power pool. The power pool operated a joint dispatch amongst the utilities for continuous economy, a type of interchange that was operated as if by one company [PJM]. Additional utilities joined in 1956, 1965, and 1981. Throughout this time, PJM was operated by a department of one member utility with contractual interchange agreements to dynamically alter the amounts of interchange and the costs of interchange amongst the participants.

In 1962, PJM installed its first online computer to control generation and dispatch generation with dynamic interchange between the member utility companies. PJM broadcasted a system lambda that each unit used as the economic solution to the overall dispatch of all companies. PJM completed its first energy management system (EMS) in 1968 and included transfer limitations between the member companies using a linear network solution technique. The EMS is the information technology system that makes it possible to monitor transmission grid operations in real time as a NETRELCO. The primary task of a NETRELCO was to study the reliability of the electric power system (Breiphol, 1990;

Billinton and Lian, 1991, Billinton and Wenyuan, 1991, Billinton and Gan, 1993, Billinton and Li, 1994). In 1996, PJM launched its first website to provide its members with current system information. PJM transitioned to an ISO in 1997. PJM includes the functions of an ICA, EMA, NETRELCO, and RTO in the competitive environment. The transition of PJM from regulated to competitive market environment is a study of the subtle changes between the two environments.

Dynamic interchange was extended for the joint operation of jointly owned units. Planning solutions in the Midwest section of the United States found that sharing the cost of unit construction and operation was economic (Podmore et al., 1979; Lee, 1988). Such units were called jointly owned units (JOUs) and the interchange between member owners became dynamically dispatched so each member could use the unit for economic operation continuously. The resulting dynamic alteration of a unit's output led to dynamic interchange between the members, mimicking a power pool with the JOUs. The ultimate JOU in the United States is the Intermountain Power Project (IPP). This JOU installation is owned by over 50 companies. Some of the companies receive the output through dynamic interchange agreements on the Utah-based alternating current (AC) transmission system, others receive the output across the high-voltage direct current (HVDC) link to California.

The Florida Energy Broker was another implementation of a pool type operation for continuous (hourly) economic dispatch of generation within the state of Florida. This organization collected unit bids and asks for auction matching in the 1980s. The resulting matches were implemented by each utility each hour resulting in significant operational savings. As such, this was the first implementation of a competitive environment using an auction mechanism (Cohen, 1982). That implementation was the basis for auction market simulation (Sheblé, 1994a,b).

The planning of utilities under such a business environment has been well documented by general textbooks and papers (Adams et al., 1972; Booth, 1972; Chao, 1983; Merril, 1991; Merril and Wood, 1990; Sullivan, 1977; Wang and McDonald, 1994; Willis, 1996; Seifi and Sepasian, 2001; Li, 2011). The production costing of generation was one of the first computerized planning tools used in both business environments (Baleriaux, 1967; Day, 1971). The application of probabilistic production costing has been one of the most thoroughly researched areas (Lee, F, 1988; Lin et al., 1989; Wang, 1989; Billinton and Lian, 1991; Billinton and Wenyuan, 1991; Billington and Li, 1992; Billington and Gan, 1993; Delson et al., 1991; Huang and Chen, 1993; Pereiran et al., 1992; Lee and Chen, 1992, 1993, 1994; Miranda, 1994; Parker and Stremel, 1996). The other area of intense research is in the area of transaction selection between utilities (Kelley et al., 1987; Winston and Gibson, 1988; Parker et al., 1989; Rau, 1989; Clayton et al., 1990; Happ, 1990; Shirmohammadi et al., 1989; Roy, 1993). The mixed business models used within the United States has spawned a number of mixed tools to provide the benefits of both environments (David and Li, 1993; Krause and McCalley, 1994; Rakic and Markovic, 1994; Post et al., 1995; Vojdani et al., 1995; Wu and Varaiya, 1995; Clayton and Mukerji, 1996; Richter and Sheblé, 1997a, 1997b, 1998). The application of competitive markets compared to traditional regulation has been analyzed but more work is required (Hobbs and Schuler, 1985; Hogan, 1977, Lerner, 1994; O'Neill and Whitmore, 1994; Oren et al., 1994; Oren, 1997; Smith, 1988; Bhattacharya et al., 2001; Loi, 2001).

15.4 Other Sections on Planning

The following sections on planning deal with the overall approach as described by Dr. H. Merrill and include sections on forecasting, power system planning, transmission planning, and system reliability. Forecasting demand is a key issue for any business entity. Forecasting for a competitive industry is more critical than for a regulated industry. Transmission planning is discussed based on probabilistic techniques to evaluate the expected advantages and costs of present and future expansion plans. Reliability of the supply is covered, including transmission reliability. The most interesting aspect of the electric power industry is the massive changes presently occurring. It will be interesting to watch as the industry adapts to regulatory changes and as the various market players find their corporate niche in this new framework.

References

Adams, N., F. Belgari, M.A. Laughton, and G. Mitra, Mathematical programming systems in generation, transmission and distribution planning, *PSCC Proceedings Paper Number 1, 1/13*, Grenoble, France, 1972.

Aggarwal, R., *Capital Budgeting Under Uncertainty*, Prentice-Hall, Englewood Cliffs, NJ, 1993.

Baleriaux, H., Simulation de l'exploitation d'un parc de machines thermiques de production d'electricite couple a des stations de pompage, *Revue E*, 5(7), 1–24, 1967.

Baughman, M.L., J.W. Jones, and A. Jacob, Model for evaluating the economics of cool storage systems, *IEEE Trans. Power Syst.*, 8(2), 716–722, May 1993.

Bently, W.G. and J.C. Evelyn, Customer thermal energy storage: A marketing opportunity for cooling off electric peak demand, *IEEE Trans. Power Syst.*, 1(4), 973–979, 1987.

Bhattacharya, K., M.H.J. Bollen, and J.E. Daalder, *Operation of Restructured Power Systems*, Springer, New York, 2001.

Billinton, R. and L. Gan, Monte Carlo simulation model for multiarea generation system reliability studies, *IEE Proc.*, 140(6), 532–538, 1993.

Billinton, R. and W. Li, A Monte Carlo method for multi-area generation system reliability assessment, *IEEE Trans. Power Syst.*, 7(4), 1487–1492, 1992.

Billinton, R. and W. Li, *Reliability Assessment of Electrical Power Systems Using Monte Carlo Methods*, Springer, New York, 1994.

Billinton, R. and G. Lian, Monte Carlo approach to substation reliability evaluation, *IEE Proc.*, 140(2), 147–152, 1991.

Billinton, R. and L. Wenyuan, Hybrid approach for reliability evaluation of composite generation and transmission systems using Monte-Carlo simulation and enumeration technique, *IEE Proc.*, 138(3), 233–241, 1991.

Binger, B.R. and E. Hoffman, *Microeconomics with Calculus*, Scott, Foresman and Company, Glenview, IL, 1988.

Booth, R.R., Optimal generation planning considering uncertainty, *IEEE Trans. PAS-91*, 70–77, Jan./Feb. 1972.

Breipohl, A.M., *Probabilistic Systems Analysis: An Introduction to Probabilistic Models, Decisions, and Applications of Random Processes*, Wiley, New York, 1990.

Bussey, L. E., *The Economic Analysis of Industrial Projects*, Prentice-Hall, Englewood Cliffs, NJ, 1981.

Chao, H.P., Peak load pricing and capacity planning with demand and supply uncertainty, *Bell J. Econ.*, 14(1), 179–190, 1983.

Chen, C.S. and J.N. Sheen, Cost benefit analysis of a cooling energy storage system, *IEEE Trans. Power Syst.*, 8(4), 1504–1510, 1993.

Chu, W., B. Chen, and C. Fu, Scheduling of direct load control to minimize load reduction for a utility suffering from generation shortage, *IEEE/PES Winter Meeting*, Columbus, OH, 1993.

Clayton, J.S., S.R. Erwin, and C.A. Gibson, Interchange costing and wheeling loss evaluation by means of incrementals, *IEEE Trans. Power Syst.*, 5(3), 759–765, 1990.

Clayton, R.E. and R. Mukerji, System planning tools for the competitive market, *IEEE Comput. Appl. Power*, 9(3), 50–55, 1996.

Cohen, L., *A Spot Market for Electricity: Preliminary Analysis of the Florida Energy Broker*, RAND Note, N-1817-DOE, February 1982.

Cohen, A.I., J.W. Patmore, D.H. Oglevee, R.W. Berman, L.H. Ayers, and J.F. Howard, An integrated system for load control, *IEEE Trans. Power Syst.*, PWRS-2(3), 645–651, 1987.

Daellenbach, H.G., *Systems and Decision Making*, John Wiley & Sons, New York, 1994.

Daryanian, B., R.E. Bohn, and R.D. Tabors, Optimal demand-side response to electricity spot prices for storage-type customers, *IEEE Trans. Power Syst.*, 4(3), 897–903, 1989.

David, A.K. and Y.Z. Li, Effect of inter-temporal factors on the real-time pricing of elasticity, *IEEE Trans. Power Syst.*, 8(1), 44–52, 1993.

Day, J.T., Forecasting minimum production costs with linear programming, *IEEE Trans. Power Appar. Syst.*, PAS-90(2), 814–823, 1971.

Dekrajangpetch, S. and G.B. Sheblé, Alternative implementations of electric power auctions, in *Proceedings of the 60th American Power Conference*, Vol. 60-1, pp. 394–398, 1998.

Delson, J.K., X. Feng, and W.C. Smith, A validation process for probabilistic production costing programs, *IEEE Trans. Power Syst.*, 6(3), 1326–1336, 1991.

Fahd, G., Optimal power flow emulation of interchange brokerage systems using linear programming, PhD dissertation, Auburn University, Auburn, AL, 1992.

Fahd, G., D.A. Richards, and G.B. Sheblé, The implementation of an energy brokerage system using linear programming, *IEEE Trans. Power Syst.*, T-PWRS, 7(1), 90–96, 1992.

Fahd, G. and G. Sheblé, Optimal power flow of interchange brokerage system using linear programming, *IEEE Trans. Power Syst.*, T-PWRS, 7(2), 497–504, 1992.

Happ, H.H., Report on Wheeling Costs, Case 88-E-238, The New York Public Service Commission, February 1990.

Hobbs, B.F. and R.E. Schuler, An assessment of the deregulation of electric power generation using network models of imperfect spatial markets, *Papers of the Regional Science Association*, 57(1), 75–89, 1985.

Hogan, W.W., A market power model with strategic interaction in electricity networks, *Energy J.*, 18(4), 107–141, 1997.

Huang, S.R. and S.L. Chen, Evaluation and improvement of variance reduction in Monte-Carlo production simulation, *IEEE Trans. Energy Convers.*, 8(4), 610–619, 1993.

Ilic, M., F. Galiana, and L. Fink, *Power Systems Restructuring: Engineering and Economics*, Kluwer Academic Publishers, Norwell, MA, 1998.

Kelley, K., S. Henderson, P. Nagler, and M. Eifert, *Some Economic Principles for Pricing Wheeled Power*, National Regulatory Research Institute, Silver Spring, MD, August 1987.

Krause, B.A. and J. McCalley, Bulk power transaction selection in a competitive electric energy system with provision of security incentives, *Proceedings of the 26th Annual North American Power Symposium*, Manhattan, KS, pp. 126–136, September 1994.

Kumar, J., Electric power auction market implementation and simulation, PhD dissertation, Iowa State University, Ames, IA, 1996.

Kumar, J. and G.B. Sheblé, A framework for transaction selection using decision analysis based upon risk and cost of insurance, *Proceedings of the 29th North American Power Symposium*, Kansas State University, Manhattan, KS, pp. 548–557, 1994.

Kumar, J. and G.B. Sheblé, A decision analysis approach to transaction selection problem in a competitive electric market, *Electric Power Syst. Res. J.*, 38(3), 209–216, 1996a.

Kumar, J. and G.B. Sheblé, Transaction selection using decision analysis based upon risk and cost of insurance, *IEEE Winter Power Meeting*, Baltimore, MD, 1996b.

Kurucz, C.N., D. Brandt, and S. Sim, A linear programming model for reducing system peak through customer load control programs, *IEEE PES Winter Meeting*, 96 WM 239-9 PWRS, Baltimore, MD, 1996.

Lai, L.L., *Power System Restructuring and Deregulation: Trading, Performance and Information Technology*, John Wiley & Sons, Chichester, U.K., 2001.

Le, K.D., R.F. Boyle, M.D. Hunter, and K.D. Jones, A procedure for coordinating direct-load-control strategies to minimize system production cost, *IEEE Trans. Power Appar. Syst.*, PAS-102(6), 1983.

Lee, F.N., Three-area joint dispatch production costing, *IEEE Trans. Power Syst.*, 3(1), 294–300, 1988.

Lee, T.Y. and N. Chen, The effect of pumped storage and battery energy storage systems on hydrothermal generation coordination, *IEEE Trans. Energy Convers.*, 7(4), 631–637, 1992.

Lee, T.Y. and N. Chen, Optimal capacity of the battery storage system in a power system, *IEEE Trans. Energy Convers.*, 8(4), 667–673, 1993.

Lee, T.Y. and N. Chen, Effect of battery energy storage system on the time-of-use rates industrial customers, *IEE Proc. Gener. Transm. Distrib.*, 141(5), 521–528, 1994.

Lee, S.H. and C.L. Wilkins, A practical approach to appliance load control analysis: A water heater case study, *IEEE Trans. Power Appar. Syst.*, 7(4), 1992.

Lerner, A.P., Monopoly and the measurement of monopoly power, *Rev. Econ. Stud.*, 1, 157–175, 1934.

Li, W., *Probabilistic Transmission System Planning*, IEEE Press, Hoboken, NJ, 2011.

Lin, M., A. Breipohl, and F. Lee, Comparison of probabilistic production cost simulation methods, *IEEE Trans. Power Syst.*, 4(4), 1326–1333, 1989.

McCalley, J., A. Fouad, V. Vittal, A. Irizarry-Rivera, R. Farmer, and B. Agarwal, A probabilistic problem in electric power system operation: The economy-security tradeoff for stability-limited systems, *Proceedings of the Third International Workshop on Rough Sets and Soft Computing*, November 10–12, 1994, San Jose, CA.

McCalley, J. and G.B. Sheblé, Competitive electric energy systems: Reliability of bulk transmission and supply, tutorial paper presented at the *Fourth International Conference of Probabilistic Methods Applied to Power Systems*, Brazil, 1994.

Merrill, H.M., Have I Ever Got a Deal for You. Economic Principles in Pricing of Services, IEEE SP 91EH0345-9-PWR, pp. 1–8, 1991.

Merril, H.M. and A.J. Wood, Risk and uncertainty in power system planning, *10th Power Systems Computation Conference, PSCC*, Graz, Austria, August 1990.

Miranda, V., Power system planning and fuzzy sets: Towards a comprehensive model including all types of uncertainties, *Proceedings of the PMAPSí94*, Rio de Janeiro, Brazil, September 1994.

Miranda, V. and L.M. Proença, A general methodology for distribution planning under uncertainty, including genetic algorithms and fuzzy models in a multi-criteria environment, *Proceedings of Stockholm Power Technology, SPT'95*, Stockholm, Sweden, June 18–22, pp. 832–837, 1995.

Mortensen, R.E. and K.P. Haggerty, Dynamics of heating and cooling loads: Models, simulation, and actual utility data, *IEEE Trans. Power Syst.*, 5(1), 253–248, 1990.

Ng, K.-H., Reformulating load management under deregulation, Master's thesis, Iowa State University, Ames, IA, May 1997.

Ng, K.-H. and G.B. Sheblé, Direct load control—A profit-based load management using linear programming, *IEEE Trans. Power Syst.*, 13(2), 688–694, 1998.

O'Neill, R.P. and C.S. Whitmore, Network oligopoly regulation: An approach to electric federalism, *Electr. Federalism Symp.*, June 24, 1993 (Revised March 16, 1994).

Oren, S.S., Economic inefficiency of passive transmission rights in congested electricity systems with competitive generation, *Energy J.*, 18(1), 63–83, 1997.

Oren, S.S., P. Spiller, P. Varaiya, and F. Wu, Nodal prices and transmission rights: A critical appraisal, University of California at Berkeley Research Report, December 1994.

Outhred, H.R., Principles of a market-based electricity industry and possible steps toward implementation in Australia, *International Conference on Advanced Power System Control, Operation and Management*, Hong Kong, December 7–10, 1993.

Parker, B.J., E. Denzinger, B. Porretta, G.J. Anders, and M.S. Mirsky, Optimal economic power transfers, *IEEE Trans. Power Syst.*, 4(3), 1167–1175, 1989.

Parker, C. and J. Stremel, A smart Monte Carlo procedure for production costing and uncertainty analysis, *Proceedings of the American Power Conference*, 58(II), 897–900, 1996.

Pereira, V., B.G. Gorenstin, and M. Morozowski Fo, Chronological probabilistic production costing and wheeling calculations with transmission network modeling, *IEEE Trans. Power Syst.*, 7(2), 885–891, 1992.

Podmore, R. et al., Automatic generation control of jointly-owned generating units, *IEEE Trans. Power Apparatus Syst.*, PAS-98 (I), 207–218, January/February, 1979.

Post, D., Electric power interchange transaction analysis and selection, Master's thesis, Iowa State University, Ames, IA, 1994.

Post, D., S. Coppinger, and G. Sheblé, Application of auctions as a pricing mechanism for the interchange of electric power, *IEEE Trans. Power Syst.*, 10(3), 1580–1584, 1995.

Rakic, M.V. and Z.M. Markovic, Short term operation and power exchange planning of hydro-thermal power systems, *IEEE Trans. Power Syst.*, 9(1), 359–365, 1994.

Rau, N.S., Certain considerations in the pricing of transmission service, *IEEE Trans. Power Syst.*, 4(3), 1133–1139, 1989.

Richter, C. and G. Sheblé, Genetic algorithm evolution of utility bidding strategies for the competitive marketplace, *1997 IEEE/PES Summer Meeting*, Berlin, Germany, PE-752-PWRS-1-05-1997, IEEE, New York, 1997a.

Richter, C. and G. Sheblé, Building fuzzy bidding strategies for the competitive generator, *Proceedings of the 1997 North American Power Symposium*, Ames, IA, 1997b.

Richter, C. and G. Sheblé, Bidding strategies that minimize risk with options and futures contracts, *Proceedings of the 1998 American Power Conference, Session 25, Open Access II-Power Marketing, Paper C*, Chicago, IL, 1998.

Roy, S., Goal-programming approach to optimal price determination for inter-area energy trading, *Intl. J. Energ. Res.*, 17, 847–862, 1993.

Rupanagunta, P., M.L. Baughman, and J.W. Jones, Scheduling of cool storage using non-linear programming techniques, *IEEE Trans. Power Syst.*, 10(3), 1279–1285, 1995.

Russel, T., Working with an independent grid in the UK—A generator's view, *Proceedings of the 24th Annual North American Power Symposium*, Manhattan, KS, pp. 270–275, September 1992.

Schweppe, F.C., M.C. Caramanis, R.D. Tabors, and R.E. Bohn, *Spot Pricing of Electricity*, Kluwer Academic Publishers, Boston, MA, 1988.

Seifi, H. and M.S. Sepasian, *Electric Power System Planning: Issues, Algorithms and Solutions*, Springer, Heidelberg, Germany, 2001.

Sheblé, G.B., Electric energy in a fully evolved marketplace, *Proceedings of the 26th Annual North American Power Symposium*, Manhattan, KS, pp. 81–90, September 1994a.

Sheblé, G.B., Simulation of discrete auction systems for power system risk management, *Proceedings of the 27th Annual Frontiers of Power Conference*, Oklahoma State University, Stillwater, OK, pp. I.1–I.9, 1994b.

Sheblé, G.B., Priced based operation in an auction market structure, Presented at the *1996 IEEE Winter Power Meeting*, Baltimore, MD, pp. 1–7, 1996.

Sheblé, G. and G. Fahd, Unit commitment literature synopsis, *IEEE Trans. Power Syst.*, 9, 128–135, 1994.

Sheblé, G. and J. McCalley, Discrete auction systems for power system management, presented at the *1994 National Science Foundation Workshop*, Pullman, WA, 1994.

Sheblé, G.B., *Auction Methods for Restructured Power System Operation*, Springer, 1999a.

Sheblé, G.B., *Computational Auction Mechanisms for Restructured Power Industry Operation*, Springer, 1999b.

Shirmohammadi, D., P.R. Gribik, T.K. Law, J.H. Malinowski, and R.E. O'Donnell, Evaluation of transmission network capacity use for wheeling transactions, *IEEE Trans. Power Syst.*, 4(4), 1405–1413, 1989.

Skeer, J., Highlights of the international energy agency conference on advanced technologies for electric demand-side management, *Proceedings of the Advanced Technologies for Electric Demand-Side Management*, International Energy Agency, Sorrento, Italy, 1991.

Smith, V.L., Electric power deregulation: Background and prospects, *Contemporary Policy Issues*, 6, 14–24, 1988.

Smith, S., Linear programming model for real-time pricing of electric power service, *Oper. Res.*, 41, 470–483, 1993.

Sullivan, R.L., *Power System Planning*, McGraw-Hill, New York, 1977.

Svoboda, A. and S. Oren, Integrating price-based resources in short-term scheduling of electric power systems, *IEEE Trans. Energy Convers.*, 9, 760–769, 1994.

Tabors, R.D., Transmission system management and pricing: New paradigms and international comparisons, Paper WM110-7 presented at the *IEEE/PES Winter Meeting, T-PWRS*, February 1994.

Vojdani, A., C. Imparto, N. Saini, B. Wollenberg, and H. Happ, Transmission access issues, presented at the *1995 IEEE/PES Winter Meeting*, 95 WM 121-4 PWRS, IEEE, New York, 1994.

Vojdani, A., C. Imparto, N. Saini, B. Wollenberg, and H. Happ, Transmission access issues, presented at the *1995 IEEE/PES Winter Meeting*, 95 WM 121-4 PWRS, IEEE, New York, 1995.

Wang, L., Approximate confidence bounds on monte carlo simulation results for energy production, *IEEE Trans. Power Syst.*, 4(1), 69–74, 1989.

Wang, C. and J.R. McDonald, *Modern Power System Planning*, McGraw-Hill, New York, 1994.

Wei, D.C. and N. Chen, Air-conditioner direct load control by multi-pass dynamic programming, *IEEE Trans. Power Syst.*, 10(1), 307–313, 1995.

Willis, H.L., *Spatial Electric Load Forecasting*, Marcel Dekker, New York, pp. 14–17, 1996.

Winston, W.E., and C.A. Gibson, Geographical load shift and its effect on interchange evaluation, *IEEE Trans. Power Syst.*, 3(3), 865–871, 1988.

Wood, A.J. and B.F. Wollenberg, *Power Generation, Operation, and Control*, 2nd edn., John Wiley & Sons, New York, 1996.

Wu, F. and P. Varaiya, Coordinated multi-lateral trades for electric power networks: Theory and implementation, University of California at Berkeley Research Report, June 1995.

16

Short-Term Load and Price Forecasting with Artificial Neural Networks[*]

Alireza Khotanzad
Southern Methodist University

16.1 Artificial Neural Networks

Artificial neural networks (ANN) are systems inspired by research into how the brain works. An ANN consists of a collection of arithmetic computing units (nodes or neurons) connected together in a network of interconnected layers. A typical node of an ANN is shown in Figure 16.1. At the input side, there are a number of so-called connections that have a weight of W_{ij} associated with them. The input denoted by X_i gets multiplied by W_{ij} before reaching node j via the respective connection. Inside the neuron, all the individual inputs are first summed up. The summed inputs are passed through a nonlinear single-input, single-output function "S" to produce the output of the neuron. This output in turn is propagated to other neurons via corresponding connections.

While there are a number of different ANN architectures, the most widely used one (especially in practical applications) is the multilayer feed-forward ANN, also known as a multilayer perceptron (MLP), shown in Figure 16.2. An MLP consists of n input nodes, h so called "hidden layer" nodes (since they are not directly accessible from either input or output side), and m output nodes connected in a feed-forward fashion. The input layer nodes are simple data distributors whereas neurons in the hidden and output layers have an S-shaped nonlinear transfer function known as the "sigmoid activation function," $f(z) = 1/1 + e^{-z}$ where z is the summed inputs.

For hidden layer nodes, the output is

$$H_j = \frac{1}{1 + \exp\left(-\sum_{i=1}^{n} W_{ij} X_i\right)}$$

---

[*] This work was supported in part by the Electric Power Research Institute and 1997 Advanced Technology Program of the State of Texas.

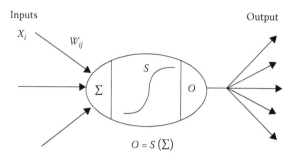

FIGURE 16.1 Model of one node of an ANN.

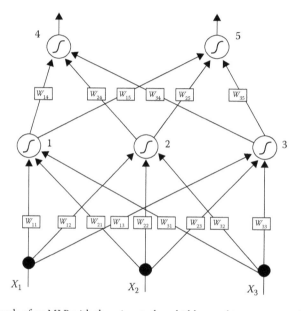

FIGURE 16.2 An example of an MLP with three input, three hidden, and two output nodes.

where H_j is the output of the jth hidden layer node, $j = 1,\ldots, h$, and X_i represents the ith input connected to this hidden node via W_{ij} with $i = 1,\ldots, n$.

The output of the kth output node is given by

$$Y_k = \frac{1}{1+\exp\left(-\sum_{j=1}^{h} W_{jk}H_j\right)}$$

where Y_k is the output of the kth output layer node with $k = h + 1,\ldots, m$, and W_{jk} representing connection weights from hidden to output layer nodes.

One of the main properties of ANNs is the ability to model complex and nonlinear relationships between input and output vectors through a learning process with "examples." During learning, known input-output examples, called the training set, are applied to the ANN. The ANN learns by adjusting or adapting the connection weights through comparing the output of the ANN to the expected output. Once the ANN is trained, the extracted knowledge from the process resides in the resulting connection weights in a distributed manner.

A trained ANN can generalize (i.e., produce the expected output) if the input is not exactly the same as any of those in the training set. This property is ideal for forecasting applications where some historical data exists but the forecast indicators (inputs) may not match up exactly with those in the history.

16.1.1 Error Back-Propagation Learning Rule

The MLP must be trained with historical data to find the appropriate values for W_{ij} and the number of required neurons in the hidden layer. The learning algorithm employed is the well-known error back-propagation (BP) rule (Rumelhart and McClelland, 1986). In BP, learning takes place by adjusting W_{ij}. The output produced by the ANN in response to inputs is repeatedly compared with the correct answer. Each time, the W_{ij} values are adjusted slightly in the direction of the correct answers by back-propagating the error at the output layer through the ANN according to a gradient descent algorithm.

To avoid overtraining, the cross-validation method is used. The training set is divided into two sets. For instance, if three years of data is available, it is divided into a two-year and a one-year set. The first set is used to train the MLP and the second set is used to test the trained model after every few hundred passes over the training data. The error on the validation set is examined. Typically this error decreases as the number of passes over the training set is increased until the ANN is overtrained, as signified by a rise in this error. Therefore, the training is stopped when the error on the validation set starts to increase. This procedure yields the appropriate number of epochs over the training set. The entire three years of data is then used to retrain the MLP using this number of epochs.

In a forecasting application, the number of input and output nodes is equal to the number of utilized forecast indicators and the number of desired outputs, respectively. However, there is no theoretical approach to calculate the appropriate number of hidden layer nodes. This number is determined using a similar approach for training epochs. By examining the error over a validation set for a varying number of hidden layer nodes, a number yielding the smallest error is selected.

16.1.2 Adaptive Update of the Weights during Online Forecasting

A unique aspect of the MLPs used in the forecasting systems described in this section is the adaptive update of the weights during online operation. In a typical usage of an MLP, it is trained with the historical data and the weights of the trained MLP are then treated as fixed parameters. This is an acceptable procedure for many applications. However, if the modeled process is a nonstationary one that can go through rapid changes, e.g., variations of electric load due to weather swings or seasonal changes, a tracking mechanism with sensitivity to the recent trends in the data can aid in producing better results.

To address this issue, an adaptive weight adjustment strategy that takes place during online operation is utilized. The MLP is initially trained using the BP algorithm; however, the trained weights are not treated as static parameters. During online operation, these weights are adaptively updated on a sample-by-sample basis. Before forecasting for the next instance, the forecasts of the past few samples are compared to the actual outcome (assuming that actual outcome for previous forecasts have become available) and a small scale error BP operation is performed with this data. This mini-training with the most recent data results in a slight adjustment of the weights and biases them toward the recent trend in data.

16.2 Short-Term Load Forecasting

The daily operation and planning activities of an electric utility requires the prediction of the electrical demand of its customers. In general, the required load forecasts can be categorized into short-term, mid-term, and long-term forecasts. The short-term forecasts refer to hourly prediction of the load for a lead time ranging from 1 h to several days out. The mid-term forecasts can either be hourly or peak load forecasts for a forecast horizon of one to several months ahead. Finally, the long-term forecasts refer to forecasts made for one to several years in the future.

The quality of short-term hourly load forecasts has a significant impact on the economic operation of the electric utility since many decisions based on these forecasts have significant economic consequences. These decisions include economic scheduling of generating capacity, scheduling of fuel purchases, system security assessment, and planning for energy transactions. The importance of accurate load forecasts will increase in the future because of the dramatic changes occurring in the structure of the utility industry due to deregulation and competition. This environment compels the utilities to operate at the highest possible efficiency, which, as indicated above, requires accurate load forecasts. Moreover, the advent of open access to transmission and distribution systems calls for new actions such as posting the available transmission capacity (ATC), which will depend on the load forecasts.

In the deregulated environment, utilities are not the only entities that need load forecasts. Power marketers, load aggregators, and independent system operators (ISO) will all need to generate load forecasts as an integral part of their operation.

This section describes the third generation of an ANN hourly load forecaster known as Artificial Neural Network Short-Term Load Forecaster (ANNSTLF). ANNSTLF, developed by Southern Methodist University and PRT, Inc. under the sponsorship of the Electric Power Research Institute (EPRI), has received wide acceptance by the electric utility industry and is presently being used by over 40 utilities across the U.S. and Canada.

Application of the ANN technology to the load forecasting problem has received much attention in recent years (Dillon et al., 1991; Park et al., 1991; Ho et al., 1992; Lee et al., 1992; Lu et al., 1993; Peng et al., 1993; Papalexopolos et al., 1994; Khotanzad et al., 1995, 1996, 1997, 1998; Mohammed et al., 1995; Bakirtzis et al., 1996). The function learning property of ANNs enables them to model the correlations between the load and such factors as climatic conditions, past usage pattern, the day of the week, and the time of the day, from historical load and weather data. Among the ANN-based load forecasters discussed in published literature, ANNSTLF is the only one that is implemented at several sites and thoroughly tested under various real-world conditions.

A noteworthy aspect of ANNSTLF is that a single architecture with the same input-output structure is used for modeling hourly loads of various size utilities in different regions of the country. The only customization required is the determination of some parameters of the ANN models. No other aspects of the models need to be altered.

16.2.1 ANNSTLF Architecture

ANNSTLF consists of three modules: two ANN load forecasters and an adaptive combiner (Khotanzad et al., 1998). Both load forecasters receive the same set of inputs and produce a load forecast for the same day, but they utilize different strategies to do so. The function of the combiner module is to mix the two forecasts to generate the final forecast.

Both of the ANN load forecasters have the same topology with the following inputs:

- 24 hourly loads of the previous day
- 24 hourly weather parameters of the previous day (temperatures or effective temperatures, as discussed later)
- 24 hourly weather parameters forecasts for the coming day
- Day type indices

The difference between the two ANNs is in their outputs. The first forecaster is trained to predict the regular (base) load of the next day, i.e., the 24 outputs are the forecasts of the hourly loads of the next day. This ANN will be referred to as the "Regular Load Forecaster (RLF)."

On the other hand, the second ANN forecaster predicts the *change* in hourly load from yesterday to today. This forecaster is named the "Delta Load Forecaster (DLF)."

The two ANN forecasters complement each other because the RLF emphasizes regular load patterns whereas the DLF puts stronger emphasis on yesterday's load. Combining these two separate forecasts results in improved accuracy. This is especially true for cases of sudden load change caused by weather fronts. The RLF has a tendency to respond slowly to rapid changes in load. On the other hand, since the DLF takes yesterday's load as the basis and predicts the changes in that load, it has a faster response to a changing situation.

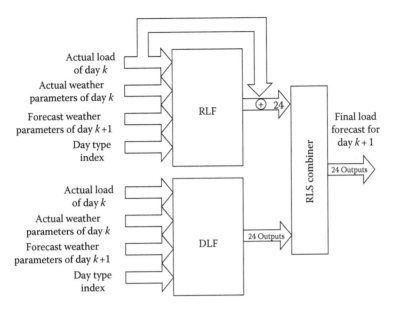

FIGURE 16.3 Block diagram of ANNSTLF.

To take advantage of the complimentary performance of the two modules, their forecasts are adaptively combined using the recursive least squares (RLS) algorithm (Proakis et al., 1992). The final forecast for each hour is obtained by a linear combination of the RLF and DLF forecasts as

$$\hat{L}_{k+1}(i) = \alpha_B(i)\hat{L}_{k+1}^{RLF}(i) + \alpha_C(i)\hat{L}_{k+1}^{DLF}(i), \quad i = 1,\dots,24$$

The $\alpha_B(i)$ and $\alpha_C(i)$ coefficients are computed using the RLS algorithm. This algorithm produces coefficients that minimize the weighted sum of squared errors of the past forecasts denoted by J,

$$J = \sum_{k=1}^{N} \beta^{N-k}[L_k(i) - \hat{L}_k(i)]^2$$

where $L_k(i)$ is the actual load at hour i, N is the number of previous days for which load forecasts have been made, and β is a weighting factor in the range of $0 < \beta \leq 1$ whose effect is to de-emphasize (forget) old data.

The block diagram of the overall system is shown in Figure 16.3.

16.2.2 Humidity and Wind Speed

Although temperature (T) is the primary weather variable affecting the load, other weather parameters, such as relative humidity (H) and wind speed (W), also have a noticeable impact on the load. The effects of these variables are taken into account through transforming the temperature value into an effective temperature, T_eff, using the following transformation:

$$T\_eff = T + \alpha * H$$

$$T\_eff = T - \frac{W * (65° - T)}{100}$$

16.2.3 Holidays and Special Days

Holidays and special days pose a challenge to any load forecasting program since the load of these days can be quite different from a regular workday. The difficulty is the small number of holidays in the historical data compared to the typical days. For instance, there would be three instances of Christmas Day in a training set of 3 years. The unusual behavior of the load for these days cannot be learned adequately by the ANNs since they are not shown many instances of these days.

It was observed that in most cases, the profile of the load forecast generated by the ANNs using the concept of designating the holiday as a weekend day, does resemble the actual load. However, there usually is a significant error in predicting the peak load of the day. The ANNSTLF package includes a function that enables the user to reshape the forecast of the entire day if the peak load forecast is changed by the user. Thus, the emphasis is placed on producing a better peak load forecast for holidays and reshaping the entire day's forecast based on it.

The holiday peak forecasting algorithm uses a novel weighted interpolation scheme. This algorithm will be referred to as "Reza algorithm" after the author who developed it (Khotanzad et al., 1998). The general idea behind the Reza algorithm is to first find the "close" holidays to the upcoming one in the historical data. The closeness criterion is the temperature at the peak-load hour. Then, the peak load of the upcoming holiday is computed by a novel weighted interpolation function described in the following.

The idea is best illustrated by an example. Let us assume that there are only three holidays in the historical data. The peak loads are first adjusted for any possible load growths. Let (t_i, p_i) designate the i-th peak-load hour temperature and peak load, respectively. Figure 16.4 shows the plot of p_i vs. t_i for an example case.

Now assume that t_h represents the peak-load hour temperature of the upcoming holiday. t_h falls in between t_1 and t_2 with the implication that the corresponding peak load, p_h, would possibly lie in the range of $[p_1, p_2] = R_1 + R_2$. But, at the same time, t_h is also between t_1 and t_3 implying that p_h would lie in $[p_1, p_3] = R_1$. Based on this logic, p_h can lie in either R_1 or $R_1 + R_2$. However, note that R_1 is common in both ranges. The idea is to give twice as much weight to the R_1 range for estimating p_h since this range appears twice in pair-wise selection of the historical data points.

The next step is to estimate p_h for each nonoverlapping interval, R_1 and R_2, on the y axis, i.e., $[p_1, p_3]$ and $[p_3, p_2]$.

For $R_1 = [p_1, p_3]$ interval:

$$\hat{p}_{h1} = \frac{p_3 - p_1}{t_3 - t_1} * (t_h - t_1) + p_1$$

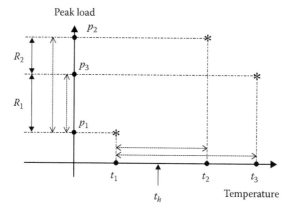

FIGURE 16.4   Example of peak load vs. temperature at peak load for a three-holiday database.

For $R_2 = [p_3, p_2]$ interval:

$$\hat{p}_{h2} = \frac{p_2 - p_3}{t_2 - t_3} * (t_h - t_3) + p_3$$

If any of the above interpolation results in a value that falls outside the respective range, R_i, the closest p_i, i.e., maximum or minimum of the interval, is used instead.

The final estimate of p_h is a weighted average of $\hat{p}_{h1}$ and $\hat{p}_{h2}$ with the weights decided by the number of overlaps that each pair-wise selection of historical datapoints creates. In this case, since R_1 is visited twice, it receives a weighting of two whereas the interval R_2 only gets a weighting coefficient of one.

$$\hat{p}_h = \frac{w_1 * \hat{p}_{h1} + w_2 * \hat{p}_{h2}}{w_1 + w_2} = \frac{2 * \hat{p}_{h1} + 1 * \hat{p}_{h2}}{2 + 1}$$

16.2.4 Performance

The performance of ANNSTLF is tested on real data from ten different utilities in various geographical regions. Information about the general location of these utilities and the length of the testing period are provided in Table 16.1.

In all cases, 3 years of historical data is used to train ANNSTLF. Actual weather data is used so that the effect of weather forecast errors do not alter the modeling error. The testing is performed in a blind fashion meaning that the test data is completely independent from the training set and is not shown to the model during its training.

One-to-seven-day-ahead forecasts are generated for each test set. To extend the forecast horizon beyond one day ahead, the forecast load of the previous day is used in place of the actual load to obtain the next day's load forecast.

The forecasting results are presented in Table 16.2 in terms of mean absolute percentage error (MAPE) defined as

$$MAPE = \frac{100}{N} \sum_{i=1}^{N} \frac{|Actual(i) - Forecast(i)|}{Actual(i)}$$

with N being the number of observations. Note that the average MAPEs over ten utilities as reported in the last row of Table 16.3 indicate that the third-generation engine is quite accurate in forecasting both hourly and peak loads. In the case of hourly load, this average remains below 3% for the entire forecast horizon of 7 days ahead, and for the peak load it reaches 3% on the seventh day. A pictorial example of one-to-seven-day-ahead load forecasts for utility 2 is shown in Figure 16.5.

TABLE 16.1   Utility Information for Performance Study

Utility	No. Days in Testing Period	Weather Variable	Location
1	141	T	Canada
2	131	T	South
3	365	T,H,W	Northeast
4	365	T	East Coast
5	134	T	Midwest
6	365	T	West Coast
7	365	T,H	Southwest
8	365	T,H	South
9	174	T	North
10	275	T,W	Midwest

TABLE 16.2   Summary of Performance Results in Terms of MAPE

Utility	MAPE OF	Days-Ahead						
		1	2	3	4	5	6	7
1	All hours	1.91	2.29	2.53	2.71	2.87	3.03	3.15
	Peak	1.70	2.11	2.39	2.62	2.73	2.94	3.10
2	All hours	2.72	3.44	3.63	3.77	3.79	3.83	3.80
	Peak	2.64	3.33	3.46	3.37	3.42	3.52	3.40
3	All hours	1.89	2.25	2.38	2.45	2.53	2.58	2.65
	Peak	1.96	2.26	2.41	2.49	2.60	2.69	2.82
4	All hours	2.02	2.37	2.51	2.58	2.61	2.65	2.69
	Peak	2.26	2.59	2.69	2.83	2.85	2.93	2.94
5	All hours	1.97	2.38	2.61	2.66	2.65	2.65	2.74
	Peak	2.03	2.36	2.49	2.37	2.49	2.51	2.55
6	All hours	1.57	1.86	1.99	2.08	2.14	2.17	2.18
	Peak	1.82	2.25	2.38	2.50	2.61	2.62	2.63
7	All hours	2.29	2.79	2.90	3.00	3.05	3.10	3.18
	Peak	2.42	2.78	2.90	2.98	3.07	3.17	3.28
8	All hours	2.22	2.91	3.15	3.28	3.39	3.45	3.50
	Peak	2.38	3.00	3.12	3.29	3.40	3.45	3.52
9	All hours	1.63	2.04	2.20	2.32	2.40	2.41	2.50
	Peak	1.83	2.25	2.36	2.51	2.54	2.64	2.78
10	All hours	2.32	2.97	3.25	3.38	3.44	3.52	3.56
	Peak	2.15	2.75	2.93	3.08	3.16	3.27	3.27
Average	All hours	2.05	2.53	2.72	2.82	2.89	2.94	2.99
	Peak	2.12	2.57	2.71	2.80	2.89	2.97	3.03

TABLE 16.3   Training and Test Periods for the Price Forecaster Performance Study

Database	Training Period	Test Period	MAE of Day-Ahead Hourly Price Forecasts ($)
CALPX	Apr 23, 1998–Dec 31, 1998	Jan 1, 1999–Mar 3, 1999	1.73
PJM	Apr 2, 1997–Dec 31, 1997	Jan 2, 1998–Mar 31, 1998	3.23

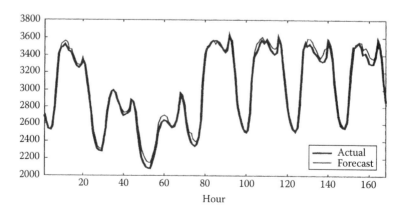

FIGURE 16.5   An example of a one-to-seven-day-ahead load forecast.

As pointed out earlier, all the weather variables (*T* or *T_*eff) used in these studies are the actual data. In online usage of the model, weather forecasts are used. The quality of these weather forecasts vary greatly from one site to another. In our experience, for most cases, the weather forecast errors introduce approximately 1% of additional error for 1–2 days out load forecasts. The increase in the error for longer range forecasts is more due to less accurate weather forecasts for three or more days out.

16.3 Short-Term Price Forecasting

Another forecasting function needed in a deregulated and competitive electricity market is prediction of future electricity prices. Such forecasts are needed by a number of entities such as generation and power system operators, wholesale power traders, retail market and risk managers, etc. Accurate price forecasts enable these entities to refine their market decisions and energy transactions leading to significant economic advantages. Both *long-term* and *short-term* price forecasts are of importance to the industry. The long-term forecasts are used for decisions on transmission augmentation, generation expansion, and distribution planning whereas the short-term forecasts are needed for daily operations and energy trading decisions. In this work, the emphasis will be on short-term hourly price forecasting with a horizon extending up to the next 24 h.

In general, energy prices are tied to a number of parameters such as future demand, weather conditions, available generation, planned outages, system reserves, transmission constraints, market perception, etc. These relationships are nonlinear and complex and conventional modeling techniques cannot capture them accurately. In a similar manner to load forecasting, ANNs could be utilized to "learn" the appropriate relationships. Application of ANN technology to electricity price forecasting is relatively new and there are few published studies on this subject (Szkuta et al., 1999).

The adaptive BP MLP forecaster described in the previous section is used here to model the relationship of hourly price to relevant forecast indicators. The system is tested on data from two power pools with good performance.

16.3.1 Architecture of Price Forecaster

The price forecaster consists of a single adaptive BP MLP with the following inputs:

- Previous day's hourly prices
- Previous day's hourly loads
- Next day's hourly load forecasts
- Next day's expected system status for each hour

The expected system status input is an indicator that is used to provide the system with information about unusual operating conditions such as transmission constraints, outages, or other subjective matters. A bi-level indicator is used to represent typical vs. atypical conditions. This input allows the user to account for his intuition about system condition and helps the ANN better interpret sudden jumps in price data that happen due to system constraints.

The outputs of the forecaster are the next day's 24 hourly price forecasts.

16.3.2 Performance

The performance of the hourly price forecaster is tested on data collected from two sources, the California Power Exchange (CALPX) and the Pennsylvania-New Jersey-Maryland ISO (PJM). The considered price data are the Unconstrained Market Clearing Price (UMCP) for CALPX, and Market Clearing Price (MCP) for PJM. The average of Locational Marginal Prices (LMP) uses a single MCP for PJM. The training and test periods for each database are listed in Table 16.3. Testing is performed in a blind fashion, meaning that the test data is completely independent from the training set and is not shown to the model during its training. Also, actual load data is used in place of load forecast.

TABLE 16.4 Results of Performance Study for the Test Period

Database	MAE of Day-Ahead Hourly Price Forecasts ($)	Sample Mean of Actual Hourly Prices ($)	Sample Standard Deviation of Actual Hourly Prices ($)
CALPX	1.73	19.98	5.45
PJM	3.23	17.44	7.67

The day-ahead forecast results are presented in the first column of Table 16.4 in terms of mean absolute error (*MAE*) expressed in dollars. This measure is defined as

$$MAE = \frac{100}{N} \sum_{i=1}^{N} \left| Actual\, Price(i) - Forecast\, Price(i) \right|$$

with N being the total number of hours in the test period.

To put these results in perspective, the sample mean and standard deviation of hourly prices in the test period are also listed in Table 16.4. Note the correspondence between MAE and the standard deviation of data, i.e., the smaller standard deviation results in a lower MAE and vice versa.

Figures 16.6 and 16.7 show a representative example of the performance for each of the databases. It can be seen that the forecasts closely follow the actual data.

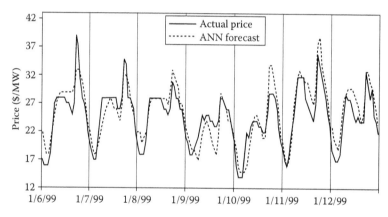

FIGURE 16.6 An example of the ANN price forecaster performance for CALPX price data.

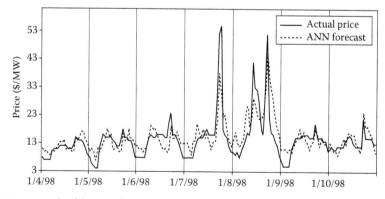

FIGURE 16.7 An example of the price forecaster performance for PJM price data.

References

Bakirtzis, A.G. et al., A neural network short term load forecasting model for the Greek power system, *IEEE Trans. Power Syst.*, 11, 2, 858–863, May 1996.

Dillon, T.S., Sestito, S., and Leung, S., Short term load forecasting using an adaptive neural network, *Int. J. Electr. Power Energy Syst.*, 13, 4, 186–192, Aug 1991.

Ho, K., Hsu, Y., and Yang, C., Short term load forecasting using a multi-layer neural network with an adaptive learning algorithm, *IEEE Trans. Power Syst.*, 7, 1, 141–149, Feb 1992.

Khotanzad, A., Afkhami-Rohani, R., Lu, T.L., Davis, M.H., Abaye, A., and Maratukulam, D.J., ANNSTLF—A neural network-based electric load forecasting system, *IEEE Trans. Neural Netw.*, 8, 4, 835–846, July 1997.

Khotanzad, A., Afkhami-Rohani, R., and Maratukulam, D., ANNSTLF—Artificial neural network short-term load forecaster-generation three, *IEEE Trans. Power Syst.*, 13, 4, 1413–1422, Nov 1998.

Khotanzad, A., Davis, M.H., Abaye, A., and Maratukulam, D.J., An artificial neural network hourly temperature forecaster with applications in load forecasting, *IEEE Trans. Power Syst.*, 11, 2, 870–876, May 1996.

Khotanzad, A., Hwang, R.C., Abaye, A., and Maratukulam, D., An adaptive modular artificial neural network hourly load forecaster and its implementation at electric utilities, *IEEE Trans. Power Syst.*, 10, 3, 1716–1722, Aug 1995.

Lee, K.Y., Cha, Y.T., and Park, J.H., Short-term load forecasting using an artificial neural network, *IEEE Trans. Power Syst.*, 7, 1, 124–132, Feb 1992.

Lu, C.N., Wu, N.T., and Vemuri, S., Neural network based short term load forecasting, *IEEE Trans. Power Syst.*, 8, 1, 336–342, Feb 1993.

Mohammed, O. et al., Practical experiences with an adaptive neural network short-term load forecasting system, *IEEE Trans. Power Syst.*, 10, 1, 254–265, Feb 1995.

Papalexopolos, A.D., Hao, S., and Peng, T.M., An implementation of a neural network based load forecasting model for the EMS, *IEEE Trans. Power Syst.*, 9, 4, 1956–1962, Nov 1994.

Park, D.C., El-Sharkawi, M.A., Marks, R.J., Atlas, L.E., and Damborg, M.J., Electric load forecasting using an artificial neural network, *IEEE Trans. Power Syst.*, 442–449, May 1991.

Peng, T.M., Hubele, N.F., and Karady, G.G., Advancement in the application of neural networks for short-term load forecasting, *IEEE Trans. Power Syst.*, 8, 3, 1195–1202, Feb 1993.

Proakis, J.G., Rader, C.M., Ling, F., and Nikias, C.L., *Advanced Digital Signal Processing*, Macmillan Publishing Company, New York, 1992, pp. 351–358.

Rumelhart, D.E. and McClelland, J.L., *Parallel Distributed Processing*, Vol. 1, MIT Press, Cambridge, MA, 1986.

Szkuta, B.R., Sanabria, L.A., and Dillon, T.S., Electricity price short-term forecasting using artificial neural networks, *IEEE Trans. Power Syst.*, 14, 3, 851–857, Aug 1999.

17

Transmission Plan Evaluation: Assessment of System Reliability

N. Dag Reppen
*Niskayuna Power
Consultants, LLC*

James W. Feltes
*Siemens Power Technologies
International*

17.1 Bulk Power System Reliability and Supply Point Reliability

Transmission systems must meet performance standards and criteria that ensure an acceptable level of quality of electric service. Service quality means continuity of supply and constancy of voltage waveform and power system frequency. Frequency is typically not an issue in large interconnected systems with adequate generation reserves. Similarly, voltage quality at the consumer connection is typically addressed at the distribution level and not by reinforcing the transmission system. This leaves continuity of power supply as the main criterion for acceptable transmission system performance.

Requirements for continuity of supply are traditionally referred to as power system reliability. Reliability criteria for transmission systems must address both local interruptions of power supply at points in the network as well as widespread interruptions affecting population centers or entire regions. Local and widespread interruptions are typically caused by different types of events and require different evaluation approaches.

Additional transmission facilities will virtually always increase reliability, but this remedy is constrained by the cost of new facilities and environmental impacts of new construction. Reliability objectives, therefore, must be defined explicitly or implicitly in terms of the value of reliable power supply to the consumer and to society at large. Reflecting the different concerns of local interruptions and widespread interruptions, reliability objectives are different for the bulk transmission system than for the local area transmission or subtransmission systems supplying electric power to electric distribution systems. These two aspects of power system reliability will be referred to as bulk power system reliability (Endrenyi et al., 1982, Parts 1 and 2) and supply point reliability.

17.1.1 Bulk Transmission Systems Reliability Is Evaluated Using Deterministic Reliability Criteria

A distinguishing characteristic of bulk transmission systems is that severe disturbances arising in them can have widespread impact. Major failures of bulk transmission systems have resulted in interruption of thousands of MW of load and interruption of service to millions of customers. Three important characteristics of reliable bulk transmission system performance are

1. Low risk of widespread shutdown of the bulk transmission system
2. Confinement of the extent of bulk transmission system shutdown when it occurs
3. Rapid restoration of operation following shutdown of the bulk transmission system

Most interconnected systems have reliability criteria and design standards that explicitly aim at limiting the risk of widespread shutdowns or blackouts. Such criteria may call for transmission reinforcements or limitations of power transfers across the system. The two other characteristics are addressed by sharpening operating command and control functions and improving control and communication facilities. Therefore, transmission system plans are typically evaluated with respect to reliability criteria that are aimed at limiting the risk of system shutdowns.

The U.S. National Electric Reliability Council (NERC), formed in response to the 1965 Northeast blackout, has developed basic design criteria aimed at reducing the risk of "instability and uncontrolled cascading" that may lead to system blackouts. The various regional reliability councils have interpreted these requirements in various ways and produced additional criteria and guides to address this problem (NERC, 1988). Deterministic criteria for bulk power systems will typically include the following requirements:

1. Test criteria for simulated tests aimed at avoiding overload cascading and instability, including voltage collapse. These test criteria specify in generic form:
   a. The system conditions to be tested: e.g., peak load conditions, lines or generators assumed out on maintenance, transfer levels
   b. The type of failure that initiates a disturbance: e.g., type and location of short circuit
   c. Assumptions to be applied regarding the operation of protection systems and other control systems
   d. The allowable limits of system response: line and transformer loading limits, high and low voltage limits, and criteria for stable operation
      The system must be reinforced to meet these criteria.
2. Requirements to test extreme contingencies such as the simultaneous outage of two or more parallel lines or the loss of entire substations. These tests are made to determine and understand the vulnerability of the system to such events. When critical extreme contingencies are identified, steps should be taken to minimize the risk of occurrence of such events.
3. Criteria and guides for protection system design to reduce the risk of critical protection initiated disturbances and for protection misoperation that may aggravate a serious system condition.

Evaluations of the system response to specified severe but rare types of failure events are labeled deterministic. The likelihood of the event specified is not considered, except in a qualitative way when the criteria were created. Since only a small subset of all potentially critical events can be tested, the tests are sometimes referred to as "umbrella" tests. A system that passes these selected tests is believed to have a degree of resiliency that will protect it not only for the specific disturbances simulated, but also for a multitude of other disturbances of similar type and severity.

17.1.2 Supply Point Reliability Is Evaluated Using Either Deterministic or Probabilistic Reliability Criteria

Reliability objectives at the local area transmission or subtransmission level focus on the reliability of supply to specific supply points as shown in Figure 17.1. Statistically, the reliability of supply may be expressed in terms of the frequency of occurrence of load interruptions, the amount of load interrupted, and the duration of the interruptions. Frequency of interruptions and MWh not served over a period such as a year are commonly used measures for the observed or predicted reliability of power supply to a particular node in a transmission system. Probabilistic reliability methods are required to predict reliability in these terms (Endrenyi, 1978; Billinton and Allan, 1984; Salvaderi et al., 1990). These methods will typically consider more likely events rather than the more extreme and very rare events that can lead to system shutdown. This is justified since system shutdown occurrences are not frequent enough to significantly impact the reliability measures calculated.

While it is practical to perform probabilistic calculations to assess supply point reliability, deterministic simulation tests are also commonly used. As a minimum, deterministic criteria call for load flow testing of all single line and single transformer outages. This is referred to as single contingency testing or N − 1 testing. For each of these outages, no line or transformer shall exceed its emergency rating, and no voltage shall violate specified high and low emergency voltage limits. Violation of these criteria calls for system reinforcements. Exceptions are typically made for supply points with low peak demand where it is judged to be too expensive to provide for redundant service. Some utilities use a peak load criterion such as 25 MW, above which redundant transmission connections to a supply point are called for.

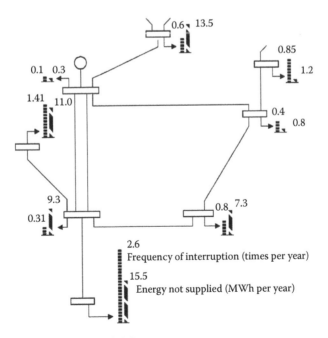

FIGURE 17.1   Prediction of supply point reliability.

17.2 Methods for Assessing Supply Point Reliability

Supply point reliability may be assessed in four different ways in order of increasing complexity:

1. *Deterministic*: System alternatives must meet criteria specifying allowable system response to specified contingencies.
2. *Probabilistic—System Trouble*: System alternatives must meet criteria specified in terms of probabilistic reliability indices reflecting risk of unacceptable system response.
3. *Probabilistic—Consumer Impact*: Same as (2), but criteria are specified in terms of consumer impact such as risk of supply interruption or risk of load curtailment.
4. *Cost/Benefit Analysis*: This approach is based on the concept that the proper level of service reliability should be defined by the balance of incremental worth of service reliability improvement and incremental cost of providing that improvement. The approach is also referred to as "effectiveness analysis" or "value-based" reliability assessment.

The limitation of the deterministic approach (1) is that it considers only the initial system problems for a few contingencies. These contingencies have typically been selected by committee based on a mixture of judgment, tradition, and experience. If the selected contingencies do not cover all important reliability concerns, the resulting system may be unreliable. If the selected contingencies put undue emphasis on severe but rare events, an unnecessarily expensive system alternative may be selected.

The probabilistic approach (2) aims at eliminating the dependency on judgment in the selection of contingencies by attempting to look at all significant contingencies. In addition, it weighs the importance of the results for each contingency according to the severity of the system problems caused by each contingency and the frequency of occurrence of each contingency.

Approach (3) looks deeper into the problem, in that it is concerned with the impact on the consumer. However, the criteria used to define an acceptable level of reliability are still judgmental. For example, how many interruptions per year would be acceptable or what percentage of total MWh demand is it acceptable to interrupt or curtail? In the cost/benefit approach (4), the criterion for acceptable reliability is implicit in the methodology used.

17.2.1 Reliability Measures—Reliability Indices

Reliability can be measured by the frequency of events having unacceptable impacts on the system or on the consumer, and by the severity and duration of the unacceptable impacts. Thus, there are three fundamental components of reliability measures:

1. Frequency of unacceptable events
2. Duration of unacceptable events
3. Severity of unacceptable events

From these, other measures, such as probability of unacceptable events, can be derived. An expectation index, such as the loss of load expectation (LOLE) index commonly used to measure the reliability of a generating system is, in its nature, a probability measure. While probability measures have proved useful in generation reliability assessment, they may not be as meaningful in assessing the reliability of a transmission system or a combined generation/transmission system. It is, for example, important to differentiate between 100 events which last 1 s and 1 event which lasts 100 s. Since probability measures cannot provide such differentiation, it is often necessary to apply frequency and duration measures when assessing the reliability of transmission systems.

Probabilistic reliability measures or indices can express the reliability improvements of added resources and reinforcements quantitatively. However, several indices are required to capture various reliability aspects. There are two major types of indices: system indices and consumer or load indices

(Guertin et al., 1978; Fong et al., 1989). The former concerns itself with system performance and system effects, the latter with the impact on the consumer. The reliability cost measure used in cost/benefit analysis may be classified as a consumer index.

17.2.2 System Indices

Indices suitable for transmission system reliability evaluation may be divided into system problem indices and load curtailment indices.

System problem indices measure frequency, duration, probability, and severity of system problems. Some examples:

- Frequency of circuit overloads (overloads/year)
- Average duration of circuit overloads (hours)
- Probability of circuit overloads

Load curtailment indices measure severity in terms of load interrupted or curtailed. The salient characteristic of these indices is that the severity of any event, regardless of the system problems resulting from the event, is expressed in terms of load curtailment. From the three fundamental reliability measures (frequency, duration, and load curtailment), a series of derived reliability indices may be defined as illustrated by the following examples.

Basic Annual Indices

- Frequency of load curtailment $F = \sum_i F_i (\text{year}^{-1})$

- Hours of load curtailment $D = \sum_i F_i D_i (\text{h year}^{-1})$

- Power curtailed $C = \sum_i F_i C_i (\text{MW year}^{-1})$

- Energy curtailed $E = \sum_i F_i D_i C_i (\text{MWh year}^{-1})$

  where
  F_i = Frequency of event i (year^{-1})
  D_i = Duration of event i (h)
  C_i = MW load curtailed for event i (MW)
  i = All events for which $C_i > 0$

Energy curtailment (E), expressed in MWh not served, is often referred to as *Energy Not Served* (ENS), *Expected Energy Not Served* (EENS), or *Expected Unserved Energy* (EUE).

Load curtailment indices are sometimes normalized to system size. Two commonly used indices are

- Power interruption index $C_N = C/CMX$ (year^{-1})
- Energy curtailment index $E_N = E/CMX$ (h year^{-1})

where CMX = peak load for system, area, or bus.

$E_N \times 60$ is referred to as system minutes, the equivalent number of minutes per year of total system shutdown during peak load conditions.

17.2.3 Cost of Interruptions to Consumers

The fact that a sudden interruption of very short duration can have a significant impact and that an outage of 4 h may have a significantly more severe impact than two outages of 2 h each, illustrates the limitations of simple aggregated reliability measures such as MWh not served. This is an important

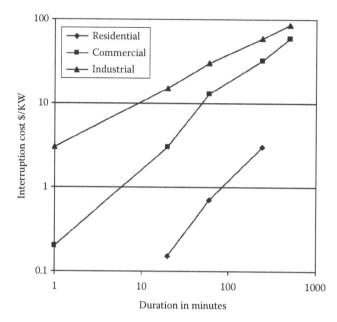

FIGURE 17.2 Illustration of customer damage functions for residential, commercial, and industrial load for process-oriented industrial load. The cost of very short duration outages may be much higher than shown here.

limitation since the various transmission reinforcement options considered may have dramatically different impacts as far as interruption durations are concerned. Since it is difficult to use a multiparameter measure when comparing reinforcement alternatives, a single aggregate measure is much preferred as long as it includes the main reliability factors of concern. The concept of cost to consumers of unreliability expressed in dollars per year has emerged as a practical measure of reliability when comparing transmission reinforcement alternatives. As a measure, reliability cost has the additional important advantage that it can be aggregated with installation cost and operating cost to arrive at a minimum "total cost" design in a cost/benefit analysis.

Conceptually, the annual reliability cost for a group of customers is the aggregated worth the customers put on avoiding load interruptions. In some cases the costs are tangible, allowing reliable dollar cost estimates; in other instances the impacts of interruptions are intangible and subjective, but still real in the eyes of the consumer. Surveys aimed at estimating what consumers would be willing to pay, either in increased rates or for backup service, have been used in the past to estimate the intangible costs of load interruptions. The results of these investigations may be expressed as *Customer Damage Functions* (CDF), as illustrated in Figure 17.2 (Mackay and Berk, 1978; Billinton et al., 1983).

Customer damage functions can be used to estimate the dollar cost of any particular load interruption given the amount of load lost and the duration of the interruption. If a customer damage function can be assigned for each supply point, then a cost of interrupted load may be determined.

17.2.4 Outage Models

Generation and transmission outages may be classified in two categories—forced outages and scheduled outages. While forced outages are beyond the control of the system operators, scheduled outages can usually be postponed if necessary to avoid putting the system in a precarious state. These two outage categories must, therefore, be treated separately (Forrest et al., 1985).

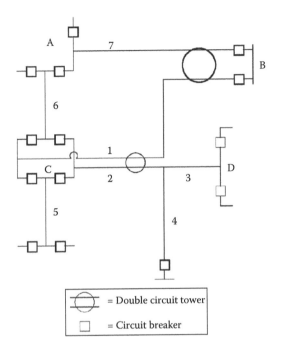

FIGURE 17.3 Sample system illustrating outage types.

17.2.4.1 Forced Outage Models

The variety and characteristics of forced outage events may be illustrated with reference to Figure 17.3. Transmission lines and transformers that can be isolated from the system by the opening of circuit breakers are referred to as "elements."

Three categories of forced outage events are recognized:

1. *Single Component Outage Event*—The outage event involves only one element. For example, a fault on circuit 1, cleared by circuit breakers in a normal manner, would only affect circuit 1.
2. *Common Mode Outage Event*—This is a multiple element outage event where a single initiating cause results in multiple element outages where the outages are not consequences of each other. For example, a single lightning stroke taking out both circuits of the double circuit line exiting substation B would be a common mode outage. This event results in the simultaneous outage of circuits 1 and 7.
3. *Substation-Related Outage Events*—This is a multiple element outage event that depends on the protection system response to a fault on a component in the substation or on an element connected to the substation. Examples of substation-related outage events are:
   a. Stuck breaker—if the breaker common to circuits 1 and 6 is stuck, a fault on either circuits 1 and 6 would result in both circuits out.
   b. Tapped circuits—a fault on circuit 2 would result in circuits 2–4 going out together.
   c. Breaker fault—if there is a fault on the breaker common to circuits 1 and 6, both circuits 1 and 6 would be outraged.
   d. Bus section fault—a fault on the bus section in substation B would outage circuits 1 and 7.

A common mode outage event may be combined with substation-related outage events. For example, a common mode failure of circuits 1 and 2 would result in an outage event encompassing circuits 1–4. Two or more independent outage events from either of the three outage categories may overlap in time, creating more complex outages. Accurate tools for the prediction of reliability measures include most if not all of these outage types.

17.3 Probabilistic Reliability Assessment Methods

Probabilistic reliability assessment tools falls in one of two categories (Endrenyi et al., 1982a,b):

1. The contingency enumeration method
2. The Monte Carlo method

In general, the contingency enumeration method is capable of looking at severe and rare events such as transmission events in great detail, but cannot practically look at many operating conditions. In contrast, the Monte Carlo methods are capable of looking at operating conditions in great detail (Noferi et al., 1975). However, from a computational standpoint, it is not possible to capture with precision the impact of infrequent but severe transmission contingencies. Thus, the two methods are capturing different aspects of the reliability problem.

17.3.1 Contingency Enumeration Approach

The contingency enumeration approach to reliability analysis includes the systematic selection and evaluation of disturbances, the classification of each disturbance according to failure criteria, and the accumulation of reliability indices. Contingency enumeration techniques are structured so as to minimize the number of disturbances that need to be investigated in detail. This is achieved by testing, to the extent possible, only those disturbances that are sufficiently severe to cause trouble and sufficiently frequent to impact the risk indices to be computed.

The contingency enumeration approach is structured as shown in Figure 17.4. For a specific predisturbance condition, a contingency is selected and tested to determine whether the contingency causes any immediate system problem such as a circuit overload or a bus voltage out of limits. If it does not, a new contingency is selected and tested.

The occurrence of a system problem may by itself be logged as a failure. However, in many cases, it will be possible to adjust generation or phase shifters to relieve overloads and to adjust generator voltages or transformer taps to bring bus voltages back within range. It is, therefore, of interest to determine whether it is possible to eliminate a system problem by such corrective actions. A failure is logged when corrective actions, short of curtailing consumer loads, are insufficient to eliminate the system problems. The severity of such system problems may be assessed by computing the amount and location of

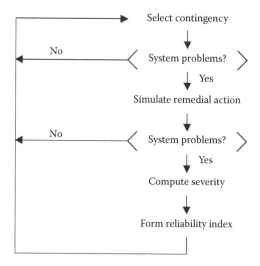

FIGURE 17.4 Contingency enumeration approach.

load curtailment necessary to eliminate the problem. In this way, it is possible to compute supply point reliability indices that measure the frequency, duration, and amount of expected load curtailment.

17.3.1.1 Monte Carlo Approach

Monte Carlo methods (Oliveira et al., 1989) may be sequential or non-sequential. The sequential approach simulates the occurrences of random events through time, recognizing the statistical properties of the various types of events. Typically, the time functions of load and planned generation schedules are established for a period of a year. Starting at the beginning of the year, a sequence of forced shutdown and restoration of transmission and generating equipment is then determined based on random sampling in accordance with the statistical characteristics of the equipment failure processes. The response of the power system during equipment outages is simulated by power flow solutions. Whenever a system condition violating predefined failure criteria is encountered, the occurrence and characteristic of this failure is recorded. At the end of 1 year of simulation, parameters describing the "observed" reliability of the system can be determined. These parameters may include frequency of equipment overload, frequency of voltage violations, MWh not served, average duration and severity of specified types of failures, etc. This process is illustrated in Figure 17.5.

One year of simulation constitutes one particular sample scenario governed by the random properties of equipment failure. In order to obtain a measure of the inherent reliability of the system, it

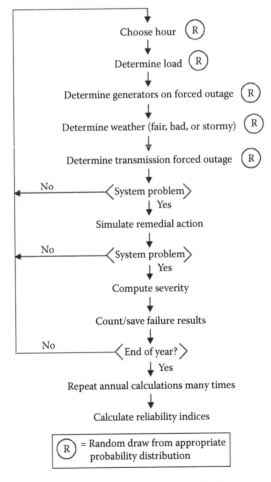

FIGURE 17.5 Possible computational sequence for Monte Carlo method.

is necessary to repeat the simulation over the annual period many times and calculate the reliability measures as the mean of the results obtained over the repeated annual simulations. For reliable systems, several hundred annual simulations may be required to obtain convergence in the reliability measures calculated.

When using the sequential approach, it is possible to model time dependencies between key variables. This allows elaborate modeling of energy-limited resources such as hydro plants and pumped hydro. It is also possible to simulate environmental effects such as the occurrence of lightning storms that may impact the failure rate of transmission equipment. A brute-force sequential Monte Carlo simulation would be prohibitively time-consuming when applied to large systems. Practical techniques rely on special sampling techniques and acceptable approximations in power system modeling.

If time dependencies are not essential, the non-sequential approach may be used. In this case, hours of simulation may be selected at random rather than in sequence. For a specific hour, a precontingency state is established including bus loads and matching generation dispatch. When it can be used, the non-sequential approach is typically much faster than the sequential approach.

17.3.2 Comparison of Contingency Enumeration and Monte Carlo Simulation

From the preceding discussion, it is clear that the Monte Carlo method differs from the contingency enumeration method in the way power system states including load, generation dispatch, and component outages are selected. The actual network solution and corrective action models used may be the same or similar for both methods. The major advantage of the Monte Carlo method is the ease with which comprehensive statistics of power system states can be included. This makes the method suitable for computing period reliability indices such as annual indices (CIGRE, 1992).

The Monte Carlo method may not be suitable for estimating the probability or frequency of occurrence of infrequent events. The contingency enumeration method may, therefore, be a more practical approach in system design. In comparing system alternatives to strengthen a local area, the Contingency Enumeration approach will provide consistent and real differences in reliability indices computed for specific situations. Unless the reliability of the alternatives are far apart, it would be very time-consuming and perhaps impractical to obtain acceptable differences in reliability by means of the Monte Carlo method. One way to mitigate this problem is to remove time-consuming calculations from the inner loop of the Monte Carlo calculations. In one approach, which is used to assess the reliability of supply to load centers, the impact of rare transmission failures are obtained from precomputed lookup tables of transmission import limits. Using this approach and various sampling techniques, several thousand years of operation can be simulated in minutes.

17.4 Application Examples

The techniques described above are presently used for transmission planning by major utilities. The following examples illustrate some of these methods. The first example uses contingency enumeration while the second example uses Monte Carlo techniques.

17.4.1 Calculation of the Reliability of Electric Power Supply to a Major Industrial Complex

Contingency enumeration techniques were used to assess the reliability of the power supply to a major manufacturing complex (Reppen et al., 1990). In this analysis, the reliability concerns encompassed system events and conditions that are capable of disturbing or shutting down all or portions of the manufacturing processes. The events of concern included initial interruptions, sustained interruptions, overloads, voltage violations, voltage collapse, and overload cascading. The reliability effects of possible system reinforcements in the immediate local power supply area and in the main grid supplying this area were evaluated by a comprehensive probabilistic reliability analysis.

The characteristics of the power system were radial feeds to the plant with provisions for automatic and manual switchover to alternative supply in case of loss of voltage on primary supply feeders, an extensive local 132 kV system, and a regional 300 and 420 kV transmission system. Contingency enumeration methods allow detailed modeling of the network, including the modeling of automatic responses of the power system to disturbances such as special relaying schemes for line tripping, generation runback, and load transfer.

Three typical categories of outages—single element outages, independent overlapping outages of multiple elements, and dependent multiple element outages—were considered in the reliability studies. The term "element" encompasses generating units as well as "transmission elements" such as transmission lines, transformers, capacitor banks, and static var devices. The reliability computations included network analysis of outages, classification of failure events according to type and severity, and calculation of reliability indices. Reliability indices representing the predicted frequency of each of the types of failure events were computed as well as load interruption and energy curtailment indices. The indices computed are referred to as annualized indices reflecting the reliability level that would be experienced if the precontingency condition considered should exist for an entire year.

The full analysis included assessment of existing power supply conditions, impact of system reconfiguration on the reliability of supply, reliability effects of system reinforcements, and impact of conditions in the main grid. Here we will concentrate on the reliability effects of system reinforcements. Two reinforcements were analyzed: construction of a new 132 kV line completing a loop at some distance from the plant and construction of a 300 kV ring connecting several of the power supply buses to the plant.

Figure 17.6 presents the results of the investigation using the energy curtailment index defined earlier. The energy curtailment index aggregates the expected loss of energy on an annualized basis. The results indicate that reinforcement A (the remote line) has no significant effect on the reliability of power supply to the plant. Reinforcement B (the 300 kV ring) provides a substantial overall improvement, although there is no significant improvement in the energy curtailment index for the winter case.

While some of the suggested means of improving reliability could have been predicted prior to the analysis, the relative effectiveness of the various actions would not be apparent without a formal reliability analysis. Performing a reliability analysis of this type gives excellent insights into the dominating failure phenomena that govern system performance. The detailed contingency information available

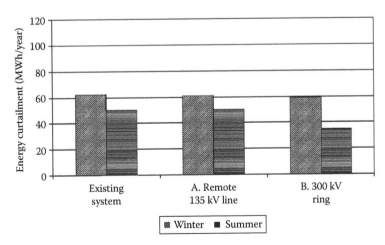

FIGURE 17.6 Benefit of system reinforcements. Annualized energy curtailment (MWh/year). Sustained interruptions only.

promotes understanding of the way systems fail, while the reliability indices computed provide the perspective necessary to make appropriate system design decisions.

17.4.2 Local Area Reliability

The second example considers the task of improving the reliability of electric power to a local area such as a city, major industrial complex, or other load center (Reppen, 1998). In simple terms, the reliability of the supply is a function of the following parameters:

- The load in the area as it fluctuates over time.
- The maximum amount of power that can be imported from the main grid. This import limit varies by maintenance and forced outage of transmission and generating equipment.
- Maximum available local generation at any particular time as it is affected by generation maintenance and forced outages.

At any particular time, load curtailment will occur if the load exceeds the maximum import capability into the area plus the maximum generation available in the area. Therefore, reliability of supply to consumers in the area can be measured in terms of statistics of load curtailment. Popular load curtailment measures include frequency of interruptions and energy not served (MWh/year). In addition, the expected annual customer interruption cost can be predicted using Monte Carlo techniques that simulate system conditions repeatedly over a time period to develop reliability measures by aggregating and averaging the impacts of individual load curtailment events. This allows the use of interruption cost functions (customer damage function) to estimate the expected annual cost of load interruptions.

Accepting calculated annual interruption costs as a realistic measure of the economic impact on the consumer, one might declare a system reinforcement alternative to be justified from a reliability standpoint if the reduction in interruption cost is greater than the net cost of investment and operation. While such a criterion may not necessarily be appropriate in all cases, it should provide a relevant benchmark in most environments.

Figure 17.7 shows key components of the power supply picture for a small city with a peak load of 210 MW. The city is supplied by two generators totaling 150 MW, and by a 138 kV double circuit transmission line from the main grid. Prime reinforcement options are as follows:

1. Add a new single circuit transmission line as indicated in Figure 17.7.
2. Add one gas turbine generator of size to be determined.
3. Add two identical gas turbine generators of size to be determined.

Key questions of interest are

1. Can the line addition be justified on the basis of savings in customer interruption cost?
2. What generator capacities will produce the same reliability improvements as the line addition?
3. What size generators can be justified on the basis of customer interruption?

Figure 17.8 shows results obtained from the Monte Carlo calculations along with the annual fixed charges for investment cost and net annual operating costs. Significant observations that can be made from Figure 17.8 are

- The additions of a new 138 kV line, a 50 MW generator, or two 25 MW generators have approximately the same interruption cost savings. However, the reason for this is that all three alternatives are an effective overkill, reducing the interruption cost to almost 0. If load is anticipated to grow, the line addition will be the better performer at a lower cost.
- One 15–25 MW generator would give much improved performance and at a cost which can be justified (marginally) based on reliability worth as expressed by the interruption cost curves.
- The line addition is clearly the most cost-effective alternative.

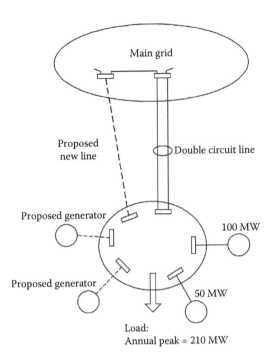

FIGURE 17.7   Power supply configuration for a small city.

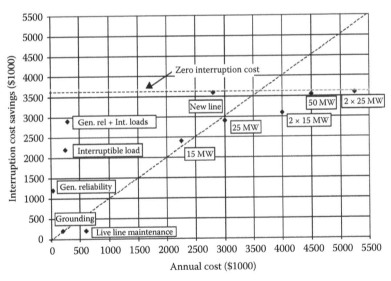

FIGURE 17.8   Annual savings in cost of interruption vs. annual combined investment and operating costs of transmission and generation reinforcements and short term measures. Measures in the upper left triangle can be justified on account of savings in interruption cost.

The first set of calculations compared the benefit of reinforcement by transmission or generation. While such additions are typically the most powerful reinforcements from the standpoint of improved reliability, they are typically also the most expensive. In addition, when dealing with small local systems, the natural or cost-effective line and generation additions are often more than what's needed to satisfy reliability needs for the next few years. This is particularly true for low load growth scenarios. Thus, there is a need for less expensive alternatives that typically will have smaller incremental reliability benefits than the addition of transmission lines and generators. The results of four such alternatives are shown in Figure 17.8:

1. Improve reliability performance of existing generators
2. Improve grounding of transmission lines
3. Introduce live line maintenance
4. Interruptible load contracts

All of the short term measures except live line maintenance can be justified as the annual interruption cost is greater than the total expense associated with the reliability improvement. Also, live line mainte-nance and transmission line grounding have too small an impact to be of interest.

The most cost effective short term measure comes from improvements in the reliability performance of the 100 MW generator, closely followed by interruptible load contracts. If both of these short term measures are taken, the reliability improvement matches that obtainable from the addition of a 25 MW gas turbine generator and at a much lower cost.

This example illustrates how it is possible to use Monte Carlo reliability calculations to predict and compare the benefit-cost trade-off of transmission and generation reinforcements and various short term measures.

References

Billinton, R. and Allan, R.N., Power-system reliability in perspective, *IEE J. Electron Power*, 30, 231–236, March 1984.

Billinton, R., Wacker, G., and Wojczynski, E., Comprehensive bibliography on electrical service interrup-tion costs, *IEEE Trans. Power Appar. Syst.*, PAS-102, 6, 1831–1837, June 1983.

CIGRE Task Force 38.03.10-1992, *Power System Reliability Analysis Volume 2 Composite Power System Reliability Evaluation*, 1992.

Discussion of Regional Council Planning Reliability Criteria and Assessment Procedures, a reference docu-ment by the North American Electric Reliability Council, June 1988.

Endrenyi, J., *Reliability Modeling in Electric Power Systems*, John Wiley & Sons, Ltd., New York, 1978.

Endrenyi, J., Albrecht, P.F., Billinton, R., Marks, G.E., Reppen, N.D., and Salvadori, L., Bulk power system reliability assessment—Why and how? Part 1: Why? IEEE Paper 82WM 147-7, presented at the *Winter Power Meeting*, New York, February 1–5, 1982a.

Endrenyi, J., Albrecht, P.F., Billinton, R., Marks, G.E., Reppen, N.D., and Salvadori, L., Bulk Power system reliability assessment—Why and how? Part 2: How? IEEE Paper 82WM 148-5, presented at the *Winter Power Meeting*, New York, February 1–5, 1982b.

Fong, C.C., Billinton, R., Gunderson, R.O., O'Neill, P.M., Raksany, J., Schneider, Jr., A.W., and Silverstein, B., Bulk system reliability—Measurement and indices, *IEEE Trans. Power Appar. Syst.*, 4, 3, 829–835, August 1989.

Forrest, D.W., Albrecht, P.F., Allan, R.N., Bhavaraju, M.P., Billinton, R., Landgren, G.L., McCory, M.F., and Reppen, N.D., Proposed terms for reporting and analyzing outages of electrical transmission and distribution facilities, *IEEE Trans. Power Appar. Syst.*, PAS-104, 2, 337–348, February 1985.

Guertin, M.B., Albrecht, P.F., Bhavaraju, M.P., Billinton, R., Jorgensen, G.E., Karas, A.N., Masters, W.E., Patton, A.D., Reppen, N.D., and Spence, R.P., Reliability indices for use in bulk power supply ade-quacy evaluation, *IEEE Trans. Power Appar. Syst.*, PAS-97, 4, 1097–1103, July/August 1978.

Mackay, E.M. and Berk, L.H., Costs of Power Interruptions to Industry Survey Results, CIGRE, Paper 3207, August 30–September 7, 1978.

Noferi, P.L., Paris, L., and Salvaderi, L., Monte Carlo Methods for power system reliability evalua-tions in transmission or generation planning, in *Proceedings, 1975 Reliability and Maintainability Symposium*, Washington, DC, 1975.

Oliveira, G.C., Pereira, M.V.F., and Cunha, S.H.F., A Technique for Reducing Computational Effort in Monte-Carlo Based Composite Reliability Evaluation, IEEE, WM 174-4 PWRS, 1989.

Overview of Planning Reliability Criteria of the Regional Reliability Councils of NERC, A reference document by the North American Electric Reliability Council, 1988.

Reppen, N.D., Balancing investments and operating costs with customer interruption costs to give increased reliability, presented at the *IEE Colloquium on Tools and Techniques for Dealing with Uncertainty*, London, U.K., January 27, 1998.

Reppen, N.D., Carlsen, T., Glende, I., Bostad, B., and Lam, B.P., Calculation of the reliability of electric power supply to a major industrial complex, presented at the *10th Power Systems Computation Conference (PSCC)*, Graz, Austria, August 19–24, 1990.

Salvaderi, L., Allan, R., Billinton, R., Endrenyi, J., Mc Gillis, D., Lauby, M., Manning, P., and Ringlee, R., State of the Art of Composite-System Reliability Evaluation, CICRE Session, paper 38-104, Paris, August 26–September 1, 1990.

18

Power System Planning

Hyde M. Merrill
Merrill Energy, LLC

Power system planning is the recurring process of studying and determining what facilities and procedures should be provided to satisfy and promote appropriate future demands for electricity. The electric power system as planned should meet or balance societal goals. These include availability of electricity to all potential users at the lowest possible cost, minimum environmental damage, high levels of safety and reliability, etc. Plans should be technically and financially feasible. Plans also should achieve the objectives of the entity doing the planning, including minimizing risk.

The *electric power system* is a force-at-a-distance energy-conversion system. It consists of three principal elements:

- Current- and voltage-producing, transmitting, and consuming hardware
- Control and protective devices
- Planning, operating, commercial, and regulatory practices and procedures

These definitions are very different from would have appeared on these pages 25 years ago. They no doubt will seem quaint 25 years hence. At this writing, the electric power industry worldwide is experiencing its most dramatic changes in two generations. These changes affect planning, but this section is intended as a practical exposition, not as a history lesson or a prophesy. We therefore will focus on how planning is or should be done today, avoiding flights of fantasy into the past or future (Sullivan, 1977; Kahn, 1988; Ringlee, 1989; Stoll et al., 1989).

Planning considers:

- Options
- Uncertainties
- Attributes

Options are the choices available to the planner. Uncertainties are parameters whose values are not known precisely or cannot be forecast without error. Attributes are measures of "goodness." Stakeholder objectives are expressed in terms of attributes. Physical, economic, and institutional realities determine how different options and uncertainties affect the attributes.

The planning problem is to identify and choose among options, in the presence of uncertainties, so as to maximize or minimize (as the case may be) the attributes.

18.1 Planning Entities

Planners generally are trained as engineers, economists, civil servants, businessmen, or mathematicians. They do power system planning for the following entities:

- Vertically integrated utilities owning generation, transmission, and distribution systems.
- Transmission companies, independent system operators (ISO), and regional transmission organizations (RTO). Transmission companies own transmission assets; the latter do not, but may have some responsibility for their planning.
- Pools or combinations of vertically integrated utilities.

Other organizations do planning studies and higher-level power sector planning. A step removed from the operation and management of the power system, their interest is in seeing that it meets society's goals:

- Various levels of government
- International development banks

Still other organizations do power system planning studies, but without system responsibility. They wish to understand the market for various services and how they might compete in it—its economics and technical requirements for entry.

- Independent power producers (IPP) or nonutility generators (NUG). These include qualifying facilities (QF as defined by the U.S. Public Utilities Regulatory Policy Act of 1978) and exempt wholesale generators (EWG as defined by the U.S. Energy Policy Act of 1992). These are subject to less stringent regulation than are utilities. They neither enjoy monopoly protection nor have an obligation to provide electricity at cost-based tariffs.
- Large industrial users.
- Commercial middlemen who buy and sell electrical energy services.
- Investors.

All of these are supported by independent purveyors of planning information. Consultants with specialized analytic skills also do planning studies.

18.2 Arenas

Planning is done in several arenas, distinguished by the planning horizon and by the types of options under consideration. These arenas include

- *Long-term vs. short-term planning.* Economists distinguish these by whether capital investment options are considered. For engineers, long-term planning has a distant horizon (perhaps 30 years for generation and half of that for transmission). Short-term planning considers about 5 years. Operations planning is for as short as a few hours and is not treated here.
- *Generation vs. transmission vs. least-cost planning.* Generation and transmission planning focus on supply options. Least-cost planning includes demand-side options for limiting or shaping load.
- *Products and services.* Some entities provide power (kW) and energy (kWh). Others plan the transmission system. Others provide for auxiliary services (voltage and power control, electrical reserves, etc.). Still others plan for diversified services like conservation and load management.

Other arenas require engineering and economic skills, but are within the purview of a book on business or policy rather than an engineering handbook.

- *Competitive markets.* Strategic planning is particularly concerned with financial and business plans in competitive markets.
- *Sector evolution.* Defining the form of the future power sector, including the relationships between competitive forces, regulation, and the broadest social objectives, is a particularly vital planning function.

18.3 The Planning Problem

18.3.1 Options

Power generation, transformer, transmission system, substation, protection, and operation and control options are discussed in other chapters of this handbook. Other options are discussed below.

18.3.1.1 Planning and Operating Standards or Criteria

Planning and operating criteria have a dual nature: they are both attributes and options. Here we will emphasize the fact that they are options, subject to change. Though they have no intrinsic value, standards or criteria are important for several reasons. Their consistent application allows independent systems to interconnect electrically in symmetrical relationships that benefit all. Criteria can also eliminate the need for planners to ask constantly, "How much reliability, controllability, etc. do I need to provide?" Criteria include

- Maximum acceptable loss-of-load probability (LOLP) or expected unserved demand, minimum required reserve margins, and similar generation planning standards
- What constitutes a single contingency (transmission systems are often designed to withstand "any" single contingency) and whether particular single contingencies are excluded because they are unlikely or expensive to forestall
- Permissible operating ranges (voltages, power flows, frequency, etc.) in the normal or preventative state, the emergency state, and the restorative state
- How criteria are to be measured or applied

Most power systems in industrialized nations are designed and operated so that

1. With all elements in service, power flows, voltages, and other parameters are within normal ranges of the equipment
2. The system remains stable after any single contingency
3. Power flows, voltages, and other operating parameters are within emergency ranges following any single contingency

For financial and economic reasons, developing countries choose weaker criteria.

18.3.1.2 Demand Management

Demand-side planning often is tied to generation planning because it affects the power and energy that the power plants will need to provide. There is no perfect classification scheme for demand-side options. Some overlapping classifications are

- Indirect load control vs. direct load control by the bulk system operator
- Power (kW) or energy (kWh) modification or both
- Type of end-use targeted

Table 18.1 shows the type of load under direct utility control in the U.S. early in the 1980s, when enthusiasm for demand-side options was especially high.

One of the most effective examples of load control was reported by a German utility. Typical off-peak winter demand was less than 70% of the peak for the same day. An indirect program promoted storage space heaters that use electricity at night, when demand is low, to heat ceramic bricks. During the day, air forced among the bricks transfers the heat to the living space. Within 5 years the program was so popular that direct control was added to avoid creating nighttime peaks. The winter daily load shape became practically flat.

18.3.1.3 Market and Strategic Options

Market and strategic options are also important. These range from buying a block of power from a neighboring utility to commodity trading in electricity futures to mergers, divestitures, and acquisitions.

TABLE 18.1 Appliances and Sectors under Direct Utility Control,
United States—1983

Appliance or Sector	Number Controlled	Percent of Total Controlled
Electric water heaters	648,437	43
Air conditioners	515,252	34
Irrigation pumps	14,261	1
Space heating	50,238	3
Swimming pool pumps	258,993	17
Other	13,710	1
Total	**1,500,891**	**100**
Residential	1,456,212	97
Commercial	29,830	2
Industrial	588	–
Agricultural	14,261	1

Source: U.S. Congress, New Electric Power Technologies: Problems and Prospects for the 1990s, OTA-E-246, Office of Technology Assessment, Washington, DC, July 1985.

18.3.2 Uncertainties

Uncertainty can seldom be eliminated. Planning and forecasting are linked so that even if the forecasts are wrong, the plans are right (Bjorklund, 1987).

18.3.2.1 Models of Uncertainty (Schweppe, 1973)

Probabilistic models, where different outcomes are associated with different probabilities, are valid if the probability structure is known. The events involved must occur often enough for the law of large numbers to apply, or else the probabilities will have little relationship to the frequencies of the outcomes. Generation planners have excellent probabilistic reliability models. (Generation and transmission reliability evaluation are treated in more detail in a separate section.)

Unknown-but-bounded (set theoretic) models are used when one or both of the conditions above are not met. For instance, transmission planners design to withstand any of a set of single contingencies, usually without measuring them probabilistically.

18.3.2.2 Demand Growth

Planners forecast the use of energy (MWh) for a period (e.g., a year) first. They divide this by the hours in the period to calculate average demand, and divide again by the projected load factor (average MW demand/peak MW demand) to forecast peak demand. Three techniques are used most often to forecast energy.

Extrapolation—Exponential growth (e.g., 4% per year) appears as a straight line on semi-log paper. Planners plot past loads on semi-log paper and use a straight edge to extrapolate to the future.

Econometric models—Econometric models quantify relationships between such parameters as economic activity and population and use of electricity. The simplest models are linear or log-linear:

$$D_i = f(P, GDP, \text{etc.}) = k_1 D_{i-1} + k_2 P_i + k_3 GDP_i + \cdots \tag{18.1}$$

where
D_i is the demand or log(demand) in period i
P_i is the population or log(population) in period i
GDP_i is the gross domestic product or log(gross domestic product) or some measure of local economic activity
k_1, k_2, k_3, etc. are coefficients

Econometric models are *developed* in a trial-and-error process. Variants of Equation 18.1 are hypothesized and least squares (regression) analysis is used to find values of coefficients that make Equation 18.1 fit historical data. Econometric models are *used* by first forecasting population, economic activity, etc. and from them calculating future energy demand using Equation 18.1.

End-use models—First, the number of households is forecast. Then the per-household penetration of various appliances is projected. The average kWh used by each appliance is estimated and is multiplied by the two previous numbers. The results are summed over all types of appliances.

Performance—Extrapolation became suspect after U.S. load forecasts in the 1970s were consistently too high. Econometric modeling is more work but is more satisfying. End-use modeling requires considerable effort but gives the most accurate forecasts of residential load.

Real drivers—One fundamental driver for per-capita load growth is the replacement by electricity of other forms of energy use. The second is the creation of new uses of energy that are uniquely satisfied by electricity.

During the decades when U.S. electric demand grew at over 7% per year, the demand for all forms of energy (of which electricity is a part) grew at about 2% per year. This obviously could not continue: the two cannot cross. The growth of electricity demand began to drop off about 1955, declining noticeably in the 1970s and thereafter. The drop in load growth was attributed to the oil crises of the 1970s. Post-1973 conservation played a part, but by then electricity had captured about all the market share it was going to get by replacement and creation of new demands for energy.

In developing countries, both fundamental drivers are limited by the ability of the electric companies to finance the necessary generation and distribution infrastructure, which is very expensive. Demand is also limited by their ability to generate. In industrialized countries, availability of capital and power plant performance are not constraining. In all countries, elasticity reduces demand if electricity is costly. This effect is much stronger in countries with low per-capita income and for energy-intensive industrial load.

18.3.2.3 Fuel and Water

In the near term, strikes, weather, and natural disasters can interrupt production or delivery. Fuel inventories and the ability to redispatch provide good hedges. In the intermediate term, government action can make fuel available or unavailable. For instance, in 1978, the U.S. Congress forbade burning natural gas by utilities, perceiving that there was a shortage. The shortage became a glut once the U.S. natural gas market was deregulated. In the long term, any single source of fuel is finite and will run out. British coal, which had fueled the industrial revolution, was shut down in the 1990s because it had been worked out.

The more important fuel uncertainties, however, are in price. For instance, Figure 18.1 shows that the price of crude oil doubled in 1974 and again in 1979. Recognizing the high variability, in 1983 the U.S. Department of Energy forecast a fuzzy band instead of a single trajectory (U.S. Dept of Energy, 1983a).

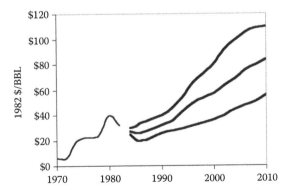

FIGURE 18.1 World oil price projections, 1983.

It is interesting that within a year of the publication of this projection, the price of oil had dropped below the low limit of the band, and it has remained there until this writing. Planners must consider extreme possibilities for all uncertainties.

Brazil, Norway, the Pacific Northwest, Quebec, and a number of developing countries are highly dependent on hydropower. Systems usually are planned and operated so that there will not be a shortfall unless one of the worst hydrological years in recorded history recurs.

18.3.2.4 Construction

Three major construction uncertainties are: How long will it take to build? How much will it cost? Will the project be completed?

A World Bank study of 41 hydro projects revealed that 37% experienced a schedule slip of 30% or more, including 17% with schedule slips of 60%–100% (Crousillat, 1989).

Figure 18.2 shows the range of actual vs. budgeted cost for a number of World Bank-financed projects. The distribution is not symmetrical—overruns are much more frequent than under-budget projects. Some of the worst cases in Figure 18.2 data occurred during periods of unexpected high inflation (Crousillat, 1989). A 1983 report projected that the cost of some 40 U.S. nuclear plants scheduled for completion by 1990 would be close to normally distributed, with the least expensive costing a bit under $2000/kW and the most expensive three times higher, at $6000/kW (U.S. Dept of Energy, 1983b).

Possibly the most expensive nuclear plant ever built, the Shoreham Plant on Long Island, was completed at a cost of some U.S. $16 billion. It was shut down by the state before producing a single kWh of commercial energy.

18.3.2.5 Technology

New technologies are generally less certain than mature technologies in their cost, construction time, and performance.

Even mature technologies may have important uncertainties. For example, transmission transfer capability is an important measure of transmission system capability. It is usually expressed as a single number, but it is actually a time-varying random variable.

18.3.2.6 Demand Management

Demand management programs are risky, in part because of uncertainty in the public's response to them. The two major uncertainties are

- What fraction of eligible customers will respond to a particular program?
- How much will the average customer change his use of electricity?

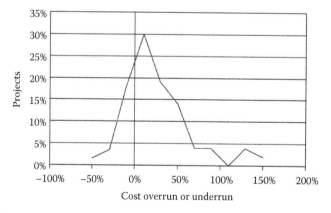

FIGURE 18.2 Budget vs. actual costs, power projects in developing countries.

These uncertainties are affected strongly by the design of the program, the incentives offered, how it is marketed, etc. Carrying out a carefully designed pilot program can reduce the uncertainty. The pilot program should be done in the region of the future commercial program.

18.3.2.7 Markets and Capital Recovery

For many years, vertically integrated utilities were guaranteed the recovery of all costs, including capital invested, plus a modest but sure profit. The customer paid this and absorbed the market uncertainties. At this writing, the regulated monopoly, cost-recovery market is being replaced in many states and countries by a more competitive market. Some market risks are being transferred from customers to utilities, power marketers, generating companies, speculators, and others.

This creates new uncertainties. For example, in competitive generation markets it is not known which potential generating units will be built. This affects both transmission and generation planning.

18.3.2.8 Regulation

For the foreseeable future, government will play a key role. The uncertainty in what governments will do propagates into uncertainties in profitability of various players, in market entry, in prices, etc.

For example, in the 1980s, U.S. state and federal governments encouraged utilities to implement demand-side programs. Program costs, and in some cases costs of foregone sales, were recovered through tariffs. The government interest later switched to competitive markets. These markets do not have such a convenient mechanism for encouraging demand-side management. As a result, demand-side programs became less attractive.

18.3.2.9 Severe Events

High-risk, low-probability events usually are not considered by standard planning practices. For example, transmission planners design so that the system will withstand any single contingency. Planning procedures, criteria, and methods generally ignore several simultaneous or near-simultaneous contingencies. The power system is not designed to withstand them—whether or not it does is happenstance.

In January 1998 an ice storm of unprecedented magnitude struck the northeastern United States and Quebec. Ice on transmission lines greatly exceeded design standards. Many towers collapsed. All lines feeding Montreal, and all lines south and east of the city, were on the ground. The government later announced a high-risk, low-probability standard: the system should be designed and operated to prevent loss of more than half of the Montreal load should such an event recur.

18.3.3 Attributes

Attributes measure "goodness" in different ways, from different perspectives. Each stakeholder has objectives; they are expressed in terms of attributes.

Customers of various kinds (residential, commercial, industrial, etc.):

- Cost of electricity
- Other costs absorbed by the customer
- Quality of service (reliability, voltage control, etc.)

Investors in various providers of energy and services

- New capital required
- Net income, earnings per share, and other measures of income
- Cash flow, coverage ratios, and other measures of cash use and replacement

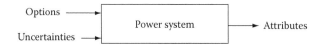

FIGURE 18.3   Options, uncertainties, and attributes.

Employees

- Security
- Promotion opportunities
- Salaries
- Healthiness and safety of working conditions

Taxpayers

- Tax revenues
- Expenditures from public funds

Neighbors (environmentalists, visitors, local inhabitants, competitors, etc.)

- Emissions or thermal discharges
- Community disruption
- Employment opportunities
- Rights-of-way and other intrusions
- Flooding
- Measures of market power

The list of attributes given above is not complete, and different attributes are important for different studies. Deciding on the planning objectives and the attributes for a given study is an important initial step in power system planning.

Some attributes are measured using complex computer models. For others, approximate or ad hoc models may be adequate or may be the best that is available. The planner calculates how the options and the uncertainties (Figure 18.3) affect the attributes.

Standards or criteria are surrogates for some attributes that are difficult or impossible to compute.

18.4 Planning Processes

18.4.1 Setting Standards or Criteria

Planning objectives often conflict. For example, maximizing reliability and minimizing environmental impacts generally conflict with minimizing costs.

Since all attributes cannot be measured in dollars, achieving the right trade-off can be a difficult socio-technico-economic-institutional problem. Doing this every day would burden the planning process. Having standards avoids having to revisit such judgments continually. For instance, once it is decided that (say) 20% generation reserve provides adequate reliability at an acceptable cost, the planner accepts 20% as a standard and designs to meet it. Testing whether a particular plan meets the reserve criterion is easy; the planner can concentrate on other issues.

Standards should be examined from time to time. If society becomes poorer or richer, its pocketbook may speak for lower or higher standards of service. Changes in technology may justify a change in standards—for instance, development of better scrubbers may make it reasonable to insist on reduced SO_2 or NOx emissions. Increased reliance on electricity may require more reliability: a proposal to shut off the power throughout the United States for 1 min to salute Edison's death was quashed. Had he died in 1900 instead of 1931, it might have been practical.

18.4.2 Assessment

18.4.2.1 Forecasts and Projections

Not all uncertainties create risk for every planning study. Those that do for a particular study are identified. Forecasts and projections are developed for these uncertainties.

18.4.2.2 System State

The state assessment begins with an evaluation of the technical and economic attributes of the present and future power system. Does it and will it satisfy established technical standards? Is it economical? Does it meet other objectives?

Chapter 8 of this handbook, "Power System Analysis and Simulation," describes the analytical tools available to planners. It also describes how these tools are used. The phenomena analyzed are described in Chapters 10 through 12.

18.4.3 Generation Planning

Chapter 2 of this handbook, "Electric Power Generation: Conventional Methods," describes generation planning options and their characteristics. The generation planner does a preliminary selection from among them, recognizing any special features of his planning problem.

Planners measure how the various options would alleviate deficiencies discovered in the assessment step. The effects of various options or combinations of options and the effects of uncertainties on other attributes are also measured.

In particular, planners compute reserve margin or other measures of reliability. They simulate the operation of the system to measure operating cost and to determine if the operation is within acceptable ranges of other parameters.

18.4.4 Transmission Planning

Traditional transmission options and new technologies are described in Chapters 3, 4, 11, 15, and 16 of this handbook. Like the generation planner, the transmission planner makes a preliminary selection based on the needs and development pattern of his system.

For instance, for technical and commercial reasons, a given system will use only a few distinct voltage classes. So a system whose existing transmission consists of 138, 345, and 765 kV equipment will rarely add new circuits at 230 or 400 kV, even though these may be popular elsewhere.

Transmission planners then identify a set of specific options and measure how these options in various combinations, along with the important uncertainties, affect the attributes. Load flow, short circuit, and stability analyses are performed to determine if voltages and currents are within acceptable bounds under various system states, and if the system will remain stable for all contingencies. How often and how much the operation of the generation system will be constrained by transmission limitations is an important consideration.

18.4.5 Least-Cost Planning

Least-cost planning is also known as integrated resource planning or integrated demand/supply planning. It considers supply-side options (generally generation options) on a level playing field with demand-side options (generally conservation, indirect load shifting, or direct load control). These options include incentives to encourage utilities and consumers to change energy consumption patterns.

As with generation planning and transmission planning, a preliminary selection weeds out options that are clearly not of interest in a particular area.

The least-cost planning process includes computing values of key attributes for various options and uncertainties.

18.4.6 Making Choices

A key question in generation, transmission, and least-cost planning is: How is one plan selected over another? A few distinctive approaches will be described.

18.4.6.1 Minimize Revenue Requirements

The planner selects the best option from the ratepayer's perspective. He selects the plan that will minimize the ratepayer's cost of electricity while satisfying reliability, environmental, and other criteria.

The ratepayer's cost—an attribute—is the revenue that the utility will have to collect to recover all operating and capital costs and to earn a commission-approved return on unrecovered investor capital:

$$RR_i(\mathbf{O},\mathbf{U}) = FC_i(\mathbf{O},\mathbf{U}) + VC_i(\mathbf{O},\mathbf{U}) \tag{18.2}$$

where
  $RR_i(\mathbf{O},\mathbf{U})$ is the revenue requirements in period i
  $FC_i(\mathbf{O},\mathbf{U})$ is the fixed costs in period i
  $VC_i(\mathbf{O},\mathbf{U})$ is the variable costs in period i
  $\mathbf{O}$ is the selection of the various options
  $\mathbf{U}$ is the realizations or values of the various uncertainties

Fixed costs are independent of how much or how little a piece of equipment is used. Depreciation (recovery of investors' capital), interest on debt, and profit (return on unrecovered capital) are typical fixed costs.

Variable costs—fuel cost—for example, are related to how much a piece of equipment is used. These costs include all system costs, not just the cost of the individual option. For instance, old plants may run less when a new plant is built. The variable cost includes fuel cost for all plants.

To apply this traditional method, the planner must know his company's return rate, which is set by the regulator. In a closely related method, the market defines the cost of electricity and Equation 18.2 is solved for the internal rate of return (IRR). The option selected is the one that maximizes the IRR, the investor's profit.

18.4.6.2 Cost-Benefit Analysis

If the benefits exceed the costs, a project is worth doing.

Typically, costs are incurred first, and benefits come later. A dollar of benefit later is not worth the same as a dollar of cost today. Present worth analysis is a way to compare dollars at different times. The basic equation is

$$P = S/(1+i)^n \tag{18.3}$$

In Equation 18.3, P is the present worth or equivalent value today of an amount S, n years in the future, with i the discount rate or annual cost of capital.

Cost-benefit analysis is also used to rank mutually exclusive projects—the one with the highest benefit/cost ratio wins.

18.4.6.3 Multi-Objective Decision Analysis

18.4.6.3.1 Utility Function Methods

Table 18.2 compares two options for a new power plant in Utah (Keeney et al., 1981). Which choice is better? The attributes are combined in a utility function of the form:

$$U(x) = k_1 Economics + k_2 Environment + \cdots + k_7 Feasibility \tag{18.4}$$

TABLE 18.2   Attributes: Nuclear Plant vs. Coal Plant

	Wellington Coal Plant	Green River Nuclear Plant
Economics ($/MWh)	60.7	47.4
Environment (corridor-miles)	532.6	500.8
Public disbenefits ($ × 000,000)	15.0	22.6
Tax revenues ($ × 000,000/year)	3.5	1.0
Health lost (equivalent years)	446.7	6.3
Public attitudes	0.33	−1.0
Feasibility	60.0	37.0

Source: Keeney, R.L. et al., Decision Framework for Technology Choice, Report EA-2153, Electric Power Research Institute, Palo Alto, CA, 1981. With permission.

In Equation 18.4, x takes on one of two values, "coal" or "nuclear." (The actual functional form for a particular study may be more complicated than Equation 18.4.) The coefficients k_i reflect the relative importance of each attribute. These coefficients convert the different attributes to a common measure. The choice that minimizes the utility function wins. In this study, for the values of coefficients selected U(coal) was $131.4/MWh; U(nuclear) was $162.9/MWh.

This approach has many variants. Uncertainties can be included, making U(x) a random variable. Work has been done to develop methods for determining the decision-maker's values (the coefficients) and risk tolerance.

18.4.6.3.2 Trade-Off Analysis

Trade-off analysis measures each attribute in its natural units, without reducing them all to a common measure, and seeks reasonable compromises at points of saturation or diminishing return. A good compromise will not necessarily optimize any of the attributes, but will come close to optimizing all of them.

For example, Figure 18.4 shows 22 plans examined in an energy strategy study. The plans in region A minimize SO₂ emissions, but are very costly. The plans in region B are cheap but have high emissions. The plans at the knee of the trade-off curve are at the point of diminishing returns for one attribute against the other. Significant reductions in emissions can be had at little cost by moving from B to the knee. Going beyond the knee toward A will not reduce SO₂ much more but will increase the cost significantly. The plans at the knee come close to minimizing both cost and emissions.

Trade-off analysis can be done graphically for two-attribute problems. More than two attributes cannot be graphed easily but can be analyzed mathematically (Crousillat et al., 1993).

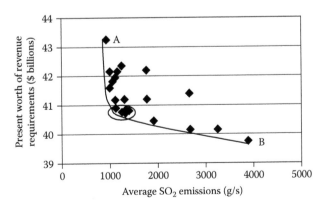

FIGURE 18.4   Trade-off: cost vs. SO₂ emissions.

18.4.6.4 Risk

Risk is the hazard due to uncertainty. Risk is also associated with decisions. Without uncertainties and alternatives, there is no risk.

System planning, engineering, and operating procedures have evolved to reduce the risks of widespread or local service interruptions. Another section of this handbook describes methods for modeling and enhancing reliability.

Not all risks are included in reliability analysis, however. Much talk about risk is directed to financial risks. Other important risks are not quantified in dollars.

One measure of risk is *robustness*, the likelihood that a particular decision will not be regretted. *Exposure* is the possible loss (in terms of an attribute) under adverse realizations of uncertainties. *Regret* is the difference between the value of an attribute for a particular set of decisions and realizations of uncertainties, and the value of the attribute for optimized decisions with perfect foreknowledge of the uncertainties.

Planners develop *hedges*, options that increase robustness or decrease exposure or regret. Building small generating units instead of large ones is an example; an insurance policy is another (De la Torre et al., 1999).

References

Bjorklund, G.J., Planning for uncertainty at an electric utility, *Public Utilities Fortnightly*, Oct. 15, 1987.

Crousillat, E., *Incorporating Risk and Uncertainty in Power System Planning*, I&ED Energy Series paper # 17, The World Bank, Washington, DC, 1989.

Crousillat, E.O., Dörfner, P., Alvarado, P., Merrill, H.M., Conflicting objectives and risk in power system planning, *IEEE Trans. Power Syst.*, 8(3), 887–893, Aug. 1993.

De la Torre, T., Feltes, J.W., Gómez, T., and Merrill, H.M., Deregulation, privatization, and competition: Transmission planning under uncertainty, *IEEE Trans. Power Syst.*, 14(2), 460–465, May 1999.

Kahn, E., *Electric Utility Planning & Regulation*, American Council for an Energy-Efficient Economy, Washington, D.C., 1988.

Keeney, R.L., Beley, J.R., Fleischauer, P., Kirkwood, C.W., Sicherman, A., Decision Framework for Technology Choice, Report EA-2153, Electric Power Research Institute, Palo Alto, CA, 1981.

Ringlee, R.J., Ed., Special section on electric utility systems planning, *Proc. IEEE*, 77(6), June 1989.

Schweppe, F.C., *Uncertain Dynamic Systems*, Prentice-Hall, Englewood Cliffs, NJ, 1973, chap. 3.

Stoll, H.G., Garver, L.J. (sic), Jordan, G.A., Price, W.H., Sigley, R.F., Jr., Szczepanski, R.S., Tice, J.B., *Least-Cost Electric Utility Planning*, John Wiley & Sons, New York, 1989.

Sullivan, R.L., *Power System Planning*, McGraw-Hill, New York, 1977.

U.S. Dept of Energy, *Energy Projections to the Year* 2010, Report DOE/PE-0029/2, Office of Policy, Planning and Analysis, Washington, DC, Oct. 1983a.

U.S. Dept of Energy, The Future of Electric Power in America: Economic Supply for Economic Growth, Report DOE/PE-0045, Office of Policy, Planning and Analysis, Washington, DC, June 1983b.

U.S. Congress, New Electric Power Technologies: Problems and Prospects for the 1990s, OTA-E-246, Office of Technology Assessment, Washington, DC, July 1985.

19

Power System Reliability

Richard E. Brown
Quanta Technology

The electric power industry began in the late 1800s as a component of the electric lighting industry. At this time, lighting was the only application for electricity, and homes had other methods of illumination if the electricity supply was interrupted. Electricity was essentially a luxury item and reliability was not an issue.

As electricity became more common, new applications began to appear. Examples include electric motors, electric heating, irons, and phonographs. People began to grow accustomed to these new electric appliances, and their need for reliable electricity increased. This trend culminated with the invention of the radio. No nonelectrical appliance could perform the same function as a radio. If a person wanted to listen to the airwaves, electricity was required. As radio sales exploded in the 1920s, people found that reliable electricity was a necessity. By the late 1930s, electricity was regarded as a basic utility (Philipson and Willis, 1999).

As electric utilities expanded and increased their transmission voltage levels, they found that they could improve reliability by interconnecting their system to neighboring utilities. This allowed connected utilities to "borrow" electricity in case of an emergency. Unfortunately, a problem on one utility's system could now cause problems to other utilities. This fact was made publicly evident on November 9, 1965. On this day, a major blackout left cities in the northeastern United States and parts of Ontario without power for several hours. Homes and businesses had become so dependent on electricity that this blackout was crippling. Action was needed to help prevent such occurrences from happening in the future.

19.1 NERC Regions

The North American Electric Reliability Corporation (NERC) was formed in 1968 as a response to the 1965 blackout. By this time, reliability assessment was already a mature field and was being applied to many types of engineered systems (Billinton and Allan, 1988; Ramakumar, 1993). NERC's mission is to promote the reliability of the North America's bulk power system (generation and transmission). It reviews past events; monitors compliance with policies, standards, principles, and guides; and assesses future reliability for various growth and operational scenarios. NERC provides planning recommendations and operating guidelines, but has no formal authority over electric utilities.

Since most of the transmission infrastructure in the United States and Canada is interconnected, bulk power reliability must look at systems larger than a single utility. The territory covered by NERC is far too large to study and manage as a whole, and is divided into 10 regions. These NERC regions are as follows: East Central Area Reliability Coordination Agreement (ECAR), Electric Reliability Council of Texas (ERCOT), Florida Reliability Coordinating Council (FRCC), Mid-Atlantic Area Council (MAAC), Mid-Atlantic Interconnected Network (MAIN), Mid-Continent Area Power Pool (MAPP), Northeast Power Coordinating Council (NPCC), Southeastern Electric Reliability Council (SERC), Southwest Power Pool (SPP), and the Western Systems Coordinating Council (WSCC). The geographic territories assigned to the 10 NERC regions are shown in Figure 19.1.

Even though there are 10 NERC regions, there are only four major transmission grids in the United States and Canada: the area associated with the WSCC, the area associated with the ERCOT, Quebec, and the eastern United States. These are usually referred to as the Western Interconnection, the ERCOT Interconnection, the Quebec Interconnection, and the Eastern Interconnection. Each of these grids is highly interconnected within their boundaries, but only has weak connections to the other grids. The geographic territories associated with these four interconnections are shown in Figure 19.1.

NERC looks at two aspects of bulk power system reliability: system adequacy and system security. A system must have enough capacity to supply power to its customers (adequacy), and it must be able to

FIGURE 19.1   NERC regions.

continue supplying power to its customers if some unforeseen event disturbs the system (security). Each of these two aspects of reliability is further discussed in the following.

19.2 System Adequacy Assessment

System adequacy is defined as the ability of a system to supply all of the power demanded by its customers (Billinton and Allan, 1988). Three conditions must be met to ensure system adequacy. First, its available generation capacity must be greater than the demanded load plus system losses. Second, it must be able to transport this power to its customers without overloading any equipment. Third, it must serve its loads within acceptable voltage levels.

System adequacy assessment is probabilistic in nature (Allan et al., 1994; Schilling et al., 1989). Each generator has a probability of being available, P_A, a probability of being available with a reduced capacity, P_R, and a probability of being unavailable, P_U. This allows the probability of all generator state combinations to be computed. A simple two-generator example is shown in Table 19.1. There are nine possible generator state combinations, and the probability of being in a particular combination is the product of the individual generator state probabilities. In general, if there are n generators and x possible states for each generator, then the number of possible generator state combinations is

$$Generator\ state\ combinations = x^n \tag{19.1}$$

In addition to generator state combinations, loading behavior must be known. Information is found by looking at historical load bus demand in recent years. For the best accuracy, 8760 h peak demand curves are used for each load bus. These correspond to hourly peak loads for a typical year. To reduce computational and data requirements, it is usually acceptable to reduce each set of 8760-h load curves to three weekly load curves (168 h each). These correspond to typical weekly load patterns for winter conditions, spring/autumn conditions, and summer conditions. Weekly load curves can be scaled up or down to represent temperatures that are above or below normal. Sample weekly load curves for a winter peaking load bus are shown in Figure 19.2.

To perform an adequacy assessment, each generation state combination is compared with all hourly loading conditions. For each combination of generation and loading, a power flow is performed. If the available generation cannot supply the loads or if any constraints are violated, the system is inadequate and certain loads must be shed. After all generation/load combinations are examined, the adequacy assessment is complete.

An adequacy assessment produces the following information for each load bus: (1) the combinations of generation and loading that result in load interruptions and (2) the probability of being in each

TABLE 19.1 Generator State Probabilities

Generator State		
Generator 1	Generator 2	Probability
Available	Available	$P_{A1} P_{A2}$
Available	Reduced	$P_{A1} P_{R2}$
Available	Unavailable	$P_{A1} P_{U2}$
Reduced	Available	$P_{R1} P_{A2}$
Reduced	Reduced	$P_{R1} P_{R2}$
Reduced	Unavailable	$P_{R1} P_{U2}$
Unavailable	Available	$P_{U1} P_{A2}$
Unavailable	Reduced	$P_{U1} P_{R2}$
Unavailable	Unavailable	$P_{U1} P_{U2}$

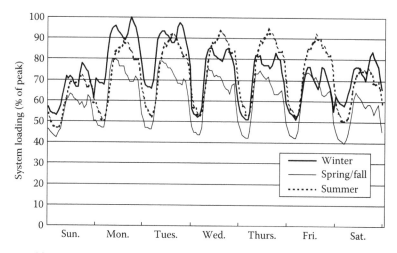

FIGURE 19.2 Weekly load curves by season.

of these inadequate state combinations. From this information, it is simple to compute the expected number of interruptions for each load bus, the expected number of interruption minutes for each load bus, and the expected amount of unserved energy for each load bus. These load bus results can then be aggregated to produce the following system indices:

- *LOLE* (loss of load expectation)—the expected number of hours per year that the system will have to shed load
- *EENS* (expected energy not served)—the expected number of megawatt hours per year that a system will not be able to supply

System adequacy assessment assumes that the transmission system is available. This may not always be the case. A classic example is the 1965 blackout, which was initiated by the unexpected loss of a transmission line. To address such events, system security assessment is required.

19.3 System Security Assessment

System security is defined as the ability of a power system to supply all of its loads in the event of one or more contingencies (a contingency is an unexpected event such as a system fault or a component outage). This is divided into two separate areas: static security assessment and dynamic security assessment.

Static security assessment determines whether a power system is able to supply peak demand after one or more pieces of equipment (such as a line or a transformer) are disconnected. The system is tested by removing a piece (or multiple pieces) of equipment from the normal power flow model, rerunning the power flow, and determining if all bus voltages are acceptable and all pieces of equipment are loaded below emergency ratings. If an unacceptable voltage or overload violation occurs, load must be shed for this condition and the system is *insecure*. If removing any single component will not result in the loss of load, the system is $N - 1$ *Secure*. If removing any X arbitrary components will not result in the loss of load, the system is $N - X$ *Secure*. N refers to the number of components on the system and X refers to the number of components that can be safely removed.

Performing a static security assessment can be computationally intensive. For example, an $N - 2$ assessment on a modest system with 5000 components (1500 buses, 500 transformers, and 3000 lines) will require more than 25 million power flows to be performed. For this reason, contingency ranking methods are often used. These methods rank each contingency based on its likelihood of resulting in

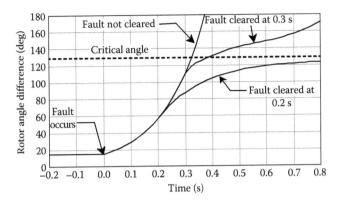

FIGURE 19.3   Dynamic security assessment.

load curtailment. Contingencies are examined in order of their contingency ranking, starting with the most severe. If a prespecified number of contingencies are tested and found to be secure, it is assumed that contingencies with less severe rankings are also secure and do not need to be examined.

Static security assessment is based on steady state power flow solutions. For each contingency, it assumes that the system protection has properly operated and the system has reached a steady state. In fact, the power system may not actually reach a steady state after it has been disturbed. Checking whether a system will reach a steady state after a fault occurs is referred to *as dynamic security assessment* (also referred to as *transient security assessment*).

When a fault occurs, the system is less able to transfer power from synchronous generators to synchronous motors. Since the instantaneous power input has not changed, generators will begin to speed up and motors will begin to slow down (analogous to the chain slipping while riding a bicycle). This increases the rotor angle difference between generators and motors. If this rotor angle exceeds a critical value, the system will become unstable and the machines will not be able to regain synchronism. After the protection system clears the fault, the rotor angle difference will still increase since the power transfer limits of the system are still less than the prefault condition. If the fault is cleared quickly enough, this additional increase will not cause the rotor angle difference to exceed the critical angle and the system will return to a synchronous state (ABB Power, 1997).

An example of a transient stability test is shown in Figure 19.3. This shows the rotor angle difference between a synchronous generator and a synchronous motor during a fault sequence. When the fault occurs, the rotor angle begins to increase. If the fault is not cleared, the rotor angle quickly exceeds the critical angle. If the fault is cleared at 0.3 s, the rotor angle still increases beyond the critical value. The system is dynamically stable for this fault if it is cleared in 0.2 s. The rotor angle will still increase after the fault occurs but will stabilize below the critical value.

A dynamic security assessment will consist of many transient stability tests that span a broad range of loading conditions, fault locations, and fault types. To reduce the number of tests required, contingency rankings (similar to static security assessment) can be used.

19.4 Probabilistic Security Assessment

Although the "N − 1 Criterion" remains popular, it has received much criticism since it treats unlikely events with the same importance as more frequent events. Using the N − 1 Criterion, large amounts of money may be spent to reinforce a system against a very rare event. From a reliability perspective, this money *might* be better spent in other areas such as replacing old equipment, decreasing maintenance intervals, adding automation equipment, adding crews, and so on. To make such value judgments, both the impact of each contingency and its probability of occurrence must be considered (Endrenyi, 1978).

This is referred to as *probabilistic security assessment*. To do this type of assessment, each piece of equipment needs at least two fundamental pieces of information: the *failure rate* of the equipment (usually denoted λ, in failures per year) and the mean time to repair of the equipment (usually denoted *MTTR*, in hours).

Performing a probabilistic security assessment is similar to a performing a standard static security assessment. First, contingencies are ranked and simulated using a power flow. If a contingency results in the loss of load, information about the number and size of interrupted loads, the frequency of the contingency, and the repair time of the contingency is recorded. This allows quantities such as *EENS* to be easily computed. If contingency *i* causes kW_i amount of kilowatts to be interrupted, then *EENS* is equal to

$$EENS = \sum_i kW_i \lambda_i MTTR_i \qquad (19.2)$$

It is important to note that this is the *EENS* due to contingencies and is separate from the *EENS* due to generation unavailability. It is also important to note that this formula assumes that $\lambda_i MTTR_i$ is small when compared to 1 year. If this is not the case, a component will experience fewer failures per year than its failure rate and the equation must be adjusted accordingly.

19.5 Distribution System Reliability

The majority of customer reliability problems stem from distribution systems. For a typical residential customer with 90 min of interrupted power per year, between 70 and 80 min will be attributable to problems occurring on the distribution system that it is connected to (Billinton and Jonnavitihula, 1996). This is largely due to radial nature of most distribution systems, the large number of components involved, the sparsity of protection devices and sectionalizing switches, and the proximity of the distribution system to end-use customers.

Since reliability means different things to different people, it is necessary to address the definition of "distribution system reliability" in more detail. In distribution systems, reliability primarily relates to equipment outages and customer interruptions:

- *Outage*—when a piece of *equipment* is deenergized
- *Momentary interruption*—when a *customer* is deenergized for less than a few minutes
- *Sustained interruption*—when a *customer* is deenergized for more than a few minutes

Customers do not, in the strictest sense, experience power outages. Customers experience power interruptions. If power is restored within a few minutes, it is considered a momentary interruption. If not, it is considered a sustained interruption. The precise meaning of "a few minutes" varies from utility to utility, but is typically between 1 and 5 min. The IEEE defines a momentary interruption based on 5 min. (Note: some references classify interruptions into four categories rather than two. Instantaneous interruptions last a few seconds, momentary interruptions last a few minutes, temporary interruptions last a few hours, and sustained interruptions last many hours.)

On a historical note, momentary interruptions used to be considered a "power quality issue" rather than a "reliability issue." It is now generally agreed that momentary interruptions are an aspect of reliability since (1) momentary interruptions can cause substantial problems to all types of customers, and (2) many trade-offs must be made between momentary interruptions and sustained interruptions during system planning, operation, and control. It can also be observed that customer voltage sags, typically considered a power quality issue, are slowly becoming a reliability issue for similar reasons.

Distribution system reliability is not dependent solely upon component failure characteristics. It is also dependent upon how the system responds to component failures. To understand this, it is necessary to understand the sequence of events that occurs after a distribution system fault.

19.6 Typical Sequence of Events after an Overhead Distribution Fault

The following is a typical sequence of events that will occur after a fault occurs on a distribution system (Brown, 2003).

1. The fault causes high currents to flow from the source to the fault location. These high currents may result in voltage sags for certain customers. These sags can occur on all feeders that have a common coupling at the distribution substation.
2. An instantaneous relay trips open the feeder circuit breaker at the substation. This causes the entire feeder to be deenergized. A pause allows the air around the fault to deionize, and then a reclosing relay will close the circuit breaker. If no fault current is detected, the fault has cleared itself and all customers on the feeder have experienced a momentary interruption.
3. If the fault persists, time overcurrent protection devices are allowed to clear the fault. If the fault is on a fused lateral, the fuse will blow and customers on the lateral will be interrupted. If the feeder breaker trips again, the reclosing relay will repeat the reclosing process a preset number of times before locking out. After the feeder breaker locks out, all customers on the feeder will be interrupted. Automated line switching and system reconfiguration will occur at this point if these capabilities exist.
4. The electric utility will receive trouble calls from customers with interrupted power. It will dispatch a crew to locate the fault and isolate it by opening up surrounding sectionalizing switches. It may also attempt to reconfigure the distribution system in an attempt to restore power to as many customers as possible while the fault is being repaired. Fault isolation can be very fast if switches are motor operated and remotely controlled, but switching usually takes between 15 and 60 min.
5. The crew repairs the faulted equipment and returns the distribution system to its normal operating state.

As can be seen, *a fault on the distribution system will impact many different customers in many different ways.* In general, the same fault will result in voltage sags for some customers, momentary interruptions for other customers, and varying lengths of sustained interruptions for other customers, depending on how the system is switched and how long the fault takes to repair.

Distribution system reliability assessment methods are able to predict distribution system reliability based on system configuration, system operation, and component reliability data (Brown et al., 1996). This ability is becoming increasingly important as the electric industry becomes more competitive, as regulatory agencies begin to regulate reliability, and as customers begin to demand performance guarantees. The most common reliability assessment methods utilize the following process: (1) they simulate a system's response to a contingency, (2) they compute the reliability impact that this contingency has on each customer, (3) the reliability impact is weighted by the probability of the contingency occurring, and (4) steps 1–3 are repeated for all contingencies. Since this process results in the reliability that each customer can expect, new designs can be compared, existing systems can be analyzed, and reliability improvement options can be explored.

19.7 Distribution Reliability Indices

Utilities typically keep track of customer reliability by using reliability indices. These are average customer reliability values for a specific area. This area can be the utility's entire service area, a particular geographic region, a substation service area, a feeder service area, and so on. The most commonly used reliability indices give each customer equal weight. This means that a large industrial customer and a small residential customer will each have an equal impact on computed indices. The most common of these *customer reliability indices* are as follows: System Average Interruption Frequency Index (SAIFI), System Average Interruption Duration Index (SAIDI), Customer Average Interruption Duration Index (CAIDI), and Average System Availability Index (ASAI) (IEEE, 2003). Notice that these indices are redundant. If SAIFI and SAIDI are known, both CAIDI and ASAI can be calculated. Formulae for these indices are as follows:

$$SAIFI = \frac{Total\ number\ of\ customer\ interruptions}{Total\ number\ of\ customers\ served} per\ year \tag{19.3}$$

$$SAIDI = \frac{\sum Customer\ interruption\ durations}{Total\ number\ of\ customers\ served}\ hours\ per\ year \tag{19.4}$$

$$CAIDI = \frac{\sum Customer\ interruption\ durations}{Total\ number\ of\ customer\ interruptions} = \frac{SAIDI}{SAIFI} hours\ per\ interruption \tag{19.5}$$

$$ASAI = \frac{Customer\ hours\ service\ availability}{Customer\ hours\ service\ demand} = \frac{8760 - SAIDI}{8760} per\ unit \tag{19.6}$$

Some less commonly used reliability indices are not based on the total number of customers served. The Customer Average Interruption Frequency Index (CAIFI) and the Customer Total Average Interruption Duration Index (CTAIDI) are based upon the number of customers that have experienced one or more interruptions in the relevant year. The Average System Interruption Frequency Index (ASIFI) and the Average System Interruption Duration Index (ASIDI) are based upon the connected kVA of customers (these are sometimes referred to as load-based indices). Formulae for these indices are as follows:

$$CAIFI = \frac{Total\ number\ of\ customer\ interruptions}{Customers\ experiencing\ one\ or\ more\ interruptions} per\ year \tag{19.7}$$

$$CTAIDI = \frac{\sum Customer\ interruption\ durations}{Customers\ experiencing\ one\ or\ more\ interruptions} hours\ per\ year \tag{19.8}$$

$$ASIFI = \frac{Connected\ kVA\ interrupted}{Total\ connected\ kVA\ served} per\ year \tag{19.9}$$

$$ASIDI = \frac{Connected\ kVA\ hours\ interrupted}{Total\ connected\ kVA\ served} hours\ per\ year \tag{19.10}$$

As momentary interruptions become more important, it becomes necessary to keep track of indices related to momentary interruptions. Since the duration of momentary interruptions is of little consequence, a single frequency-related index, the Momentary Average Interruption Frequency Index (MAIFI), is all that is needed. MAIFI, like SAIFI, weights each customer equally (there is currently no load-based index for momentary interruptions). The formula for MAIFI is as follows:

$$MAIFI = \frac{Total\ number\ of\ customer\ momentary\ interruptions}{Total\ number\ of\ customers\ served} per\ year \tag{19.11}$$

The precise application of *MAIFI* varies. This variation is best illustrated by an example. Assume that a customer experiences three recloser operations followed by a recloser lockout, all within a period of 1 min. Some utilities would not count this event as a momentary interruption since the customer experiences a sustained interruption. Other utilities would count this event as three momentary interruptions and one sustained interruption. Similarly, if a customer experiences three recloser operations within a period of 1 min with power being restored after the last recloser, some utilities would count the event as three momentary interruptions and other utilities would count the event as a single momentary interruption. The IEEE defines a momentary interruption as lasting less than five minutes and a "momentary event" as a grouping of one or more momentary interruptions occurring within a five minute interval.

19.8 Storms and Major Events

When electric utilities compute reliability indices, they often exclude interruptions caused by "storms" and "major events." The definition of a major event varies from utility to utility, but a typical example is when more than 10% of customers experience an interruption during the event. The event starts when the notification of the first interruption is received and ends when all customers are restored service. The IEEE defines a major event based on a statistical approach called 2.5 Beta (IEEE, 2003).

In nonstorm conditions, equipment failures are independent events—the failure of one device is completely independent of another device. In contrast, major events are characterized by common-mode failures. This means that a common cause is responsible for all equipment failures. The result is that many components tend to fail at the same time. This puts a strain on utility resources, which can only handle a certain number of concurrent failures (Brown et al., 1997). The most common causes of major events are wind storms, ice storms, and heat waves.

Wind storms refer to linear winds that blow down trees and utility poles. The severity of wind storms is dependent upon sustained wind speed, gust speed, wind direction, and the length of the storm. Severity is also sensitive to vegetation management and the time elapsed since the last wind storm. Since a wind storm will tend to blow over all of the weak trees, a similar storm occurring a few months later may have little impact. A U.S. map showing wind speeds for the worst expected storm in 50 years is shown in Figure 19.4.

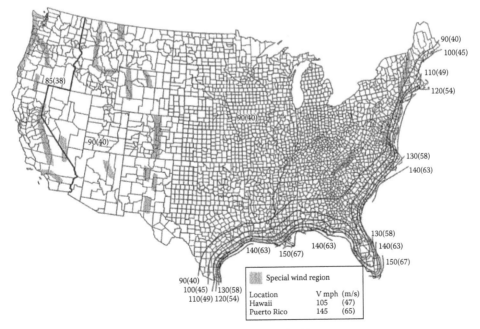

FIGURE 19.4 Fifty-year wind storm (sustained wind speed in miles/h).

Ice storms refer to ice buildup on conductors. This has four major effects: (1) it places a heavy physical load on the conductors and support structures, (2) it increases the cross-sectional area that is exposed to the wind, (3) ice can break off and cause a conductor to jump into the phase wires located above it, and (4) galloping. Galloping occurs when ice buildup assumes a teardrop shape and acts as an airfoil. During high winds, this can cause conductors to swing wildly and with great force. Ice can also cause problems by accumulating in trees, causing limbs to break off, and causing entire trunks to fall over into power lines.

Heat waves are extended periods of exceedingly hot weather. This hot weather causes electricity demand to skyrocket due to air-conditioning loads. At the same time, conductors cannot carry as much electricity since they cannot transfer heat as effectively to their surroundings. This combination of heavy loading and conductor de-rating can cause overhead wires to become overloaded and sag to dangerous levels. Overloaded cables will cause insulation to lose life. In a worst-case scenario, the maximum power transfer capabilities of the system can be approached, resulting in a voltage collapse condition. Humidity exacerbates the impact of heat waves since it causes air conditioners to consume more energy.

19.9 Component Reliability Data

For a reliability model to be accurate, component reliability data must be representative of the system being modeled. Utilities recognize this and are increasing their efforts to keep track of component failure rates, failure modes, repair times, switching times, and other important reliability parameters. Unfortunately, reliability statistics vary widely from utility to utility and from country to country. The range of equipment reliability data that can be found in published literature is shown in Table 19.2.

Because component reliability is very system specific, it is beneficial to calibrate reliability models to historical reliability indices. In this process, component reliability parameters are adjusted until historical reliability indices match computed reliability indices (Brown and Ochoa, 1998). The amount that each parameter is adjusted should depend on the confidence of the original value and the sensitivity of the reliability indices to changes in this value. To illustrate, consider an overhead distribution system. A reliability model of this system is created using component reliability data from published literature. Unfortunately, the reliability indices that the model produces do not agree with the historical performance of the system

TABLE 19.2   Equipment Reliability Data

Component	Failure Rate (year^{-1})	MTTR (h)
Substation Equipment		
Power transformers	0.015–0.07	15–480
Circuit breakers	0.003–0.02	6–80
Disconnect switches	0.004–0.16	1.5–12
Air-insulated buswork	0.002–0.04	2–13
Overhead Equipment		
Transmission lines[a]	0.003–0.140	4–280
Distribution lines[a]	0.030–0.180	4–110
Switches/fused cutouts	0.004–0.014	1–4
Pole mounted transformer	0.001–0.004	3–8
Underground Equipment		
Cable[a]	0.005–0.04	3–30
Padmount switches	0.001–0.01	1–5
Padmount transformers	0.002–0.003	2–6
Cable terminations/joints	0.0001–0.002	2–4

[a] Failure rates for lines and cable are per mile.

over the past few years. To fix this, the failure rate and repair times of overhead lines (along with other component parameters) can be adjusted until predicted reliability matches historical reliability.

19.10 Utility Reliability Problems

To gain a broader understanding of power system reliability, it is necessary to understand the root causes of system faults and system failures. A description of major failure modes is now provided.

19.10.1 Underground Cable

A major reliability concern pertaining to underground cables is electrochemical treeing. Treeing occurs when moisture penetration in the presence of an electric field reduces the dielectric strength of cable insulation. When the dielectric strength is degraded sufficiently, transients caused by lightning or switching can result in dielectric breakdown. Electrochemical treeing usually affects extruded dielectric cable such as cross-linked polyethylene (XLPE) and ethylene-propylene rubber (EPR), and is largely attributed to insulation impurities and bad manufacturing. To reduce failures related to electrochemical treeing, a utility can install surge protection on riser poles (transitions from overhead to underground), can purchase tree-retardant cable, and can test cable reels before accepting them from the manufacturer.

Existing cable can be tested and replaced if problems are found. One way to do this is to apply a DC voltage withstand test (approximately three times nominal RMS voltage). Since cables will either pass or not pass this test, information about the state of cable deterioration cannot be determined. Another popular method for cable testing is to inject a small signal into one end and check for reflections that will occur at partial discharge points. Other methods are measuring the power factor over a range of frequencies (dielectric spectroscopy), analyzing physical insulation samples in a lab for polymeric breakdown (degree of polymerization), and using cable indentors to test the hardness of the insulation.

Not all underground cable system failures are due to cable insulation. A substantial percentage occurs at splices, terminations, and joints. Major causes are due to water ingress and poor workmanship. Heat shrink covers can be used to waterproof these junctions and improve reliability.

The last major reliability concern for underground cable is dig-ins. This is when excavation equipment cuts through one or more cables. To prevent dig-ins, utilities should encourage the public to have cable routes identified before initiating site excavation. In extreme cases where high reliability is required, utilities can place cable in concrete-encased duct banks.

19.10.2 Transformer Failures

Transformers are critical links in power systems, and can take a long time to replace if they fail. Through-faults cause extreme physical stress on transformer windings, and are the major cause of transformer failures. Overloads rarely result in transformer failures, but do cause thermal aging of winding insulation.

When a transformer becomes hot, the insulation on the windings slowly breaks down and becomes brittle over time. The rate of thermal breakdown approximately doubles for every 10°C. Ten degree Celcius is referred to as the "Montsinger Factor" and is a rule of thumb describing the Arrhenius theory of electrolytic dissociation. Because of this exponential relationship, transformer overloads can result in rapid transformer aging. When thermal aging has caused insulation to become sufficiently brittle, the next fault current that passes through the transformer will mechanically shake the windings, a crack will form in the insulation, and an internal transformer fault will result.

Extreme hot-spot temperatures in liquid-filled transformers can also result in failure. This is because the hot spot can cause free bubbles that reduce the dielectric strength of the liquid. Even if free bubbles are not formed, high temperatures will increase internal tank pressure and may result in overflow or tank rupture.

Many transformers are fitted with load tap changers (LTCs) for voltage regulation. These mechanically moving devices have historically been prone to failure and can substantially reduce the reliability

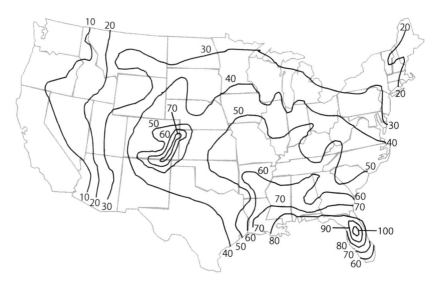

FIGURE 19.5 Number of thunderstorm days per year.

of a transformer (Willis, 1997). Manufacturers have addressed this problem and new LTC models using vacuum technology have succeeded in reducing failure rates.

19.10.3 Lightning

A lightning strike occurs when the voltage generated between a cloud and the ground exceeds the dielectric strength of the air. This results in a massive current stroke that usually exceeds 30,000 A. To make matters worse, most strokes consist of multiple discharges within a fraction of a second. Lightning is the major reliability concern for utilities located in high keraunic areas (Burke, 1994). An isokeraunic map for the United States is shown in Figure 19.5.

Lightning can affect power systems through direct strikes (the stroke contacts the power system) or through indirect strikes (the stroke contacts something in close proximity and induces a traveling voltage wave on the power system). Lightning can be protected against by having a high system BIL (basic impulse level) by using shield wires, by using surge arrestors to clamp voltages across equipment, and by having a low impedance ground. Direct strikes are virtually impossible to protect against on a distribution system.

19.10.4 Tree Contact

Trees continuously grow, can fall over onto conductors, can drop branches onto conductors, can push conductors together, and can serve as gateway for animals. This is why many utilities spend more on tree trimming than on any other preventative maintenance activity.

When a tree branch bridges two conductors, a fault does not occur immediately. This is because a moist tree branch has a substantial resistance. A small current begins to flow and starts to dry out the wood fibers. After several minutes, the cellulose will carbonize, resistance will be greatly reduced, and a short circuit will occur. Branches brushing against a single phase conductor typically *do not* result in system faults.

Faults due to tree contact can be reduced by using tree wire. This is overhead wire with an insulated jacket similar to cable. Tree wire can be effective, but faults tend to result in conductor burndown since they will not motor (move themselves along the conductor) like faults on bare conductor.

19.10.5 Birds

Birds are the most common cause of animal faults on both transmission systems and air-insulated substations. Different types of birds cause different types of problems, but they can generally be classified as nesting birds, roosting birds, raptors, and woodpeckers.

Nesting birds commonly build their homes on transmission towers and in substations. Nesting materials can cause faults, and bird excrement can contaminate insulators. Nesting birds also attract predators such as raccoons, snakes, and cats. These predators can be a worse reliability problem than the birds themselves.

Roosting birds use electrical equipment to rest on or to search for prey. They can be electrocuted by bridging conductors with their wings, and their excrement can contaminate insulators. To prevent birds from roosting, anti-roosting devices can be placed on attractive sites. For locations that cater to thousands of roosting birds, more extreme deterrent methods such as pyrotechnics can be used.

Raptors are birds of prey such as eagles, hawks, ospreys, owls, and vultures. Reliability problems are similar to other roosting and nesting birds, but special consideration may be required since most raptors are protected by the federal government.

Woodpeckers peck holes in wood with their beaks as they search for insects. This does not harm trees (the bark regenerates), but can cause devastating damage to utility poles. This can be prevented by using steel poles, by using repellent, or by tricking a woodpecker into believing that there is already a resident woodpecker (woodpeckers are quite territorial).

19.10.6 Squirrels

Squirrels are a reliability concern for all overhead distribution systems near wooded areas. Squirrels will not typically climb utility poles, but will leap onto them from nearby trees. They cause faults by bridging grounded equipment with phase conductors. Squirrel problems can be mitigated by cutting down nearby access trees or by installing animal guards on insulators.

19.10.7 Snakes

Snakes are major reliability concerns in both substations and underground systems. They can squeeze through very small openings, can climb almost anything, and have the length to easily span phase conductors. Snakes are usually searching for food (birds in substations and mice in underground systems), and removing the food supply can often remove the snake problem. Special "snake fences" are also available.

19.10.8 Insects

It is becoming more common for fire ants to build nests in pad-mounted equipment. Their nesting materials can cause short circuits, the ants can eat away at conductor insulation, and they make equipment maintenance a challenge.

19.10.9 Bears, Bison, and Cattle

These large animals do not typically cause short circuits, but degrade the structural integrity of poles by rubbing on guy wires. Bears can also destroy wooden poles by using them as scratching posts, and black bears can climb wooden utility poles. These problems can be addressed by placing fences around poles and guy wire anchors.

19.10.10 Mice, Rats, and Gophers

These rodents cause faults by gnawing through the insulation of underground cable. They are the most common cause of animal-related outages on underground equipment. To make matters worse, they will attract snakes (also a reliability problem). Equipment cabinets should be tightly sealed to prevent these small animals from entering. Ultrasonic devices can also be used to keep rodents away (ultrasonic devices will not keep snakes away).

19.10.11 Vandalism

Vandalism can take many different forms, from people shooting insulators with rifles to professional thieves stealing conductor wire for scrap metal. Addressing these reliability problems will vary greatly from situation to situation.

19.11 Reliability Economics

When a power interruption occurs, both the utility and the interrupted customers are inconvenienced. The utility must spend money to fix the problem, will lose energy sales during the interruption, and may be sued by disgruntled customers. From the customer perspective, batch processes may be ruined, electronic devices may crash, production may be lost, retail sales may be lost, and inventory (such as refrigerated food) may be ruined.

When a customer experiences an interruption, there is an amount of money that it would be willing to pay to have avoided the interruption. This amount is referred to as the customer's incurred cost of poor reliability, and consists of a base cost plus a time-dependent cost. The base cost is the same for all interruptions, relates to electronic equipment shutdown and interrupted processes, and is equivalent to the cost of a momentary interruption. The time-dependent cost relates to lost production and extended inconvenience, and reflects that customers would prefer interruptions to be shorter rather than longer.

The customer cost of an interruption varies widely from customer to customer and from country to country. Other important factors include the time of year, the day of the week, the time of day, and whether advanced warning is provided. Specific results are well documented by a host of customer surveys (Billinton et al., 1983; IEEE Std. 493-1990; Tollefson et al., 1991, 1994). For planning purposes, it is useful to aggregate these results into a few basic customer classes: commercial, industrial, and residential. Since larger customers will have a higher cost of reliability, results are normalized to the peak kW load of each customer. Reliability cost curves for typical U.S. customers are shown in Figure 19.6.

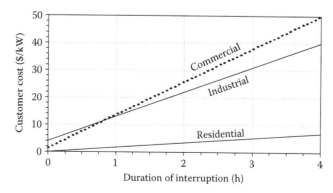

FIGURE 19.6 Typical U.S. customer interruption costs (1999 dollars).

Average customer cost curves tend to be linear and can be modeled as an initial cost plus a first order, time-dependent cost. Specific customer cost curves may be extremely nonlinear. For example, a meat packing warehouse depending upon refrigeration may be unaffected by interruptions lasting many hours. At a certain point, the meat will begin to spoil and severe economic losses will quickly occur. After the meat spoils, additional interruption time will harm this particular customer much more.

19.12 Annual Variations in Reliability

Power system reliability varies from year to year. In a lucky year, a system may have a SAIDI of 30 min. The next year, this exact same system may experience a SAIDI of 8 h. This type of variation is inevitable and must be considered when comparing reliability indices. It is also important to note that the variance of reliability indices will tend to be less for areas serving more customers. Individual customer reliability will tend to be the most volatile, followed by feeder reliability, substation reliability, regional reliability, and so forth.

The importance of annual reliability variance will grow as utilities become subject to performance-based rates and as customer reliability guarantees become more common. These types of contracts expose utilities to new risks that must be understood and managed. Since performance-based contracts penalize and reward utilities based on reliability, annual variations must be understood for fair contracts to be negotiated and managed.

Contractual issues concerning service reliability are becoming important as the electric industry becomes more competitive. Customers can choose between suppliers, and realize that there is a trade-off between reliability and rates. Some customers will demand poor reliability at low rates, and other customers will demand high reliability at premium rates. To address the wide variation in customer needs, utilities can no longer be suppliers of energy alone, but must become suppliers of both energy and reliability. Power system reliability is now a *bona fide* commodity with explicit value for utilities to supply and explicit value that customers demand.

References

Allan, R.N., Billinton, R., Breipohl, A.M., and Grigg, C.H., Bibliography on the application of probability methods in power system reliability evaluation, *IEEE Trans. Power Syst.*, 9, 1, 41–49, Feb. 1994.

Billinton, R. and Allan, R.N., *Reliability Assessment of Large Electric Power Systems*, Kluwer Academic Publishers, Dordrecht, the Netherlands, 1988.

Billinton, R. and Allan, R., *Reliability Evaluation of Engineering Systems: Concepts and Techniques*, 2nd edn., Plenum Press, New York, 1992.

Billinton, R. and Jonnavitihula, S., A test system for teaching overall power system reliability assessment, *IEEE Trans. Power Syst.*, 11, 4, 1670–1676, Nov. 1996.

Billinton, R., Wacker, G., and Wojczynski, E., Comprehensive bibliography on electrical service interruption costs, *IEEE Trans. Power Appar. Syst.*, PAS-102, 6, 1831–1837, June 1983.

Brown, R.E., *Electric Power Distribution Reliability*, Second Edition, CRC Press, 2009.

Brown, R.E., Gupta, S., Christie, R.D., Venkata, S.S., and Fletcher, R.D., Distribution system reliability analysis using hierarchical Markov modeling, *IEEE Trans. Power Delivery*, 11, 4, 1929–1934, Oct. 1996.

Brown, R.E., Gupta, S., Christie, R.D., Venkata, S.S., and Fletcher, R.D., Distribution system reliability: Momentary interruptions and storms, *IEEE Trans. Power Delivery*, 12, 4, 1569–1575, Oct. 1997.

Brown, R.E. and Ochoa, J.R., Distribution system reliability: Default data and model validation, *IEEE Trans. Power Syst.*, 13, 2, 704–709, May 1998.

Burke, J.J., *Power Distribution Engineering*, Marcel Dekker, Inc., New York, 1994.

Electrical Transmission and Distribution Reference Book, ABB Power T&D Company, Inc., Raleigh, NC, 1997.

Endrenyi, J., *Reliability in Electric Power Systems*, John Wiley & Sons, Ltd., New York, 1978.

IEEE Std. 493-1990, *IEEE Recommended Practice for the Design of Reliable Industrial and Commercial Power Systems*, 1990.

IEEE Standard 1366–2003, *Guide for Electric Power Distribution Reliability Indices*, 2003.

Philipson, L. and Willis, H.L., *Understanding Electric Utilities and De-regulation*, Marcel Dekker, Inc., New York, 1999.

Ramakumar, R., *Engineering Reliability: Fundamentals and Applications*, Prentice-Hall, Inc., Englewood Cliffs, NJ, 1993.

Schilling, M.T., Billinton, R., Leite da Silva, A.M., and El-Kady, M.A., Bibliography on composite system reliability (1964–1988), *IEEE Trans. Power Syst.*, 4, 3, 1122–1132, Aug. 1989.

Tollefson, G., Billinton, R., and Wacker, G., Comprehensive bibliography on reliability worth and electric service consumer interruption costs 1980–1990, *IEEE Trans. Power Syst.*, 6, 4, 1508–1514, Nov. 1991.

Tollefson, G., Billinton, R., Wacker, G., Chan, E., and Aweya, J., A Canadian customer survey to assess power system reliability worth, *IEEE Trans. Power Syst.*, 9, 1, 443–450, Feb. 1994.

Willis, H.L., *Power Distribution Planning Reference Book*, Marcel Dekker, Inc., New York, 1997.

20

Probabilistic Methods for Planning and Operational Analysis

Gerald T. Heydt
Arizona State University

Peter W. Sauer
University of Illinois at Urbana-Champaign

20.1 Uncertainty in Power System Engineering

Probabilistic methods are mathematical techniques to formally consider the impact of uncertainty in models, parameters, or data. Typical uncertainties include the future value of loading conditions, fuel prices, weather, and the status of equipment. Methods to consider all possible values of uncertain data or parameters include such techniques as interval analysis, minimum/maximum analysis, and fuzzy mathematics. In many cases, these techniques will produce conservative results because they do not necessarily incorporate the "likelihood" of each value of a parameter in an expected range, or they might be intentionally designed to compute "worse case" scenarios. These techniques are not discussed further here. Instead, this chapter presents two techniques that have been successfully applied to power system planning and operational analysis as noted by the "Application of Probability Methods" subcommittee of the IEEE Power Engineering Society in Rau et al. (1994). They are

- Monte Carlo simulation: analysis in which the system to be studied is subjected to pseudorandom operating conditions, and the results of many analyses are recorded and subsequently statistically studied. The advantage is that no specialized forms or simplifications of the system model are needed.
- Analytical probability methods: analysis in which the system to be studied is represented by functions of several random variables of known distribution.

Rau et al. (1994) also noted that probabilistic load flow methods would be well suited to evaluate loadability limits and transfer capabilities under uncertainties created by industry restructuring. They also noted the needs to include uncertainty analysis in cost/worth studies and security assessment. This chapter focuses on those two analysis tools and illustrates the differences between the Monte Carlo simulation method and the analytical probability method. There has been renewed interest in the subject of uncertainty in generation with the incorporation of renewable resources. Mainly these resources are wind energy, solar energy, and concentrated solar energy. In each of these cases, uncertainty in the supply power occurs because of the variability of the wind and the stochastic nature of wind forecasts, as well as the variability of insolation (solar intensity) due to weather and man-made degradation to atmospheric transmissivity. One approach in these areas is to "bracket" the variability of the generation (e.g., high wind case, high solar case versus low wind case, low solar case). Another is to model the probability density function of the wind or solar generation, and then use various algorithms that process the probability density functions of the sources to obtain the probability density functions of the line loads and bus voltages (i.e., a "stochastic load flow study"). The stochastic analysis described makes available such information as the probability of insufficient generation, the probability of high line loads (e.g., overload), and the probability of high bus voltage (this may occur during light load conditions with high penetration of solar generation at unity power factor common to residential solar photovoltaic generation systems).

20.2 Deterministic Power Flow Studies

Conventional, deterministic power flow studies (load flow studies) are one of the most widely used analysis tools for both planning and operations (Heydt, 1996). They can imprecisely be described through three features:

- Power flow analysis is the computation of steady-state conditions for a given set of loads P (active power) and Q (reactive power) and a given system configuration (interconnecting lines/transformers).
- Active-power generation and bus voltage magnitudes are specified at the generator buses.
- Power flow analysis gives a solution of system states, mainly bus voltage phase angles (δ) and magnitudes (V).

Additional complexities such as phase-shifting and tap-changing-under-load (TCUL) transformers and other devices are often added to this basic description. And of course many other "solution" quantities such as generator reactive power output and line power flows are often added.

The main method used for power flow analysis is the Newton–Raphson (NR) method, a method to find the zeros of $f(X) = 0$ where f is a vector-valued function of P and Q summations at each bus (Kirchhoff's current law in terms of power), and X is the system state vector (primarily δ and V). There is normally a natural weak coupling between real power and voltage magnitude (P and V) as well as reactive power and voltage angle (Q and δ). This weak coupling can be exploited in the NR method to speed up solutions. Figures 20.1 and 20.2 are pictorials of this decoupling.

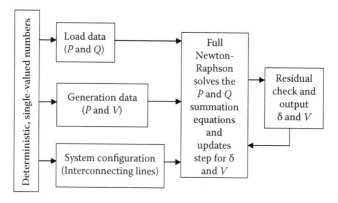

FIGURE 20.1   Conventional NR power flow study.

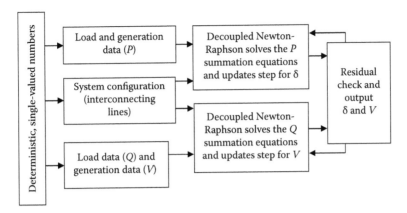

FIGURE 20.2 Decoupled NR power flow study.

This decoupling can be utilized to increase the speed of computing exact solutions, and also can be used to create superfast approximate linear solutions. In both cases, the inputs to the power flow solution algorithms shown are deterministic. When there is uncertainty in the input data, this can be considered using the two methods introduced earlier—Monte Carlo and analytical probability. These two methods are discussed next in the context of power flow analysis.

20.3 Monte Carlo Power Flow Studies

In Monte Carlo power flow studies, multiple power flow studies are run using different sets of input data. The various sets of input data are typically obtained from random number generators to correspond to some desired statistical distribution. For example, the true future total load of a power system (and its allocation to system buses) might be unknown, but an expected value might be known together with a variance with specified statistical distribution (i.e., uniform or normal). This problem could be a problem with a scalar random variable (random total load at one time instant with known deterministic allocation to system buses), or it could be a problem with a vector of random variables (random total load at one time instant plus random allocation to system buses). In the scalar case, this would require numerous power flow studies using numerous random samples for the total load. In the vector case, this would require numerous power flow studies using numerous random samples for the total load (perhaps one statistical distribution) plus numerous random samples for the factors that allocate the total load to the system buses (perhaps another statistical distribution). This could perhaps be defined more easily by simply considering each system bus load to be a random variable. This would require a decision of which "multivariate" statistical distribution to use (i.e., all bus load values are independent and normally distributed, or all bus loads are fully correlated and normally distributed, or something in between). In most cases, the Monte Carlo power flow studies are done for "variations" from the expected value of the random variables. As such, the expected value of the variation in bus load from the expected value would be zero. The first Monte Carlo power flow study is usually the case where all variations are equal to their expected value (zero). This is also often called the "base case." Figure 20.3 shows the Monte Carlo simulation method for considering uncertainty in the input data.

The statistical description of the distribution is typically a histogram. This is a plot of the value of the output variable on the horizontal axis and the number of times that value occurs on the vertical axis. The horizontal values are grouped into "ranges" of possible values (i.e., all values from 0.1 to 0.2, 0.2 to 0.3, etc.). These histograms are similar in concept to the probability density function of analytical methods. There are also numerical descriptions (moments) of the statistical distribution that can be computed—i.e., mean, variance, skewness, kurtosis, etc. One problem with the Monte Carlo method is in determining the number of samples that are sufficient to properly represent the variation in the

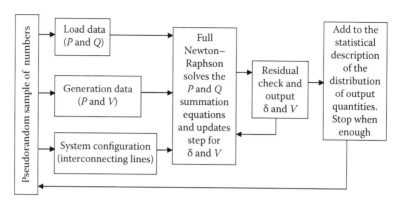

FIGURE 20.3 Monte Carlo power flow study.

uncertain parameters. For example, are 100 samples sufficient to represent all possible variations in the total system load? How about 1000? One way to look at this is—"The deterministic solution is just one sample, so anything more than one is better than the deterministic solution." But, what is important in the choice of samples is the specification of what is most likely to represent the most likely possible values of the uncertain data. This normally requires a large number of samples. Another problem with Monte Carlo methods is the large computational burden required to study a large number of samples.

Multiple power flow studies can also be used to compute the variation of system bus voltages (and all other output quantities) throughout the day, week, or year as the load varies. For example, if there was an interest in computing the line flows for every hour of the day for 1 year (assuming all load and generation data are known for each hour), this would require 8760 power flow studies (1/h). It is important to realize that although this is not really a Monte Carlo simulation as the variation of the load data is not from a random phenomenon (i.e., daily, weekly, and seasonal variations in load are somewhat predictable), it is computationally the same as running 8760 "samples" and observing the distribution of the output quantities. The "statistical distribution" would give an indication of the percentage of time throughout the year during which a line flow was in a certain range. If there was an interest in considering the uncertainty in the total load variation throughout the year, some mechanism for specifying the "variations" from the base case would need to be given for each hour of everyday. Presumably, these variations would take the form of random samples created from some known statistical distribution. A Monte Carlo simulation of this could easily result in the need for over one million power flow studies. And, if the uncertainty of the data was extended to random samples of load at each bus for each hour, the result could be the need for over one billion power flow studies.

20.4 Analytical Probabilistic Power Flow Studies

Because it is very difficult to perform analytical probability analysis for large nonlinear systems, the most probabilistic power flow studies exploit two properties of power systems that lead to a linear, smaller model. First, the decoupling that was discussed above allows the real power and voltage angle to be completely separated from the reactive power and voltage magnitude in computation. This not only reduces the size of the problem, but also reduces the accuracy and loses voltage information. Second, the fact that "small" changes in bus real power injections result in "small" changes in bus voltage angles leads to a linear approximation as follows:

$$\Delta \delta_{bus} = T_1 \Delta P_{bus} \tag{20.1}$$

So, for an initial power flow solution giving the bus voltage angle vector δ^0_{bus} for an initial bus real power injection vector, P^0_{bus}, an approximate power flow solution for a problem with $P^1_{bus} = P^0_{bus} + \Delta P_{bus}$ is given by $\delta^1_{bus} = \delta^0_{bus} + T_1 \Delta P_{bus}$.

It is important to point out that the vector of bus real power injection changes does not include the injection change at the swing (or slack) bus of the power flow study. And, the vector of bus voltage angles does not include an entry for the swing (or slack) bus of the power flow study. This is because the swing (or slack) bus is chosen as the fixed-angle reference for the power flow study and also provides the balance in real power for any specified condition. In this manner, the vector of bus real power injection changes could include a single entry and the result of this injection change would automatically be offset by a nearly equal and opposite injection change at the swing bus.

This linear analysis can be extended to approximate the resulting change in real power line flows as

$$\Delta P_{\text{line}} = T_2 \Delta \delta_{\text{bus}} \tag{20.2}$$

This combination of linear approximations gives the traditional "generation shift distribution factor," which is also the "power transfer distribution factor" for power transfers from a bus to or from the swing bus (Heydt, 1996):

$$\Delta P_{\text{line}} = T_2 T_1 \Delta P_{\text{bus}} = T \Delta P_{\text{bus}} \tag{20.3}$$

This relationship is the heart of the linear power flow, which enables analytical probabilistic methods. One reason a linear computation is important is that the linear sum of jointly normal random variables is also jointly normal (Papoulis, 1965). This makes the computation of the statistical distribution of the output variables very easy to compute from the specified statistical distribution of the input variables. For example, if there is some uncertainty in the value of the loads, an initial power flow study could be done using the expected value of the loads. This would be a deterministic power flow study. To account for the uncertainty in load, an analytical probabilistic power flow could be done as follows:

Step 1. Specify the statistical distribution of the variation of load (ΔP_{bus})—for a scalar (total load) this might be a zero mean, normal distribution with some specified variance. For a vector of zero-mean jointly normal load variations, the statistical distribution would be specified by also giving the covariance between each load variation. A covariance of zero would mean that the two uncertain jointly normal variations are statistically "independent." A covariance of ±1.0 would mean that the two jointly normal variations are fully correlated (linearly related). The covariance matrix C_{load} contains the complete statistical description of all the load variations through the variances (diagonal) and covariances (off diagonal).

Step 2. Compute the statistical distribution of the variation of line real power flows. When the variations in loads are assumed to be zero-mean jointly normal, the complete description of the statistical distribution of the variation of line real power flows is found easily as (Papoulis, 1965)

$$C_{\text{line}} = T_3 C_{\text{load}} T_3^t \tag{20.4}$$

where "t" denotes matrix transposition. The mean value of the variation in line real power flows is taken as zero because the mean value of the bus real power injection variations was assumed zero and the two are assumed to be linearly related. Now, the fact that the variations in line real power flows are not really linearly related to the bus real power injection variations means that these statistical distributions will not be exact. Figure 20.4 shows how the analytic probabilistic power flow study is done.

Stochastic power flow is another term that has been used for analytical probabilistic power flow. Since stochastic processes are statistical processes involving a number of random variables depending on a variable parameter (usually time), this terminology has been adopted as equivalent. One of the first publications on this method was of Borkowska (1974). The subject continues to attract interest (Vorsic et al., 1991).

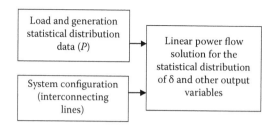

FIGURE 20.4   Analytic probabilistic power flow study.

20.5 Applications for Available Transfer Capability (ATC)

In order to realize open access to electric power transmission networks and promote generation competition and customer choice, the Federal Energy Regulatory Commission requires that ATC be made available on a publicly accessible open access same-time information system (OASIS). The ATC is defined as a measure of the transfer capability, or available "room" in the physical transmission network for transfers of power for further commercial activity, over and above already committed uses (NERC, 1996). The ATC is defined by NERC as the total transfer capability (TTC) minus the transmission reliability margin (TRM), capacity benefit margin (CBM), and existing power flows:

$$ATC = TTC - TRM - CBM - \text{Existing power flows} \qquad (20.5)$$

where the TTC is the total amount of power that can be sent from bus A to bus B within a power network in a reliable manner, the TRM is the amount of transmission transfer capability necessary to ensure network security under uncertainties, and the CBM is the amount of TTC reserved by load serving entities to ensure access to generation from interconnected systems to meet generation reliability requirements. Neglecting TRM and CBM reduces the computation of ATC to

$$ATC = TTC - \text{Existing power flows.} \qquad (20.6)$$

The ATC between two buses (or groups of buses) determines the maximum additional power that can be transmitted in an interchange schedule between the specified buses. ATC is clearly determined by load flow study results and transmission limits. As an illustration, consider Figure 20.5 in which 1000 MW is transmitted from bus A to bus B. If the line rating is 1500 MW and if no parallel paths exist from A to B, the ATC from A to B is 500 MW.

In a real power system, the network and the computation of the ATC are much more complicated as parallel flows make the relationship between transfers and flows less obvious. In addition, the ATC is uncertain due to the uncertainty of power system equipment availability and power system loads. An evaluation of the stochastic behavior of the ATC is important to reduce the likelihood of congestion. The ATC is determined by power flow studies subject to transmission system limits. These phenomena are

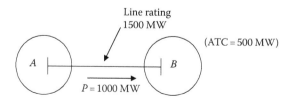

FIGURE 20.5   An example of ATC.

nonlinear in behavior. As discussed above, linearization can be used to estimate line flows in a power flow study model. This section presents a method of finding the stochastic ATC using a linear transformation of line power flows into ATC.

Increasing the transfer power increases the loading in the network. At some point, operational or physical limits to various elements are reached that prevent further increase. The largest value of transfer that causes no limit violations is used to compute the TTC and ATC. Limits are affected by the power injections at both buses A and B. This effect can be found analytically by finding the distribution factors of the lines and other components. In this context, a "distribution factor" refers to the power transfer distribution factor for line $i-j$ and bus k, where an equal and opposite injection is automatically made at the swing bus.

The complexity of the ATC calculation is drastically reduced by linearizing the power flow study problem and considering only thermal transmission limits. The linearization is most accurate when only small deviations from the point of linearization are encountered. The use of thermal limits is justified for short transmission circuits. The transmission system linearization in this case is done using power transfer distribution factors, which are discussed above. Starting with a base case power flow solution, the addition of a power transfer from bus A to bus B impacts the flow in the line between bus i and bus j as

$$P_{ij}^1 = P_{ij}^0 + (T_{ij,A} - T_{ij,B})P_{AB} \tag{20.7}$$

where P_{AB} is the power transfer from bus A to bus B (assumed to be positive). When $(T_{ij,A} - T_{ij,B})$ is positive, the transfer that results in rated power flow on line $i-j$ is

$$P_{AB,ij} = \frac{\left(P_{ij}^{\text{Rated}} - P_{ij}^0\right)}{(T_{ij,A} - T_{ij,B})} \tag{20.8}$$

When $(T_{ij,A} - T_{ij,B})$ is negative, the transfer that results in rated power flow on line $i-j$ is

$$P_{AB,ij} = \frac{\left(-P_{ij}^{\text{Rated}} - P_{ij}^0\right)}{(T_{ij,A} - T_{ij,B})} \tag{20.9}$$

Each line will have a value of $P_{AB,ij}$ that represents the maximum value of power transfer from bus A to bus B without overloading line ij. The minimum of all these values would be the ATC for the AB transfer.

Once this minimum is determined, the question is, "What if there is uncertainty in the initial bus and line loadings?" As discussed above, there are at least two ways to incorporate uncertainty in this computation. The first way is the Monte Carlo method that specifies a statistical distribution of the uncertain parameter (perhaps total load) and utilizes pseudorandom number samples to perform brute force repeated ATC solutions. This could involve full nonlinear power flow studies, or faster linear power flow studies. The collection of ATC solutions that come from a large sample of total load values gives the statistical distribution of the ATC, perhaps as a histogram. If the critical line that determines the ATC of the AB transfer does not change with the uncertainty of the initial bus and line loadings, then the possible variation in ATC can be easily computed from the statistical distribution of the initial line flow. For example, if the uncertainty of initial line flows is normally distributed with zero mean and some variance, then the ATC from A to B is also normally distributed with zero mean, as it is a linear function of the initial flow in the critical line. The variance is also directly available because of the linear transformation.

TABLE 20.1   Comparison of Monte Carlo and Analytical Methods for ATC

"From" Bus	"To" Bus	Analytical Mean	Analytical Variance	Monte Carlo Mean	Monte Carlo Variance
1	9	8.5	4.5	7.147	5.008
1	10	4.177	2.3005	4.7766	2.3356
1	11	2.7144	0.7038	2.764	1.1549
2	9	7.8480	4.5208	7.318	5.6584

20.6 An Example of Stochastic ATC

The following example illustrates the basic principles of stochastic ATC. The IEEE 14-bus system was studied with the base case bus and line variables as given in Christie (1999) and Pai (1979). For the uncertainty analysis by Monte Carlo methods, the bus loads were generated on the computer using 100 samples from a random number generator. The bus real power loads were assumed to be statistically independent with a normal distribution that had a mean equal to the base case and a standard deviation equal to 10% of the base case. Bus load power factor was maintained at the base case value. A comparison of the Monte Carlo results with the analytical method is given in Table 20.1.

The Monte Carlo results required 100 full nonlinear power flow studies of ATC, whereas the analytical results required a simple direct multiplication. Additional results for this example are available in Stahlhut et al. (2005).

20.7 An Example of Expected Financial Income from Transmission Tariffs

A second illustration of the two methods for considering uncertainty deals with the estimation of transmission tariff revenue because of unknown future loading. This illustration was reported also in Westendorf (2005) and Stahlhut et al. (2005). The work is presented in three parts. The first includes a study of transmission tariff revenue based on an estimated hourly forecast for 1 year. This "base case" solution provides the expected revenue for the year for that forecasted loading level. The second part includes a Monte Carlo simulation of possible variations in revenue for a forecast error of known statistical distribution. This is a repeat of the base case solution for 100 different possible cases (per hour) sampled from a normal distribution function—using total load as the random variable. This error in total load was then "distributed" to the individual buses using the base case percentage allocation. The third part included an analytical solution to estimate the expected statistical distribution of the transmission tariff revenue using linear power flow techniques.

This analysis used the IEEE 14-bus test system (Pai, 1979; Christie, 1999) and computed revenue from transmission tariffs using an assumed tariff of $0.04 per MW-mile for each transmission line. An annual income was also computed by summing the hourly incomes for the year. Because the transmission tariff is a function of the length of each transmission line, an estimate of the transmission line lengths was necessary. The total impedance of a transmission line is dependent on the length of the line. The line length is directly proportional to the transmission line impedance. The method for calculating the lengths of the transmission lines in the IEEE 14-bus test system based on the total line impedance of each line was developed as follows. A power base for the system was assumed to be $S_{base} = 100$ MVA. Using the nominal voltages, V_{base}, of the transmission lines and the power base of the system, each transmission line reactance value, the length of the line was calculated by assuming a conversion factor of 0.7 Ω/mile. Secondly, lines containing transformers (i.e., lines 8–10) were considered to be zero-length lines.

A load forecast taking into account hourly variations was developed for this investigation by using publicly accessible approximate historical load data for each hour of everyday for the year 2004 from

TABLE 20.2 Comparison of Monte Carlo and Analytical Methods for Expected Revenue for 1 Day

Time	Mean Total Revenue ($)	Base Case Total Revenue ($)	Simulated Standard Deviation $\sigma_{revenue}$ ($)	Analytical Standard Deviation $\sigma_{revenue}$ ($)
100	390.66	390.96	12.49	9.96
200	382.23	382.95	11.28	9.76
300	378.27	381.11	12.82	9.72
400	383.13	384.06	12.29	9.79
500	394.60	393.56	13.23	10.03
600	427.99	426.22	14.47	10.83
700	478.56	479.04	14.23	12.13
800	510.09	510.85	16.94	12.91
900	505.21	506.51	14.03	12.81
1000	501.30	501.86	13.83	12.69
1100	497.15	496.49	13.93	12.56
1200	490.09	491.41	12.79	12.43
1300	494.68	490.73	13.83	12.42
1400	488.98	488.53	15.66	12.36
1500	484.86	483.91	15.24	12.25
1600	484.97	484.78	14.77	12.27
1700	502.80	504.46	14.16	12.76
1800	537.99	538.33	17.39	13.59
1900	543.86	544.99	16.76	13.75
2000	536.59	538.02	19.71	13.58
2100	527.99	527.33	14.94	13.32
2200	500.62	502.73	15.19	12.71
2300	464.40	466.21	13.47	11.81
2400	427.75	430.47	14.41	10.93

the PJM Web site (PJM, 2005). The integrated hourly load data for the PJM-E area were scaled to match the system load of 259 MW of the IEEE 14-bus test system. Using the hourly load data, power flows were conducted to determine the megawatt flows on the transmission lines allowing the calculation of the each line revenue for the hour. The total revenue generated for each hour was calculated by summing all of the line revenues calculated from the power flows. While standard nonlinear power flow methods were used to compute the Monte Carlo solutions, standard linear power flow methods were used to calculate the analytical solutions.

The Monte Carlo simulations and the analytical methods were compared. Both methods provided results for the distribution of the system revenue generated for each hour. Although an entire year was analyzed, Table 20.2 lists the results from both methods for just 1 day.

Column 1 is the time of day studied. Column 2 is the expected revenue using the Monte Carlo simulation, which created deviations from the base case using a zero-mean random number generator (100 samples per hour). Column 3 shows the revenue from the base case. The fact that these two columns are quite close is due to the fact that the deviations were sampled from a zero-mean number generator and the system behavior is nearly linear for deviations of this size. The standard deviation of the sampled distribution was 3%. This was designed to produce a sample load forecast with expected variations between plus and minus 10% of the base case. Column 4 shows the Monte Carlo sample standard deviation, and Column 5 shows the analytical standard deviation found using the assumed linear nature of the calculation and the assumed normal distribution of the forecast error.

The error in standard deviations can be attributed to several things. The linear power flow method is an approximation that is used in the analytical method but not in the Monte Carlo simulation method. The Monte Carlo method only used 100 random samples to create the sample standard deviation. This is not a large number of samples and contributes to the discrepancy in the two methods.

20.8 Wind Energy Resources and Stochastic Power Flow Studies

Wind is caused by regions on Earth that have different air pressure. Perez (2007) and Pavia (1986) are brief samples of the extensive meteorological literature in this area. The wind speed is often characterized as a random process, and one particular model used frequently is the Weibull distribution (see Figure 20.6) because this distribution is one sided (i.e., the random variable is always non-negative), and actual wind data appear to behave in this fashion. The Weibull probability density $f(x)$ is

$$f_x(x) = \begin{cases} \dfrac{kx^{k-1}e^{-(x/\lambda)k}}{\lambda^k} & \text{for } x \geq 0 \\ 0 & \text{otherwise} \end{cases}$$

where
   λ is the scale of the density
   k is termed the shape parameter

The actual *power* contained in wind is proportional to the cube of the wind speed under some conditions, and therefore the available wind power, if viewed as a random process, does not possess a Weibull probability density. The power output of a wind turbine, however, is not simply characterized because the cubic behavior of the wind power with wind speed only holds below a certain threshold of wind speed (typically in the <50 m/s range)—and there are also nonlinear confounding factors of air temperature, wind direction, and humidity. To illustrate the complexity of the problem, this discussion suggests that wind power is indeed a random process, but one cannot rely on a simple statistical model such as a Weibull density as an accurate representation of wind power.

Having modeled the probability density function of the wind resource (active power P), it is then possible to use algorithms to calculate the probability density function of the line flows and bus voltage magnitudes in the interconnected power system. While this has been done under laboratory and experimental conditions, a full-scale commercialized application software package is still lacking.

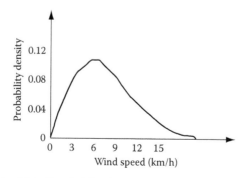

FIGURE 20.6   A representative Weibull probability density curve to model wind speed.

20.9 Solar Photovoltaic Energy Resources and Stochastic Power Flow Studies

Similar to the preceding discussion with regard to wind resources, solar photovoltaic resources also can be modeled probabilistically, and a probability density function of the solar generation can be estimated. This generation has two components, one being completely deterministic. The deterministic component comes from the "equation of time," which gives the sun's position for every day of the year for any place on Earth. This equation is known to a high precision and basically the equation is a trigonometric expression. The second component is probabilistic due to weather conditions, cloud cover, and man-made smoke and haze. There are estimates of the probability density function of the stochastic component; these vary in accuracy and, invariably, are obtained for specific locations. As in the case of wind resource calculations, it is possible to process the probability density functions of the solar insolation in order to obtain the statistics of system line power flows and bus voltages. Commercial software packages that encompass all elements of solar energy uncertainty are lacking.

20.10 Conclusions

Power system planning will always have a need to consider the uncertainty of future conditions and the impact that these uncertainties have on technical and financial issues. When uncertainties can be assumed small, linear power flow methods that are needed for analytical techniques can significantly reduce computational effort. A significant element in stochastic power flow analysis is the appearance of wind, solar photovoltaic, and concentrated solar energy resources in power systems. These resources have significant uncertainty, and stochastic power flow studies will be needed to analyze the impact of these renewable resources on system operating parameters.

References

Borkowska, B., Probabilistic load flow, *IEEE Trans. Power Appar. Syst.*, PAS-93(3), 1974, 752–759.

Christie, R., Power system test archive, 1999, http://www.ee.washington.edu/research/pstca/pf14/ pg_tca-14bus.htm

Heydt, G.T., *Computer Analysis Methods for Power Systems*, Macmillan, New York, 1996.

Pérez, I.A., Sánchez, M.L., and Ángeles García, M. Weibull wind speed distribution: Numerical considerations and use with solar data, *J. Geophysical Research-Atmospherics*, October 2007, V.112, record locator D20112, doi: 10.1029/2006JD008278, 2007.

NERC, Available transfer capability definitions and determination, North American Electric Reliability Council, Princeton, NJ, June 1996.

Pai, M.A., *Computer Techniques in Power System Analysis*, Tata McGraw Hill, New Delhi, India, 1979.

Papoulis, A., *Probability, Random Variables, and Stochastic Processes*, McGraw-Hill Book Co., New York, 1965.

Pavia, E.G. and J.J. O'Brien, Weibull statistics of wind speed over the ocean, *J. Climate and Applied Meteorology*, 25, 1324–1332, 1986.

PJM Interconnection, Hourly load data, 2005, http://www.pjm.com/markets/jsp/loadhryr.jsp

Rau, N., Grigg, C.H., and Silverstein, B., Living with uncertainty: R&D trends and needs in applying probability methods to power system planning and operation, *IEEE PES Rev.*, November 1994, pp. 24–25.

Stahlhut, J., Feng, G., Hedman, K., Westendorf, B., Heydt, G., Sauer, P., and Sheble, G., Uncertain power flows and transmission expansion planning, Accepted for presentation at *the 2005 North American Power Symposium*, Ames, IA, October 23–25, 2005.

Vorsic, J., Muzek, V., and Skerbinek, G., Stochastic load flow analysis, in *Proceedings of the Electrotechnical Conference*, Ljubljana, Slovenia, 2, 1445–1448, May 22–24, 1991.

Westendorf, B., Stochastic transmission revenues, MSEE thesis, University of Illinois at Urbana-Champaign, Urbana, IL, May 2005.

21

Engineering Principles of Electricity Pricing

Lawrence J. Vogt
Mississippi Power Company

Electricity is produced by both utilities and nonutility entities, and it is sold in retail transactions with end-use customers and in wholesale transactions with other utilities for ultimate resale. Pricing of these transactions is market based and/or cost based. For example, in some jurisdictions electricity producers are permitted to compete in retail markets in which customers are allowed to choose an energy service provider and pay market-based prices for electricity. However, the electricity must then be transported from the producers to the customers by means of transmission and distribution (T&D) facilities that are owned and operated by the local utility. T&D power delivery service is typically regulated by a government agency under cost-based pricing. In many other cases, local utilities are fully integrated and have exclusive jurisdictional service rights; thus, their production and T&D functions are fully regulated under cost-based pricing.

Electricity pricing is a multidisciplinary function involving accounting, economics, engineering, finance, and regulatory law. The costs of building, operating, and maintaining a power system arise from meeting the capacity, energy, reliability, and power quality needs of a diverse base of electricity consumers. Engineering principles play a particularly key role in developing effective cost-based prices for electricity by taking into account these technical requirements of providing electric services [1].

21.1 Electricity Pricing Overview

To adequately capture the cost-related attributes of electric service through pricing, while preserving equity among customers, three fundamental cost/pricing components are recognized:

- *Energy component*: variable costs associated with a customer's requirements for a volume of kWh, for example, generation fuels

- *Demand component*: fixed costs associated with a customer's maximum kW or kVA load requirements, for example, T&D line and transformer capacity
- *Customer component*: fixed costs that are independent of a customer's energy or capacity requirements, for example, electric service meter

These components are represented as fundamental *unit prices* or *rates*, and they are combined and often modified in a variety of fashions to form the basic billing mechanism or *rate structure* for the various rate classes.

The form of rate structure is varied for different customer segments in order to reflect the distinct load characteristics and cost-of-service differences of each of the groups. Typically, simple rate structures based solely on kWh consumption are utilized to recover energy-related and demand-related costs of service for small watt-hour metered residential and commercial customer rate classes. The composite energy/demand charges may be flat (same price per kWh for all kWh) or blocked, as exemplified in the following.

Residential/small commercial rate example:

Monthly Rate for Secondary Service

Declining block rate structure version	Inverted block rate structure version
Customer charge: $20.00 per month	*Customer charge*: $20.00 per month
Energy/demand charges:	*Energy/demand charges*:
First 150 kWh @ 5.6¢/kWh	First 300 kWh @ 3.0¢/kWh
Next 350 kWh @ 4.8¢/kWh	Next 450 kWh @ 4.5¢/kWh
Next 500 kWh @ 4.1¢/kWh	Next 750 kWh @ 6.5¢/kWh
Excess kWh @ 3.7¢/kWh	Excess kWh @ 7.5¢/kWh

In contrast, large demand metered commercial and industrial customer rate structures are much more complex because of wide variations in the cost to serve and the load characteristics between individual customers within a given rate class. In particular, maximum demands and load factors (LFs) often vary significantly between such customers due to unique end-use energy requirements and daily operational cycles. Typically, rate structures for demand metered commercial and industrial customer rate classes include an explicit demand charge along with either a flat or blocked energy charge configuration. When the LFs of customers within a given rate class vary widely, an *hours use of demand* (kWh per kW) energy charge structure is often utilized, as exemplified in the following.

Large commercial/industrial rate example:

Monthly Rate for Secondary Service

Customer charge		$250.00 per month
Demand charges		
First 30 kW	@	$5.25 per kW
Over 30 kW	@	$4.95 per kW
Energy charges:		
First 200 kWh per kW		
First 6,000 kWh	@	4.0¢ per kWh
Over 6,000 kWh	@	3.0¢ per kWh
Next 250 kWh per kW		
First 10,000 kWh	@	2.0¢ per kWh
Over 10,000 kWh	@	1.0¢ per kWh
Over 450 kWh per kW		
All kWh	@	0.5¢ per kWh

Establishment of cost-based rate structures and their unit prices is accomplished by first defining homogenous groups (classes) of customers, including residential, lighting, and various subgroups of commercial and industrial, based on load characteristics (primarily demand requirements and LF) and electric service methodology (including delivery voltage level and single-phase service vs. three-phase service). Next the total costs of providing electric service to customers are assigned to these groups and subgroups using *cost causation* principles. The total costs assigned to a given group are then apportioned to the customers within that group on the basis of monthly customer LFs or kWh per kW usage. This intra-group cost assignment represents an extension of the basic class cost-of-service study, and its results, along with other information, provide guidance for setting practical rate structures and associated prices for each of the customer rate groups or classes. The following sections focus on some of the essential engineering principles that are utilized in the production of an electric cost-of-service study and the subsequent rate design for retail electric service.

21.2 Electric Cost-of-Service Study

Cost-of-service studies are data intensive, and the numerous input requirements, as shown in Table 21.1, call for information from across many of the operational areas of the utility. The cost-of-service study requires both monetary and technical operating data and information. Cost and revenue figures for the study are

TABLE 21.1 Major Cost-of-Service Study Input Data Requirements

Rate base (investment) items	Income taxes
• Electric plant in service	• Federal
• Intangible plant	• State
• General plant	• Tax rates
• Common plant	• Deductible interest
• Construction work in progress	
• Plant held for future use	*Financial information*
• Fuel stock	• Cost of capital
• Materials and supplies	• Capital structure
• Working capital	
• Prepayments	*Revenues*
• Accumulated depreciation[a]	• Revenues from sales
• Property insurance and other operating reserves[a]	• Late charges and forfeited discounts
	• Miscellaneous service fees
• Accumulated deferred income taxes[a]	• Leases and rentals
• Investment tax credits[a]	• Nonterritorial sales credits
• Customer advances[a]	• Revenue credits
• Customer deposits[a]	
	Customer information
Expenses	• Class designation
• Fuel burned	• Service voltage level
• Purchased power	• Billing determinants
• Operations expenses	
• Maintenance expenses	*Class information*
• Customer accounts, assistance, and sales expenses	• Hourly load shapes
	• Unbilled sales
• Salaries and wages	
• Depreciation expense	*System information*
• Administrative and general	• Generation output
• Taxes other than income taxes	• Territorial purchases
• Miscellaneous fees	• T&D System loss characteristics
• Interest on customer deposits	• Substation diagrams
• Amortization of nonrecurring expenses	• Distribution circuit characteristics
	• Current distribution
	• Equipment costs

[a] Rate base deductions.

acquired from corporate financial operating reports, which summarize data entries from various accounting systems (i.e., mass property accounting, revenue accounting, etc.). Having cost information booked in a configuration, such as the Federal Energy Regulatory Commission's (FERC) Uniform System of Accounts, facilitates the organization of the data elements within the actual cost-of-service study. Technical data are acquired from the utility's customer information system (CIS) and various engineering sources.

21.3 Cost-of-Service Study Framework

The cost-of-service study is fundamental for establishing cost-based rates for electric service. Stated from a financial point of view, the cost-of-service study is a methodology for measuring the earnings position, or profitability, of the various classes of electric service. A key result of the study is the quantification of an annual test period rate of return yield for each class under the existing rates for electric service. A rudimentary framework of the cost-of-service study process is presented in Figure 21.1. The process consists of four principal cost analysis steps:

1. *Functionalization of costs*: Basic cost elements are organized in accordance with major operating functions of the power system and its supporting business functions. These power system expenditures are assigned to function based on the voltage level at which the costs are incurred.
2. *Classification of costs*: The levelized functional costs are then categorized as being energy related, demand related, or customer related. Since plant costs are associated with specific equipment and

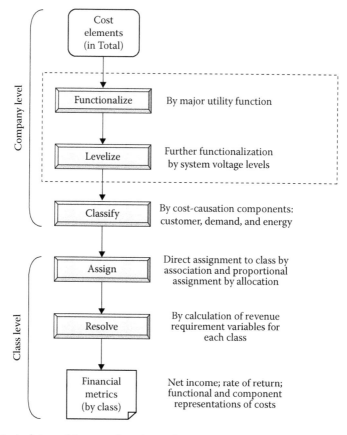

FIGURE 21.1 Principal steps of the cost-of-service analysis process. Functionalization and classification of rate base and expenses are executed for total company costs. The cost assignment process then apportions the total company costs to each of the classes where rates of return and other metrics can be determined on a class basis.

facilities, some distribution system cost elements are related to both the customer and demand cost components and must be separated accordingly.

3. *Assignment of costs*: The levelized functional cost components are then proportionally assigned to each customer class using (a) class kWh usage to allocate the energy cost components, (b) class load responsibilities to allocate the demand cost components, and (c) the numbers of customers in each class to allocate the customer-related cost components. Cost elements that are incurred exclusively for a specific class of electric service (or even a single customer) are directly assigned to that class (or to the class in which the specific customer is included).

4. *Quantification of results*: Allocated expenses and taxes are deducted from class revenues in order to determine class net income amounts, which are then divided by the allocated rate base to determine class rates of return. Other results, such as customer, demand, and energy cost component summaries, can be produced for rate design guidance and support purposes.

As noted in Figure 21.1, the functionalization and classification steps represent analyses of cost data for the utility as a whole. The accounting data and the engineering inputs interact in the cost assignment step in accordance with the cost causation criterion. The completed cost-of-service study indicates results on both a class and a total company basis.

The cost functionalization step is reasonably straightforward as the Uniform System of Accounts defines a utility's operational cost structure in terms of six principal functions: production, transmission, distribution, customer accounts, customer service and information, and sales. A series of accounts under each major function provide further information, and utilities typically create subaccounts that yield additional detail for cost of service. For example, FERC Account 368—Line Transformers contains the book costs and counts not only for distribution line transformers but also for distribution capacitors, voltage regulators, and cutouts. Levelization, which is a subfunctionalization step, arranges costs according to system voltage levels. For example, the FERC prescribes a 5-level system for cost-of-service studies as follows:

Level 1: production, including generator step-up (GSU) transformers
Level 2: transmission, including subtransmission and associated substations
Level 3: distribution substations, defined by level 2 to distribution voltage conversion
Level 4: primary distribution lines, including primary line equipment and facilities
Level 5: distribution line transformers, including secondary distribution lines and facilities

In the cost classification step, the functionalized costs are classified as fixed or variable in terms of the fundamental cost components: customer, demand, and energy. Customer-related costs are fixed as they are independent of energy or load. Demand-related costs are fixed as they relate to peak load conditions, and a fixed amount of capacity is installed to meet maximum demands even though a customer's actual demand can vary considerably from hour to hour or even minute to minute. Energy-related costs are variable as they relate to the production and delivery of kWh as energy is consumed. In general, the production, transmission, and distribution substation plant functions (levels 1, 2, and 3) are demand related. The distribution system (levels 4 and 5) is partially demand related and partially customer related. A key analysis that separates distribution level costs between the demand and customer cost components is referred to as the *minimum distribution system*. Level 1 production fuels and some other O&M expenses are energy related. Classification further aligns the functionalized costs in preparation for assignment to jurisdiction or classes using customer, demand, and energy allocation factors.

The cost assignment step distributes the functionalized and classified total company costs to the specified customer or rate classes. Much of the system functions are jointly used by different classes; thus, an allocation of costs is required at each voltage level. Allocation factors are based on cost causation criteria, which are driven by various technical characteristics of the power system. For instance, all customers do not peak at the same time due to load diversity throughout the system. Thus, coincidence of load is a major issue, which must be considered in the development of rational demand allocation factors. In addition, demand and energy losses are key issues to be considered in the formulation of the

TABLE 21.2   Example Cost-of-Service Study Results for Major Customer Classes

	Total $(000s)	Residential	Commercial	Industrial	Lighting
Revenue	755,015	291,936	258,687	194,681	9,710
Expenses	517,789	199,914	170,093	141,328	6,454
NOI$_{BIT}$	237,226	92,021	88,594	53,354	3,257
Income tax	148,229	60,204	52,825	33,104	2,096
NOI	88,997	31,818	35,769	20,250	1,160
Rate base	1,139,845	493,163	374,767	257,280	14,636
ROI	7.81%	6.45%	9.54%	7.87%	7.93%
ROE	7.80%	4.86%	11.57%	7.94%	8.06%

ultimate demand and energy allocators to ensure cost allocation equity among the customer groups, which are served at various voltage levels.

Once all investment items (i.e., the rate base), O&M expenses, and income taxes are allocated to the classes, various class financial metrics and other cost-of-service study results can be quantified. An example summary of cost-of-service study results is shown in Table 21.2. Financial performance measures include the return on investment (ROI) and the return on equity (ROE) under present rates. The ROI for the total utility or for a given customer group j is determined by the following:

$$ROI_j = \frac{REV_{PRES_j} - EXP_j - IT_j}{RB_j} \tag{21.1}$$

where

ROI_j is the return on investment for group j in percent
REV_{PRES_j} is the group j annual revenue under present rates
EXP_j are O&M expenses assigned to group j
IT_j are income taxes assigned to group j
RB_j is the rate base (investment) assigned to group j

The numerator in Equation 21.1 is referred to as the *net operating income* (NOI), while the result of $(REV_{PRES_j} - EXP_j)$ is referred to as the *NOI before income tax* (NOI$_{BIT}$).

For an investor-owned business, common equity (common stock) is a component of the overall cost of capital, which also includes debt (bonds) and preferred stock. Like ROI, the ROE can also be measured on both a total and a customer or rate class basis using the results of the cost-of-service study. The ROE for the total utility or for a given customer group j is determined by the following:

$$ROE_j = \frac{NOI_j - [RB_j \times (D + PS)]}{RB_j \times EQ_{CAP}} \times 100 \tag{21.2}$$

where

ROE_j is the return on equity for group j in percent
NOI_j is the net operating income for group j
RB_j is the rate base for group j
D is the cost rate for long-term debt (bonds) in percent
PS is the cost rate for preferred stock in percent
EQ_{CAP} is the common equity capitalization in percent

The common equity capitalization ratio represents the amount of the utility's common equity relative to its total amount of capital.

As noted in Table 21.2, the ROI and ROE values for the commercial customer class are significantly higher than the total company values, whereas the same values for the residential class are significantly lower than the total company values. ROI and ROE values of both the industrial class and the lighting class are fairly close to the total company values. Such information is useful when allocating rate revenue increases or decreases to the customer or rate classes of electric service, particularly when the objective is to move the class rate of return values closer to the total company values.

The cost-of-service study also provides information that is vital for establishing rate structures and unit prices for each of the rate schedules for electric service. In particular, the basic customer, demand, and energy cost components derived from the cost-of-service study allocations can be cast as component revenue requirements, which include O&M expenses, income taxes, and an ROI. These data serve as inputs to the rate design process by providing a starting point for establishing rates that are intended to recover costs on a cost causation basis.

The following three sections address key cost-of-service analyses that specifically rely on engineering principles and methodologies. These sections include the minimum distribution system analysis, analysis of load diversity, and analysis of demand and energy losses. These analyses ensure a cost causation approach to cost classification and cost assignment.

21.4 Minimum Distribution System Analysis

The concept of a minimum distribution system recognizes that the costs of the primary and secondary distribution system have both customer-related and demand-related attributes. As discussed previously, the customer cost component is associated with no-load conditions, whereas the demand cost component is associated with load conditions, that is, equipment and facilities have the capacity to meet peak loads. Some devices can be categorized as being either customer related or demand related, while some individual devices are related to both components.

A direct approach to identifying and quantifying the customer and demand cost components is to evaluate each major distribution item in terms of its mission. For example, a distribution system protection scheme consists of a mix of circuit breakers, reclosers, and fused cutouts that are coordinated to minimize customer outages in the event of faults on the system. A simple feeder schematic including protective devices is shown in Figure 21.2.

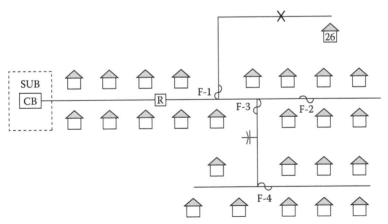

FIGURE 21.2 A schematic of a primary feeder system. A fault occurs on the primary tap line, which connects customer #26 to the main feeder. Under a temporary fault condition, customers downstream from the recloser will experience a momentary outage, but service is restored to all customers. Under a permanent fault condition, the same momentary outage will occur to all customers, but only customer #26 will lose service as fuse F-1 blows to clear the fault from the main feeder. A switched capacitor located on the line between fuses F-3 and F-4 provides feeder voltage support under heavy loading conditions.

The protective devices work together to preserve service to as many customers as possible under various fault conditions. In essence, the protection scheme safeguards the "voltage path," which connects each customer on the circuit to the source, that is, the substation. Consider a case where all customer load is removed from the circuit, and a fault occurs as before; the protection scheme will operate exactly as it did under load in an effort to maintain the voltage path to as many customers as possible. The protection scheme's mission is independent of load or demand and thus is a customer-related function. Furthermore, other facilities that provide a connection of a customer to the source have a customer-related attribute.

Consider again the circuit in Figure 21.2 during peak demand periods. Under light loading conditions, the capacitor is switched off. However, once high load currents create an unacceptable voltage drop, the capacitor connects to the circuit to raise the voltage along the feeder profile, thereby releasing capacity. The capacitor's mission is to support voltage under load and is thus dependent on load. Absent customer load, the capacitor is not required at all, even as the circuit is energized to create a voltage path. The capacitor is independent of the voltage path; thus, it is a demand-related component.

The missions of a number of other distribution system devices have both customer-related and demand-related attributes. Primary and secondary conductors clearly create a path interconnecting customers to a source, but they also have capacity to carry load. Line transformers are integral to the path as they connect the secondary conductor system to the primary conductor system, but they also provide capacity to serve customers' loads.

When a single device has both customer-related and demand-related attributes, its total cost must be allocated. The *minimum-intercept* or *zero-intercept methodology* provides a rational basis for separating the cost of a device between its customer and demand components. The zero-intercept methodology is a weighted linear regression of the unit costs of standard ratings or sizes of a specific device, such as a single-phase overhead line transformer, plotted as a function of its capacity-related characteristic, which would be kVA for a line transformer. The objective of the regression analysis is to determine the y-intercept. The y-intercept represents that portion of a device's total cost that is associated with zero capacity and thus the customer-related component. The unit costs must be weighted by the numbers of devices because of the uneven distribution of the various ratings or sizes of the devices in service. The slope and the y-intercept of the weighted linear regression equation are given by the following:

$$m = \frac{\left[N \times \sum_{i=1}^{\tau} n_i X_i Y_i \right] - \left[\sum_{i=1}^{\tau} n_i X_i \times \sum_{i=1}^{\tau} n_i Y_i \right]}{\left[N \times \sum_{i=1}^{\tau} n_i X_i^2 \right] - \left[\sum_{i=1}^{\tau} n_i X_i \right]^2} \qquad (21.3a)$$

$$b = \frac{\sum_{i=1}^{\tau} n_i Y_i}{N} - m \times \left[\frac{\sum_{i=1}^{\tau} n_i X_i}{N} \right] \qquad (21.3b)$$

where
  m is the slope
  b is the y-intercept
  X_i is the unit size
  Y_i is the unit cost
  n_i is the number of units of a particular size or rating
  N is the total number of all units (i.e., Σn_i)
  τ is the total number of sizes or ratings

It is not critical for the regression analysis to have unit cost data points for all transformers in service; however, a range of the smaller size units in each category is important to ensure data linearity and reasonable *y*-intercept results. In addition, the data points of a particular regression should represent the same input and output voltages (e.g., 15 kV, 208Y/120 volt pad mount transformers). Given the wide range of characteristics (voltages, one vs. two bushings, etc.) within a major transformer category, the regression analysis can be conducted using the most common or representative transformer type in each major category with the results then applied to the population of transformers in each respective category.

A problem arises often when embedded (book) costs are used in the regression analysis because of wide variations in equipment vintage. Over time, equipment ratings have generally trended upward due to higher distribution voltages and increased customer loads. However, the book records may reflect a significant number of small size devices in service that have relatively low average unit costs due to the older age of their installations. Concurrently, larger size devices have much higher average unit costs as they were more recently installed. This temporal disparity in unit costs can distort the regression results and even produce a negative *y*-intercept. An alternative approach overcomes such incongruities as it is based on an estimate of the cost required to "rebuild" all of the devices in service all at once using current unit construction costs applied to the full inventory of devices. The *y*-intercept from a regression using current unit costs in lieu of embedded unit costs can then be applied to determine the customer component proportion of the total "rebuild" cost. This ratio is then applied to the book cost in order to estimate the customer component on an embedded cost basis.

An example of zero-intercept regression results for single-phase and three-phase overhead (pole mount) and underground (pad mount) line transformers is shown in Figure 21.3. The *y*-intercept (where kVA = 0) represents the no-load customer cost component for each of the four transformer types.

Once the *y*-intercept unit cost is calculated for a given category of devices, the total cost of all such devices (units) in that category is apportioned between the customer cost component and the demand cost component by

$$\text{Customer cost} = y\text{-intercept} \times \text{Total number of units} \qquad (21.4a)$$

$$\text{Demand cost} = \text{Total category cost} - \text{Customer cost} \qquad (21.4b)$$

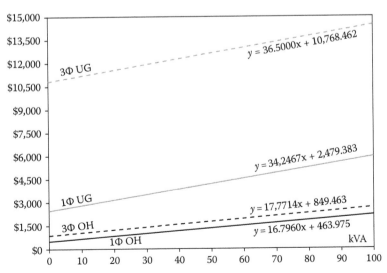

FIGURE 21.3   Zero-intercept regression analysis for four categories of distribution line transformers.

TABLE 21.3   Example Results of a Level 5 Line Transformer Classification Analysis

Transformer Type	Number of Units	*y*-Intercept Unit Costs $/Unit	Customer Cost $(000s)	Total Cost $(000s)	Demand Cost $(000s)
1φ pole mount	98,728	463.975	45,807	109,961	64,154
3φ pole mount	1,135	849.463	964	6,339	5,375
1φ pad mount	6,188	2,479.383	15,342	24,946	9,604
3φ pad mount	1,310	10,768.462	14,107	41,220	27,113
Total	107,361		$76,220	$182,466	$106,246

As shown in Table 21.3, the *y*-intercepts from the regression examples shown in Figure 21.3 are multiplied by the total number of units of line transformers, by major category, to compute the respective customer-related costs. The demand-related amounts are then calculated as the difference between the total cost of all transformers and the customer cost.

Conductors and poles also fall into the category of distribution facilities that must be classified by both customer and demand components. Separate regressions are needed for overhead and underground and primary and secondary distribution conductors. Separate regressions are needed for wood, concrete, and steel distribution poles. The regression analyses are similar to line transformers, but the independent variables that represent capacity are different. For example, conductor capacity is typically addressed by using conductor ampacity, although conductor size (MCM) also serves as a very good proxy for capacity.

21.5 Analysis of Load Diversity

The assignment of costs depends on customer loads being quantified at each level of the system. Due to load diversity, a customer's load ratio share of a line transformer is vastly different than the customer's load ratio share of generation or transmission. Load diversity both between individual customers and between customer groups varies considerably throughout the power system. This effect occurs as a result of the interaction of the customer's load with other customer loads at that particular service level and at the higher levels of the system. Thus, the cost causation effect of an individual customer's load varies when viewed at the point of power delivery and at all other voltage higher within the system.

Monthly load shapes for an example commercial customer, an entire commercial class of customers, and a total system are illustrated in Figure 21.4. As shown in the upper chart of Figure 21.4, the system peak demand is established on the third Friday of the month at a magnitude of about 1,600 MW. In the center chart of Figure 21.4, the commercial class peak for the month is established on Friday of the following week at a magnitude of about 230 kW. Because of the time difference between the occurrence of the system peak and the commercial class peak, the value of the class maximum load is referred to as the class *non-coincident peak* (NCP). The magnitude of the commercial class demand at the time of the system peak is about 200 kW, and this value is referred to as the class *coincident peak* (CP). As shown in the lower chart of Figure 21.4, an individual customer within the commercial class is found to establish a monthly maximum demand of approximately 37 kW during the third Tuesday of the month. However, the customer's demand at the time of the commercial class peak is found to be less than 5 kW. Furthermore, the customer's demand at the time of the system peak is also less than 5 kW as its operation is observed to have shut down to a minimum load level just prior to the system peak hour. A comparison of the three load shapes illustrates the nature of load diversity throughout the system. The example customer's peak demand is significantly diverse with respect to both the peak load of the commercial class and the peak load of the system.

Demand-related cost drivers are functions of load diversity. The cost to serve a customer at the local level is based on sizing the capacity of local facilities to meet the customer's peak demand, which could

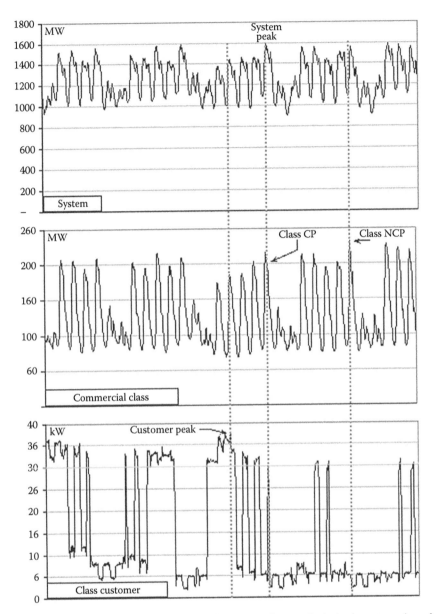

FIGURE 21.4 An example of load diversity during the course of a month. A single customer's peak demand (lower chart) does not occur at the same time as either the class peak (center chart) or the system peak (upper chart). Load diversity characteristics are key to providing an adequate amount of capacity at each functional level of the power system.

occur at any hour of the day and on any day of the year. In this example, line transformer capacity that can properly handle the kVA load associated with the customer's 37 kW peak demand would be required. As observed in Figure 21.4, the customer's load at the time of either the commercial class peak or the system peak would not provide an accurate representation of the customer's service transformer capacity requirements.

In contrast, a customer's peak demand does not provide a true representation of that customer's production level capacity requirements. The example commercial customer actually requires less than 5 kW of production capacity, which is a small fraction of its peak demand of 37 kW. While some customers may establish their peak demands simultaneously with the system peak, many other customers' peaks occur in other hours. Production capacity is sized to meet the diversified loads of all customers as a whole, that is, where load diversity is greatest. Since costs are typically allocated to rate or customer classes, class CP load values are highly indicative of production cost responsibilities assignable to the classes.

Building production capacity to serve an extreme load circumstance that may happen only occasionally would not be the most economical decision when other operational options exist for providing capacity. For example, supplemental capacity might be acquired from tie-line interchange with neighboring utility systems, and/or interruptible service contracts with large customers might be invoked during system critical loading conditions. An understanding of the typical system load diversity characteristics is a key factor in not only providing adequate capacity to serve load but also determining the capacity-related cost responsibilities of the customer or rate classes.

Figure 21.4 provides just a single month's view of load diversity. During another month, the example customer may indeed peak at the same time as the class peak, system peak, or both as the class peak might also occur at the time of system peak in any given month. A more comprehensive view of load diversity at the upper functional levels of the system is provided by viewing the interactions of load over multiple months or a whole year. Compared to a single month CP, an average of multiple monthly class peaks relative to the average of multiple monthly system peaks is more appropriate for determining the typical amount of load diversity at the production and transmission levels of the system. A "12-CP" methodology is utilized frequently for the allocation of Level 1 demand–related production plant costs as it yields a very high load diversity characteristic to the demand allocation factors. The 12-CP allocation factor for each customer or rate class is determined by the following equation:

$$DAF_{CP_j} = \frac{\sum_{i=1}^{12} CP_{ij}}{\sum_{i=1}^{12} SP_i} \times 100 \tag{21.5}$$

where
DAF_{CP_j} is the 12-CP allocation factor for the jth class, as a percent
CP_{ij} is the coincident load of the jth class in the ith month, in MW
SP_i is the system monthly peak load in the ith month, in MW

The result of Equation 21.5 is a class allocation factor representing an average month of a test year. Averaging across the 12 monthly system and class peaks captures the diversity in weather-sensitive end-use load devices in utility systems that experience appreciable seasonal peak load cycles.

Coincidence factor (CF) values, which correspond to the major functional levels of the power system, are plotted for example residential, commercial, and industrial customer classes in Figure 21.5. A substantial difference in LF exists among the three classes. The correlation between LF and load diversity is evidenced by the difference in shapes of the class CF profiles. Higher LF class loads are more coincident with system peaks than are lower LF class loads. With the exception of a 100% LF customer, this relationship holds for groups (classes) of customers as opposed to individual customers since even a solitary low LF customer could peak simultaneously with the total system load.

As discussed previously, a 12-CP allocation factor is utilized often for allocation of level 1 production demand costs. At lower system levels, other allocation factors based on load diversities that are characteristic of each level are more reflective of cost causation principles. For example, the average of monthly peak season class CP values may be more typical of transmission and/or distribution substation load diversities.

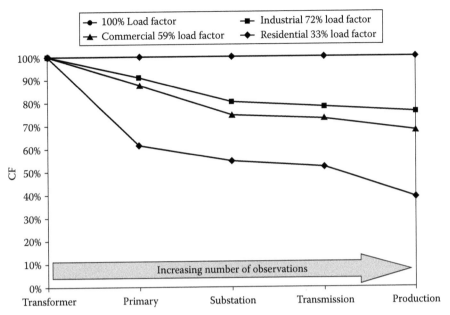

FIGURE 21.5 A plot showing the effect of LF on the coincidence of load at various power system levels. The number of observations (customer demands), upon which the calculations of the class CFs are based, is a key aspect of the level of load diversity realized. The CFs for the residential, commercial, and industrial classes decrease at different rates when moving from the service transformer to the generator due to differences in class LFs. The decrease is more pronounced with lower LFs; however, at 100% LF, a load is unconditionally coincident with the peak at every level of the system regardless of the number of observations considered.

Load diversity at the primary distribution system level is well represented by class NCP loads since the times at which feeders peak vary widely based on each feeder's particular load mix. Secondary distribution customer NCP values, that is, individual customer maximum demands, are more appropriate for representing load diversity at the line transformer and secondary voltage level. The customer NCP allocation factor is based on the maximum annual demands of customers for each secondary voltage customer or rate class and is determined by the following:

$$DAF_{MAX_j} = \frac{\sum_{k=1}^{n} D_{jk}}{\sum_{j=1}^{N} \sum_{k=1}^{n} D_{jk}} \times 100 \qquad (21.6)$$

where
DAF_{MAX_j} is the maximum load allocation factor for the jth class, as a percent
D_{jk} is the maximum annual demand of the kth secondary voltage customer of the jth customer or rate class, in MW
n is the total number of secondary voltage customers in each class
N is the total number of secondary voltage customer or rate classes

An example of load diversified demand allocation factors used to assign costs to customer classes is shown in Table 21.4. A given allocator is indicated by a series of percentage-based factors, which represent the cost causation share of the total cost element for each applicable class.

TABLE 21.4 Example Customer Class Demand Cost Allocation Factors

Allocation Factor	Total MW	Customer Class			
		Residential	Commercial	Industrial	Lighting
Demand					
• *Level 1—prod* (12-CP)	1,646	652	588	427	—
		39.6254%	34.4077%	25.9669%	—
• *Level 2—trans* (5-CP)	1,857	842	588	427	—
		45.3550%	31.6674%	22.9777%	—
• *Level 3—subs* (1-CP)	1,757	874	572	311	—
		49.7345%	32.5479%	17.7176%	—
• *Level 4—pri* (NCP)	2,085	985	717	368	15
		47.2216%	34.3832%	17.6537%	0.7415%
• *Level 5-A—xfmr* (MAX-A)	3,147	1738	1228	164	17
		55.2195%	39.0272%	5.2168%	0.5364%
• *Level 5-B—sec* (MAX-B)	3,050	1689	1184	160	16
		55.3877%	38.849%	5.2403%	0.5370%

21.6 Analysis of Demand and Energy Losses

Electrical losses throughout the T&D system represent a major factor in the development of energy and demand cost allocators. Energy sales are recorded by revenue meters located at each customer's point of service. Since customers receive service at different voltage levels, the adjustment of the metered kWh units to incorporate total system line and transformation losses ensures that allocations of the fuel and variable O&M costs incurred at the production level of the system are accomplished in a fair and equitable manner. In other words, prior to allocation, class sales at the meters are transformed to their equivalent share of energy production at level 1 generator buses. For cost allocation purposes, energy-related losses are assessed on an annual basis.

Demand-related losses are assessed on a system peak load basis, since they are related to system capacity requirements. Class demands at the service points are adjusted for losses up through the system to the level or levels at which they will be applied as demand-related cost allocators. For example, a CP type of allocator used for assigning production plant costs would require that the associated class demands be adjusted for total system line and transformation losses in order to reflect their equivalent share of production capacity at level 1. In contrast, secondary voltage class NCP demands would first be adjusted for secondary line and line transformer losses to make them equivalent with primary voltage class NCP demands. The loss-adjusted secondary voltage class demands would then be further adjusted, along with the primary voltage class demands, to incorporate primary feeder losses for development of a level 4 NCP set of allocation factors. Core, or no-load, losses are constant in time as they represent a steady-state condition caused simply by energizing the transformers. Core losses are independent of a transformer's loading conditions. On the other hand, load losses are a function of load current, that is, i^2R, and are present in both the windings of transformers and line conductors.

Figure 21.6 illustrates the i^2R aspect of losses. The load-related losses in a given hour are proportional to the square of the load in that hour. As a result, the losses at peak loading conditions are proportionately higher relative to the system load than during off-peak loading conditions. At 5:00 AM, when system load is at its minimum for the day, the associated load losses are slightly more than 4% of the load in that hour. However, at the 5:00 PM system peak hour, losses represent nearly 10% of the system load. The magnitude of the no-load losses is constant in every hour.

A plot comparing a unitized system load with its associated load-related losses, on an annual load duration basis, is shown in Figure 21.7. Compared to a system LF of 55% (i.e., the unitized average load), the i^2R nature of losses results in a 29% LF for the losses. The LF of the losses is a key component for

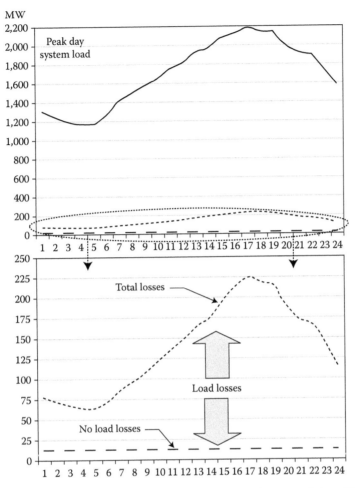

FIGURE 21.6 Peak day profile of system load and the associated T&D losses consisting of load- and no-load-related losses.

modeling line conductor and transformer load–related energy and demand losses. The LF of the load-related losses is a function of the square of the load, as determined by the following equation:

$$LF_{Loss} = \frac{1}{T} \sum_{i=1}^{T} \left(\frac{L_i}{L_{Max}} \right)^2 \tag{21.7}$$

where
LF_{Loss} is the annual LF of the load losses, in percent
L_i is the load in the *i*th hour, in MW
L_{Max} is the maximum hourly load occurring during the year, in MW
T is the number of hours in the year

For the cost-of-service study, system losses are modeled for each voltage level. A simplified schematic diagram for an entire power system is used as a framework for modeling peak demand and annual energy flows from the territorial inputs (generators and intersystem tie lines) to the customer delivery points. Transformer and line losses are accounted for at each voltage level. Each voltage level is analyzed as a node with power flowing in from various sources, including the higher voltage subsystems, and with

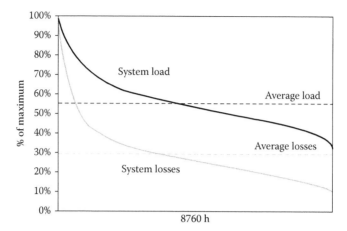

FIGURE 21.7 Annual unitized load duration curves of system load and the associated load-related losses. The LF of the losses is much less than the LF of the load itself as the average losses relative to the peak system losses are much lower. The LFs are equal to the unitized average values.

power flowing out in terms of level sales, level losses, and outputs to lower voltage subsystems. The sum of the input and output power flows for each node is equal to zero.

A detailed schematic diagram of a 34.5 kV subsystem is shown in Figure 21.8. Power flows into this 34.5 kV node, which is indicated by the encircling dashed line, from the 138 and 69 kV subsystems at points "A" and "B." In addition, one or more generators are connected to the 34.5 kV system by means

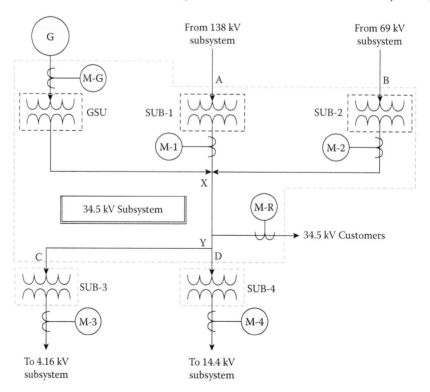

FIGURE 21.8 A nodal model of a 34.5 kV subsystem. Power inputs to the node are shown at "A," "B," and the generator "G." Power outputs are shown at "C," "D," and the 34.5 kV customers. Implicit outputs to be determined include transformer losses and line losses occurring along the 34.5 kV lines, which are indicated by the line segment "X" to "Y."

of 13–34.5 kV GSU transformers. Note that the GSU and the input substations (SUB-1 and SUB-2) represent the total of all such voltage transformations. For instance, SUB-2 may represent the composite of just a few or even dozens of 69 to 34.5 kV substations that are scattered throughout the power system territory. The input substations are metered at the low voltage buses of the transformers (M-1 and M-2), while the generators are metered at their output terminals (M-G).

Sales to customers taking service at 34.5 kV are metered for billing purposes (M-R). In reality, customer delivery points are distributed along the lines; however, for modeling losses for the cost-of-service study, sales can be treated as a single nodal output from the composite 34.5 kV line represented by "X" to "Y." The 34.5 kV level outputs to the two lower voltage subsystems are indicated as "C" and "D;" however, these outputs are metered at the low voltage buses of SUB-3 and SUB-4 (M-3 and M-4), which are located a voltage transformation beyond the edge of the 34.5 kV node.

As indicated in Figure 21.8, metered load data are essential to determining the power flows into and out of the node. Differences in power flows are key to quantifying the energy and demand losses. Basic watt-hour metering adequately captures the annual energy flows, but interval metering is necessary to determine the demands during the time of system peak. While large commercial and industrial customers typically are metered with interval recorders for billing purposes, peak demands for residential and smaller commercial and industrial customers might have to be estimated from load research sample data.

Transformer data are also required for the loss analysis, including factory results of full-load-loss and no-load-loss tests. Individual test results are best for substation power transformers, but typical results may be sufficient for line transformers. Installed transformer MVA capacities are needed as well, including substation OA/FA/FOA ratings. Transformer demand and energy losses are determined from the factory test data. Since the no-load-related transformer loss data are constant under any transformer loading condition, the associated demand and energy no-load losses for a given transformer are determined simply by the following:

$$D_{TNL} = NLL \tag{21.8a}$$

$$E_{TNL} = NLL \times T \tag{21.8b}$$

where
D_{TNL} are peak hour demand-related transformer no-load losses in MW
NLL is the no-load loss rating of the transformer in MW
E_{TNL} are annual energy-related transformer no-load losses in MWh
T is the number of hours in the year

The no-load-related demand and energy losses of individual transformers are summed to get the total for each of the different voltage transformations. For the 34.5 kV subsystem shown in Figure 21.8, total no-load-loss values are calculated for each of the four substation transformers and for the GSU transformers.

Load-related transformer demand and energy losses are functions of transformer loading conditions. The demand load losses of a transformer at the time of system peak are determined by the following:

$$D_{TLL} = FLL \times \left(\frac{L_{@Peak}}{T_{CAP} \times PF} \right)^2 \tag{21.9a}$$

where
D_{TLL} are peak hour demand-related transformer load losses in MW
FLL is the full-load loss rating of the transformer in MW
T_{CAP} is the transformer capacity in MVA
PF is the power factor of the transformer load
$L_{@Peak}$ is the transformer load coincident with the system peak in MW

The quantity contained within the brackets of Equation 21.9a represents the percent loading of the transformer at the time of system peak.

A transformer's energy load loss across the year is a function of its demand load loss as determined by the following:

$$E_{TLL} = \frac{D_{TLL}}{(L_{@Peak}/L_{Max})^2} \times T \times LF_{Loss} \tag{21.9b}$$

where

E_{TLL} are annual energy-related transformer load losses in MWh
D_{TLL} are peak hour demand-related transformer load losses in MW
$L_{@Peak}$ is the transformer load coincident with the system peak in MW
L_{MAX} is the transformer maximum annual load
T is the number of hours in the year
LF_{Loss} is the annual LF of the losses in percent

The denominator of Equation 21.9b adjusts the transformer's demand load losses at the time of system peak to the transformer's demand losses at the time of its peak loading condition. In other words, this adjustment takes into account load diversity between the system peak and the transformer peak. Without this adjustment, the annual energy load losses would be understated.

21.6.1 Methodology for Evaluating Losses

The loss characteristics of the power system are evaluated by modeling the power flows from the generators to the loads. Generally, the analysis begins at the highest voltage levels and ends with the lowest voltage levels. However, each voltage level (node) can be analyzed as a subsystem and then consolidated with the other voltage subsystems. The 34.5 kV subsystem schematic diagram in Figure 21.8, along with the input data provided in the following, will serve as the basis for this example. Note that even at three decimal places, some rounding difference will be observed.

Transformer Data	Capacity (MVA)	NLL (MW)	FLL (MW)
GSU	19.412	0.008	0.038
SUB-1	70.185	0.100	0.301
SUB-2	10.910	1.015	0.042
SUB-3	20.605	0.029	0.091
SUB-4	7.035	0.002	0.008

Meter Data	Annual MWh	MW @ Peak	LF[a] (%)	LF_{Loss}[a] (%)	CF[a] (%)
M-G	24,900	18.261	15.57	15.44	100.00
M-1	251,170	29.772	77.63	58.24	78.53
M-2	37,850	4.498	75.63	58.24	78.53
M-3	89,405	18.913	53.96	30.52	100.00
M-4	27,438	6.662	47.00	23.94	100.00
M-R	192,000	25.710			

[a] Load factor, loss factor, and coincidence factors determined from hourly loads.

Other data
Annual hours = 8760
Power factor @ peak = 98%

Calculation of 34.5 kV subsystem inflow @ "A"	MW	MWh
SUB-1 metered demand and energy (M-1)	29.772	251,170
SUB-1 NL demand losses (Equation 21.8a)	0.100	
SUB-1 NL energy losses (Equation 21.8b)		879
SUB-1 demand load losses (Equation 21.9a)	0.056	
SUB-1 energy load losses (Equation 21.9b)		467
Total inflow from 138 kV subsystem	29.929	252,515

In the same manner, the demand and energy inflows from the 69 kV subsystem to SUB-2 @ "B," the demand and energy outflows to the 4.16 kV subsystem (SUB-3) @ "C," and the demand and energy outflows to the 14.4 kV subsystem (SUB-4) @ "D" are determined to be the following:

	MW	MWh
34.5 kV subsystem inflow @ "B"	4.522	38,153
34.5 kV subsystem outflow @ "C"	6.672	27,472
34.5 kV subsystem outflow @ "D"	19.027	89,886

Note that the sum of the demands and energies for "C" and "D" represents the 34.5 kV subsystem outflows at "Y," which are 25.699 MW and 117,358 MWh.

Generation is also connected to the 34.5 kV subsystem. The output of the GSU is determined by subtracting the calculated transformer no-load and load-related demand and energy losses from the generator terminal demand and energy meter readings (M-G), which are determined to be 18.218 MW and 24,781 MWh. The total input to the 34.5 kV lines @ "X" can now be determined:

	MW	MWh
SUB-1 metered demand and energy (M-1)	29.772	251,170
SUB-2 metered demand and energy (M-2)	4.498	37,950
GSU output:	18.218	24,781
Total input to the 34.5 kV lines @ "X"	52.488	313,901

While line losses are distributed along the lengths of the conductors and demand and energy sales are also distributed based on customer delivery points, both the load-related losses and sales can be represented as single nodal outputs. The line losses between "X" and "Y" can be determined, as residual values, given the known inputs and outputs:

	MW	MWh
Total input to the 34.5 kV lines @ "X"	52.488	313,901
Less: total output of the 34.5 kV lines @ "Y"	25.699	117,358
Less: metered demand and energy sales (M-R)	25.710	192,000
Total line losses	1.079	4,543

The results of the calculated losses and power flows through the example 34.5 kV subsystem are organized in Table 21.5. While the initial analysis of system losses proceeded by evaluating power flows from

TABLE 21.5   Total Power Flow through an Example 34.5 kV Subsystem

	MW @ Peak	Loss Factor	Branch Factor	Annual MWh	Loss Factor	Branch Factor
34.5 kV subsystem outflows						
34.5–4.16 kV	6.672			27,472		
34.5–14.4 kV	19.027			89,886		
34.5 kV sales	25.710			192,000		
Sub-total outflow	51.409			309,358		
34.5 kV line losses	1.079	2.100%		4,542	1.468%	
	52.488			313,900		
Transformer outflows	52.488			313,900		
69–34.5 kV subs	4.498		8.570%	37,950		12.090%
138–34.5 kV subs	29.772		56.722%	251,170		80.016%
34.5 kV GSU subs	18.218		34.708%	24,781		7.895%
Transformer load losses	0.100			584		
69–34.5 kV subs	0.009	0.190%		71	0.186%	
138–34.5 kV subs	0.056	0.189%		467	0.186%	
34.5 kV GSU subs	0.035	0.191%		47	0.190%	
Transformer no-load losses	0.124			1084		
69–34.5 kV subs	0.015			133		
138–34.5 kV subs	0.100			879		
34.5 kV GSU subs	0.008			72		
Total 34.5 kV losses	*1.303*	*2.472%*		*6,210*	*1.968%*	
Total 34.5 kV outflow	52.712			315,568		
34.5 kV subsystem inflows						
From 69 kV subsystem	4.522		8.579%	38,153		12.090%
From 138 kV subsystem	29.929		56.779%	252,515		80.019%
From 34.5 kV generation	18.261		34.642%	24,900		7.891%
Total 34.5 kV inflow	52.712			315,568		

the generators and the upper level voltage subsystems to the loads and to the lower voltage subsystems (a downward direction), the loss analysis model developed for the cost-of-service study works in reverse (an upward direction).

To complete the model framework, (a) load-related loss factors for lines and transformers and (b) power flow branching factors are determined. For example, the 34.5 kV line losses were calculated to be 1.079 MW at peak load and 4,542 MWh across the year. Thus, the respective line loss factors as a percentage of the subsystem demand and energy outflows are 2.100% (1.079 MW/51.409 MW = 0.020988) and 1.468% (4,542 MWh/309,358 MWh = 0.014683). Similarly, transformer load–related loss factors are determined by the ratio of transformer loss to transformer outflow. For example, the load-related loss factor for the 138–34.5 kV substation transformers is 0.189% (0.056 MW/29.772 MW = 0.001894).

Branching factors indicate how the system power flows split between different voltage levels. For example, the total power inflow to the 34.5 kV subsystem from 34.5 kV generation and the higher voltage subsystems is 52.712 MW at peak load. The inflow from the 138 kV subsystem alone is 29.992 MW or 56.7789% of the total subsystem inflow. Due to no-load and load-related losses, the power flow out of the 138 kV transformers to the 34.5 kV lines is 29.722 MW. Since the total 34.5 kV subsystem transformer outflow to the 34.5 kV lines is 52.488 MW at peak load, the 138–34.5 kV transformer output represents 56.7219% of the total transformer outflow.

All information needed to model power flows and calculate losses for each of the customer or rate classes has now been determined. The model integrates all of the voltage subsystems together into a single, uniform analysis. Given the determination of the various system loss and branching factors, the only input required to determine the results for a given customer or rate class is that class's annual energy sales and load at the system peak.

21.7 Electric Rate Design

Rate design is the process of establishing a rate structure for a given class of electric service, along with its constituent price components, that is capable of achieving a specified revenue requirement at a commensurate level of risk. The cost-of-service study is a key resource for the rate design function, as it provides electric service costs on a customer, demand, and energy component basis for the major power system functions of production, transmission, distribution (primary and secondary), customer accounts, customer assistance, and sales. Such information, along with the associated customer, demand, and energy billing units of the rate classes, provides a basis from which to develop elementary rates for electric service.

21.8 Cost Curve Development

A cost curve bridges the gap between the cost-of-service study and rate design. Basic unit costs are translated into prices within a potentially complex rate structure. Cost curve development can be thought of as an intra-rate cost allocation methodology that further encompasses the characteristics of load diversity relative to the costs of electric service across all levels of energy and demand usage for a specified rate class. Like the cost-of-service study, the more intricate part of the cost curve allocation process revolves around the demand cost components. The cost curve serves as an essential model for designing and evaluating alternative demand rate structures, particularly when displayed in graphical form.

A flowchart describing the input data and analytical processes needed to develop a cost curve for a given demand rate class is shown in Figure 21.9. Input data include the cost-of-service study component revenue requirements for the given rate. Another input is an hours use of demand bill frequency, which is based on the subject rate's kWh usage and actual kW maximum demands. The frequency is a distribution of customer monthly maximum demands as a function of LF in the form of kWh per kW. Since customer maximum demands represent customer NCP values (kW_{MAX}), they must be diversified in order to determine the equivalent class kW_{NCP} and kW_{CP} values. The relationship that exists between CFs and LFs is used to diversify the customer maximum demands. The unit costs of service for an example rate class consisting of 12,000 commercial and industrial secondary three-phase customers having an annual energy use of 2,100,000 MWh are calculated as follows:

Unit customer cost = $6,885,417/(12,000 customers × 12 months)
= $47.8154 per bill

Unit demand costs:

- Production/transmission = $67,560,485/4,099,734 kW_{CP}
  = $16.4792 per kW_{CP}
- Substation/primary lines = $9,451,570/5,273,832 kW_{NCP}
  = $1.7922 per kW_{NCP}
- Transformers/sec. lines = $4,372,917/6,888,228 kW_{MAX}
  = $0.6348 per kW_{MAX}

Unit energy costs:

- Nonfuel = $19,637,444/2,100,000,000 kWh
  = $0.009351 per kWh
- Fuel/purchased power = $51,292,168/2,100,000,000 kWh
  = $0.024425 per kWh

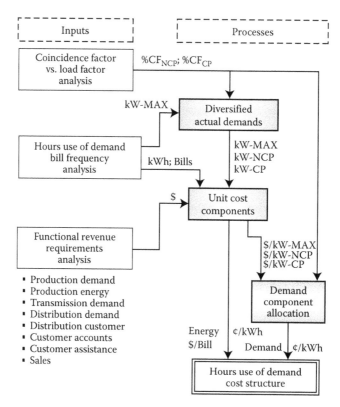

FIGURE 21.9 Cost curve development process. Three major analytical steps are implemented to produce a cost curve for a given rate class. The cost curve is of a form similar to an hours use of demand rate structure, and it can be easily plotted and compared to the associated rate structure that is currently used for billing.

21.8.1 Coincidence Factor–Load Factor Relationship

The relationship between the coincidence of load, measured in terms of a CF, and customer monthly LF is a key cost causation driver and thus a central model for development of a rate class cost curve. In essence, the relationship shows that CF has a nonlinear relationship to LF, as illustrated by the CF–LF curve plotted in Figure 21.10. Specifically, the relationship is in the form of a third-order polynomial.

The CF–LF curve in Figure 21.10 is based on the conventional calculation of the CF, which is the ratio of the maximum demand of a group of individual loads to the sum of the maximum demands of the individual loads. In this context, the individual loads are customers having a common value of LF, and several LF groups are required to accurately plot the CF–LF curve across the entire LF spectrum. When considering an LF group's position with respect to the total system load, the resulting maximum demand of the customer group represents an NCP type of factor. In other words, a group peak may or may not occur simultaneously with the system peak demand. The cost-of-service study regards an NCP type of factor as being highly indicative of the cost causation of the demand-related component of the primary distribution system. Thus, the conventional CF–LF relationship, designated here as CF_{NCP}, can be utilized as an intra-rate cost allocator for the primary distribution demand costs assigned to a particular rate class.

The conventional CF–LF relationship is not a plausible indicator of cost causation for production and transmission system demand-related costs since load diversity is much greater at these levels of the system than at the distribution system level. A strong indicator of the cost causation driver of production and transmission is a CP type of demand cost allocation factor. Consequently, an adapted CF is

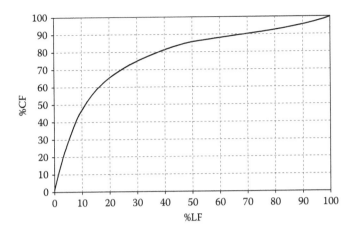

FIGURE 21.10 The nonlinear relationship between CF and LF. The *x*-axis represents monthly LFs of individual customers. The *y*-axis represents the CFs of groups of customers having the same monthly LFs. The CF–LF relationship was first identified by Constantine W. Bary; thus, it is often referred to as the Bary Curve.

required for the development of a modified CF–LF relationship in order to properly allocate these costs in an intra-rate cost curve development. The conventional NCP and the modified CP CFs are given by the following:

$$CF_{NCP} = \frac{D_{MAX}}{\sum_{i=1}^{n} d_i} \quad CF_{CP} = \frac{D_{SYS}}{\sum_{i=1}^{n} d_i} \tag{21.10}$$

where

CF_{NCP} is an LF group's NCP type of CF
CF_{CP} is an LF group's CP type of CF
D_{MAX} is an LF group's maximum demand
D_{SYS} is an LF group's demand at the time of system peak
d_i is the maximum demand of the *i*th customer in an LF group
n is the number of customers in the LF group

The significant demand-related cost causation driver of local facilities is customer maximum demands. In essence, a single customer's maximum demand is fully coincident with itself under all conditions (i.e., 1 kW/1 kW = 1 kW), since diversity of a single load is undefined. Thus, the CF based on customer maximum demands is represented as 100% at all levels of LF, and it can be designated as CF_{MAX}. A chart comparing CF_{CP}, CF_{NCP}, and CF_{MAX} curves is shown in Figure 21.11. As shown, a kWh per kW scale has been added to the *x*-axis of the CF–LF chart. A cost curve for a rate class is in a form similar to a multi-step hours use of demand rate structure; thus, it is more practical to work with kWh per kW units than with LF percentages. The average month is 730 h; thus, 730 kWh per kW is equivalent to a monthly LF of 100%.

21.8.2 Allocation of Unit Demand Cost Components

The unit demand costs developed previously for the example rate class consist of the following functional components: $16.4792 per kW_{CP} for production and transmission, $1.7922 per kW_{NCP} for distribution substations and primary lines, and $0.6348 per kW_{MAX} for line transformers and secondary lines. The CF–LF curves are used to prorate the unit demand costs for production and transmission and for distribution

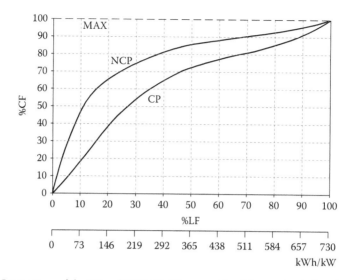

FIGURE 21.11 Comparison of the CP and NCP CF–LF curves. The CP curve lies below the NCP curve due to greater customer load diversity at the time of the system peak as opposed to the time of an LF group's peak. The MAX curve represents a customer's demand at the local facilities level where diversity is nonexistent for all practical purposes at maximum loading conditions. Also shown is a second x-axis scale having units of kWh per kW, which facilitates the chart's use for the development of cost curves.

TABLE 21.6 Allocation of Unit Demand Costs

HUD (kWh/kW)	CF_{MAX} (%)	CF_{NCP} (%)	CF_{CP} (%)	$ Per kW$_{MAX}$	$ Per kW$_{NCP}$	$ Per kW$_{CP}$	$ Per kW$_{TOTAL}$
0	100.00	0.00	0.00	0.0000	0.0000	0.0000	0.0000
20	100.00	16.69	4.42	0.6348	0.2992	0.7288	1.6628
50	100.00	35.96	11.93	0.6348	0.6445	1.9661	3.2454
100	100.00	55.92	25.71	0.6348	1.0021	4.2361	5.8731
150	100.00	65.98	39.32	0.6348	1.1825	6.4804	8.2978
200	100.00	72.64	50.61	0.6348	1.3018	8.3395	10.2762
250	100.00	77.83	59.36	0.6348	1.3947	9.7828	11.8123
300	100.00	81.77	66.03	0.6348	1.4655	10.8814	12.9817
350	100.00	84.73	71.04	0.6348	1.5186	11.7065	13.8599
400	100.00	86.94	74.82	0.6348	1.5582	12.3292	14.5222
450	100.00	88.64	77.80	0.6348	1.5886	12.8207	15.0441
500	100.00	90.07	80.42	0.6348	1.6142	13.2521	15.5012
550	100.00	91.47	83.10	0.6348	1.6393	13.6946	15.9688
600	100.00	93.09	86.29	0.6348	1.6682	14.2194	16.5225
650	100.00	95.15	90.40	0.6348	1.7053	14.8975	17.2376
700	100.00	97.91	95.88	0.6348	1.7547	15.8002	18.1898
≥730	100.00	100.00	100.00	0.6348	1.7922	16.4792	18.9062
Unit costs:				0.6348	1.7922	16.4792	

substations and primary distribution lines, as shown in Table 21.6. The prorated unit demand costs are keyed to various hours use of demand levels. The prorated unit costs are applicable to a customer's peak demand, based on that customer's hours use of demand, since the unit costs themselves account for the characteristics of load diversity between a customer's peak demand and its effective contribution to the system peak and the LF group peak loads. The three unit demand cost components are summed to obtain a total unit cost assessment as a function of hours use of demand. Equation 21.11 is used to develop the average demand cost on a ¢ per kWh basis between any 2 kWh per kW points of usage:

$$E_{AVG} = \frac{\Delta D_{TOT}}{\Delta HUD} \times 100 \qquad (21.11)$$

where
E_{AVG} is the average demand cost in ¢ per kWh
ΔD_{TOT} is the difference in unit demand cost in \$ per kW
ΔHUD is the difference in hours use of demand in kWh per kW

Generally, increments of 50 kWh per kW are sufficient enough to graph the average cost curve, particularly on a hyperbolic scale of hours use of demand. For example, the average cost between 100 and 150 kWh per kW, based on Table 21.6 data, is calculated as follows:

$$[(\$8.2978/kW - \$5.8731/kW)/(150\ kWh/kW - 100\ kWh/kW)] \times 100 = 4.8494¢\ per\ kWh$$

All or a portion of fuel-related costs may be embedded in the basic rate structure. Assuming for this example that all fuel-related costs are to be recovered through a separate rate mechanism, only the non-fuel energy cost component, that is, \$0.009351 per kWh, needs to be added to the average demand costs to obtain the total average demand and energy cost as a function of kWh per kW. The total cost structure is thus defined by the following:

Customer cost: \$47.8154 per bill

Demand/energy cost:

kWh/kW	¢/kWh	kWh/kW	¢/kWh
0	0.0000	400	2.2597
20	9.2493	450	1.9790
50	6.2103	500	1.8492
100	6.1905	550	1.8704
150	5.7845	600	2.0424
200	4.5308	650	2.3655
250	4.0074	700	2.8394
300	3.2739	≥730	3.3232
350	2.6913		

The hours use–based cost structure is plotted in Figure 21.12. The combined demand and energy cost structure is plotted both with and without the customer cost component. Since the customer cost component is truly fixed, that is, not a function of either kW or kWh, plotting of the cost curve with the customer cost component requires an assumption of demand so that its average cost per kWh can be calculated as a function of kWh per kW. A 50 kW value was assumed for this plot as it is approximately equal to the average monthly kW per customer for this example rate class. The y-intercept at $x = 20$ kWh per kW is approximately 14¢ per kWh. A cost curve based on a demand higher than 50 kW would have a comparable intercept that is somewhat less than 14¢ per kWh because of the relatively higher amount of kWh that exists, even at low hours use. The opposite would occur with demands that are less than 50 kW. While this cost curve could serve as a billing rate, it is rather complicated. Rate design utilizes the cost curve as a guide for developing more practical rate structures.

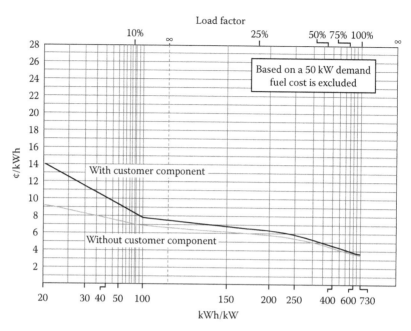

FIGURE 21.12 Demand rate cost curves. A cost curve in ¢ per kWh is plotted as a function of the hours use of demand using a hyperbolic scale for the *x*-axis. The demand and energy cost structure is plotted both with and without the customer cost component. A demand of 50 kW is assumed for plotting the curve with the customer component. The selection of different demand assumptions would result in a family of curves when the customer cost component is included.

21.9 Rate Design Methodology

The design of a practical rate structure, which tracks the cost curve in a reasonable manner, could be approached in different ways. For this example, a three-step rate design is illustrated for the hours use ranges of 0–200 kWh per kW, 200–400 kWh per kW, and over 400 kWh per kW. To simplify the rate design process, the design initially addresses only the kWh charges (the finalization of the customer charge is considered during the fine tuning process). Thus, the combined demand and nonfuel energy revenue target is $101,022,417 (the total base rate revenue target with the customer-related cost is $107,907,834). The demand and energy-related cost curve, without the customer cost component, is shown in Figure 21.13.

The 0–200 kWh per kW price step is considered first. The average cost is found to be 6.808¢ per kWh at 100 kWh per kW and 5.850¢ per kWh at 200 kWh per kW. By inspection of the cost curve, it is reasonable to set the initial unit price of the first 200 kWh per kW step in the middle of these values, that is, 6.33¢ per kWh. Without inclusion of the customer charge, the first hours use of demand price of the rate structure will display as a flat line when graphed, as observed in Figure 21.13.

Now consider the tail step (i.e., 400 kWh per kW) of the concept rate structure. By inspection of Figure 21.13, a straight-line extrapolation of the cost curve based on the range of 400–730 kWh per kW indicates an average rate at infinity of approximately 2.3¢ per kWh. This rate at infinity would also represent the value of the tail step price. Thus, initial values for the first and last steps of the rate structure have been selected. The revenue produced by each of these 2 h use steps can then be determined by applying the two prices to their associated kWh quantities, which results in $82,131,266.

The remainder of the combined demand and energy revenue amount is to be collected by the middle 200–400 kWh per kW step. The unit price for this middle step is determined by dividing its revenue responsibility by its associated kWh, which yields a result of 2.84¢ per kWh.

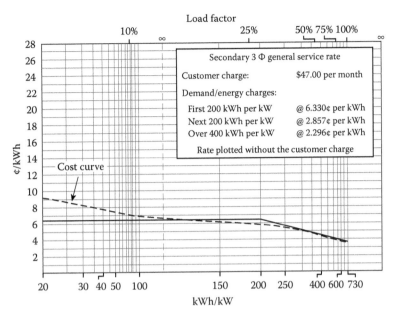

FIGURE 21.13 Hours use of demand rate design. A three-step hours use of demand rate designed to track the cost curve closely between 100 kWh per kW and 730 kWh per kW. The plots exclude the customer charge. The hours use of demand rate structure was originally developed by Arthur Wright, a British engineer.

Due to rounding, when the initial kWh prices are applied back to the billing determinants, a revenue target overcollection of $12,622 is calculated. The customer-related cost component is now incorporated for fine tuning to the total base rate revenue target. The kWh prices for the final rate design are shown in Figure 21.13, and the rate curve is plotted in comparison to the cost curve.

As can be observed in Figure 21.13, the rate curve does not track the cost curve below 100 kWh per kW. If low LF customers are served under this rate, then another rate design could be considered by adding an explicit demand charge to the hours use of demand structure. The addition of such a fixed charge will increase the average rate per kWh in the lower LF range. Reducing the first 200 kWh per kW price by ½¢ would free up revenue, in the amount of $6,096,856, for the explicit demand charge. The demand charge is then determined by dividing this revenue amount by the sum of the associated customer maximum demands, which yields a result of $0.8851/kW. After fine tuning of the charges to achieve the target revenue, the modified hours use of demand rate structure is plotted in comparison to the cost curve in Figure 21.14. Although the rate structure has been made somewhat more complex, it tracks the cost curve very closely across the whole spectrum of usage.

Alternatively, a simple rate structure with an explicit demand charge and a flat energy charge (as well as the customer charge) can be developed. The CF–LF relationship was utilized previously to develop a cost curve based on diversified demands. A cost curve, which is not based on diversified demands, can also be developed from the cost-of-service study and demand unit data for the example general service rate. In this case, the total demand-related revenue requirements for production, transmission, distribution substations, primary distribution lines, line transformers, and secondary distribution lines are summed and then divided by the total of the customer maximum demands. Thus, the nondiversified cost curve is represented by $11.8151 per kW_{MAX} and $0.009351 per kWh. As illustrated in Figure 21.15, the nondiversified cost structure has been designated as the "fundamental Hopkinson" rate design as its demand charge consists of only demand-related costs while its energy charge consists of only the non-fuel energy–related costs. By shifting a portion of demand cost recovery responsibility from the demand charge to the energy charge, that is, by applying the principle of rate tilt, the average rate per kWh is observed to decrease on the left side of the chart and to increase on the right side of the chart. The point

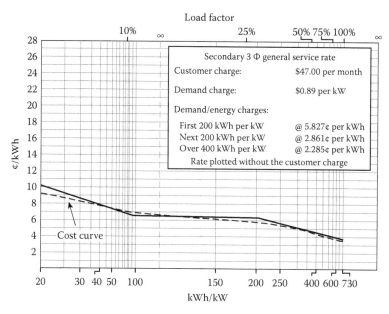

FIGURE 21.14 Modified hours use of demand rate design. By adding an explicit demand charge to a basic hours use of demand rate structure, the average rate is elevated at low hours use of demand since the demand charge is a fixed charge and thus provides the same declining average rate effect as caused by a customer charge. This design is referred to as a Wright-Hopkinson rate structure. Like Wright, Dr. John Hopkinson was a British engineer.

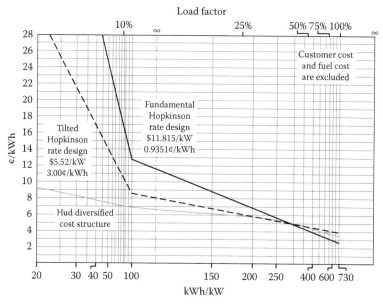

FIGURE 21.15 Hopkinson rate design. The nondiversified cost curve is designated as the "fundamental Hopkinson" rate design. By tilting the fundamental Hopkinson rate, a portion of the demand cost recovery is transferred to the kWh price. In so doing, the average rate per kWh of the "tilted Hopkinson" rate structure rate shifts downward on the left and upward on the right. The pivot point is 305 h use of demand, that is, the average hours use of the rate class.

of pivot is located at 305 kWh per kW, which is also the hours use of demand at which the fundamental Hopkinson rate and the tilted Hopkinson rate intersect. Furthermore, 305 kWh per kW represents the average hours use of demand for the example rate class. Any degree of applied rate tilt will cause the average rate curve to pivot at this point. With full rate tilt, the average rate would be plotted as a straight line across the entire chart, that is, a uniform rate per kWh; thus, the Hopkinson rate structure would be transformed into a nondemand, flat energy rate structure.

Note in Figure 21.15 how the tilted Hopkinson rate tracks the HUD diversified cost curve generally over the range of 175–500 kWh per kW. Further rate tilt could be applied so as to make the Hopkinson rate coincide with the HUD diversified cost curve at 100 kWh per kW. Then the Hopkinson rate would track the HUD diversified cost curve generally over the range of 100–275 kWh per kW. However, at hours use above about 275 kWh per kW, the average rate would rise well above the HUD diversified cost curve.

Selection of a particular rate structure design is guided to a great extent by the load characteristics of the customers served under each particular rate class of service. Rate classes of high LF customers are best served under the fundamental Hopkinson rate structure. A rate class of medium to high LF customers would be well served under a basic Wright rate structure. If a rate class serves customers with a wide distribution of LFs from low to high, the Wright–Hopkinson design would be most practical.

Reference

1. Vogt, L. J. 2009. *Electricity Pricing: Engineering Principles and Methodologies*. Boca Raton, FL: Taylor & Francis.

22

Business Essentials

Richard E. Brown
Quanta Technology

22.1 Introduction

There are many challenging power engineering problems to be solved. New customers must be served. Old equipment must be maintained. New technologies must be assessed and adopted. To solve these challenges, power engineers find themselves responsible for planning, engineering, system analysis, system design, equipment specification, maintenance management, operations, and a host of other functions. Whatever their role, power engineers make many decisions. Some of these decisions result from extensive and careful analyses. Others are made quickly during everyday activities. In virtually all cases, decisions have cost and other implications. Some options are cheap. Others options are expensive. Some options spend as little as possible now. Others options spend money now in order to save money later. Some options result in high safety margins. Other options are more risky. With so many choices, it is valuable for power engineers to understand the criteria for deciding which decisions are best from a business perspective.

Typical businesses prefer engineering decisions that result in higher profits. A cheaper engineering solution may produce higher profits if a resulting lower price causes an increase in sales. It is also possible that a more expensive engineering solution will produce higher profits if the resulting higher quality product can command a premium price. In both cases, the business objective is clear: while acting legally and ethically, maximize profits whenever possible. Utilities are a bit different. Investor-owned utilities need to be profitable, but have their profits essentially controlled by regulators. Government-owned utilities will be more political when making decisions. Member-owned utilities have yet another perspective with regards to revenues and costs.

In any case, it is helpful for power engineers to understand the basics of business. Like engineering, business has a large number of words and phrases with precise meanings. Many of these words and phrases represent simple concepts, and can be easily learned and understood. Although some terms represent difficult or confusing concepts, this section will (hopefully) remove much of the mystery and serve as a reference when necessary.

The breadth of subject matter covered in this section is necessarily extensive so that the reader can become exposed to the essentials of business. The consequence is that each topic can only be given brief treatment. These and other business topics are addressed more fully the book *Business Essentials for Utility Engineers* (CRC Press).

22.2 Accounting

Although many perceive accounting to be arcane and esoteric, it can be boiled down to one simple equation: the assets of a company must be equal to the claims on these assets. This *basic accounting equation* is most commonly represented as the following:

$$\text{Assets} = \text{Liabilities} + \text{Owner's Equity}$$

An *asset* is something of value, and a company's assets are the sum of all things of value that a company owns. A *liability* is an obligation of the company, such as a loan or an unpaid bill. When the value of all liabilities is subtracted from the value of all assets, the residual value is left for the owners. This residual value is called *owner's equity*.

To illustrate the basic accounting equation in action, a simple example is now provided. Imagine a group of investor purchasing and running a small utility. These investors initially sell 10 million shares of common stock at $10 per share. The sale of stock raises $100 million in cash, which is referred to as *paid-in capital*. The raised cash, an asset, corresponds to $100 million in owner's equity.

The stock transaction described earlier is reflected in the *balance sheet* of the company. The balance sheet can be thought of as an expanded version of the accounting equation, and describes all assets, liabilities, and owner's equity. The balance sheet of the company (all values in millions of dollars) after the sale of common stock is

Assets		Liabilities		Owner's Equity	
100	Cash			100	Common stock
100	**Assets**	**0**	**Liabilities**	**100**	**Owner's equity**

Transaction: Issue $10 million shares of common stock at $10 per share.

The aforementioned table is organized into columns corresponding to assets, liabilities, and owner's equity. Each individual item in these categories is listed, starting from the top, in nonbold font. The total amount of each category is listed at the bottom in bold font.

After the first transaction of the company, the balance sheet confirms that assets are equal to liabilities plus owner's equity. In a more practical sense, the company now has (1) cash on hand and (2) stockholders that expect a return on their investment. The company finds that it needs additional funds to start business operations. It therefore sells $50 million in bonds. The bond issuance raises another $50 million in cash, and results in a corresponding $50 million liability. The balance sheet is now

Assets		Liabilities		Owner's Equity	
150	Cash	50	Bonds	100	Common stock
150	**Assets**	**50**	**Liabilities**	**100**	**Owner's equity**

Transaction: Issue $50 million in bonds.

The bond transaction causes the *cash account* to increase from $100 to $150 million. To balance the increase in assets, a $50 million liability is recorded so that the total assets of $150 million equal the liabilities plus owner's equity.

The company now negotiates to buy a small utility system for $90 million and pays in cash. The effect is to reduce the cash account by $90 million and to add a new asset worth the same amount. Total assets have not changed, and total assets still equal liabilities plus owner's equity; the balance sheet still balances.

Assets		Liabilities		Owner's Equity	
60	Cash	50	Bonds	100	Common stock
90	Utility system				
150	**Assets**	**50**	**Liabilities**	**100**	**Owner's equity**

Transaction: Purchase a $90 million utility system.

The company is now a utility. In its first month, it provides utility services to its customers and then bills them $15 million for these services. The customers have not yet paid these bills, but their legal obligation to pay these bills is an asset to the company. This type of asset is typically recorded in a category called *accounts receivable*.

When a customer is obligated to pay for services rendered, the result is an increase in profits, also called *earnings*. Since the owners of the company have rights to these earnings, an increase in accounts receivable is balanced by an increase in an owner's equity account called *retained earnings*.

Assets		Liabilities		Owner's Equity	
60	Cash	50	Bonds	100	Common stock
90	Utility system			15	Retained earnings
15	Accounts receivable				
165	**Assets**	**50**	**Liabilities**	**115**	**Owner's equity**

Transaction: Bill $15 million to customers.

Notice that the aforementioned balance sheet still balances. Assets are worth $165 million and the sum of liabilities and owner's equity is worth the same amount.

In the process of providing services to its customers, the utility incurs expenses of $5 million, which it owes to a variety of contractors and outsourcing companies. The utility has not yet paid its bills, but must record the obligation to pay these bills as a liability. Unpaid obligations are typically recorded in a category called *accounts payable*. An increase in accounts payable results in a decrease in earnings, which is recorded as a decrease in the retained earnings account.

Assets		Liabilities		Owner's Equity	
60	Cash	50	Bonds	100	Common stock
90	Utility system	5	Accounts payable	10	Retained earnings
15	Accounts receivable				
165	**Assets**	**55**	**Liabilities**	**110**	**Owner's equity**

Transaction: Incur $5 million in expenses.

The company now pays the $5 million that it owes in bills. These payments come out of cash accounts, and are offset by a reduction in accounts payable. Since the company is transferring cash out of the company, total assets are lower. Since the company no longer has unpaid bills, total liabilities are also lower.

The balance sheet reflects clearly the difference between a company with many unpaid bills and a company with few unpaid bills. The new balance sheet is

Assets			Liabilities		Owner's Equity	
55	Cash	50	Bonds	100	Common stock
90	Utility system	0	Accounts payable	10	Retained earnings
15	Accounts receivable				
160	**Assets**	**50**	**Liabilities**	**110**	**Owner's equity**

Transaction: Pay $5 million in unpaid bills.

Customers now pay $10 million of their unpaid bills. This is not the total amount owed, but has the effect of increasing cash by $10 million and reducing accounts receivable by the same amount. There is no net effect on total assets.

Assets			Liabilities		Owner's Equity	
65	Cash	50	Bonds	100	Common stock
90	Utility system	0	Accounts payable	10	Retained earnings
5	Accounts receivable				
160	**Assets**	**50**	**Liabilities**	**110**	**Owner's equity**

Transaction: Customers pay $10 million of their bills.

The utility system will not live forever. To account for this, the value of the utility system on the balance sheet is reduced over time. This reduction in value is called *depreciation*. Depreciation is covered in more detail later, but for now the assumption is a utility system depreciation amount of $1 million. This reduces the *book value* of the utility system by $1 million, and is treated as an expense that lowers retained earnings. The updated balance sheet is

Assets			Liabilities		Owner's Equity	
65	Cash	50	Bonds	100	Common stock
89	Utility system	0	Accounts payable	9	Retained earnings
5	Accounts receivable				
159	**Assets**	**50**	**Liabilities**	**109**	**Owner's equity**

Transaction: Utility system depreciates by $1 million.

In addition to operating expenses and depreciation expenses, a utility is obligated to pay interest payments to bond holders. In this case, the utility pays $1 million in interest. These payments reduce the cash account and retained earnings accordingly. The interest payment does not affect the face value of the bonds, and therefore does not affect the amount of liabilities in the bond account. If the utility paid off part of the principle of the bonds, the cash account (an asset) and the bond account (a liability) would both be reduced.

The updated balance sheet after the $1 million in bond interest payments are made is

Assets			Liabilities		Owner's Equity	
64	Cash	50	Bonds	100	Common stock
89	Utility system	0	Accounts payable	8	Retained earnings
5	Accounts receivable				
158	**Assets**	**50**	**Liabilities**	**108**	**Owner's equity**

Transaction: Utility pays $1 million in interest to bond holders.

From an accounting perspective, this utility is performing well. It is both profitable and generating cash. In order to transfer some of these profits to its owners, the utility now decides to distribute $2 million of retained earnings to common stockholders. This amount is taken from the cash account and is called a *dividend*. Since there are 10 million shares of common stock, the dividend corresponds to 20 cents per share. The resulting balance sheet is

Assets		Liabilities		Owner's Equity	
62	Cash	50	Bonds	100	Common stock
89	Utility system	0	Accounts payable	6	Retained earnings
5	Accounts receivable				
156	**Assets**	**50**	**Liabilities**	**106**	**Owner's equity**

Transaction: Issue $2 million in dividends.

22.2.1 More on Assets, Liabilities, and Owner's Equity

An asset is not necessarily a tangible thing, and can include items such as owed money, patents, and prepaid insurance. Assets that can be touched and felt are called *tangible assets* and others are called *intangible assets*. *Goodwill* is a special type of nonmonetary intangible asset. It refers to a premium paid above market value for an asset, typically during the acquisition of a company.

Another way to classify assets is based on the expected length of time until the asset is used up. Assets that are expected to be used up within 1 year are called *current assets*. Typical current assets include cash, cash equivalents, accounts receivable, inventory, and short-term investments. Assets that are not expected to be used up within 1 year are called *noncurrent assets* or *long-term assets*. Typical long-term assets include land, buildings, equipment, utility infrastructure, and long-term investments. Tangible long-term assets are also called *fixed assets*.

Like assets, liabilities are grouped according to their timeframe into *current liabilities* and *noncurrent* or *long-term liabilities*. Common current liabilities include unpaid wages, unpaid bills, unpaid interest, declared but unpaid dividends, and prepayments for services. Common long-term liabilities include long-term debt (e.g., bonds and bank loans), long-term leases, and employee pension obligations.

Working capital is the net amount of financial capital tied up in daily business operations. Mathematically, working capital is equal to current assets minus current liabilities.

Capital employed is defined as total assets minus current liabilities, and represents the total amount of capital used to finance a utility.

As shown in the accounting example, owner's equity can be divided into *paid-in capital* and *retained earnings*. Paid-in capital is the amount of money raised through the issuance of common stock. Retained earnings are everything else, and are equal to cumulative revenue minus cumulative expenses minus cumulative dividends. Since paid-in capital is generated by selling common stock, it is sometimes called *common stock equity*.

22.2.2 Amortization and Depreciation

When a utility purchases an expensive piece of equipment, the cost of this equipment is spread over the expected useful life. This process is called *amortization*. When amortization is applied to a tangible asset, it is called *depreciation*.

Consider the purchase of an expensive construction vehicle for $1 million that is expected to last for 20 years. Accounting rules require that the total cost of the vehicle be distributed across each of these 20 years. Typically, this is done by allocating an equal amount to each year. For this example, $50,000 of the $1 million is treated as a *depreciation expense* for a period of 20 years; each year the vehicle is depreciated by

$50,000 until the initial value has been depreciated. In accounting records, the current asset value is equal to the initial value minus accumulated depreciation expenses, and is referred to as the *book value* of the asset.

22.2.3 Financial Statements

When people speak of financial statements, they are typically referring to the *income statement*, the *balance sheet*, and the *statement of cash flows*. The income statement describes a utility's revenues and expenses over a specific period of time. The balance sheet describes a utility's assets, liabilities, and owner's equity at a specific point in time. The statement of cash flows describes changes in the cash account over a specific period of time.

An income statement, also called and *earnings statement* or a *profit and loss statement* (income, earnings, and profits are all synonymous), describes the profitability of a utility over a specific period of time, such as a month, 3 months (quarter), or a year. It does this by presenting revenue and expenses in different categories so that income can be presented at different levels. Income statements start with company revenue, and successively subtract expense categories until there are no more expenses to subtract. This process generally results in operating income, earnings before interest and taxes (EBIT), earnings before interest and taxes depreciation and amortization (EBITDA), income before taxes (IBT), net operating profit after taxes (NOPAT), and net income. The relationship of revenue and income measures to each other is shown in Figure 22.1.

A balance sheet, such as the tables shown in the accounting example, shows the financial status of a utility at a specific point in time. Information is typically categorized into assets, liabilities, and owner's equity. When looking at a balance sheet, it is important to remember that the value of assets (for the most part) are recorded at historical cost, and might not represent the value of the asset today.

Because of differences in accounting treatment, income statements and balance sheets are somewhat subjective. But cash is objective. Identical utilities could have different balance sheets, but cash flow would be the same. Most importantly, the statement of cash flows is very difficult to manipulate.

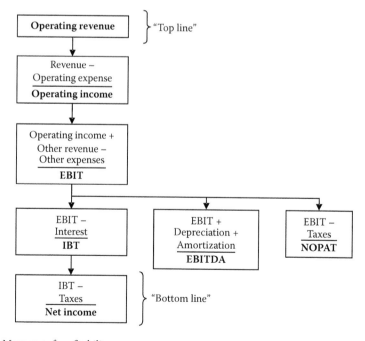

FIGURE 22.1 Measures of profitability.

An important concept in cash analysis is that of *free cash flow*. This is typically defined as the amount of cash generated by a company that is available for distribution to owners as dividends. In the long run, net income must be equal to free cash flow. Because they are equal in the long run, net income is a good first approximation of free cash flow. To make this approximation better, several adjustments must be made. First, all noncash expenses must be added back. Next, changes in working capital must be considered. Last, all capital expenditures must be subtracted. Free cash flow corresponds to the following equation:

Net income

\+ Noncash expenses

\+ Decrease in working capital

− Capital expenditures

= Free cash flow

The free cash flow of a utility is recorded in the statement of cash flows, which is regarded by many financial analysts as the most important financial statement. This typically includes a section on cash from operations (net income with adjustments for noncash expenses and changes in working capital), cash from investments (capital purchases and capital sales), and cash from financing (debt purchase, debt retirement, stock issuance, stock repurchase, dividends). The statement of cash flow concludes with the amount of cash at the beginning of the period, the change in cash, and the corresponding amount of cash at the end of the period.

It should be understood that this section discusses *financial accounting*, which is what appears in annual reports and is submitted to government agencies like the Security and Exchange Commission.

22.3 Finance

The subject of *engineering economics* typically deals with the time value of money and net present value (NPV) calculations. In the business world, these topics are included in the subject of *finance*. As a verb, finance means raising money for an investment. This section presents a summary of finance from an engineering and technical management perspective.

22.3.1 Time Value of Money

An amount of money received today is more valuable than the same amount of money received sometime in the future. Consider putting $100 today into a savings account that offers 5% interest per year. After 1 year, the $100 has turned into $105. In the terminology of finance, the *present value* of $100 is has a *future value* of $105 in 1 year. After 2 years, the amount increases by an additional 5% to $110.50. The increase in the first year is $5.00 and the increase in the second year is $5.50. Left alone, the amount of increase will grow, resulting in an exponential increase in account value. This exponential increase is caused by *compound interest*—interest being earned on all previous interest payments.

It is common to consider a number of cash flow events for a single analysis. For example, a project may consist of a number of costs requiring cash outflows at various times. Once completed, the project may also result in revenue from customer payments at various times. The sum of all positive present values minus the sum of all negative present values is called *net present value*. When cash inflows are from revenue and cash outflows are from expenses, NPV is equal to the following:

$$NPV = \sum_{n=0}^{\infty} (R_n - E_n)(1+r)^{-n}$$

where
    NPV represents net present value
    R_n represents revenue in year n
    E_n represents expense in year n
    r represents interest rate

There are several types of cash flow streams that commonly arise in an NPV analysis. For this reason, it is convenient to provide a quick method of NPV calculation. The first type of cash flow stream is a *perpetuity*, which corresponds to a constant amount of cash at the beginning of each year, starting in Year 1. The second type of common cash flow stream is a *growing perpetuity*, which provides payments that increase by a fixed percentage each year, g.

The perpetuity equations are simple to remember and convenient to use. However, cash flows are typically not expected to last forever, but end after a specific number of years, n. These *annuities* have formulae similar to perpetuities but with an additional term that essentially subtracts out the perpetuity values in distant years.

Cash flow diagrams and their associated NPV equations for perpetuities and annuities are shown in Figure 22.2. These are the most commonly used cash flow streams, and can be combined to represent many cash flow scenarios.

The generic term for the "r" value used in time value of money calculations is the *discount rate*. It is possible that a discount rate corresponds to a specified interest rate, but often this is not the case. Factors that are typically considered when choosing a discount rate are inflation, time frame, and risk exposure.

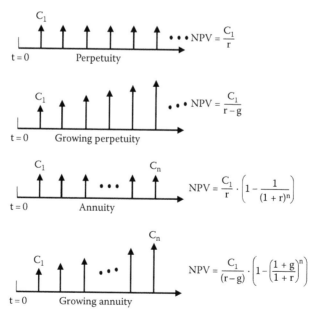

FIGURE 22.2 Common NPV calculations.

It is often necessary to determine the average cost of raising cash for a company. This is done by considering the interest rates for all debt (such as bonds and bank loans) and the expected return for stockholders. This *weighted average cost of capital,* or *WACC* (pronounced "wack"), represents is the average profit that a company must achieve in order to meet all investor expectations.

22.3.2 Capital Structure

Financial capital is money available for funding a business. The sources of financial capital are owners and lenders. Financial capital provided by owners is called owner's equity, often shortened to equity. Financial capital provided by lenders is called debt. All sources of equity and debt are called the *capital structure* of a company. Both the equity (e.g., common stock) and debt (i.e., bonds) of a company are typically valued by markets. The value of the total capital structure of a company is equal to the value of equity plus the value of debt.

If a company goes bankrupt, available funds are paid out to sources of capital based on *seniority.* A debt that must repaid before another is said to be *senior.* A debt that must repaid after another is said to be *subordinate* to the senior debt. Debts that have the same priority are said to be *pari passu.* A common classification of seniority is the following: senior secured debt (highest seniority), senior unsecured debt subordinated debt, convertible debt preferred stock (mezzanine), and common stock (lowest seniority).

Senior secured debt gets paid off first. The term *secured* means that the debt is guaranteed by the assets of the company. This is sometimes called *mortgage-backed debt.* Unsecured debts, also called *debentures,* are not secured by assets and only by the general credit worthiness of the issuer. Senior unsecured debt gets paid off after secured debt, followed by various levels of subordinate debt. Next to be paid is debt that can be converted into preferred or common stock. Mezzanine capital is senior only to common stock; in this case the mezzanine capital is preferred stock. Anything left after all senior obligations are fulfilled goes to common stockholders.

Capital structure is typically measured as amount of total debt divided by the total company value (debt plus owner's equity). This is called a company's *debt ratio,* and can vary from 0% for companies without any debt, to more than 50% for companies with very low financial risk such as regulated utilities. Alternatively, capital structure can be measured by dividing total debt by owner's equity, resulting in the *debt-to-equity ratio.*

22.4 Financial Risk

Financial risk relates to the predictability of financial outcomes. Less risk is preferable to high risk in the sense that investors will pay more for more predictable outcomes. The tools of financial risk analysis allow risk-based valuations to be treated quantitatively.

Mathematical analyses based on expected values and average outcomes are insufficient for financial risk analysis. Financial risk analysis is not interested in averages. Rather, it is interested in knowing the predictability of outcomes, which requires knowledge of all possible outcomes and their likelihood of occurrence. This type of analysis can either be backward-looking or forward-looking. A backward-looking financial risk analysis examines historical data and uses the tools and techniques of statistics. A forward-looking financial risk analysis creates predictive models and uses the tools and techniques of probability theory.

22.4.1 Diversification

The essence of diversification can be summed up by the old adage, "don't put all of your eggs in one basket." Good investors do not care whether the value of a particular stock goes up or down, but are extremely interested in whether their overall portfolio value goes up or down.

Some stocks do better than others primarily due to random chance and/or factors not knowable through public information. These uncertainties are specific to each company and are called *idiosyncratic risks* which can be diversified away. Investing in one small startup company is risky, since it may succeed and it may not succeed. Investing in many small startup companies is much less risky since a certain percentage will fail and a certain percentage will succeed. In a developed market, idiosyncratic risk can be effectively eliminated by purchasing 30–40 assets with similar idiosyncratic risk characteristics.

Not all risk is idiosyncratic. There are some risk factors that will impact all investments at the same time. This is similar to the engineering concept of a *common mode* root cause. Something external is impacting many things at once. Whereas idiosyncratic risks are diversifiable, common mode risks are *nondiversifiable*. Another common term for nondiversifiable risk is *systematic risk*. Purchasing a large number of disparate investments does not impact nondiversifiable risk because all of the investments will be impacted by the common mode events in a similar way. For example, higher interest rates will reduce the value of all stock prices in a similar way. Lower consumer confidence will have a similar effect on the overall market. Some systematic risk will be specific to certain industries (such as the development of a substitute product or service), while others will be broader in scope.

22.4.2 Portfolio Theory

Portfolio theory assumes that rational investors are interested in maximizing investment returns for a given level of risk. Investment return is defined as the expected value of future returns divided by current price. Investment risk is defined as the standard deviation of this distribution.

Portfolio theory builds on diversification theory; all of the mathematics and derivations assume a fully diversified portfolio with no idiosyncratic risk, with lognormally distributed stock prices. Portfolios are created by buying and selling stock through the stock market. Portfolio theory can be easily extended to include any type of asset such as bonds, real estate, and commodities. Terms that are used synonymously are investment, company, firm, security, and asset.

Consider a large number of assets, each with an expected return and risk, allowing them to be plotted as points on a graph. Random portfolios can now be generated by selecting random mixes of asset with different asset weights. The risk and return of each portfolio can be calculated, allowing them also to be plotted on the graph. This requires a correlation matrix, or the assumption that returns are not correlated. Some of these portfolios will have returns that are higher than all other points with equal or lower risk. The set of all such portfolios is called the *efficient frontier*. Rational investors will only purchase a portfolio located on the efficient frontier because otherwise they could achieve higher returns for the same level of risk. The point of the efficient frontier with the lowest level of risk is called the *minimum variance portfolio*, which is equivalent to the lowest possible standard deviation. The efficient frontier is shown in Figure 22.3.

When a portfolio is on the efficient frontier, the trade-off between risk and return is binding. Higher portfolio returns cannot be achieved without increasing risk. Similarly, portfolio risk cannot be reduced

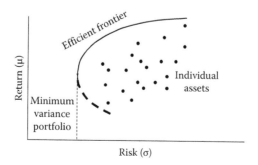

FIGURE 22.3 The efficient frontier.

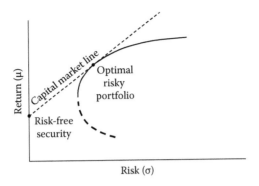

FIGURE 22.4 The capital market line.

without accepting lower levels of return. Portfolio theory is extended by consideration of a risk-free security—a security with a highly predictable return. An example of a nearly risk-free security is a U.S. treasury bill, which has nominal returns only exposed to the risk of a collapse of the U.S. government.

A risk-free security is graphically represented in Figure 22.4. There is a single point on the efficient frontier where a tangent line can be drawn through the risk-free security. This point is called the *optimal risky portfolio*, and the line is called the *capital market line*. The capital market line represents the best possible risk versus return potential for combinations of risky assets and risk-free securities.

The optimal risky portfolio represents the weighted sum of every asset in the market, with weights being proportional to market value. A portfolio with these characteristics is called the *market portfolio*. A reasonable estimate of the stock market portfolio can be achieved through the purchase of a mutual fund that tracks a broad market index such as the S&P 500 or the Russell 3000.

22.4.3 Capital Asset Pricing Model

The capital asset pricing model (CAPM, pronounced "Cap M") provides a methodology for determining what the expected returns of stockholders should be. Like portfolio theory, CAPM assumes that investors are fully diversified and are only exposed to systematic risk. It also recognizes that rational investors will want to both invest in the market portfolio with the expected market return (r_m) and invest in risk-free securities with the risk-free rate of return (r_f).

Stock price risk is most commonly measured by comparing it to the overall risk of the market using covariance. The covariance of market movement and stock price movement, when divided by market variance, is called *beta* (β). Beta is mathematically defined as

$$\beta = \frac{\text{Covariance (S,M)}}{\sigma_m^2}$$

where
S represents individual stock prices
M represents overall market prices
σ_m^2 represents variance of market price

The beta of a company describes how its price moves relative to the overall market. If $\beta = 1$, a stock market increase (or decrease) of 1% results in the company stock tending to increase (or decrease) by 1%. If beta is greater than 1, stock price movement tends to be greater than the overall market. For example, if $\beta = 2$, a stock market increase by 1% results in the company stock tending to increase by 2%. If beta is less than one, stock price movement tends to be less than the overall market. For example, if $\beta = 0.5$, a stock market increase of 1% results the company stock tending to increase by one half of a percent.

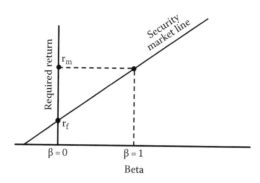

FIGURE 22.5 The security market line.

For a fully diversified portfolio, it can be mathematically shown that the required return on a stock is a linear function of beta. This relationship states that the required return of a stock is the following:

$$r_s = r_f + \beta(r_m - r_f)$$

where

r_s represents required return on a stock
r_f represents risk-free rate
r_m represents expected market return
$(r_m - r_f)$ represents market risk premium

CAPM states that the required return on a stock is equal to the risk free rate plus a risk premium. If the stock has the same risk as the overall market ($\beta = 1$), the risk premium is equal to the market risk premium. If the stock is twice as risky as the market ($\beta = 2$), the risk premium is twice the market risk premium. If the stock is half as risky as the market ($\beta = 0.5$), the risk premium is half the market risk premium.

Plotting required return versus beta results in the *security market line*. The security market line is easily drawn. One point corresponds to a beta of zero and the risk-free rate. A second point corresponds to a beta of one and the expected market return. The security market line is drawn through these two points, and represents the required expected return for a security with any specified beta. The security market line is shown in Figure 22.5.

22.4.4 Financial Options

The world of finance is replete with contracts and other legal documents that have monetary value. There are innumerable possibilities such as stocks, bonds, and bank loans. Legal documents like these that have monetary value are called *financial instruments*. Financial instruments are broadly categorized as either *cash instruments* or *derivative instruments*. Cash instruments do not relate to hard currency per se. Rather, the monetary value of cash instruments is directly determined by the market. Examples include stocks and bonds.

Unlike cash instruments, the monetary value of derivative instruments is derived from the underlying value of one or more cash instruments (hence the name). The most common derivative instruments are options, futures, and swaps. For example, the owner of an option on a utility stock does not have any ownership rights to the utility. Rather, the value of the option rises and falls as the value of the utility stock rises or falls. It is a pure financial mechanism. Today, new and complex derivative instruments are constantly being developed through *financial engineering*.

A swap is an agreement between two parties to exchange a stream of cash flows. Consider a loan with a floating interest rate. The holder of the loan can purchase an *interest rate swap* that will exchange the floating interest rate payments with fixed payments. Purchasing the swap removes the risk of interest rate fluctuations. Other common types of swaps include foreign currency swaps, credit swaps, commodity swaps and equity swaps.

A call option is the right to buy a share of stock (or other cash instrument) at a specified price, called the *strike price*, at a future date. Consider a utility with stock trading at $50 per share. Now consider a call option with a strike price of $55 with an expiration of 1 year. Before the option expires, the holder has the right, but not the obligation, to *exercise* the option and purchase a share of stock for $55. If the utility stock is trading below the strike price, it does not make sense to exercise the call option since a share of stock can be obtained for less money through a direct purchase. If the utility stock is trading above the strike price, the call option is *in the money*; the holder can exercise the option and obtain the stock for a price below current market value. If the stock were trading at $60, the holder could obtain a share of stock for the strike price of $55, immediately sell the share back to the market at $60, and retain the difference of $5.

Investors can both buy and sell call options. A purchased call option is called a *long call*. A pure investment in a call option is a bet that the price will increase above the strike price in the future. A sold call option is called a *short call*. A pure sale of a call option is a bet that the price will not increase above the strike price in the future.

A put option is the right to sell a share of stock at a specified strike price before a future expiration date. Consider a utility with stock trading at $50 per share. Now, consider a put option with a strike price of $45 with an expiration of 1 year. Before the option expires, the holder can exercise the option and sell a share of stock for $45. If the utility stock is trading above the strike price, it does not make sense to exercise the put option since a share of stock can be sold for more money in the market. If the utility stock is trading below the strike price, the put option is in the money; the holder can exercise the option and sell the stock for a price above current market value. If the stock were trading at $40, the holder could purchase a share of stock on the market for $40, immediately exercise the option and sell the share for $45, and retain the difference of $5.

Like call options, investors can both buy and sell put options. A purchased put option is called a *long put*. A pure investment in a put option is a bet that the price will decrease below the strike price in the future. A sold call option is called a *short put*. A pure sale of a put option is a bet that the price will not decrease below the strike price in the future.

There are two common ways to treat the expiration date of an option. An *American option* allows the option to be exercised at any date up to and including the expiration date. A *European option* only allows the option to be exercised on the expiration date.

A mathematical way to value financial options is the famous *Black–Scholes Option Pricing Model*, which provides a closed-form equation for the price of a call option. The Black–Scholes equation is easily computed, and is mathematically expressed as follows:

$$P_{call} = P_{stock} \cdot N(d_1) - P_{strike} \cdot e^{-r \cdot t} \cdot N(d_2)$$

$$d_1 = \frac{1}{\sigma\sqrt{t}} \left(\ln\left(\frac{P_{stock}}{P_{strike}}\right) + t \cdot \left(r + \frac{\sigma^2}{2}\right) \right)$$

$$d_2 = d_1 - \sigma\sqrt{t}$$

where
   P is the price
   t is the time until option expiration
   r is the risk-free interest rate
   σ is the volatility of underlying stock (standard deviation as a % of mean)
   N is the cumulative standard normal distribution function

Black–Scholes allows the price of a call option to be computed directly. The price of a put option is calculated by first computing the price of the corresponding call option and then using the put-call parity relationship as follows:

$$P_{put} = P_{call} + PV(Cash) - P_{stock}$$

where PV(Cash) is the present value of cash, equal to the value of the underlying asset, invested at the risk-free rate for the term of the option.

22.5 Financial Ratios

Since companies come in many sizes, it is often difficult to compare financial performance using raw accounting numbers. For example, it is difficult to know whether a company with $100 million in earnings is doing better than a company with $500 million in earnings without knowing the size each company. It is equally difficult to compare the stock price of different companies without knowing the total number of issued shares. To help with the interpretation of accounting numbers, financial analysts often use financial ratios.

There are a large number of financial ratios. Some are commonly used and others are rarely used. This section does not attempt to be comprehensive and address every obscure ratio that will occasionally be encountered. Rather, it focuses on the more commonly seen ratios.

Profitability ratios reflect how much money a company is making compared to some measure of company size. There are many different profitability ratios using different measures of profit for the numerator and different size measures for the denominator. Some of the more common profitability ratios are now presented:

$$\text{Operating profit margin} = \frac{\text{Operating income}}{\text{Operating revenue}}$$

Operating profit margin, often referred to as *operating margin*, is the average amount of profit made per sale considering all operating expenses. If operating margin is positive, the company is fundamentally profitable not considering nonoperational expenses such as interest payments, taxes, and other possible nonoperational items:

$$\text{Net profit margin} = \frac{\text{Net income}}{\text{Total revenue}}$$

Net profit margin, often referred to as *net margin*, is the average amount of profit made per sale considering all revenue sources and all expenses. Net margin represents the amount of profit that is left for common shareholders, and can be either kept as retained earnings or distributed as dividends. Net profit margin is best used to compare companies within the same industry. Some healthy companies can have very low net margins but very high revenues (e.g., Wal-Mart). Other healthy companies can have very high net margins but relatively low revenues (e.g., Rolex):

$$\text{Return on assets (ROA)} = \frac{\text{Net income}}{\text{Assets}}$$

Return on assets (ROA) is a measure of how effectively a company is utilizing its assets to generate profits for its shareholders. Since assets are a good measure of the overall amount that has been invested in a utility, ROA provides a good measure of profitability normalized by size. The problem with ROA is that it normalizes profits available for shareholders by investments made by both shareholders and lenders (recall that assets are equal to liabilities plus owner's equity). Therefore, a highly leveraged company will have a relatively low ROA (since it has high interest payments), even though shareholder returns will be higher precisely due to this leverage. Because of this problem, ROA should be used with caution when comparing companies with differing capital structures:

$$\text{Return on capital employed (ROCE)} = \frac{\text{EBIT}}{\text{Assets} - \text{Current liabilities}}$$

Return on capital employed (ROCE, pronounced "Rocky") is a measure of how efficiently a company is utilizing net invested capital to generate profits for all stakeholders (i.e., interest, taxes, and dividends). Of the ratios examined so far, ROCE is probably the best for comparing the profitability of different companies. It is somewhat insensitive to capital structure and tax situation, since it includes both interest payments and taxes in the numerator. It is also a good measure of capital efficiency, since the denominator subtracts current liabilities (which reduce capital requirements) from assets:

$$\text{Return on equity (ROE)} = \frac{\text{Net income}}{\text{Owner's equity}}$$

Return on equity (ROE) is a measure of how efficiently a company is utilizing equity investments to generate profits for shareholders. It is the best measure of "bottom line" profitability for the book value of shareholder's equity, which represents retained earnings plus the original paid-in capital from the issuance of common stock. It is important to remember that the market value of common stock (i.e., market capitalization) will be different from the book value of common stock:

$$\text{Payout ratio} = \frac{\text{Dividend payments}}{\text{Net income}}$$

Payout ratio is a measure of how much net income is distributed as dividends (the rest being kept as retained earnings). Companies that do not distribute any dividends will have a payout ratio of zero. Companies with a negative net income that still distribute dividends will have, confusingly, a negative payout ratio. Investors expect mature companies to have stable net incomes and stable payout ratios. Since the stock price of a company is based on expected future dividends, payout ratio is an important ratio that investors consider when valuing companies.

A common criticism of profitability ratios is that they are based on accounting values and not market values. Analysts and investors in agreement with this criticism tend to use market ratios, which are ratios based wholly or partly on values determined by the market:

$$\text{Price-to-earnings (P/E)} = \frac{\text{Stock price} \times \text{Shares}}{\text{Net income}}$$

Price-to-earnings (P/E, pronounced "P to E," or "Pee Eee Ratio") is a measure of how the share price of a stock compared to per-share earnings. Equivalently, it compares market capitalization (share price times shares outstanding) to net income. Price-to-earnings is one of the most important financial indicators of a company, especially when compared to industry peers. A high P/E indicates a high stock price, most likely due to strong financial performance and high earnings growth expectations. A low P/E indicates a low stock price, most likely due to weak financial performance and low earnings growth expectations. Low P/E ratios may indicate a bargain, since very low P/E stocks have historically outperformed very high P/E stocks:

$$\text{Price-to-book (P/B)} = \frac{\text{Stock price} \times \text{Shares}}{\text{Owner's equity}}$$

Price-to-book (P/B) is a measure of how market capitalization compares to owner's equity as shown on the balance sheet. Market capitalization is equal to the trading price of common stock multiplied by shares outstanding. Owner's equity is equal to paid-in capital plus retained earnings. A price-to-book ratio greater than one shows that the market values the company more than what investors have provided in equity:

$$\text{Dividend yield} = \frac{\text{Dividends}}{\text{Stock price} \times \text{Shares}} \times 100\%$$

Dividend yield is a measure of dividends per share compared to stock price. For example, if a company's stock is trading at $50 per share and it distributes $5 per share in dividends, the dividend yield is 10%. Dividend yield is an important measure for mature companies with stable dividend payments. This is certainly true for utilities, where many utility stock owners count on dividend payments to supplement other sources of personal income:

$$\text{Earnings per share (EPS)} = \frac{\text{Net income}}{\text{Number of common shares}}$$

Earnings per share (EPS) is, as its name implies, the amount of net earnings divided by the number of outstanding common shares. Earnings per share is difficult to compare across companies since the number of common shares outstanding may not be proportional to company size. Regardless, it is common for earnings targets and earnings reports to be reported on a per-share basis. It is also common for earnings results to be stated relative to the target, such as "the utility beat its quarterly earnings target by two cents a share."

22.6 Asset Management

Asset management is a business approach designed to align the management of asset-related spending to corporate goals. The objective is to make all infrastructure-related decisions according to a single set of stakeholder-driven criteria. The outcome is a set of spending decisions capable of delivering the greatest stakeholder value from the investment dollars available.

Asset management is a well-defined term that has a long history in the financial community. When an investment banker is asked about asset management services, the answer is very specific and will be similar to the following: financial asset management is a process to make financial investment decisions so that returns are maximized while satisfying risk tolerance and other investor requirements.

Power system assets are different from financial assets. They deteriorate with age. They require periodic inspection and maintenance. They are part of an integrated system. Once installed, they cannot easily be taken out of service and sold. The list goes on. With all of these differences, it is expected that power system asset management will not be identical to financial asset management, and will likely be more complicated. With this point in mind, the definition of power system asset management becomes the following:

Power system asset management—Making data-driven power system investment decisions so that life cycle costs are minimized while satisfying performance, risk tolerance, budget, and other operational requirements.

Stated simply, asset management is a corporate strategy that seeks to balance performance, cost, and risk. Achieving this balance requires the alignment of corporate goals, management decisions, and engineering decisions. It also requires the corporate culture, business processes, and information systems capable of making rigorous and consistent spending decisions based on asset-level data. The result is a multi-year investment plan that maximizes shareholder value while meeting all performance, cost, and risk constraints. With these points in mind, the goals of asset management become the following: (1) balance cost, performance, and risk; (2) align spending decisions with corporate objectives; and (3) base spending decisions on asset-level data.

It is becoming more common for asset management companies to separate the asset management function from asset ownership and asset operations. The asset owner is responsible for setting financial, technical, and risk criteria. The asset manager is responsible for translating these criteria into a multi-year asset plan. The asset service provider is responsible for executing these decisions and providing feedback on actual cost and benefits. This decoupled structure allows each asset function to have focus: owners on corporate strategy, asset managers on planning and budgeting, and service providers on operational excellence.

Asset management is also about process. Instead of a hierarchical organization where decisions and budgets follow the chain of command into functional silos, asset management is a single process that links asset owners, asset managers, and asset service providers in a manner that allows all spending decisions to be aligned with corporate objectives and supported by asset data.

A conceptual diagram of an asset management organization is shown in Figure 22.6. The center circle represents primary functions. The outer ring represents the asset management process that links the primary functions. The primary inputs for the asset manager are corporate objectives from the asset owner and data from the asset service provider. The asset manager is then responsible for developing a

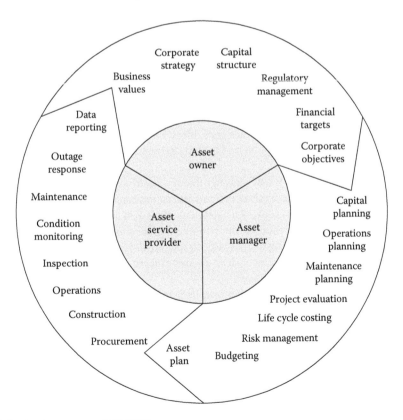

FIGURE 22.6 Asset management structure.

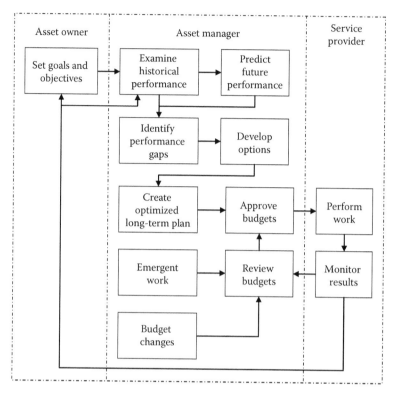

FIGURE 22.7 Example asset management process.

multi-year asset plan that is able to achieve all corporate objectives for the least life cycle cost. Once a plan is developed, the asset manager translates the short-term portion of the asset plan into a budget and work packages that are contracted out to the asset service provider.

Figure 22.7 shows a typical asset management system from an organizational process perspective. The asset manager is given high level goals and objectives from the asset owner. The asset manager then examines current and potential performance gaps, identifies an optimal long-term asset plan, creates budgets based on the short-term component of the long-term plan, and contracts with the service provider to perform the work.

In many ways, an asset manager is a hybrid engineer and businessperson. An asset manager is able to understand the business objectives of power systems and communicate with senior management using business language and business concepts. An asset manager is also able to understand the engineering issues of the power system and communicate with operational staff using engineering language and engineering concepts. An asset manager takes pride in spending as little as possible so that profits can be maximized. An asset manager also takes pride in good engineering solutions that result in the achievement of all technical performance objectives.

A large amount of spending decisions will always be made during everyday operations. It is therefore important for all power engineers to shift their mind-set and become familiar with basic business concepts and jargon. Asset managers will need to thoroughly learn these business skills and will apply them extensively. But in order to achieve the next level in business performance, all power engineers should learn the essentials of business and supplement their engineering prowess with business acumen.

IV

Power Electronics

R. Mark Nelms

R. Mark Nelms is a professor and chair of the Department of Electrical and Computer Engineering at Auburn University, Auburn, Alabama, where he has taught electric circuit analysis for over 25 years. He received his BEE (1980) and MS (1982) in electrical engineering from Auburn University and his PhD in electrical engineering from Virginia Polytechnic Institute and State University, Blacksburg, Virginia in 1987. He is a member of the American Society for Engineering Education and the Institute of Electrical and Electronics Engineers, Inc. (IEEE). His awards include being named an IEEE Fellow "for technical leadership and contributions to applied power electronics" in 2004 and the William F. Walker Merit Teaching Award from the Samuel Ginn College of Engineering at Auburn University in 2004. His research interests are in power electronics, energy conversion, and power systems. In addition, he is a registered professional engineer in the State of Alabama.

23

Power Semiconductor Devices

Kaushik
Rajashekara
Rolls-Royce Corporation

Z. John Shen
*University of
Central Florida*

The modern age of power electronics began with the introduction of thyristors in the late 1950s. Now there are several types of power devices available for high-power and high-frequency applications. The most notable power devices are gate turn-off thyristors (GTOs), power bipolar junction transistors (BJTs), power MOSFETs, insulated-gate bipolar transistors (IGBTs), and integrated gate-commutated thyristors (IGCT). Power semiconductor devices are the most important functional elements in all power conversion applications. The power devices are mainly used as switches to convert power from one form to another. They are widely used in utility power applications such as power quality conditioning, renewable power source integration, high-voltage DC transmission (HVDC), and flexible AC transmission systems (FACTS) as well as motor drives, uninterrupted power supplies, power supplies, induction heating, and many other power conversion applications. A review of the basic characteristics of these power devices is presented in this section.

23.1 Thyristor and Triac

The thyristor, also called a silicon-controlled rectifier (SCR), is basically a four-layer three-junction *pnpn* device. It has three terminals: anode, cathode, and gate. The device is turned on by applying a short pulse across the gate and cathode. Once the device turns on, the gate loses its control to turn off the device. The turn-off is achieved by applying a reverse voltage across the anode and cathode. The thyristor symbol and its volt-ampere characteristics are shown in Figure 23.1. There are basically two classifications of thyristors: phase-controlled thyristors and fast-switching thyristors. The difference between a phase-controlled and fast-switching thyristor is the low turn-off time (on the order of a few microseconds) for the latter. The phase-controlled thyristors are slow type and are used in natural commutation (or phase-controlled) applications. Fast-switching thyristors are used in forced commutation applications such as DC–DC choppers and DC–AC inverters. The phase-controlled thyristors are

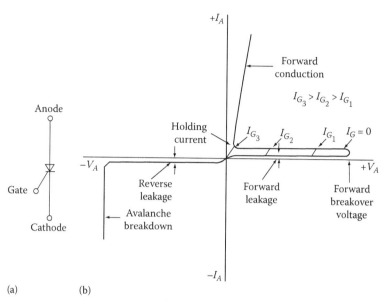

(a)   (b)

FIGURE 23.1 (a) Thyristor symbol and (b) volt-ampere characteristics. (From Bose, B.K., *Modern Power Electronics: Evaluation, Technology, and Applications*, IEEE Press, New York, p. 5. © 1992.)

turned off by forcing the current to zero using an external commutation circuit. This requires additional commutating components, thus resulting in additional losses in the inverter.

Thyristors are highly rugged devices in terms of transient currents, *di/dt,* and *dv/dt* capability. The forward voltage drop in thyristors is about 1.5–2 V, and even at a current level of several thousand amperes. Thyristors are available up to a current rating of 7000 A and a voltage rating of 10 kV, providing the highest power rating among all power semiconductor devices. A high-power thyristor is typically fabricated on an entire silicon wafer up to 150 mm in diameter, and housed in a press pack package with very low thermal resistance and low electrical parasitic impedance. When assembled in series and parallel connections, thyristors can deliver much higher voltage and current capabilities. One such example is the thyristor valves used in HVDC converters that handle a voltage level of several hundreds of kV. A thyristor's high-voltage and current capability can be attributed to the very high concentration of excess electrons and holes in the semiconductor switch in its forward conduction state or the so-called conductivity modulation. A semiconductor device is categorized as a *bipolar* type when its conduction current is made of both electron and hole currents. A thyristor is a *bipolar* type power semiconductor device with a very high level of conductivity modulation. However due to the nature of the excess electron and hole plasma, it takes a relatively long time to turn on and off the thyristor, resulting in a high switching power loss at high operating frequencies. Because of this, the maximum switching frequencies possible using thyristors are limited in comparison with other power devices considered in this section.

Thyristors have I^2t withstand capability and can be protected by fuses. The nonrepetitive surge current capability for thyristors is about 10 times their rated root mean square (rms) current. They must be protected by snubber networks for *dv/dt* and *di/dt* effects. If the specified *dv/dt* is exceeded, thyristors may start conducting without applying a gate pulse. In DC-to-AC conversion applications, it is necessary to use an antiparallel diode of similar rating across each main thyristor.

A triac is functionally a pair of converter-grade thyristors connected in antiparallel. The triac symbol and volt-ampere characteristics are shown in Figure 23.2. Because of the integration, the triac has poor reapplied *dv/dt*, poor gate current sensitivity at turn-on, and longer turn-off time. Triacs are mainly used in phase control applications such as in AC regulators for lighting and fan control and in solid-state AC relays.

easondoo

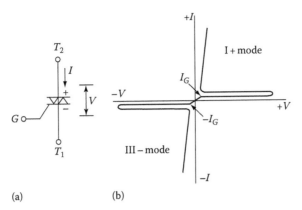

FIGURE 23.2 (a) Triac symbol and (b) volt-ampere characteristics. (From Bose, B.K., *Modern Power Electronics: Evaluation, Technology, and Applications*, IEEE Press, New York, p. 5. © 1992.)

23.2 Gate Turn-Off Thyristor

The GTO is a power thyristor that can be turned on by a short pulse of gate current and turned off by a reverse gate pulse. This reverse gate current amplitude is dependent on the anode current to be turned off. Hence there is no need for an external commutation circuit to turn it off. Because turn-off is provided by bypassing carriers directly to the gate circuit, its turn-off time is short, thus giving it more capability for high-frequency operation than thyristors. The GTO symbol and turn-off characteristics are shown in Figure 23.3.

GTOs have the I^2t withstand capability and hence can be protected by semiconductor fuses. For reliable operation of GTOs, the critical aspects are proper design of the gate turn-off circuit and the snubber circuit. A GTO has a poor turn-off current gain of the order of four to five. For example, a 2000 A peak current GTO may require as high as 500 A of reverse gate current. Also, a GTO has the tendency to latch at temperatures above 125°C. GTOs are available up to about 6500 V, 4000 A.

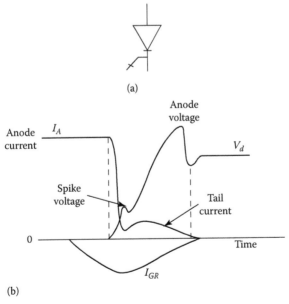

FIGURE 23.3 (a) GTO symbol and (b) turn-off characteristics. (From Bose, B.K., *Modern Power Electronics: Evaluation, Technology, and Applications*, IEEE Press, New York, p. 5. © 1992.)

23.3 Reverse-Conducting Thyristor and Asymmetrical Silicon-Controlled Rectifier

Normally in inverter applications, a diode in antiparallel is connected to the thyristor for commutation/freewheeling purposes. In reverse-conducting thyristors (RCTs), the diode is integrated with a fast-switching thyristor in a single silicon wafer or chip. Thus, the number of power devices could be reduced. This integration brings forth a substantial improvement of the static and dynamic characteristics as well as its overall circuit performance.

The RCTs are designed mainly for converter applications such as traction drives or wind power converters. The antiparallel diode limits the reverse voltage across the thyristor to 1–2 V. Also, because of the reverse recovery behavior of the diodes, the thyristor may see very high reapplied *dv/dt* when the diode recovers from its reverse voltage. This necessitates the use of large RC snubber networks to suppress voltage transients. As the range of application of thyristors and diodes extends into higher frequencies, their reverse recovery charge becomes increasingly important. High reverse recovery charge results in high power dissipation during switching.

The asymmetrical silicon-controlled rectifier (ASCR) has similar forward blocking capability to a fast-switching thyristor, but it has a much lower reverse blocking capability. It has an on-state voltage drop of about 25% less than a fast-switching thyristor of a similar rating. The ASCR features a fast turn-off time; thus, it can work at a higher frequency than an SCR. Since the turn-off time is down by a factor of nearly 2, the size of the commutating components can be halved. Because of this, the switching losses will also be low.

Gate-assisted turn-off techniques are used to even further reduce the turn-off time of an ASCR. The application of a negative voltage to the gate during turn-off helps to evacuate stored charge in the device and aids the recovery mechanisms. This will, in effect, reduce the turn-off time by a factor of up to 2 over the conventional device.

23.4 Power Bipolar Junction Transistor

Power BJTs are used in applications ranging from a few to several hundred kilowatts and switching frequencies up to about 10 kHz. They may be used in utility power systems as relay gear drivers or in auxiliary power supplies. Power BJTs used in power conversion applications are generally *npn* type. The power transistor is turned on by supplying sufficient base current, and this base drive has to be maintained throughout its conduction period. It is turned off by removing the base drive and making the base voltage slightly negative (within $-V_{BE(\max)}$). The saturation voltage of the device is normally 0.5–2.5 V and increases as the current increases. Hence, the on-state losses increase more than proportionately with current. A power BJT is a *bipolar* type power semiconductor device but with a conductivity modulation level below that of a thyristor. Because of relatively larger switching times, the switching loss significantly increases with switching frequency for power BJTs. Power BJTs can block only forward voltages. The reverse peak voltage rating of these devices is as low as 5–10 V. Power BJTs do not have I^2t withstand capability. In other words, they can absorb only very little energy before device failure. Therefore, they cannot be protected by semiconductor fuses, and thus an electronic protection method has to be used.

To eliminate high base current requirements, Darlington configurations are commonly used. They are available in monolithic or in isolated packages. The basic Darlington configuration is shown schematically in Figure 23.4. The Darlington configuration presents a specific advantage in that it can considerably increase the current switched by the transistor for a given base drive. The $V_{CE(sat)}$ for the Darlington is generally more than that of a single transistor of similar rating with corresponding increase in on-state power loss. During switching, the reverse-biased collector junction may show hot-spot breakdown effects that are specified by reverse-bias safe operating area (RBSOA) and forward-bias safe operating area (FBSOA). Modern devices with highly interdigitated emitter base geometry force more uniform current distribution and therefore considerably improve the secondary breakdown performance. Normally, a well-designed switching aid network constrains the device operation well within the SOAs.

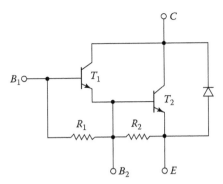

FIGURE 23.4 Two-stage Darlington transistor with bypass diode. (From Bose, B.K., *Modern Power Electronics: Evaluation, Technology, and Applications*, IEEE Press, New York, p. 6. © 1992.)

23.5 Power MOSFET

Power MOSFETs are widely used in power supplies, relay drivers, or other auxiliary circuits in utility power systems. Due to the power rating limitation, power MOSFETs are not used in the main power conversion stage of the utility power systems. They are essentially voltage-driven rather than current-driven devices, unlike bipolar transistors.

The gate of a MOSFET is isolated electrically from the source by a layer of silicon oxide. Hence, the gate drive circuit is simple and power loss in the gate control circuit is practically negligible. Although in steady state the gate draws virtually no current, this is not so under transient conditions. The gate-to-source and gate-to-drain capacitances have to be charged and discharged appropriately to obtain the desired switching speed, and the drive circuit must have a sufficiently low output impedance to supply the required charging and discharging currents. The circuit symbol of a power MOSFET is shown in Figure 23.5.

Power MOSFETs are majority carrier devices, and there is no minority carrier storage time. Hence, they have exceptionally fast rise and fall times. They are essentially resistive devices when turned on, while bipolar transistors present a more or less constant $V_{CE(sat)}$ over the normal operating range. Power dissipation in MOSFETs is $Id^2 R_{DS(on)}$, and in bipolars it is $I_C V_{CE(sat)}$. At low currents, therefore, a power MOSFET may have a lower conduction loss than a comparable bipolar device, but at higher currents, the conduction loss will exceed that of bipolars. Also, the $R_{DS(on)}$ increases with temperature.

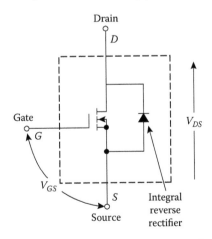

FIGURE 23.5 Power MOSFET circuit symbol. (From Bose, B.K., *Modern Power Electronics: Evaluation, Technology, and Applications*, IEEE Press, New York, p. 7. © 1992.)

An important feature of a power MOSFET is the absence of a detrimental secondary breakdown effect, which is present in a bipolar transistor, and as a result, it has an extremely rugged switching performance. In MOSFETs, $R_{DS(on)}$ increases with temperature, and thus the current is automatically diverted away from the hot spot. The drain body junction appears as an antiparallel diode between source and drain. Thus, power MOSFETs will not support voltage in the reverse direction. Although this inverse diode is relatively fast, it is slow by comparison with the MOSFET. Recent devices have the diode recovery time as low as 100 ns. Since MOSFETs cannot be protected by fuses, an electronic protection technique has to be used.

With the advancement in MOS technology, ruggedized MOSFETs are replacing the conventional MOSFETs. The need to ruggedize power MOSFETs is related to device reliability. If a MOSFET is operating within its specification range at all times, its chances for failing catastrophically are minimal. However, if its absolute maximum rating is exceeded, failure probability increases dramatically. Under actual operating conditions, a MOSFET may be subjected to transients—either externally from the power bus supplying the circuit or from the circuit itself due, for example, to inductive kicks going beyond the absolute maximum ratings. Such conditions are likely in almost every application, and in most cases are beyond a designer's control. Rugged devices are made to be more tolerant for over-voltage transients. Ruggedness is the ability of a MOSFET to operate in an environment of dynamic electrical stresses, without activating any of the parasitic BJTs. The rugged device can withstand higher levels of diode recovery *dv/dt* and static *dv/dt*.

23.6 Insulated-Gate Bipolar Transistor

The IGBT is a switching transistor controlled by a voltage applied to its gate terminal. Device operation and structure are similar to that of a power MOSFET. The principal difference is that the IGBT relies on conductivity modulation to reduce on-state conduction losses. The IGBT has high input impedance and fast turn-on speed like a MOSFET, but exhibits an on-state voltage drop and current-carrying capability comparable to that of a bipolar transistor while switching much faster. IGBTs have a clear advantage over MOSFETs in high-voltage applications where conduction losses must be minimized. Since the initial introduction of the IGBT into market in the mid 1980s, the semiconductor industry has made great technological advancement in improving device performance and reducing fabrication cost. IGBTs are available up to a current rating of several hundred amperes and a voltage rating of 6.5 kV, and widely used in power quality conditioning, renewable power source integration, motor drives, uninterrupted power supplies, power supplies, induction heating and many other power conversion applications.

Like the power MOSFET, the IGBT does not exhibit the secondary breakdown phenomenon common to bipolar transistors. However, care should be taken not to exceed the maximum power dissipation and specified maximum junction temperature of the device under all conditions for guaranteed reliable operation. The on-state voltage of the IGBT is heavily dependent on the gate voltage. To obtain a low on-state voltage, a sufficiently high gate voltage must be applied.

In general, IGBTs can be classified as punch-through (PT) and nonpunch-through (NPT) structures, as shown in Figure 23.6. In the PT IGBT, an N+ buffer layer is normally introduced between the P+ substrate and the N− epitaxial layer, so that the whole N− drift region is depleted when the device is blocking the off-state voltage, and the electrical field shape inside the N− drift region is close to a rectangular shape. Because a shorter N− region can be used in the punch-through IGBT, a better trade-off between the forward voltage drop and turn-off time can be achieved. PT IGBTs are available up to about 1200 V.

High-voltage IGBTs are realized through a NPT process. The devices are built on an N− float-zone (FZ) wafer substrate which serves as the N− base drift region. Experimental NPT IGBTs of up to about 6.5 kV have been reported in the literature. NPT IGBTs are more robust than PT IGBTs, particularly under short circuit conditions. But NPT IGBTs have a higher forward voltage drop than the PT IGBTs.

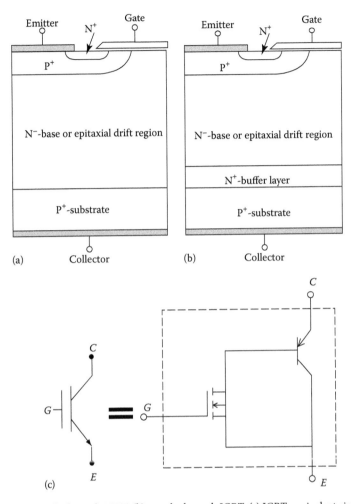

FIGURE 23.6 (a) Nonpunch-through IGBT, (b) punch-through IGBT, (c) IGBT equivalent circuit.

The PT IGBTs cannot be as easily paralleled as MOSFETs. The factors that inhibit current sharing of parallel-connected IGBTs are (1) on-state current unbalance, caused by $V_{CE(sat)}$ distribution and main circuit wiring resistance distribution, and (2) current unbalance at turn-on and turn-off, caused by the switching time difference of the parallel connected devices and circuit wiring inductance distribution. The NPT IGBTs can be paralleled because of their positive temperature coefficient property.

Trench gate structures were introduced into the IGBT to reduce the MOS channel resistance and the so-called JFET resistance in the mid 1990s. It was found that the IGBT structure with a deep trench gate and relatively wide cell pitch considerably enhances the electron injection efficiency at the emitter side by minimizing the back injection of holes into the P-base. The large cell pitch in this type of IGBT leads to a slight penalty in the MOS channel resistance, but is more than compensated by the significant reduction of the N-base conduction resistance. Alternatively, an N-type layer can be added under the P-base to block the hole back injection current and enhance electron injection. These two carrier enhancement approaches are widely adopted in the state-of-the-art IGBTs, providing an optimum stored excess carrier profile similar to that of an ideal PiN diode and hence a reduction in forward voltage. Thin wafer punch-through IGBTs represent the latest development in IGBT technology which features a short N-base and an N-buffer layer. The difference between this new type of IGBT and the conventional PT-IGBT is that the former uses a shallow P-emitter and no carrier lifetime killing techniques very much like the conventional NPT-IGBT.

23.7 Integrated Gate-Commutated Thyristor

The IGCT is a gate-controlled turn-off switch which turns off like a transistor but conducts like a thyristor with the lowest conduction losses. The fundamental difference between a conventional GTO and the IGCT lies in the very low inductance gate driver system, inherent to the IGCT. Ultra-low inductance has been achieved through the development of a new optimized housing and integrated gate driver concept. In the IGCT, the entire anode current is commutated from cathode to gate in a very short time. Since the *npn*-transistor is inactive thereafter, the *pnp*-transistor is deprived of base current, and turns off. The IGCT, therefore, turns off in a transistor mode, thus completely eliminating the current filamentation problems inherent in conventional GTOs. Additional advantages are a dramatic reduction of storage time to less than 2 μs, and a reduction in fall time to around 1 μs. Thus, the series connection of IGCTs is facilitated, compared to GTOs, by the very low dispersion associated with these times. The key to achieving "hard" turn-off of this nature is the duration of the time interval in which it occurs. The user thus only needs to connect the device to a 28–40 V power supply and optical fiber for on/off control. Because of the topology in which it is used, the IGCT produces negligible turn-on losses. In addition, IGCT enables operation at higher frequencies than formerly obtained by other high-power semiconductor devices (Figure 23.7).

The IGCTs are available up to about 6500 V, 4000 A, and is the power switching device of choice for demanding high-power applications such as medium voltage drives (MVD), wind power converters, STATCOMs, dynamic voltage restorers (DVR), solid state breakers, DC traction line boosters, traction power compensators, and interties.

The current and future power semiconductor devices developmental direction is shown in Figure 23.8. High-temperature operation capability and low forward voltage drop operation can be obtained if silicon is replaced by silicon carbide, gallium nitride, or other wide bandgap (WBG) semiconductor materials for producing power devices. The silicon carbide has a higher bandgap than silicon. Hence, higher breakdown voltage devices could be developed. Silicon carbide devices have excellent switching characteristics and stable blocking voltages at higher temperatures. But WBG devices are still in the very early stages of development.

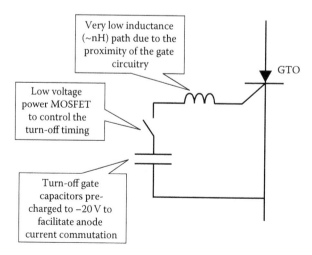

FIGURE 23.7   Equivalent circuit for IGCT.

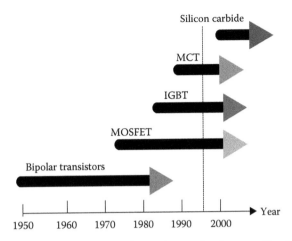

FIGURE 23.8 Current and future power semiconductor devices development direction. (From Huang, A.Q., Recent developments of power semiconductor devices, in *VPEC Seminar Proceedings*, Blacksburg, VA, pp. 1–9, September 1995. With permission.)

References

Bose, B.K., *Modern Power Electronics: Evaluation, Technology, and Applications*, IEEE Press, New York, 1992.

Huang, A.Q., Recent developments of power semiconductor devices, in *VPEC Seminar Proceedings*, Blacksburg, VA, September 1995, pp. 1–9.

Mohan, N. and Undeland, T., *Power Electronics: Converters, Applications, and Design*, John Wiley & Sons, New York, 1995.

Wojslawowicz, J., Ruggedized transistors emerging as power MOSFET standard-bearers, *Power Technics Magazine*, January 1988, 29–32.

Further Information

Bird, B.M. and King, K.G., *An Introduction to Power Electronics*, Wiley-Interscience, New York, 1984.

Sittig, R. and Roggwiller, P., *Semiconductor Devices for Power Conditioning*, Plenum, New York, 1982.

Williams, B.W., *Power Electronics, Devices, Drivers and Applications*, John Wiley, New York, 1987.

24

Uncontrolled and Controlled Rectifiers

Mahesh M. Swamy
*Yaskawa America
Incorporated*

24.1 Introduction

Rectifiers are electronic circuits that convert bidirectional voltage to unidirectional voltage. This process can be accomplished either by mechanical means employing commutators or by static means employing semiconductor devices. Static means of rectification is more efficient and reliable compared to rotating commutators. In this chapter, we will discuss rectification of electric power for industrial and commercial use. In other words, we will not be discussing small signal rectification that generally involves low power and low voltage signals. Static power rectifiers can be classified into two broad groups. They are (a) uncontrolled rectifiers and (b) controlled rectifiers. Uncontrolled rectifiers make use of power semiconductor diodes while controlled rectifiers make use of thyristors (SCRs), gate turn Off thyristors (GTOs), and MOSFET controlled thyristors (MCTs).

Rectifiers, in general, are widely used in power electronics to rectify single-phase as well as three-phase voltages. DC power supplies used in computers, consumer electronics, and a host of other applications typically make use of single-phase rectifiers. Industrial applications include, but are not limited to industrial drives, metal extraction processes, industrial heating, power generation and transmission, etc. Most industrial applications of large power rating typically employ three-phase rectification processes.

Uncontrolled rectifiers in single-phase and in three-phase circuits will be discussed in Section 24.2. Section 24.3 will focus on controlled rectifiers. Application issues regarding uncontrolled and controlled rectifiers will be briefly discussed within each section. Section 24.4 will conclude this chapter.

24.2 Uncontrolled Rectifiers

Simplest uncontrolled rectifier use can be found in single-phase circuits. There are two types of uncontrolled rectification. They are (a) half-wave rectification and (b) full-wave rectification. Half-wave and full-wave rectification techniques have been used in single-phase as well as in three-phase circuits. As mentioned earlier, uncontrolled rectifiers make use of diodes. Diodes are two terminal semiconductor devices that allow flow of current in only one direction. The two terminals of a diode are known as the anode and the cathode.

24.2.1 Mechanics of Diode Conduction [4]

Anode is formed when a pure semiconductor material, typically Silicon, is doped with impurities that have fewer valence electrons than Silicon. Silicon has an atomic number of 14, which according to *Bohr's* atomic model means that the *K and L* shells are completely filled by 10 electrons and the remaining 4 electrons occupy the *M shell*. The *M shell* can hold a maximum of 18 electrons. In a Silicon crystal, every atom is bound to four other atoms, which are placed at the corners of a regular tetrahedron. The bonding, which involves sharing of a valence electron with a neighboring atom, is known as covalent bonding. When a group-3 element (typically boron, aluminum, gallium, and indium) is doped into the Silicon lattice structure, three of the four covalent bonds are made. However, one bonding site is vacant in the Silicon lattice structure. This creates vacancies or *holes* in the semiconductor. In the presence of either thermal field or an electrical field, electrons from neighboring lattice or from external agency tend to migrate to fill this vacancy. The vacancy or *hole* can also be said to move toward the approaching electron thereby creating a mobile hole and hence current flow. Such a semiconductor material is also known as lightly doped semiconductor material or *p type*. Similarly, cathode is formed when Silicon is doped with impurities that have higher valence electrons than Silicon. This would mean elements belonging to group 5. Typical doping impurities of this group are phosphorus, arsenic, and antimony. When a group 5 element is doped into the Silicon lattice structure, it oversatisfies the covalent bonding sites available in the Silicon lattice structure, creating excess or loose electrons in the valence shell. In the presence of either thermal field or an electrical field, these loose electrons easily get detached from the lattice structure and are free to conduct electricity. Such a semiconductor material is also known as heavily doped semiconductor material or *n type*.

The structure of the final doped crystal even after the addition of *acceptor* impurities (group 3) or *donor* impurities (group 5) remains electrically neutral. The available electrons balance the net positive charge and there is no charge imbalance.

When a *p*-type material is joined with an *n*-type material, *p-n* junction is formed. Some loose electrons from the *n*-type material migrate to fill the holes in the *p*-type material and some holes in the *p*-type migrate to meet with the loose electrons in the *n*-type material. Such a movement causes the *p*-type structure to develop a slight negative charge and the *n*-type structure to develop some positive charge. This slight positive and negative charges in the *n*-type and *p*-type areas, respectively, prevent further migration of electrons from *n*-type to *p*-type and holes from *p*-type to *n*-type areas. In other words, an energy barrier is automatically created due to the movement of charges within the crystalline lattice structure. Keep in mind that the combined material is still electrically neutral and no charge imbalance exists.

When a positive potential greater than the barrier potential is applied across the *p-n* junction, then electrons from the *n*-type area migrate to combine with the holes in the *p*-type area and vice versa. The *p-n* junction is said to be *forward biased*. Movement of charge particles constitutes current flow. Current is said to flow from the anode to the cathode when the potential at the anode is higher than the potential at the cathode by a minimum threshold voltage also known as the junction barrier voltage. The magnitude of current flow is high when the externally applied positive potential across the *p-n* junction is high.

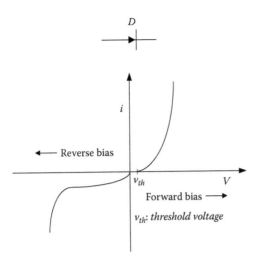

FIGURE 24.1 Typical *v-i* characteristic of a semiconductor diode and its symbol.

When the polarity of the applied voltage across the *p-n* junction is reversed compared to the case described previously, then the flow of current ceases. The holes in the *p*-type area move away from the *n*-type area and the electrons in the *n*-type area move away from the *p*-type area. The *p-n* junction is said to be *reverse biased*. In fact, the holes in the *p*-type area get attracted to the negative external potential and similarly the electrons in the *n*-type area get attracted to the positive external potential. This creates a depletion region at the *p-n* junction and there is almost no charge carriers flowing in the depletion region. This phenomenon brings us to an important observation that a *p-n* junction can be utilized to force current to flow only in one direction depending on the polarity of the applied voltage across it. Such a semiconductor device is known as a *diode*. Electrical circuits employing diodes to convert ac voltage to unidirectional voltage across a load are known as *rectifiers*. The voltage-current characteristic of a typical power semiconductor diode along with its symbol is shown in Figure 24.1.

24.2.2 Single-Phase Half-Wave Rectifier Circuits

A single-phase half-wave rectifier circuit employs one diode. Typical circuit, which makes use of a half-wave rectifier, is shown in Figure 24.2. A single-phase ac source is applied across the primary windings of a transformer. The secondary of the transformer consists of a diode and a resistive load. This is typical since many consumer electronic items including computers utilize single-phase power.

Typically, the primary side is connected to a single-phase ac source, which could be 120 V, 60 Hz, 100 V, 50 Hz, 220 V, 50 Hz, or any other utility source. The secondary side voltage is generally stepped

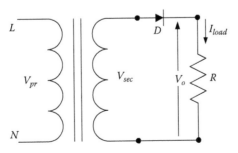

FIGURE 24.2 Electrical schematic of a single-phase half-wave rectifier circuit feeding a resistive load. Average output voltage is V_o.

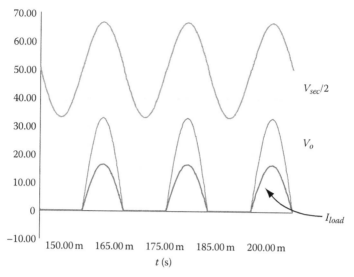

FIGURE 24.3 Typical waveforms at various points in the circuit of Figure 24.2.

down and rectified to achieve low dc voltage for consumer applications. The secondary voltage, the voltage across the load resistor and the current through it is shown in Figure 24.3. For a purely resistive load, $V_o = \sqrt{2} * V_{sec}/\pi$.

When the voltage across the anode-cathode of diode $D1$ in Figure 24.2 goes negative, the diode does not conduct and no voltage appears across the load resistor R. The current through R follows the voltage across it. The value of the secondary voltage is chosen to be 24Vac and the value of R is chosen to be 2Ω. Since, only one-half of the input voltage waveform is allowed to pass onto the output, such a rectifier is known as *half-wave* rectifier. The voltage ripple across the load resistor is rather large and in typical power supplies, such ripples are unacceptable. The current through the load is discontinuous and the current through the secondary of the transformer is unidirectional. The ac component in the secondary of the transformer is balanced by a corresponding ac component in the primary winding. However, the dc component in the secondary does not induce any voltage on the primary side and hence it is not compensated for. This dc current component through the transformer secondary can cause the transformer to saturate and is not advisable for large power applications. In order to smoothen the output voltage across the load resistor R and to make the load current continuous, a smoothing filter circuit comprising either a large dc capacitor or a combination of a series inductor and shunt dc capacitor is employed. Such a circuit is shown in Figure 24.4 and the resulting waveforms are shown in Figure 24.5.

It is interesting to see that the voltage across the load resistor has very little ripple and the current through it is smooth. However, the value of the filter components employed is large and is generally not economically feasible. For example, in order to get a voltage waveform across the load resistor R that has

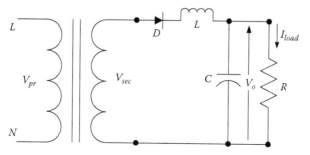

FIGURE 24.4 Modified circuit of Figure 24.2 employing smoothing filters.

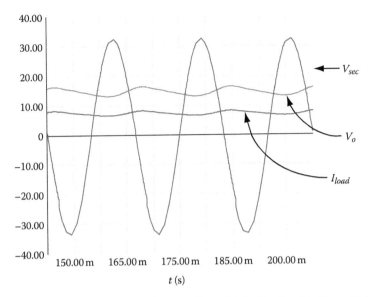

FIGURE 24.5 Voltage across load resistor R and current through it for the circuit in Figure 24.4.

less than 25% peak-peak voltage ripple, the value of inductance that had to be used is 3 mH and the value of the capacitor is 20,000 μF. Increasing the value of inductance does reduce the peak-to-peak ripple across the load. However, the voltage drop across the inductor increases and the average voltage across the resistor reduces significantly.

24.2.3 Full-Wave Rectifiers [1]

In order to improve the performance without adding bulky filter components, it is a good practice to employ full-wave rectifiers. The circuit in Figure 24.2 can be easily modified into a full-wave rectifier. The transformer is changed from a single secondary winding to a center-tapped secondary winding. Two diodes are now employed instead of one. The new circuit is shown in Figure 24.6.

The waveforms for the circuit of Figure 24.6 are shown in Figure 24.7. The voltage across the load resistor is a full-wave rectified voltage. The current has subtle discontinuities but it can be improved by employing smaller size filters. For a purely resistive load, $V_o = 2 * \sqrt{2} * V_{sec}/\pi$.

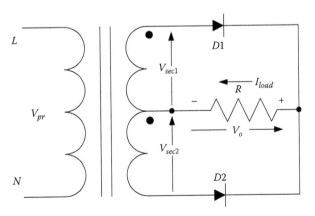

FIGURE 24.6 Electrical schematic of a single-phase full-wave rectifier circuit. Average output voltage is V_o. For balanced operation, $V_{sec1} = V_{sec2} = V_{sec}$.

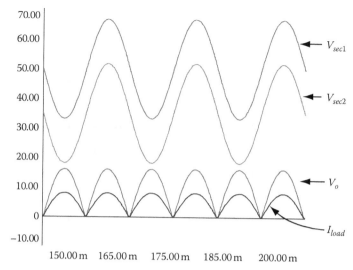

FIGURE 24.7   Typical waveforms at various points in Figure 24.6. Vscale: 1/10 of actual value.

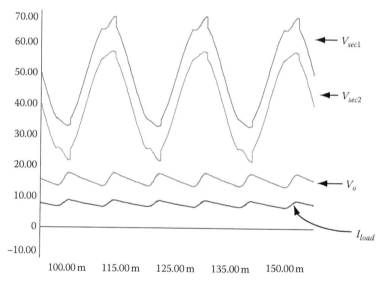

FIGURE 24.8   Voltage across the load resistor and current through it with only capacitor filter. Vscale: 1/10 of actual value.

A typical filter for the circuit of Figure 24.6 may include only a capacitor. The waveforms are shown in Figure 24.8. Adding a capacitor filter distorts the secondary voltage due to discontinuous and pulsating current flowing through the secondary windings.

Another way of reducing the size of the filter components is to increase the frequency of the supply. In many power supply applications similar to the one used in computers, a high frequency ac supply is achieved by means of switching. The high frequency ac is then level translated via a ferrite core transformer with multiple secondary windings. The secondary voltages are then rectified employing a simple circuit as that shown in Figure 24.4 or 24.6 with much smaller filters. The resulting voltage across the load resistor is then maintained to have a peak-peak voltage ripple of less than 1%.

Full-wave rectification can be achieved without the use of center-tap transformers. Such circuits make use of four diodes in single-phase circuits and six diodes in three-phase circuits. The circuit configuration is typically referred to as the H-bridge circuit. A single-phase full-wave H-bridge topology

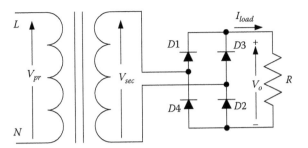

FIGURE 24.9 Schematic representation of a single-phase full-wave H-bridge rectifier.

is shown in Figure 24.9. The main difference between the circuit topology shown in Figures 24.6 and 24.9 is that the H-bridge circuit employs four diodes while the topology of Figure 24.6 utilizes only two diodes. The *VA* rating of the center-tap transformer, however, is higher than that of the transformer needed for H-bridge rectifier. The voltage and current stresses in the diodes of Figure 24.6 is also greater than that in the diodes of Figure 24.9.

In order to highlight the basic difference in the two topologies, it will be interesting to compare the component ratings for the same power output. To make the comparison easy, let both topologies employ very large filter inductors such that the current through *R* is constant and ripple free. Let this current through *R* be denoted by I_{dc}. Let the power being supplied to the load be denoted by P_{dc}. The output power and the load current are then related by the following expression:

$$P_{dc} = I_{dc}^2 * R \tag{24.1}$$

The rms current flowing through the first secondary winding in the topology in Figure 24.6 will be $I_{dc}/\sqrt{2}$. This is because the current through a secondary winding flows only when the corresponding diode is forward biased. This means that the current through the secondary winding will flow only for one-half cycle. If the voltage at the secondary is assumed *V*, the *VA* rating of the secondary winding of the transformer in Figure 24.6 will be given by

$$VA_1 = V * \frac{I_{dc}}{\sqrt{2}}$$

$$VA_2 = V * \frac{I_{dc}}{\sqrt{2}} \tag{24.2}$$

$$VA = VA_1 + VA_2 = \sqrt{2} * V * I_{dc}$$

This is the secondary-side *VA* rating for the transformer shown in Figure 24.6.

For the isolation transformer shown in Figure 24.9, let the secondary voltage be *V* and the load current be of a constant value, I_{dc}. Since, in the topology of Figure 24.9, the secondary winding carries the current I_{dc} when diodes *D*1 and *D*2 conduct as well as when diodes *D*3 and *D*4 conduct, the rms value of the secondary winding current is I_{dc}. Hence, the *VA* rating of the secondary winding of the transformer shown in Figure 24.9 is given by $VA = V * I_{dc}$ and is less than that needed for the topology of Figure 24.6. Note that the primary *VA* rating for both cases remain the same since in both cases, the power being transferred from the source to the load remains the same.

When diode *D*2 in the circuit of Figure 24.6 conducts, the secondary voltage of the second winding V_{sec2} (=*V*) appears at the cathode of diode *D*1. The voltage being blocked by the diode *D*1 can thus reach two times the peak secondary voltage (=$2 * V_{pk}$) (Figure 24.7). In the topology of Figure 24.9, when diodes *D*1 and *D*2 conduct, the voltage V_{sec} (=*V*), which is same as V_{sec2}, appears across *D*3 and *D*4 in series.

This means that the diodes have to withstand only one time the peak of the secondary voltage, V_{pk}. The rms value of the current flowing through the diodes in both topologies is the same. Hence, from the diode voltage rating as well as from the secondary VA rating points of view, the topology of Figure 24.9 is better than that of Figure 24.6. Further, the topology in Figure 24.9 can be directly connected to a single-phase ac source since it does not need a center-tapped transformer. The voltage waveform across the load is similar to that shown in Figures 24.7 and 24.8.

In many industrial applications, the topology shown in Figure 24.9 is used along with a dc filter capacitor to smoothen the ripples across the load resistor. The load resistor is simply a representative of the active part of the load. It could be an inverter system or a high-frequency resonant link. In any case, the diode rectifier bridge would see a representative load resistor. For the same output power and the same peak-to-peak ripple voltage across the load, the dc filter capacitor in case of single-phase source will need to be much larger compared to that for a three-phase source connected to a six diode rectifier bridge circuit.

When the rectified power is large, it is advisable to add a dc link inductor. This can reduce the size of the capacitor to some extent and reduce the current ripple through the load. When the rectifier is turned ON initially with the capacitor at zero voltage, large amplitude of charging current will flow into the filter capacitor through a pair of conducting diodes. The diodes $D1 \sim D4$ should be rated to handle this large surge current. In order to limit the high inrush current, it is a normal practice to add a charging resistor in series with the filter capacitor. The charging resistor limits the inrush current but creates a significant power loss if it is left in the circuit under normal operation. Typically, a contactor is used to short-circuit the charging resistor after the capacitor is charged to a desired level. The resistor is thus electrically non-functional during normal operating conditions. A typical arrangement showing a single-phase full-wave H-bridge rectifier system for an inverter application is shown in Figure 24.10. The charging current at time of turn ON is shown in a simulated waveform in Figure 24.11. Note that the contacts across the soft-charge resistor are closed under normal operation. The contacts across the soft-charge resistor are initiated by various means. The coil for the contacts could be powered from the input ac supply and a timer or it could be powered ON by a logic controller that senses the level of voltage across the dc bus capacitor or senses the rate of change in voltage across the dc bus capacitor. A simulated waveform depicting the inrush with and without a soft-charge resistor is shown in Figure 24.11a and b, respectively.

The value of soft-charge resistor used is $6\,\Omega$. The dc bus capacitor is about $1200\,\mu F$. To show typical operation, at start-up, there is no load and resistor R represents only the bleed-off resistor of approximately $4.7\,k\Omega$ present across the capacitor. The peak value of the charging current for this case is observed to be approximately 50 A.

Next, simulation result is given in Figure 24.11b for the case with no soft charge resists. The current is limited by the system impedance and by the diode forward resistance. The peak current is seen to be about 175 A for the same parameters as that chosen for the simulation shown in Figure 24.11a. Note that the dc bus capacitor gets charged to a value almost twice the input peak voltage and this is due to the resonance type of condition set in by the input impedance of the ac source that includes the leakage inductance of the input transformer. The high voltage across the dc bus takes a long time to bleed off through the

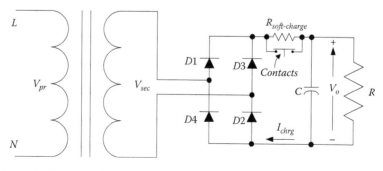

FIGURE 24.10 Single-phase H-bridge circuit with soft-charge resistor-contactor arrangement.

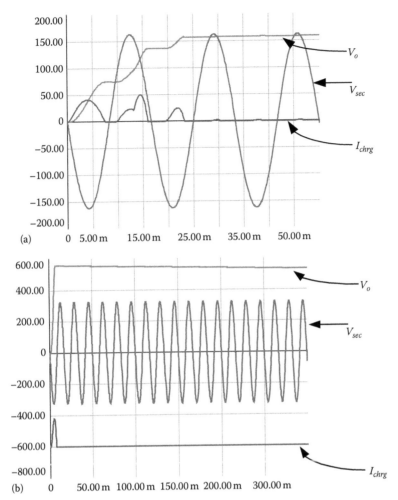

FIGURE 24.11 (a) Charging current and voltage across capacitor for the circuit of Figure 24.10. Vscale: 1/2 of actual value. (b) Charging current and voltage across capacitor for no soft-charge resistor. Vscale: Actual voltage.

bleed resistor. The high voltage can damage the inverter IGBTs if the load is an inverter and hence charging the dc bus capacitor without soft-charge resistors is not recommended and should always be avoided.

For larger power applications, typically above 1.5 kW, it is advisable to use a power source with a higher voltage. In some applications, two of the three phases of a three-phase power system are used as the source powering the rectifier of Figure 24.9. The line-line voltage could be either 240 or 480 Vac. Under those circumstances, a load of up to 4 kW can be powered using single-phase supply before adopting a full three-phase H-bridge configuration. Beyond 4 kW, the size of the capacitor becomes too large to achieve a peak-peak voltage ripple of less than 10%. Further, the strain that the pulsating current will put on the power system can become unacceptable. Hence, it is advisable to employ three-phase rectifier configurations for loads typically rated at 4 kW and higher.

24.2.4 Three-Phase Rectifiers (Half Wave and Full Wave)

Similar to the single-phase case, there exist half-wave and full-wave three-phase rectifier circuits. Again, similar to the single-phase case, the half-wave rectifier in the three-phase case also has dc components in the source current. The source has to be large enough to handle this. It is thus not advisable to use three-phase half-wave rectifier topology for large power applications. The three-phase half-wave

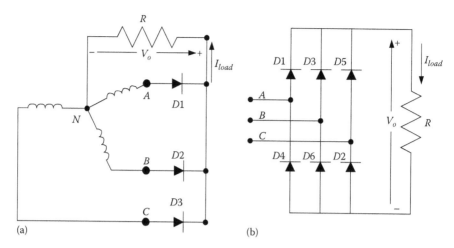

FIGURE 24.12 Schematic representation of three-phase rectifier configurations. (a) Half-wave rectifier needing a neutral point, *N*; and (b) full-wave rectifier.

rectifier employs three diodes while the full-wave H-bridge configuration employs six diodes. Typical three-phase half-wave and full-wave topologies are shown in Figure 24.12.

In the half-wave rectifier shown in Figure 24.12a, the shape of the output voltage and current through the resistive load is dictated by the instantaneous value of the source voltages, *A*, *B*, and *C*. These source voltages are phase shifted in time by 120 electrical degrees, which corresponds to approximately 5.55 ms for a 60 Hz system. This means that if *A* phase reaches its peak value at time t_1, *B* phase will achieve its peak 120 electrical degrees later (t_1 + 5.55 ms), and *C* will achieve its peak 120 electrical degrees later than *B* (t_1 + 5.55 ms + 5.55 ms). Since all three phases are connected to the same output resistor *R*, the phase that provides the highest instantaneous voltage is the phase that appears across *R*. In other words, the phase with the highest instantaneous voltage reverse biases the diodes of the other two phases and prevents them from conducting which consequently prevents those phase voltages from appearing across *R*. Since a particular phase is connected to only one diode in Figure 24.12a, only three pulses, each of 120° duration, appears across the load resistor, *R*. Typical output voltage across *R* and current through it for the circuit of Figure 24.12a is shown in Figure 24.13a.

A similar explanation can be provided to explain the voltage waveform across a purely resistive load in the case of a three-phase full-wave rectifier shown in Figure 24.12b. The output voltage that appears across *R* is the highest instantaneous line-line voltage and not simply the phase voltage. Since there are six such intervals, each of 60 electrical degrees duration in a given cycle, the output voltage waveform will have six humps in one cycle—Figure 24.13b. Since a phase is connected to two diodes (diode pair), each phase conducts current out and into itself thereby eliminating dc component in one complete cycle.

The waveform for a three-phase full-wave rectifier with a purely resistive load is shown in Figure 24.13b. Note that the number of humps in Figure 24.13a is only three in one ac cycle while the number of humps in Figure 24.13b is six in one ac cycle. Further, the peak-to-peak ripple in the voltage as well as in the current is significantly lower in the full-bridge configuration compared to the half-bridge configuration.

In both the configurations shown in Figure 24.12, the load current does not become discontinuous due to three-phase operation. Comparing this to the single-phase half-wave and full-wave rectifier, the output voltage ripple is much lower in three-phase rectifier systems compared to single-phase rectifier systems. Hence, with the use of moderately sized filters on the DC side, three-phase full-wave rectifiers can be operated at hundred to thousands of kilowatts. The only limitation would be the size of the diodes used and power system harmonics, which will be discussed next. Since there are six humps in the output voltage waveform per electrical cycle, the three-phase full-wave rectifier shown in Figure 24.12b is also known as a six-pulse rectifier system.

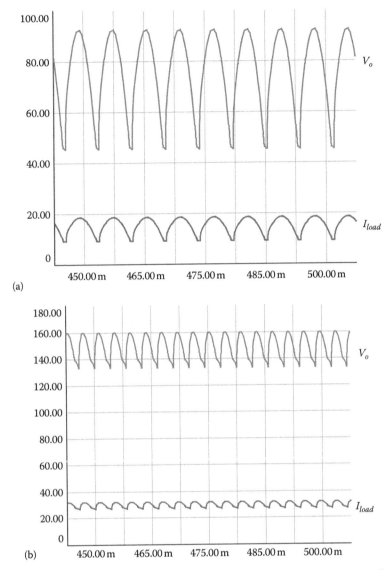

FIGURE 24.13 (a) Typical output voltage across a purely resistive load and current through it for the half-wave rectifier shown in Figure 24.12a. Vscale: 0.5 of actual value. (b) Typical output voltage across a purely resistive load and current through it for the full-wave rectifier shown in Figure 24.12b. Vscale: 0.5 of actual value.

24.2.5 Average Output Voltage

In order to evaluate the average value of the output voltage for the two rectifiers shown in Figure 24.12, the output voltages in Figure 24.13a and b have to be integrated over a cycle. For the circuit shown in Figure 24.12a, the integration yields the following:

$$V_O = \frac{3}{2\pi} \int_{\pi/6}^{5\pi/6} \sqrt{2}\, V_{L-N} \sin(\omega t)\, d(\omega t)$$

$$V_O = \frac{3 * \sqrt{3} * \sqrt{2} * V_{L-N}}{2 * \pi}$$

(24.3)

Similar operation can be performed to obtain the average output voltage for the circuit shown in Figure 24.12b. This yields

$$V_O = \frac{3}{\pi} \int_{\pi/3}^{2\pi/3} \sqrt{2}\, V_{L-L} \sin(\omega t)\, d(\omega t)$$

$$V_O = \frac{3 * \sqrt{2} * V_{L-L}}{\pi} = \frac{3 * \sqrt{2} * \sqrt{3} * V_{L-N}}{\pi}$$

(24.4)

In other words, the average output voltage for the circuit in Figure 24.12b is twice that of the circuit in Figure 24.12a.

24.2.6 Influence of Three-Phase Rectification on the Power System

Events over the last several years have focused attention on certain types of loads on the electrical system that result in power quality problems for the user and utility alike. Equipment which has become common place in most facilities including computer power supplies, solid state lighting ballast, adjustable speed drives (ASDs), and uninterruptible power supplies (UPSs) are examples of nonlinear loads. Nonlinear loads are loads in which the current waveform does not have a linear relationship with the voltage waveform. In other words, if the input voltage to the load is sinusoidal and the current is nonsinusoidal then such loads will be classified as nonlinear loads because of the nonlinear relationship between voltage and current. Nonlinear loads generate voltage and current harmonics, which can have adverse effects on equipment that are used to deliver electrical energy to them. Examples of power delivery equipment include power system transformers, feeders, circuit breakers, etc. Power delivery equipments are subject to higher heating losses due to harmonic currents consumed by nonlinear loads to which they are connected. Harmonics can have a detrimental effect on emergency or standby power generators, telephones and other sensitive electrical equipment. When reactive power compensation in the form of passive power factor improving capacitors is used with nonlinear loads, resonance conditions can occur that may result in even higher levels of harmonic voltage and current distortion thereby causing equipment failure, disruption of power service, and fire hazards in extreme conditions.

The electrical environment has absorbed most of these problems in the past. However, the problem has now reached a magnitude where Europe, the United States, and other countries have proposed standards to responsibly engineer systems considering the electrical environment. IEEE 519-1992 [5] and IEC 555 have evolved to become a common requirement cited when specifying equipment on newly engineered projects. Various harmonic filtering techniques have been developed to meet these specifications. The present IEEE 519-1992 document establishes acceptable levels of harmonics (voltage and current) that can be introduced into the incoming feeders by commercial and industrial users. Where there may have been little cooperation previously from manufacturers to meet such specifications, the adoption of IEEE 519-1992 and other similar world standards now attract the attention of everyone.

24.2.7 Why VFDs Generate Harmonics?

The current waveform at the inputs of a variable frequency drive (VFD) is not continuous. It has multiple zero crossings in one electrical cycle. The dc bus capacitor draws charging current only when it is discharged due to the motor load. The charging current flows into the capacitor when the input rectifier is forward biased, which occurs when the instantaneous input voltage is higher than the dc voltage across the dc bus capacitor. The pulsed current drawn by the dc bus capacitor is rich in harmonics because it is discontinuous as seen in Figure 24.1.

The voltage harmonics generated by VFDs are due to the flat-topping effect caused by weak ac source charging the dc bus capacitor without any intervening impedance. The distorted voltage waveform gives rise to voltage harmonics and this is of a more important concern than current harmonics. The reason is simple. Voltage is shared by all loads and it affects all loads connected in an electrical system. Current distortion has a local effect and pertains to only that circuit that is feeding the nonlinear load. Hence, connecting nonlinear loads like VFDs to a weak ac system requires more careful consideration than otherwise.

The discontinuous, nonsinusoidal current waveform as shown in Figure 24.1 can be mathematically represented by sinusoidal patterns of different frequencies having a certain amplitude and phase relationship among each other. By adding these components, the original waveform can be reconstructed. The amplitude of the various sinusoidal components that need to be used to reconstruct a given nonsinusoidal waveform is expressed in terms of a mathematical expression called total harmonic distortion. The total harmonic current distortion is defined as: $THD_I = \sqrt{\sum_{n=2}^{n=\infty} I_n^2 / I_1}$; I_1 is the rms value of the fundamental component of current; and I_n is the rms value of the nth harmonic component of current.

The reason for doing this is that it is easier to evaluate the heating effect caused by continuous sinusoidal waveforms of different frequencies and corresponding amplitudes than to estimate the heating effects caused by discontinuous nonsinusoidal waveforms.

The order of current harmonics produced by a semiconductor converter during normal operation is termed as characteristic harmonics. In a three-phase, six-pulse converter with *no dc bus capacitor*, the characteristic harmonics are nontriplen odd harmonics (e.g., 5th, 7th, 11th, etc.). In general, the characteristic harmonics generated by a semiconductor converter is given by

$$h = kq \pm 1 \tag{24.5}$$

where
  h is the order of harmonics
  k is any integer
  q is the pulse number of the semiconductor converter (six for a six-pulse converter)

When operating a six-pulse rectifier-inverter system with a dc bus capacitor (voltage source inverter or VSI), harmonics of orders other than those given by the previous equation may be observed. Such harmonics are called noncharacteristic harmonics. Though of lower magnitude, these also contribute to the overall harmonic distortion of the input current. The per unit value of the characteristic harmonics present in the theoretical current waveform at the input of the semiconductor converter is given by $1/h$ where h is the order of the harmonics. In practice, the observed per unit value of the harmonics is much greater than $1/h$. This is because the theoretical current waveform is a rectangular pattern made up of equal positive and negative halves, each occupying 120 electrical degrees. The pulsed discontinuous waveform observed commonly at the input of a VFD (Figure 24.14) digresses greatly from the theoretical waveform.

24.2.8 Harmonic Limit Calculations Based on IEEE 519-1992 [5]

The IEEE 519-1992 relies strongly on the definition of the point of common coupling or PCC. The PCC from the power utility point of view will usually be the point where power comes into the establishment (i.e., point of metering). However, the IEEE 519-1992 document also suggests that, "within an industrial plant, the point of common coupling (PCC) is the point between the nonlinear load

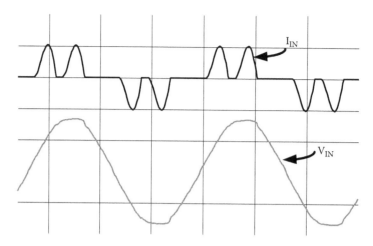

FIGURE 24.14   Typical pulsed current waveform as seen at input of a VFD.

and other loads" [1]. This suggestion is crucial since many plant managers and building supervisors feel that it is equally if not more important to keep the harmonic levels at or below acceptable guidelines within their facility. In view of the many recently reported problems associated with harmonics within industrial plants [2], it is important to recognize the need for mitigating harmonics at the point where the nonlinear load is connected to the power system. This approach would minimize harmonic problems, thereby reducing costly downtime and improving the life of electrical equipment. If harmonic mitigation is accomplished for individual nonlinear loads or a group of nonlinear loads collectively, then the total harmonics at the point of the utility connection will in most cases meet or better the IEEE recommended guidelines. In view of this, it is becoming increasingly common for project engineers and consultants to require nonlinear equipment suppliers to adopt the procedure outlined in IEEE 519-1992 to mitigate the harmonics to acceptable levels at the point of the offending equipment. For this to be interpreted equally by different suppliers, the intended PCC must be identified. If not defined clearly, many suppliers of nonlinear loads would likely adopt the PCC to be at the utility metering point, which would not benefit the plant or the building but rather the utility.

Having established that it is beneficial to adopt the PCC to be the point where the nonlinear load connects to the power system, the next step is to establish the short circuit ratio. Short circuit ratio calculations are key in establishing the allowable current harmonic distortion levels. For calculating the short circuit ratio, the available short circuit current at the input terminals of the nonlinear load needs to be determined. If the short circuit value available at the low-voltage side of the utility transformer feeding the establishment (building) is known, and the cable and other series impedances in the electrical circuit between the low-voltage side of the transformer and the input to the nonlinear load are known, then the available short circuit at the nonlinear load can be calculated. *In practice, it is common to assume the same short circuit current level as at the secondary of the utility transformer feeding the nonlinear load.* The next step is to compute the fundamental value of the rated input current into the nonlinear load. In case the nonlinear load is a VFD operating an induction motor, the NEC amp rating for induction motors can be used to obtain this number. NEC amps are fundamental amps that a motor draws when connected directly to the utility supply. An example is presented here to recap the previous procedure.

A 100 hp ASD/motor combination connected to a 480-V system being fed from a 1500-kVA, 3 ph transformer with an impedance of 4% is required to meet IEEE 519-1992 at its input terminals. The rated current of the transformer is: $1500 * 1000/(\sqrt{3} * 480)$, which is calculated to be 1804.2 A. The short circuit current available at the secondary of the transformer is equal to the rated current divided by the per unit impedance of the transformer. This is calculated to be: 45,105.5 A. The short circuit ratio, which is defined as the ratio of the short circuit current at the PCC to the fundamental value of the nonlinear current, is computed next. NEC rating for a 100 hp, 460 V induction motor is 124 A. Assuming that the short circuit current at the VFD input is practically the same as that at the secondary of the utility transformer, the short circuit ratio is calculated to be: 45,105.5/124 which equals 363.75. On referring to the IEEE 519-1992 Table 10.3 [1], the short circuit ratio falls in the 100–1000 category. For this ratio, the total demand distortion (TDD) at the point of VFD connection to the power system network is recommended to be 15% or less. For reference, Table 10.3 [5] is reproduced hereafter.

Current Distortion Limits for General Distribution Systems (120 through 69,000 V)

| | Maximum Harmonic Current Distortion in Percent of I_L | | | | | |
| | Individual Harmonic Order (Odd Harmonics) | | | | | |
I_{sc}/I_L	<11	$11 \leq h \leq 17$	$17 \leq h \leq 23$	$23 \leq h \leq 35$	$35 \leq h$	*TDD*
<20[a]	4.0	2.0	1.5	0.6	0.3	5.0
20 < 50	7.0	3.5	2.5	1.0	0.5	8.0
50 < 100	10.0	4.5	4.0	1.5	0.7	12.0
100 < 1000	12.0	5.5	5.0	2.0	1.0	15.0
>1000	15.0	7.0	6.0	2.5	1.4	20.0

Even harmonics are limited to 25% of the odd harmonic limits mentioned earlier.

[a] All power generation equipment is limited to these values of current distortion, regardless of actual I_{sc}/I_L, where I_{sc} is the maximum short circuit current at PCC and I_L is the maximum demand load current (fundamental frequency) at PCC. *TDD* is *total demand distortion* and is defined as the harmonic current distortion in percent of maximum demand load current. The maximum demand current interval could be either a 15 min or a 30 min interval.

24.2.9 Harmonic Mitigating Techniques

Various techniques of improving the input current waveform are discussed hereafter. The intent of all techniques is to make the input current more continuous to reduce the overall current harmonic distortion. The different techniques can be classified into three broad categories:

1. Passive techniques
2. Active techniques
3. Hybrid technique—combination of passive and active techniques

There are three different options in the passive configuration. They are as follows:

1. Addition of inductive impedance—line reactors and/or dc link chokes
2. Capacitor based harmonic filters—tuned as well as broadband type
3. Multi-pulse techniques (12 pulse, 18 pulse, etc.)

This chapter will concentrate only on the passive techniques. Each of the mentioned passive option will be briefly discussed with their relative advantages and disadvantages.

24.2.10 Addition of Inductive Impedance

24.2.10.1 Three-Phase Line Reactors

A line reactor makes the current waveform less discontinuous resulting in lower current harmonics. Since the reactor impedance increases with frequency, it offers larger impedance to the flow of higher order harmonic currents.

On knowing the input reactance value, the expected current harmonic distortion can be estimated. A table illustrating the expected input current harmonics for various amounts of input reactance is shown as follows:

Percent Harmonics Versus Total Line Impedance

Harmonic	Total Input Impedance							
	3%	4%	5%	6%	7%	8%	9%	10%
5th	40	34	32	30	28	26	24	23
7th	16	13	12	11	10	9	8.3	7.5
11th	7.3	6.3	5.8	5.2	5	4.3	4.2	4
13th	4.9	4.2	3.9	3.6	3.3	3.15	3	2.8
17th	3	2.4	2.2	2.1	0.9	0.7	0.5	0.4
19th	2.2	2	0.8	0.7	0.4	0.3	0.25	0.2
%THID	44.13	37.31	34.96	32.65	30.35	28.04	25.92	24.68
True rms	1.09	1.07	1.06	1.05	1.05	1.04	1.03	1.03

Input reactance is determined by the series combination of impedance of the ac reactor, input transformer (building/plant incoming-feed transformer), and power cable. By adding all the inductive reactance upstream, the effective line impedance can be determined and the expected harmonic current distortion can be estimated from the previous chart. The effective impedance value in % is based on the actual loading and is

$$Z_{eff(pu)} = \frac{\sqrt{3} * 2 * \pi * f * L_T * I_{act(fnd.)}}{V_{L-L}} * 100 \qquad (24.6)$$

where

$I_{act(fnd.)}$ is the fundamental value of the actual load current
V_{L-L} is the line-line voltage

L_T is the total inductance of all reactance upstream. The effective impedance of the transformer as seen from the nonlinear load is

$$Z_{eff,x-mer} = \frac{Z_{x-mer} * I_{act(fnd.)}}{I_r} \qquad (24.7)$$

where

$Z_{eff,x-mer}$ is the effective impedance of the transformer as viewed from the nonlinear load end
Z_{x-mer} is the nameplate impedance of the transformer
I_r is the nameplate rated current of the transformer

The reactor also electrically separates the dc bus voltage from the ac source so that the ac source is not clamped to the dc bus voltage during diode conduction. This feature reduces flat topping of the ac voltage waveform caused by many VFDs when operated with weak ac systems.

However, introducing ac inductance between the diode input terminal and the ac source causes overlap of conduction between outgoing diode and incoming diode in a three-phase diode rectifier system. The overlap phenomenon reduces the average dc bus voltage. This reduction depends on the duration of the overlap in electrical degrees, which in turn depends on the value of the intervening inductance used and the current amplitude. The duration of overlap in electrical degrees is commonly represented by μ. In order to compute the effect quantitatively, a simple model can be assumed. Assume that the line comprises inductance L in each phase. Let the dc load current be I_{dc} and let it be assumed that this current does not change during the overlap interval. The current through the incoming diode at start is zero and by the end of the overlap interval, it is I_{dc}. Based on this assumption, the relationship between current and voltage can be expressed as

$$v_{ab} = \sqrt{2} * V_{L-L} * \sin(\omega t) = 2 * L * \left(\frac{di}{dt}\right)$$

$$\sqrt{2} * V_{L-L} * \int\limits_{(\pi/3)}^{(\pi/3)+\mu} \sin(\omega t)d(t) = 2 * L * \int\limits_{0}^{I_{dc}} di \tag{24.8}$$

$$I_{dc} = \frac{\sqrt{2} * V_{L-L} * (\cos(\pi/3) - \cos(\pi/3 + \mu))}{2\omega L} = \frac{\sqrt{2} * V_{L-L} * \sin(\pi/3 + \mu/2) * \sin(\mu/2)}{\omega L}$$

For small values of overlap angle μ, $\sin(\mu/2) = \mu/2$ and $\sin(\pi/3 + (\mu/2)) = \sin(\pi/3)$. Rearranging the preceding equation yields

$$\mu = \frac{2 * \sqrt{2} * \omega L * I_{dc}}{V_{L-L} * \sqrt{3}} \tag{24.9}$$

From the preceding expression, the following observations can be made:

1. If the inductance L in the form of either external inductance or leakage inductance of transformer or lead length is large, the overlap duration will be large.
2. If the load current, I_{dc} is large, the overlap duration is large.

The average output voltage will reduce due to the overlap angle as mentioned before. In order to compute the average output voltage with a certain overlap angle, the limits of integration have to be changed. This exercise yields the following:

$$V_O = \frac{3}{\pi} \int\limits_{\mu+(\pi/3)}^{\mu+(2\pi/3)} \sqrt{2} V_{L-L} \sin(\omega t)d(\omega t) \tag{24.10}$$

$$V_O = \frac{3 * \sqrt{2} * V_{L-L} * \cos(\mu)}{\pi} = \frac{3 * \sqrt{2} * \sqrt{3} * V_{L-N} * \cos(\mu)}{\pi}$$

Thus, it can be seen that the overlap angle contributes to the reduction in the average value of the output dc bus voltage. Unfortunately, higher values of external inductive reactance, increases the overlap angle, which in turn reduces the average output voltage as seen from the previous equation.

24.2.10.2 DC Link Choke

Based on the earlier discussion, it can be noted that any inductor of adequate value placed in between the ac source and the dc bus capacitor of the VFD will help in making the input current waveform more continuous. Hence, a dc link choke, which is electrically present after the diode rectifier bridge and before the dc bus capacitor, can be used to reduce the input current harmonic distortion. The dc link choke appears to perform similar to the three-phase line inductance. However, on analyzing the behavior of the dc link choke, it can be seen that the dc link choke behaves similar to the input ac line inductor only from current distortion point of view but has a completely different influence on the average output voltage.

An important difference is that the dc link choke is after the diode rectifier block and so they do not contribute to the overlap phenomenon discussed earlier with regards to external ac input reactors. Hence, there is no dc bus voltage reduction similar to the way as experienced when ac input reactors are used. The dc link choke increases the diode conduction duration. There is a critical dc link choke inductance value, which when exceeded will result in complete 60° conduction of a diode pair. Any value of dc link inductance beyond this critical value is of no further importance and thus introducing a very large dc link inductor will have no further benefit. A larger inductance value will only be associated with a higher winding resistance and cause marginally extra voltage drop across the winding resistance, resulting in higher power loss without altering the input current distortion or the average output dc voltage significantly. The critical dc link choke that is needed to achieve complete 60° conduction is derived next.

It is assumed that the source is ideal with zero impedance. The forward voltage drop across the conducting diodes is also neglected. When no dc link inductance is used, the dc bus charges up to the peak of the input ac line and since the source is assumed ideal and the voltage drop across the diode is neglected, the average dc bus voltage remains at the peak of the input ac line-line voltage even under loaded condition. This assumption is not true and corrections for this will be made later. The critical inductance is that value that will result in the average dc bus voltage to drop from its peak value to the average 3 ph rectified value. The voltage across the dc link inductor absorbs the difference. Mathematically, the following is true:

$$L_{cr} * \frac{\Delta i}{\Delta t} = V_m - V_{3\text{-}ph\text{-}avg} = V_m - \frac{3 * V_m}{\pi}$$

$$L_{cr} = \frac{\pi - 3}{\pi} * V_m * \frac{\Delta t}{\Delta i} = \frac{\pi - 3}{\pi} * V_m * \frac{T/6}{I_{dc}}$$

(24.11)

where

L_{cr} is the critical value of the dc link choke
I_{dc} is the load current

The change in current Δi in Equation 24.11 is the difference from noload condition to rated load condition. Hence, Δi is the rated average dc link current that flows continuous for a 60° conduction interval. T is the period of the input ac supply. In Equation 24.11, it should be pointed out that if continuous current conduction for 60° duration is desired at a lower value of dc load current, a dc link inductor of a large value is required.

From the expression for the critical dc link inductance, it is seen that the value depends on the load condition, frequency of the input ac supply and the peak value of the input ac line-line voltage, V_m. It is also interesting to note that the value of the critical dc link inductor for a 240 V system for the same load is 1/4th the value for a 480 V system.

24.2.10.3 AC Reactor versus DC Link Choke

Often, application engineers are asked the question regarding the choice between input ac reactors and dc link chokes. As mentioned, the dc link choke appears to have an advantage over the input ac reactor from size, cost, and performance points of view. The ac reactor increases overlap duration of the diodes and reduces the overall dc link voltage, whereas the dc link choke does not have such effect on the dc bus. Variation of dc bus voltage with respect to different values of ac reactor and dc link choke have been studied and plotted in Figure 24.15. The dc bus voltage is seen to keep reducing for increasing values of input ac reactor, while it remains flat when dc link choke of value greater than the critical value is employed.

In spite of some shortcomings, the ac reactor does provide reasonable attenuation to switching noise riding on the input ac line due to disturbance in the input ac source. The reason for the disturbance could be switching in and out of power factor correcting capacitors, often done by utilities to improve the overall power factor or the disturbance caused by lightning and other weather related events. Hence, an optimal solution would be to use a small percent input ac reactor (0.02 pu or lower) and a standard dc link choke, as shown in Figure 24.16.

On knowing the input reactance (both ac reactor and dc choke values), the expected current harmonic distortion can be estimated. A graph illustrating the expected input current harmonics for various amounts of reactance is shown in Figure 24.17.

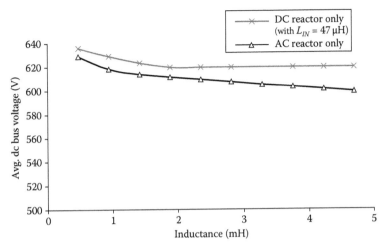

FIGURE 24.15 Variation of dc bus voltage with increasing value of ac reactor/dc link choke for a 460 V, 7.5 hp VFD.

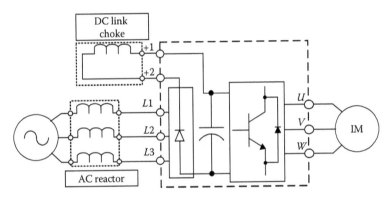

FIGURE 24.16 Commonly used inductive filters.

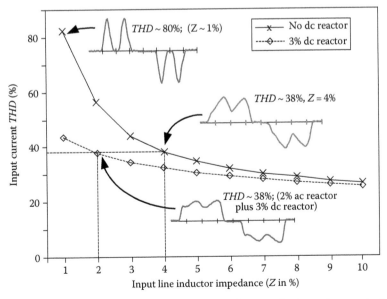

FIGURE 24.17 Performance with ac line reactor and dc link choke. Note the slope of current in the two cases.

24.2.11 Capacitor-Based Passive Filters

Passive filters consist of passive components like inductors, capacitors, and resistors arranged in a pre-determined fashion either to attenuate the flow of harmonic components through them or to shunt the harmonic component into them. Passive filters can be of many types. Some popular ones are series passive filters, shunt passive filters, and low-pass broadband passive filters. Series and shunt passive filters are effective only in a narrow proximity of the frequency at which they are tuned. Low-pass broadband passive filters have a broader bandwidth and attenuate a larger range of harmonics above their cutoff frequency.

24.2.11.1 Series Passive Filter

One way to mitigate harmonics generated by nonlinear loads is to introduce a series passive filter (Figure 24.18) in the incoming power line so that the filter offers high impedance to the flow of harmonics from the source to the nonlinear load. Since the series passive filter is tuned to a particular frequency, it offers high impedance at only its tuned frequency. Depending on the physical property of L and C chosen, typically there exists a narrow band around the tuned frequency where the impedance remains high.

Series passive filters have been used more often in 1 ph applications where it is effective in attenuating the third-harmonic component. The series pass filter is generally designed to offer low impedance at the fundamental frequency. A major drawback of this approach is that the filter components have to be designed to handle the rated load current. Further, one filter section is not adequate to attenuate the entire harmonic spectrum present in the input current of a nonlinear system. Multiple sections may be needed to achieve this, which makes it bulky and expensive.

24.2.11.2 Shunt Passive Filter

The second and more common approach is to use a shunt passive filter, as shown in Figure 24.19. The shunt passive filter is placed across the incoming line and is designed to offer very low impedance to current components corresponding to its tuned frequency. Another way of explaining the behavior of a shunt filter is to consider the energy flow from source to the nonlinear load via the shunt filter. Energy at fundamental frequency flows into the shunt passive filter and the energy at the filter's tuned frequency flows out of the shunt filter since it offers lower impedance for flow of energy at its tuned

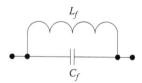

FIGURE 24.18   Single-phase representation of a series filter configuration.

FIGURE 24.19   Single-phase representation of a shunt-tuned filter configuration.

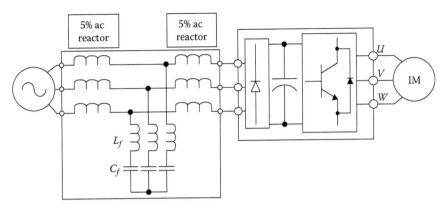

FIGURE 24.20 Fifth harmonic tuned filter with input and output reactors.

frequency compared to the source. In other words, the harmonic component needed by the nonlinear load is provided by the shunt filter rather than the ac source.

The fundamental frequency energy component flowing into the shunt filter is the reason for leading VARs and can cause overvoltage at the filter terminals. This can create problems with VFDs that are vulnerable to higher than normal voltage and under light-load condition can encounter overvoltage trips. Similar to the series tuned filter, the shunt-tuned filter is effective only at and around its tuned frequency and only one section of the filter alone is inadequate to provide for all the harmonic energy needed by a typical nonlinear load (VFD). Multiple sections are needed, which makes them bulky and expensive.

The commonly used 3 ph shunt filter sections comprise individual sections tuned for the 5th, the 7th, and perhaps a high-pass section typically tuned near the 11th harmonic. Unfortunately, if care is not taken, the shunt filter will try to provide the harmonic energy needed by all nonlinear loads connected across its terminals. In this process, it can be overloaded and be damaged if unprotected. In order to avoid import of harmonics, it is important to use series line reactors, which impede the harmonic energy flow from other sources into the shunt-tuned filter sections.

A popular type of passive filter (Figure 24.20) comprises a shunt filter tuned to the fifth harmonic along with series impedance to limit import of harmonics from other sources. One more reactor is placed in between the filter section and the VFD to further reduce the current distortion. This type of filter is bulky, expensive, inefficient, and can cause dc bus overvoltage. All passive filters are associated with circulating current that cause unnecessary power loss. Circulating current in capacitor filters causes high voltage at VFD input terminals. Passive filters can also cause system resonance. Given these disadvantages, it is best to avoid them.

The addition of extra line inductance can aggravate the overvoltage condition experienced by inverter drive systems at light-load conditions. The over voltage tolerance margin would be compromised and the VFD could be more vulnerable to fault out on over voltage thereby causing nuisance trips.

24.2.11.3 Low-Pass Broadband Filter

The low-pass broadband filter is similar to the circuit configuration of Figure 24.20. To improve the filtering performance, the inductor L_f is removed and placed in place of the 5% input ac reactor. By removing L_f from the shunt path, the filter configuration changes from tuned type to broadband type. One advantage of the low-pass broadband harmonic filter is that unlike the shunt and series type filters, the broadband filter need not be configured in multiple stages or sections to offer wide spectrum filtering. In other words, one filter section achieves the performance close to the combined effect of a fifth, seventh, and a high-pass shunt-tuned filter section. A typical broadband filter section is shown in Figure 24.21. The series inductor L_f offers high impedance to limit import and export of harmonics from and to other nonlinear loads on the system [6].

However, by removing L_f from the shunt branch and moving it to the series branch, aggravates the overvoltage problem experienced by VFDs. Autotransformers have been used in the past to address this problem.

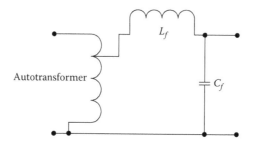

FIGURE 24.21   Broadband filter section with autotransformer.

Since the overvoltage is a function of the load current, the correction offered by autotransformer works only at one operating point at best and is inadequate to handle wide range of operating conditions. In addition, the size and cost of the total filter configuration becomes high and less appealing. The leading VAR problem is not resolved and in fact has been found to interfere with power measurement and monitoring systems. These unfavorable features are serious enough to limit use of such filters for VFD applications.

24.2.12 Multi-Pulse Techniques

As discussed earlier, the characteristic harmonics generated by a semiconductor converter is a function of the pulse number for that converter. Higher the pulse number, lower is the total harmonic distortion since the order of the characteristic harmonics shifts to a higher value. Pulse number is defined as the number of diode-pair conduction intervals that occur in one electrical cycle. In a three-phase, six-diode bridge rectifier, the number of diode-pair conduction intervals is six and such a rectifier is known as a six-pulse rectifier. By using multiple six-pulse diode rectifiers in parallel and phase shifting the input voltage to each rectifier bridge by a suitable value, multi-pulse operation can be achieved.

24.2.12.1 12-Pulse Techniques

A 12-pulse rectifier operation can be achieved by using two six-pulse rectifiers in parallel with one rectifier fed from a power source that is phase shifted with respect to the other rectifier by 30 electrical degrees. The 12-pulse rectifier will have the lowest harmonic order of 11. In other words, the fifth, and the seventh harmonic orders are theoretically nonexistent in a 12-pulse converter. Again, as mentioned in Section 24.2.7, the amplitude of the characteristic harmonic is typically proportional to the inverse of the harmonic order. In other words, the amplitude of the 11th harmonic in a 12-pulse system will be 1/11 of the fundamental component and the amplitude of the 13th harmonic will be 1/13 of the fundamental component. There are many different ways of achieving the necessary phase shift to realize 12-pulse operation. Some popular methods are

1. Three winding isolation transformer
2. Hybrid 12-pulse method
3. Autotransformer method

24.2.12.1.1 *Three Winding Isolation Transformer Method*

A three winding isolation transformer has three different sets of windings. One set of winding is typically called the primary, while the other two sets are called secondary windings. The primary winding can be connected in delta or in wye configuration. One set of secondary winding is connected in delta while the other set is connected in wye configuration. This arrangement automatically yields a 30° phase shift between the two sets of secondary windings. A traditional 12-pulse arrangement using a three winding isolation transformer is shown in Figure 24.22. The realization of 12-pulse operation in the circuit of Figure 24.19 is discussed next.

The current flowing out of the secondary windings, viewed independently, is similar to that observed in a six-pulse rectifier. However, since the voltages are phase shifted by 30 electrical degrees, the currents

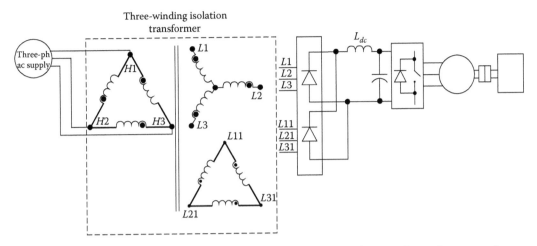

FIGURE 24.22 Typical schematic of a 12-pulse configuration using a three winding isolation transformer. DC link choke improves performance.

are also phase shifted by the same amount. In other words, if i_1 is the fundamental current through one set of secondary windings, and i_2 is the 30° phase-shifted current in the other set of secondary windings, then i_1 and i_2 can be expressed as follows:

$$i_1 = I_m * \sin(\omega t)$$

$$i_2 = I_m * \sin\left(\omega t - \frac{\pi}{6}\right)$$

(24.12)

The fifth harmonic component of the current in one of the secondary windings will be phase shifted with respect to its corresponding phase in the other set. However, it should be noted that the phase shift will get multiplied by the harmonic number as well. The fifth harmonic component in the two sets of windings can be represented as follows:

$$i_{5(1)} = I_{5m} * \sin(5 * \omega t)$$

$$i_{5(2)} = I_{5m} * \sin\left(5 * \omega t - \frac{5 * \pi}{6} - \frac{\pi}{6}\right) = -I_{5m} * \sin(5 * \omega t)$$

(24.13)

Similar expressions can be written for the seventh harmonic currents in each set of the secondary windings. From the previous expressions, it can be said that the flux pattern formed by the fifth and seventh harmonic components by one set of secondary windings are theoretically equal and opposite to the fifth and seventh harmonic flux components produced by the second set of secondary windings. Consequently, there is no fifth and seventh harmonic component reflected on to the primary windings and so the fifth and seventh harmonic components do not theoretically exist in the input ac supply feeding the primary windings.

Based on the previous explanation, it can be said that in a three winding isolation transformer arrangement, magnetic flux coupling plays an important role in assuring the elimination of low order current harmonics. Any departure from the ideal scenario assumed earlier will yield suboptimal flux cancellation and higher total current harmonic distortion. Leakage flux and the primary magnetizing flux create nonideal conditions and are responsible for the existence of noncharacteristic harmonics in the input current of a typical 12-pulse system. Minor winding imbalance between the two sets of secondary windings also contributes to suboptimal performance.

Advantages

Some important advantages of the three winding isolation transformer configuration to achieve 12-pulse operation is listed as follows:

- 12-pulse operation yields low total current harmonic distortion.
- Three winding arrangement yields isolation from the input ac source, which has been seen to offer high impedance to conducted EMI.
- It offers in-built impedance due to leakage inductance of transformer. This smoothes the input current and helps further reduce the total current harmonic distortion.
- It is ideally suited for voltage level translation. If the input is at a high voltage (3.2 or 4.16 kV), and the drive system is rated for 480 V operation, it is ideal to step down and to achieve the benefits of 12-pulse operation.

Disadvantages

In spite of its appeal, the three winding isolation transformer configuration has a few shortcomings listed as follows:

- The three winding transformer has to be rated for full power operation, which makes it bulky and expensive.
- Leakage inductance of the transformer will cause reduction in the dc bus voltage, which will require the use of taps in the primary winding to compensate for this drop. Addition of taps will increase cost.
- Due to minor winding mismatch, leakage flux, and nontrivial magnetizing current, the total current harmonic distortion can be higher than expected.
- Needs the VFD to be equipped with two six-pulse rectifiers that increases the cost of the VFD.

24.2.12.1.2 Hybrid 12-Pulse Method

One disadvantage of the three winding arrangement mentioned earlier is its size and cost. On reexamining the circuit of Figure 24.22, it can be noted that one set of winding does not have any phase shift with respect to the primary winding. This is important because it allows one 6-pulse rectifier circuit to be directly connected to the ac source via some balancing inductance to match the inductance in front of the other 6-pulse rectifier circuit to achieve 12-pulse operation.

The resulting scheme has one six-pulse rectifier powered via a phase shifting isolation transformer, while the other six-pulse rectifier is fed directly from the ac source via matching impedance. Such a 12-pulse arrangement is called a hybrid 12-pulse configuration and is shown in Figure 24.23. The phase

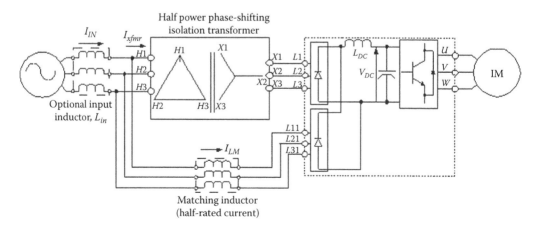

FIGURE 24.23   Schematic of a typical hybrid 12-pulse arrangement.

shifting transformer feeding one of the two six-pulse rectifiers is sized to handle half the rated power. Similarly, the matching inductor is sized to carry only half the rated current. This arrangement results in the overall size of the transformer and matching inductor combination to be smaller and less expensive than the three winding arrangement.

Advantages

Some important advantages of the hybrid 12-pulse is listed as follows:

- Size and cost of the hybrid 12-pulse configuration are much less than the three winding arrangement.
- 12-pulse operation is achieved with low total current harmonic distortion.
- Unlike three winding method, in this method the current (instead of flux in the core) in the two bridges are combined at the source to cancel the low order harmonics. Leakage flux and winding mismatch problems are reduced by adjusting the matching inductor to effectively cancel the fifth and seventh harmonic currents.

Disadvantages

The hybrid version also has some important disadvantages that need to be pointed out. They are as follows:

- The impedance mismatch between the leakage inductance and the external matching inductance can never be accomplished for all operating conditions because the leakage inductance is a function of current through the transformer while the external inductance is in the form of self inductance, which is constant till its rated current value.
- In order to minimize the effect of mismatch, an input ac line inductor may need to be used sometimes to comply with the harmonic levels recommended in IEEE 519(1992).
- The use of extra inductance ahead of the transformer-inductor combination can cause extra voltage drop that cannot be compensated for.
- The arrangement of Figure 24.23 cannot be used where level translation is needed.
- The advantage of high impedance to conducted EMI as offered by the three winding arrangement is reduced on using the hybrid arrangement of Figure 24.23.
- Similar to the three winding configuration, this method also requires the VFD to have two six-pulse rectifiers.

It should be pointed out that in spite of the previously listed shortcomings the hybrid 12-pulse method is gaining in popularity primarily because of size and cost advantage. The transformer leakage inductance and the external matching inductance are matched to perform at rated current so that low harmonic distortion is achieved at rated operating conditions. Tests conducted at 75 hp along with the associated harmonic spectrum are shown in Figure 24.24.

24.2.12.1.3 Autotransformer Method

The phase shift necessary to achieve multi-pulse operation can also be achieved by using autotransformers. Autotransformers do not provide any isolation between the input and output but can be used to provide phase shift. Autotransformers are typically smaller compared to regular isolation transformers because they do not need to process the entire power. Majority of the load current passes directly from the primary to the secondary terminals and only a small amount of *VA* necessary for the phase shift is processed by the autotransformer. This makes them small, inexpensive, and attractive for use in multi-pulse systems.

Though autotransformers are appealing for multi-pulse applications, they are not well suited for single VFD load. In all ac to dc rectification schemes, the diode pair that has the highest voltage across the input terminals conducts to charge the dc bus. When parallel rectifiers are used as in multi-pulse

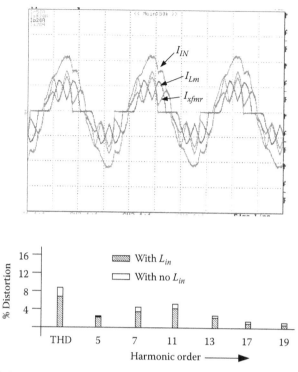

FIGURE 24.24 75 hp hybrid 12-pulse waveforms. *THD* = 6.7% with 5% input ac reactor and 8.8% with no input ac reactor.

techniques, it is important to maintain sharing of current among the multi-pulse rectifiers. If current sharing is compromised, then the amplitudes of lower order harmonics between the two rectifiers in a 12-pulse scheme will not cancel completely and this will result in poor harmonic performance. By electrically isolating one rectifier from the other in the two schemes discussed thus far, acceptable 12-pulse performance was possible. However, when autotransformers are employed, such isolation is lost and current from one set of phase-shifted windings can flow into the other set thereby compromising the equal distribution of current between the phase-shifted sets of windings. One way to force the rectifiers to share correctly is to introduce inter-phase transformer (IPT) in between the outputs of the two diode rectifier units as shown in Figure 24.25. A zero sequence blocking transformers (ZSBT) in between the rectifier and one of the phase-shifted outputs of the autotransformer also helps in reducing noncharacteristics triplen harmonics from flowing into the ac system. The autotransformer of Figure 24.8 has phase-shifted outputs of ±15°.

The use of ZSBT and IPT makes the overall system bulky and expensive and the choice of autotransformer less appealing. However, in many cases, the VFD is not equipped with two rectifier units and so none of the 12-pulse schemes can be really used in such cases. In such applications, if multiple VFDs are being employed and they can be paired into approximately equal ratings then the delta-fork autotransformer shown in Figure 24.25 can be effectively implemented. Instead of isolating the two diode rectifier units in one VFD, it is possible to use two different VFDs operating two independent loads of approximately equal rating and supplying them power from the phase-shifted outputs of the delta-fork transformer. This type of matched pair possibilities exists in a given system and is ideal for VFDs that do not have two independent six-pulse rectifier units. One such scheme of distributing the load between the phase-shifted outputs of a delta-fork autotransformer is shown in Figure 24.26. This arrangement has been seen in the field to achieve low total current harmonic distortion even with load imbalance in the neighborhood of 20%–25%.

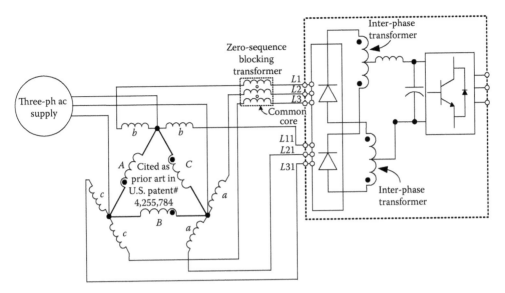

FIGURE 24.25 Delta-fork autotransformer with ZSBT and IPT for 12-pulse applications.

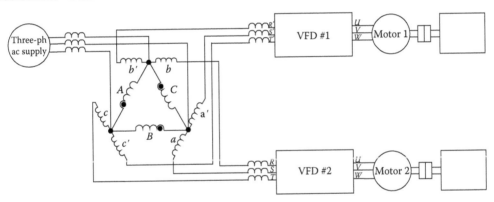

FIGURE 24.26 Use of low-cost autotransformer for 12-pulse operation in case of isolated and balanced loads.

Advantages

Some important advantages of the autotransformer connection shown in Figure 24.22 are listed as follows:

- VFDs do not need to have multiple rectifier units to achieve benefits of 12-pulse operation.
- Size and cost of autotransformer is small and unlike the circuit of Figure 24.21, there is no need for IPTs and ZSBTs.
- The three-phase input ac reactor in front of each VFD helps in making the current more continuous. These may be replaced by dc link chokes.

Disadvantages

The circuit of Figure 24.22 has a few shortcomings and the reader should be aware of these. They are

- Better harmonic performance is achieved if the loads are balanced. Since the loads are independent, many times it is not possible to guarantee balance and this may reduce the overall harmonic performance.
- Input ac line inductors or dc link chokes may be necessary to get better harmonic performance.
- VFDs need to be isolated to prevent crosscurrent flow between the two sets of windings and to assure good sharing.

24.2.12.2 18-Pulse Techniques

Harmonic distortion concerns are serious when the power ratings of the VFD load increases. Large power VFDs are gaining in popularity due to their low-cost and impressive reliability. The use of large power VFDs increases the amplitude of low order harmonics that can impact the power system significantly. In many large power installations, current harmonic distortion levels achievable using 12-pulse technique is insufficient to meet the levels recommended in IEEE 519(1992). In view of this, lately, quite a lot of interest has been shown in developing 18-pulse VFD systems to achieve much superior harmonic performance compared to the traditional 12-pulse systems.

18-pulse systems have become economically feasible due to the recent advances in autotransformer techniques that help reduce the overall cost and achieve low total current harmonic distortion. As mentioned earlier, when employing autotransformers, care should be taken to force the different rectifier units to share the current properly. The 18-pulse configuration lends itself better in achieving this goal compared to the 12-pulse scheme. Some popular 18-pulse autotransformer techniques are discussed next.

For 18-pulse operation, there is need for three sets of three-phase ac supply that are phase shifted with respect to each other by 20 electrical degrees. Traditionally, this is achieved using a four winding isolation transformer that has one set of primary windings and three sets of secondary windings. One set of secondary winding is in phase with the primary winding, while the other two sets are phase shifted by +20 electrical degrees and –20 electrical degrees with the primary. This arrangement yields three phase-shifted supplies that allow 18-pulse operation as shown in Figure 24.27.

The use of dc link choke as shown in Figure 24.27 is optional. The leakage inductance of the transformer may be sufficient to smooth the input current and improve the overall current harmonic distortion levels.

The primary disadvantage of the scheme shown in Figure 24.27 is that the phase shifting isolation transformer is bulky and expensive. A common disadvantage with all 18-pulse schemes is that all of

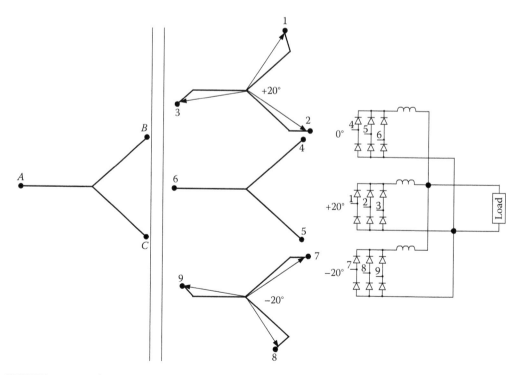

FIGURE 24.27 Schematic representation of 18-pulse converter circuit fed from phase-shifted isolation transformer.

them need three independent three-phase rectifier units. Many VFD manufacturers do not provide this feature and the additional rectifier units needed may have to be provided external to the VFD.

Instead of using ±20° phase-shifted outputs from isolation transformer for 18-pulse operation, a 9-phase supply where each phase lags the other by 40 electrical degrees can be used. Autotransformers have been developed that implement this idea and are widely used [3]—Figure 24.28. Due to the nature of autotransformers, the size, weight, and cost can be reduced compared to the conventional technique shown in Figure 24.27.

Figure 24.28a shows a nine-phase ac supply using wye-fork with a tertiary delta winding to circulate triplen harmonics. The size of the autotransformer is big and there is need for additional series

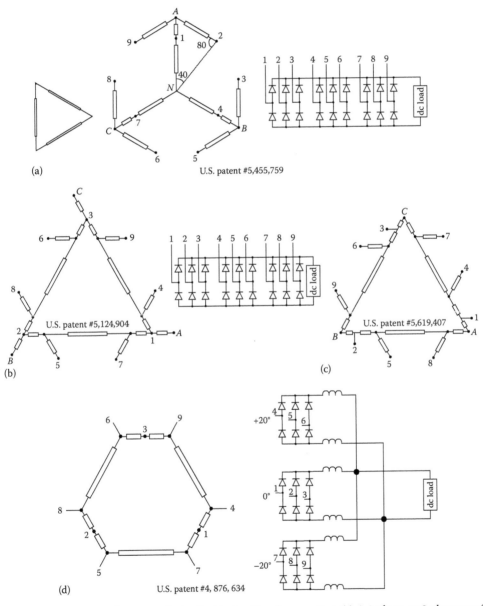

FIGURE 24.28 Autotransformer methods of achieving 18-pulse operation. (a) A 3-phase to 9-phase wye fork with closed delta autotransformer. (b) A differential delta autotransformer with three sets of outputs phase shifted to 40 electrical degrees. (c) A 3-phase to 9-phase differential data type autotransformer. (d) A polygon autotransformer with three sets of windings phase displaced by 20 electrical degrees.

impedance to smoothen the input ac currents. The rating of the transformer is about 70% of the rating of the load. If the series inductance is not used, then the output dc voltage is about 4.3% higher than that achieved when a standard six-pulse rectifier is used.

Figure 24.28b shows a nine-phase ac supply using delta-fork that does not require additional delta winding. In this configuration, the average dc output voltage is about 14% higher than that obtained using a standard six-pulse rectifier scheme. This can potentially stress the dc bus capacitors and the IGBTs in the inverter section of a VFD. In order to overcome this, additional teaser windings are used as shown. These windings not only add cost and increase the overall rating of the transformer, they also cause imbalance that results in higher than normal circulating currents in the delta windings, which need to be accommodated. The harmonic performance is good but the overall size is large with rated current flow through the teaser windings.

In order to overcome the 14% higher average dc bus voltage observed in the previous configuration, a modification of the configuration was proposed in the patent cited in Figure 24.28c. The harmonic performance is equally good and the average dc bus voltage is equal to that observed in six-pulse rectifiers. Similar to the previous configuration, the stub winding currents are high and the teaser winding needs to carry rated load current making the overall transformer big in size and expensive to wind.

In autotransformer configurations using stub and/or teaser windings shown in Figure 24.28a through c, the overall size and rating of the autotransformer is higher than the optimal value. The use of stub windings typically results in poor utilization of the core and involves more labor to wind the coils. Polygon type of autotransformer is better than stub type autotransformer from size and core utilization points of view. A polygon type autotransformer is shown in Figure 24.28d. It should be pointed out that the configuration of Figure 24.28d needs the use of IPTs and input ac inductors to achieve low total current harmonic distortion. The reason is that the outputs are not equally spaced to achieve a nine-phase ac supply as in the previous configurations. The polygon autotransformer of Figure 24.28d provides ±20° phase-shifted outputs to achieve 18-pulse operation.

One of the widely used 18-pulse autotransformer configurations is that shown in Figure 24.29. This configuration is a modified version of the configuration shown in Figure 24.28a and was proposed by the same author. In the configuration of Figure 24.29, the delta-connected tertiary winding is included in the wye fork. This construction is called the windmill construction. Initially the windmill structure was present in each phase and the size of the transformer was still big. The kVA rating was about 60%. By intelligently removing the windmill structure from two of the three phases, it was shown that the performance remained equally good. By adopting the modified structure of Figure 24.29, the kVA rating of the autotransformer was reduced from 60% to 55%.

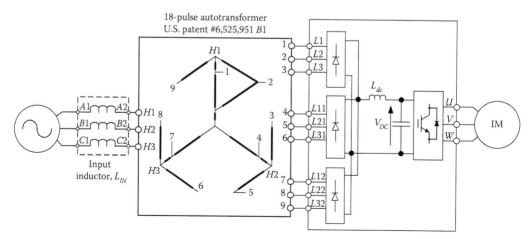

FIGURE 24.29 Schematic of the modified windmill construction of the 18-pulse autotransformer configuration used with VFDs.

In all the 18-pulse autotransformer methods, the change of current from one conducting diode pair to the other is quite sudden. Though the rms current rating may not exceed the current rating of the diode, attention should be given to the *di/dt* of the current through the diodes. Since the use of autotransformer method of 18-pulse operation is recent, there is not much statistical data available to comment on the *di/dt* issue with diodes when used in conjunction with 18-pulse autotransformer techniques.

24.2.12.3 Summary of Drawbacks with Autotransformers

From the discussion on autotransformers thus far, some important shortcomings of the autotransformer based topology are summarized as follows:

1. The leakage and magnetizing inductances of many auto transformers in the market is far lower than that in isolation transformers. Powering up an autotransformer typically results in an inrush current that is much higher than that observed in systems with isolation transformer. This requires careful fuse selection and coordination so that nuisance trips are avoided and fuse protection is still available.
2. In all the 18-pulse autotransformer methods, the change of current from one conducting diode pair to the other is quick. Though the rms current rating may not exceed the current rating of the diode, attention should be given to the *di/dt* of the current through the diodes. One solution is to use additional inductors in between the autotransformer and the input rectifier to lower the *di/dt*. This makes the overall scheme bulky and expensive. The rectangular current through the windings also increases losses, prompting the need to use fans to keep the size of the transformer small.
3. Due to the sudden change in current and lack of sufficient leakage inductance in autotransformers, such topologies require significant input impedance (shown as L_{IN} in Figure 24.29) to smooth the current and reduce the overall input current distortion. All the autotransformer configurations discussed here do not operate well without a significant amount of input inductance ahead of the autotransformer.
4. Autotransformer techniques utilize complex winding structures, either of the stub type or the polygon type. These transformers are labor intensive to manufacture and result in poor core utilization. Even the polygon type shown in Figure 24.28d is labor intensive to wind.
5. Autotransformer topologies that convert a three-phase system to a nine-phase output create an aberration in the dc bus ripple content of a VFD. When one or two of nine output phases have a bad rectifier, the increase in dc bus ripple is hardly noticeable and this reduces the chance for detection of failure. The power flow is now shared by existing rectifiers that can eventually fail.

Given the previous shortcomings, it is clear that there is room for improvement in multi-pulse rectification schemes. Similar to the idea shown in Figure 24.23, a new 18-pulse scheme that has two 6-pulse rectifiers powered via a phase shifting isolation transformer, and a third 6-pulse rectifier fed directly from the ac source via a matching impedance was proposed in [8]. Such an 18-pulse arrangement will be referred to as hybrid 18-pulse configuration and is shown in Figure 24.30. The phase shifting zig-zag isolation transformer feeding two of the three six-pulse rectifiers is sized to handle 2/3 the rated load power. Similarly, the matching inductor is sized to carry only 1/3 the rated input load current. This arrangement results in the overall size of the transformer and matching inductor combination to be smaller and less expensive than the four winding arrangement of Figure 24.27. As an example, a 50 hp conventional 18-pulse transformer without input inductor L_{IN} was recently quoted to be 42″(H) × 36″(W) × 24″(D) with an estimated weight of 880 lb, while the hybrid 18-pulse structure (without L_{IN}) to handle the same load was quoted to be 24″(H) × 26″(W) × 14″(D) with an estimated weight of 550 lb. The topology requires a matching inductor to perform comparably, which will add 12 lb, for a total weight of 562 lb. Both these structures are naturally cooled.

The phase shift in the transformer shown in Figure 24.30 is achieved by winding extra teaser windings on appropriate limbs of a transformer. The teaser windings are marked "T" with subscript denoting the phase that they are wound on. For example, T_{H21} denotes a teaser winding that is wound

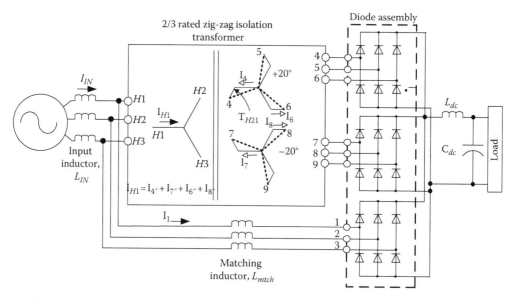

FIGURE 24.30   Schematic of proposed hybrid 18-pulse topology [8].

on the H2 winding of the primary side of the isolation transformer and is used in the first set of secondary winding to yield a phase shift of +20°.

The required phase shift could have been achieved using an autotransformer, but due to the reasons mentioned earlier, it is not the choice topology.

24.2.12.4 Harmonic Mitigation Technique Summary

This section discussed generation of current harmonics by nonlinear loads and the IEEE-519-1992 standard to limit the quantity of these harmonics. A methodology of applying this standard to a practical industrial site has been described. Different harmonic mitigating techniques presently available in the industry have been highlighted. Multi-pulse techniques to achieve low total current harmonic distortion have been discussed. Relative advantages and disadvantages of the techniques presented have also been discussed. Based on the materials presented in this report, the following important conclusions can be drawn:

1. Passive techniques involving capacitors are associated with circulating current, leading power factor, and high dc bus voltage at light load condition and hence should be avoided as far as possible. They are also associated with the possibility of causing network resonance and hence if they are installed, care should be taken to monitor resonance conditions and avoid them.
2. DC link chokes are a better alternative than ac line reactors for harmonic mitigation since they do not cause additional voltage drop if the value is greater than the critical inductance.
3. To handle transients and surges on the ac line, a combination of small value of ac inductance and dc link choke is preferred.

Multi-pulse techniques offer the best passive solution to handle harmonics. The hybrid 18-pulse technique and the hybrid 12-pulse technique are attractive for medium power applications, while distributing the load on phase-shifted outputs of an autotransformer for small power application is an interesting alternative.

24.2.12.5 Active Harmonic Compensation

Most passive techniques discussed earlier aim to cure the harmonic problems once nonlinear loads have created them. However, motor-drive manufacturers are developing rectification techniques that do not generate low order harmonics. These drives use active front ends. Instead of using diodes as rectifiers, the active front end ASDs make use of active switches like IGBTs along with antiparallel diodes.

In such active front-end rectifiers, power flow becomes bidirectional. The input current can be wave shaped and made sinusoidal to have low values of low-order harmonics.

Apart from the active front ends, there also exists shunt active filters used for actively introducing a current waveform into the ac network which when combined with the harmonic current, results in an almost perfect sinusoidal waveform.

Interesting and effective combinations of passive tuned and active components have been proposed by many researchers and quite a few of them are reportedly in use in the steel, rail, and power utility industries. Such topologies are commonly referred to as hybrid structures and have been extensively researched by authors of reference [7].

Most active filter topologies are cost-effective in high power ratings but require high initial investment. Hybrid filters also have large bandwidth and good dynamic response. Control is accomplished using digital signal processing (DSP) chips. The hybrid structures also need current and voltage sensors and corresponding analog to digital (A/D) converters.

Manufacturers of smaller power equipment like computer power supplies, lighting ballast, etc., have successfully employed single-phase active circuits, employing boost converter topologies.

Detailed discussions on active harmonic compensation and related topics can be found in literature. Since it is beyond the scope of this section, the topic on active harmonic compensating circuits is not dealt in detail here.

24.3 Controlled Rectifiers

Controlled rectifier circuits make use of controlled switches. One such device is the "thyristor." A thyristor is a four-layer (p-n-p-n), three-junction device that conducts current only in one direction similar to a diode. The junction marked $J3$ in Figure 24.26 is utilized as the control junction and consequently the rectification process can be initiated at will provided the device is favorably biased and the load is of favorable magnitude. The operation of a thyristor can be explained by assuming it to be made up of two transistors connected back-to-back as shown in Figure 24.31.

Let α_1 and α_2 be the ratio of collector to emitter currents of transistors Q1 and Q2, respectively. In other words, $\alpha_1 = I_{c1}/I_{e1}$; $\alpha_2 = I_{c2}/I_{e2}$; Also, from Figure 24.26, $I_{e1} = I_{e2} = I_A$ where I_A is the anode current flowing through the thyristor. From transistor theory, the value of I_{e2} is equal to $I_{c2} + I_{b2} + I_{lkg}$; where I_{lkg} is the leakage current crossing the $n1$-$p2$ junction. From Figure 24.26, $I_{b2} = I_{c1}$. Hence the anode current can be rewritten as

$$I_A = I_{c1} + I_{c2} + I_{lkg} \tag{24.14}$$

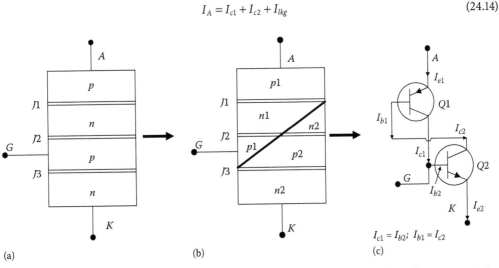

(a) (b) (c)

FIGURE 24.31 Virtual representation of a thyristor to explain its operation. (a) Semiconductor layer representation for a thyristor, (b) step towards deriving an equivalent model for a thyristor, and (c) equivalent model of a thyristor.

Substituting the collector currents by the product of ratio α and emitter current, the anode current becomes

$$I_A = (\alpha_1 * I_{e1}) + (\alpha_2 * I_{e2}) + I_{lkg}$$

$$I_A = (\alpha_1 + \alpha_2)I_A + I_{lkg} \tag{24.15}$$

$$I_A = \frac{I_{lkg}}{1 - (\alpha_1 + \alpha_2)}$$

If the ratio of the collector current to base current (gain) of the transistors is assumed to be β_1 and β_2 respectively, then the relationship between to β_1, β_2 and α_1, α_2 can be written as

$$\alpha_1 = \frac{\beta_1}{1 + \beta_1}; \quad \alpha_2 = \frac{\beta_2}{1 + \beta_2} \tag{24.16}$$

Substituting for α_1 and α_2 in the expression for I_A yields the following expression:

$$I_A = \frac{(1 + \beta_1)(1 + \beta_2)I_{lkg}}{1 - \beta_1\beta_2} \tag{24.17}$$

If the values of α_1 and α_2 are low (low gains) then the anode current is low and comparable to the leakage current. Under this condition, the thyristor is said to be in its OFF state. However, if the effective gain of the transistor is such that the product of the gains are close to 1 (i.e., sum of the ratios of α_1 and α_2 are close to 1), then there is large increase in anode current and the thyristor is said to be in conduction. External circuit conditions can be changed to influence the product of the gains ($\beta_1\beta_2$). Some techniques of achieving this are briefly discussed next.

1. *Increasing applied voltage*: On applying a voltage across the anode to cathode terminals of the thyristor (anode being more positive than the cathode), junctions J1 and J3 in Figure 24.26 are forward biased while junction J2 is reverse biased. The thyristor does not conduct any current and is said to be in a blocking state. On increasing the applied voltage, minority carriers in junction J2 (i.e., holes in $n1$, $n2$ and electrons in $p1$, $p2$) start acquiring more energy and hence start to migrate. In the process, these holes could dislodge more holes. Recombination of the electrons and holes also occur which creates more motion. If the voltage is increased beyond a particular level, the movement of holes and electrons becomes great and junction J2 ceases to exist. The product of the gains of the two transistors in the two-transistor model is said to achieve values close to unity. This method of forcing current to flow through the thyristor is not recommended since junction J2 gets permanently damaged and the thyristor ceases to block forward voltage. Hence this method is a destructive method.

2. *High dv/dt*: As explained earlier, junction J2 is the forward blocking junction when a forward voltage is applied across anode to cathode of a thyristor. Any *p-n* junction behaves like a depletion region when it is reverse biased. Since J2 is reverse biased, this junction behaves like a depletion region. Another way of looking at a depletion region is that the boundary of the depletion region has abundant holes and electrons while the region itself is depleted of charged carriers. This characteristic is similar to that of a capacitor. If the voltage across the junction (J2) changes very abruptly, then there will be rapid movement of charged carriers through the depleted region. If the rate of change of voltage across this junction (J2) exceeds a predetermined value, then the movement of charged carriers through the depleted region is so high that junction J2 is again annihilated. After this event, the thyristor is said to have lost its capability to block forward voltage and even a small amount of forward voltage will result in significant current flow, limited only by the load impedance. This method is destructive too and is hence not recommended.

3. *Temperature*: Temperature affects the movement of holes and electrons in any semiconductor device. Increasing the temperature of junction J2 will have a very similar effect. More holes and electrons will begin to move causing more dislodging of electrons and holes from neighboring lattice. If a high temperature is maintained, this could lead to an avalanche breakdown of junction J2 and again render the thyristor useless since it would no longer be able to block forward voltage. Increasing temperature is yet another destructive method of forcing the thyristor to conduct.

4. *Gate current injection*: If a positive voltage is applied across the gate to cathode of a thyristor, then junction J3 would be forward biased. Charged carriers will start moving. The movement of charged carriers in junction J3 will attract electrons from n2 region of the thyristor (Figure 24.26). Some of these electrons will flow out of the gate terminal but there would be ample electrons that could start crossing junction J2. Since electrons in p2 region of junction J2 are minority carriers, these can cause rapid recombination and help increase movement of minority carriers in junction J2. By steadily increasing the forward biasing potential of junction J3, the depletion width of junction J2 can be controlled. If a forward biasing voltage is applied across anode to cathode of the thyristor with its gate to cathode favorably biased at the same time, then the thyristor can be made to conduct current. This method achieves conduction by increasing the leakage current in a controlled manner. The gain product in the two-transistor equivalent is made to achieve a value of unity in a controlled manner and the thyristor is said to turn ON. This is the only recommended way of turning ON a thyristor. When the gate-cathode junction is sufficiently forward biased, the current through the thyristor depends on the applied voltage across the anode-cathode and the load impedance. The load impedance and the externally applied anode-cathode voltage should be such that the current through the thyristor is greater than a minimum current known as *latching current*, I_l. Under such a condition, the thyristor is said to have *latched ON*. Once it has latched ON, the thyristor remains ON. In other words, even if the forward biasing voltage across the gate-cathode terminals is removed, the thyristor continues to conduct. Junction J2 does not exist during ON condition. The thyristor reverts to its blocking state only when the current through it falls below a minimum threshold value known as *holding current*, I_h. Typically, holding current is lower than latching current ($I_h < I_l$). There are two ways of achieving this. They are either (1) increase the load impedance to such a value that the thyristor current falls below I_h or (2) apply reverse biasing voltage across the anode-cathode of the thyristor. An approximate *v-i* characteristic of a typical thyristor and its symbol are shown in Figure 24.32.

Since the thyristor allows flow of current only in one direction like a diode and the instant at which it is turned ON can be controlled, the device is a key component in building a controlled rectifier unit. The diode in all the circuits discussed so far can be replaced with the thyristor. Because of its controllability, the instant at which the thyristor conducts can be delayed to alter the average and rms output voltages. By doing so, the output voltage and output power from the rectifier can be controlled. Rectifiers that employ thyristors are thus also known as silicon controlled rectifiers or SCR.

A typical single-phase *R-L* rectifier circuit with one thyristor as the rectifier is shown in Figure 24.33. The figure also shows the relevant circuit waveforms. The greatest difference between this circuit and its diode counterpart is also shown for comparison. Both circuits conduct beyond π radians due to the presence of the inductor *L* since the average voltage across an inductor is zero. If the value of the circuit components and the input supply voltage are the same in both cases, the duration for which the current flows into the output *R-L* load depends on the values of *R* and *L*. In the case of the diode circuit it does not depend on anything else while in the case of the thyristor circuit, it also depends on the instant the thyristor is given a gate trigger.

From Figure 24.33, it is important to note that the energy stored in the inductor during conduction interval can be controlled in the case of thyristor in such a manner that it reduces the conduction interval and thereby alters (reduces) the output power. Both the diode and the thyristor show reverse recovery phenomenon. The thyristor similar to the diode can block reverse voltage applied across it repeatedly, provided the voltage is less than its breakdown voltage.

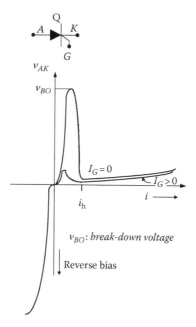

FIGURE 24.32 *v-i* characteristic of a thyristor along with its symbol.

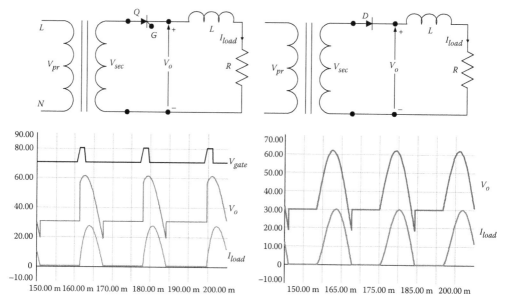

FIGURE 24.33 Comparing a single thyristor rectifier circuit with a single diode rectifier circuit. Note that the thyristor conduction is delayed deliberately to bring out the differences.

24.3.1 Gate Circuit Requirements

The trigger signal should have voltage amplitude greater than the minimum gate trigger voltage of the thyristor being turned ON. It should not be greater than the maximum gate trigger voltage, either. The gate current should likewise be in between the minimum and maximum values specified by the thyristor manufacturer. Low gate current driver circuits can fail to turn ON the thyristor. The thyristor is current controlled switch and so the gate circuit should be able to provide the needed turn ON gate current into the thyristor. Unlike the bipolar transistor, the thyristor is not an amplifier and so the gate current requirement does not absolutely depend on the voltage and current rating of the thyristor. Sufficient gate trigger current will turn ON the thyristor and current will flow from the anode to the cathode, provided the thyristor is favorably biased and the load is such that the current flowing is higher than the latching current of the thyristor. In other words, in single-phase ac to dc rectifier circuits, the gate trigger will turn ON the thyristor only if it occurs during the positive part of the ac cycle (Figure 24.33). Any trigger signal during the negative part of the ac cycle will not turn ON the thyristor and thyristor will remain in blocking state. Keeping the gate signal ON during the negative part of the ac cycle does not typically damage a thyristor.

24.3.2 Single-Phase H-Bridge Rectifier Circuits with Thyristors [1–3]

Similar to the diode H-bridge rectifier topology, there exists SCR–based rectifier topologies. Because of their unique ability to be controlled, the output voltage and hence the power can be controlled to desired levels. Since the triggering of the thyristor has to be synchronized with the input sinusoidal voltage in an ac to dc rectifier circuit, soft charging of the filter capacitor can be achieved. In other words, there is no need for employing soft-charge resistor and contactor combination as is required in single-phase and three-phase ac to dc rectifier circuits with dc bus capacitors.

In controlled ac to dc rectifier circuits, it is important to discuss control of resistive, inductive, and resistive–inductive load circuits. DC motor control falls into the resistive–inductive load circuit. DC motors are still an important part of the industry. However, the use of dc motor in industrial application is declining rapidly. Control of dc motors is typically achieved by controlled rectifier circuits employing thyristors. Small motors of less than 3-kW (approximately 5 hp) rating can be controlled by single-phase SCR circuits while larger ratings require three-phase versions. A typical single-phase H-bridge SCR–based circuit for the control of a dc motor is shown in Figure 24.29. Typical output waveforms are shown in Figure 24.34. The current in the load side can be assumed continuous due to the large inductance of the armature of the dc motor.

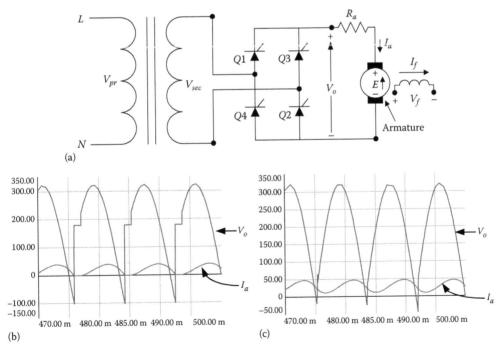

FIGURE 24.34 Single-phase dc motor control circuit for controlling a separately excited dc motor. R_a indicates equivalent armature resistance and E is the back emf. Typical waveforms are shown in (a) topological representation, (b) discontinuous mode of operation, and (c) continuous mode of operation.

In Figure 24.34a, V_f is the field voltage, which is applied externally and generally is independent of the applied armature voltage. Such a dc motor is known as a separately excited motor. I_a is the armature current while I_f is the field current. By altering the instant of turn ON of the thyristors, the average output voltage can be altered. A dc motor typically generates a back emf that is dependent on speed. In Figure 24.34b, discontinuous mode of operation is depicted. When the trigger instant is delayed appreciably from the start of the sinusoidal supply voltage waveform, the average output voltage reduces, resulting in discontinuous current flow, which is characterized by zero output current durations as seen in Figure 24.34b. During the instant the current goes to zero, the output voltage is simply the back emf, E. When current starts flowing again, the input voltage is fed into the output and so the output has portions of the input voltage.

On advancing the instant of triggering the thyristors, the output current can be made continuous as seen in Figure 24.34c. The output voltage increases and since the thyristors conduct for the entire duration, the input wave shape appears at the output. Since the output voltage can be controlled, the armature current can be effectively controlled. Since the torque produced by a dc motor is directly proportional to the armature current, the torque developed can thus be controlled:

$$T = K\varphi I_a \tag{24.18}$$

K is the motor constant and depends on the number of armature conductors, number of poles, and type of winding employed in the dc machine. ϕ is flux produced by the field and is proportional to the field current, I_f. Hence, the torque produced by a dc machine can be rewritten as

$$T = K(K_1 I_f) I_a \tag{24.19}$$

By keeping the field current constant, the torque then becomes directly proportional to the armature current, which is controlled by controlling the output voltage of the ac to dc controlled rectifier. In the circuit shown in Figure 24.34, it is important to note that the current I_a, cannot flow in the opposite direction. Hence, the motor cannot generate negative torque. In order to make the motor run in the opposite direction, the direction of the field has to be changed. Speed control within the base speed can also be accomplished by controlling the armature voltage as follows:

$$E = K\varphi\omega = K(K_1 I_f)\omega \tag{24.20}$$

ω is the speed of the armature in radians/s. The back emf, E is the difference between the output dc voltage of the ac to dc controlled rectifier and the drop across the equivalent armature resistance. Hence, E can be rewritten as

$$E = V_a - (I_a R_a); \quad \omega = \frac{V_a - (I_a R_a)}{KK_1 I_f} \tag{24.21}$$

For control of speed above base speed, the field current has to be reduced. Hence, it can be shown that controlling the armature current controls torque and speed below base speed while controlling the field current achieves speed control above base speed. Because of the large inductance of the armature circuit, the current through it can be assumed continuous for practical operating region. The average output voltage of a single-phase ac to dc rectifier circuit for continuous current operation is given by (referring to Figure 24.34c:

$$V_O = \frac{1}{\pi} \int_{\alpha}^{\pi+\alpha} \left(\sqrt{2} * V_{rms}\right) d(\omega t) = \frac{2 * \sqrt{2} * V_{rms} * \cos(\alpha)}{\pi} \tag{24.22}$$

Equation 24.22 is derived for continuous current condition. By controlling the triggering angle, α, the average value of the output voltage, V_O can be controlled. If armature current control is the main objective (to control output torque), then the controller of Figure 24.34 can be configured with a feedback loop. The measured current can be compared with a set reference and the error can be used to control the triggering angle, α. Since the output voltage and hence the armature current is not directly proportional to α but to $\cos(\alpha)$, the previous method will yield a nonlinear (cosinusoidal) relationship between the output voltage and control angle, α. However, the error signal $\cos(\alpha)$ instead of α can be chosen to be the control parameter. This would then yield a linear relationship between the output voltage and cos of control angle, α.

It is important to note from the equation for the output average voltage that the output average voltage can become negative if the triggering angle is greater than 90 electrical degrees. This leads us to the topic of regeneration. AC to DC controlled rectifiers employing thyristors and having large inductance on the dc side can be made to operate in the regeneration mode by simply delaying the trigger angle. This is quite beneficial in hoist applications as explained hereafter.

When a load on a hoist needs to be raised, electrical energy is supplied to the motor. The voltage across the motor is positive and the current through the armature is positive. Positive torque is generated and the load is raised. When the load is being brought down, the motor rotates in the opposite direction that results in a negative value of back emf. The current through the thyristors cannot go negative so the motor is still developing positive torque tending to raise the load and prevent it from running away due to the gravitational pull. The negative back emf is supported by advancing the gating angle to be greater than 90 electrical degrees so that the voltage across the armature of the motor is negative but remains slightly more positive than the back emf, the difference causing positive current flow into the motor. The large inductance of the motor helps to maintain the positive direction of current through the armature. From electrical energy flow point of view, the product of current through the motor and the voltage across it is negative meaning that the motor is in regeneration mode.

The kinetic energy due to the motors motion is converted to electrical energy and this produces considerable braking torque. The electrical energy is fed back to the source via the input thyristors. Converting kinetic energy to electrical energy has the desired braking effect and such conversion is known as regenerative braking.

The preceding application describes two-quadrant operation. Cranes and elevators employed in hoist operation are required to operate in all four quadrants (Figure 24.30). Using only one H-bridge rectifier allows two-quadrant operation—quadrants, one and four or quadrants two and three. For achieving four-quadrant operation, two H-bridge rectifiers are needed, as shown in Figure 24.31. The four different quadrants of operation are described next for a crane/hoist operation.

In the first quadrant, the motor develops positive torque and motor runs in the positive direction meaning, speed is positive—product of torque and speed is power and so positive electric power is supplied to the motor from the ac to dc rectifier.

When the crane with a load is racing upward, close to the end of its travel, the ac to dc controlled rectifier is made to stop powering the motor. The rectifier practically is switched off. The inertia of the load moving upward generates a voltage in the form of a back emf. This voltage is fed into a second rectifier bridge arranged in the opposite direction (Figure 24.35). The second bridge is turned ON to let the generated voltage across the upwardly mobile motor to flow into the utility thereby converting the inertial motion to electric power. In the second quadrant, speed remains positive but torque becomes negative, since the current through the motor flows in the opposite direction into the second rectifier bridge arrangement (Figure 24.35). The product of speed and torque is negative meaning that the motor behaves like a generator during this part of the travel.

Third quadrant operation occurs at the beginning of the lowering action. Both torque and speed are negative and so the product of torque and speed is positive. Power is applied to the motor to overcome static friction and allow the rotating parts of the mechanism to move the load downward. In this case, the direction of armature current through the motor is opposite to that in quadrant one and the electrical power needed by the motor is supplied by the second rectifier bridge arrangement (Figure 24.36).

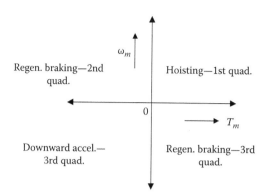

FIGURE 24.35 Four-quadrant operation of a crane or hoist.

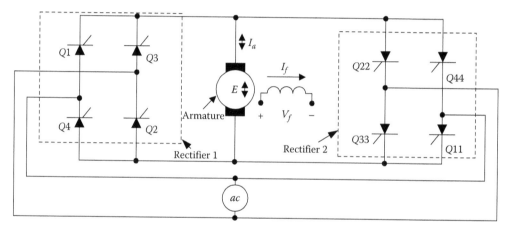

FIGURE 24.36 Two rectifier bridge arrangements for four-quadrant operation of dc motor.

The mechanical load and motor arrangement goes into the fourth quadrant of operation for the larger part of the downward motion. This is the duration during which, the motor resists the tendency of the load to accelerate downward by developing positive torque. Since motion is downward, speed is negative and the product of torque and speed is negative. This means the motor behaves like a generator. Due to the downward motion of the motor, the back emf is negative but the current is positive and so from electrical energy flow point of view, the power is negative, meaning that the motor is in regeneration mode.

Since the thyristors cannot conduct in the opposite direction, a new rectifier section arranged in an opposite manner had to be provided to enable the four-quadrant operation needed in cranes and hoists. The method by which unidirectional electrical power was routed to the bidirectional ac utility lines is known as inversion (opposite to rectification). Since no external means of switching OFF the thyristors was employed, the process of inversion is achieved by natural commutation provided by the ac source. Such an inverter is known as line commutated inverter.

24.3.3 Three-Phase Controlled AC to DC Rectifier Systems

The observations made so far for the single-phase controlled ac to dc rectifiers can be easily extended to three-phase versions. An important controlled rectification scheme that was not mentioned in the single-phase case is the semiconverter circuit. In Figure 24.37, if the thyristors Q2 and Q4 are replaced by diodes (*D2* and *D4*), then the circuit of Figure 24.37 is converted into a semiconverter circuit. Such a circuit does not have the potential to provide regeneration capability and hence is of limited use. However, in dual converter applications, especially in three-phase versions, there are a few instances where a

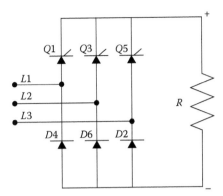

FIGURE 24.37 A typical three-phase semiconverter.

semiconverter can be employed to reduce cost. A typical three-phase semiconverter circuit will consist of three thyristors and three diodes arranged in an H-bridge configuration as shown in Figure 24.37.

Three-phase dual converter schemes similar to the single-phase version shown in Figure 24.36 are still employed to operate large steel mills, hoists, and cranes. However, the advent of vector controlled ac drives has drastically changed the electrical landscape of the modern industry. Most dc motor applications are being rapidly replaced by ac motors with field oriented control schemes. DC motor application in railway traction has also seen significant reduction due to the less expensive and more robust ac motors.

However, there are still a few important applications where three-phase controlled rectification (inversion) is the most cost-effective solution. One such application is the regenerative converter module that many inverter-drive manufacturers provide as an optional equipment to customers with overhauling loads. Under normal circumstance, during motoring mode of operation of an ac drive, the regenerative unit does not come into the circuit. However, when the dc bus voltage tends to go higher than a predetermined level due to overhauling of the load, the kinetic energy of the load is converted into electrical energy and is fed back into the ac system via a six-pulse thyristor-based inverter bridge. One such scheme is shown in Figure 24.38.

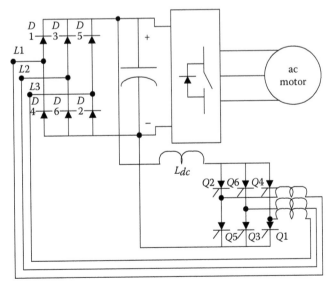

FIGURE 24.38 Use of six-pulse thyristor bridge in the inverter mode to provide regeneration capability to an existing ac drive system.

24.3.4 Average Output Voltage

In order to evaluate the average value of the output voltage for a three-phase full bridge converter, the process of integrating the output voltage similar to the one in Figure 24.39b has to be undertaken. For the circuit shown in Figure 24.39b, where the diodes are replaced by thyristors, the integration yields the following:

$$V_O = \frac{3}{\pi} \int_{\alpha+(\pi/3)}^{\alpha+(2\pi/3)} \sqrt{2}\, V_{L-L} \sin(\omega t)\, d(\omega t)$$

$$V_O = \frac{3 * \sqrt{2} * V_{L-L} * \cos(\alpha)}{\pi} = \frac{3 * \sqrt{2} * \sqrt{3} * V_{L-N} * \cos(\alpha)}{\pi}$$

(24.23)

The average output voltage for the circuit in Figure 24.39b with the diodes being replaced by thyristors is only different in the *cosine* of the triggering angle, α. If the triggering angle is zero, the circuit performs similar to a three-phase diode rectifier and the average output voltages become the same.

24.3.5 Use of Thyristors for Soft Charging DC Bus of Voltage Source Inverters

Though thyristor-based inverters may have been replaced by better controlled and higher speed semiconductor switches, the thyristors are still the work horse in many rectifier applications.

As discussed earlier, variable frequency drives (VFDs) with diode rectifier front end are typically equipped with a resistor-contactor arrangement to limit the inrush current into the dc bus capacitors, thereby providing a means for soft charging the dc bus capacitors. Because of the mechanical nature of the magnetic contactor typically used in VFDs, there exists a concern for reliability. In addition, during a brownout condition, typically the contactor remains closed and when the voltage recovers, the ensuing transient is often large enough to possibly cause unfavorable influence to surrounding components in the VFD. Many researchers and application engineers have thought about this problem and have worked in resolving this dilemma in a cost-effective manner.

There have been suggestions of replacing the magnetic contactor (MC in Figure 24.39a) with a semiconductor switch, as shown in Figure 24.39b. The semiconductor switch shown in Figure 24.39b is typically a thyristor and requires intelligent control. Unfortunately, during steady state operation, it is associated with power loss and reduces the overall efficiency of the VFD. Thyristor controlled semiconverters, similar to the one shown in Figure 24.37 are also effective in soft charging the dc bus of VFDs. The logic used for controlling the three thyristors in Figure 24.37 are also effectively used to eliminate problems during brownout conditions and is perhaps a very effective way for soft charging the dc bus of VFDs. However, the only drawback of the topology shown in Figure 24.37 is the need for a logic controller and additional gate driver circuits that can add cost to the VFD. Yet another

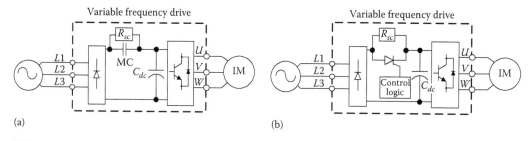

(a)   (b)

FIGURE 24.39 Soft-charge circuit configuration, (a) with magnetic contactor (MC) and (b) with thyristor.

technique to soft charge the dc bus capacitor that employs thyristors is discussed in this section. The features of the circuit discussed are as follows:

- No mechanical contactors.
- Should be able to handle brownout conditions in an efficient manner.
- Autonomous operation (without any control logic) to handle various power supply conditions.
- Unit should be compact and economical.

24.3.5.1 Thyristor Assist Clamp Circuit [9]

The proposed topology, shown in Figure 24.40, meets most of the desired features. A dc link inductor with a resistor assist circuit is employed to soft charge the dc bus capacitor. The assist resistor also has a series thyristor [9].

24.3.5.2 Principle of Operation

When ac power is applied to the circuit shown in Figure 24.40, an inrush current begins to flow, assuming that the dc bus capacitor has no initial stored voltage. The inrush current is divided into two distinct paths. The first path is through the resistor-thyristor (TH2) combination and the second path is through the dc link inductor, L_{dc}. The current through the resistor-thyristor path is initially higher and quicker than that through L_{dc} since the inductor delays the build up of current through it. The dc bus capacitor starts to charge, with the resistor-thyristor combination providing as much charging as possible. The second charging path, through L_{dc}, creates a resonant circuit. Due to the nature of LC circuit, the voltage across the dc bus capacitor C tends to increase over and above the peak value of the applied input ac voltage. At this time, the thyristor across L_{dc}, TH1, experiences a forward bias and turns ON. The turning ON of TH1, causes the voltage across the inductor to start falling and eventually turns OFF thyristor TH2 in series with the assist resistor, by reverse biasing it. The inductor voltage linearly ramps to zero and gets clamped by TH1. The voltage across the dc bus capacitor stops increasing and eventually discharges into its discharge resistor to a level dictated by the input voltage condition.

The important aspect of the resistor-assist circuit cannot be overlooked since the charging current flowing through L_{dc} is reduced due to the parallel resistor assist circuit. This reduces the stored energy in L_{dc}. It also lowers the saturation current requirement and makes the inductor physically smaller. Due to the LC nature of the circuit, the voltage across the capacitor is still higher than the peak value of the input voltage. The clamping circuit consisting of TH1 assures that the dc bus voltage is clamped to an acceptable value.

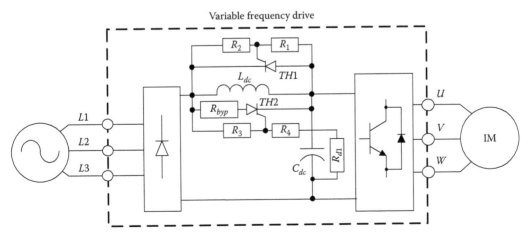

FIGURE 24.40 Proposed circuit for soft charging the dc bus capacitor, employing two thyristors.

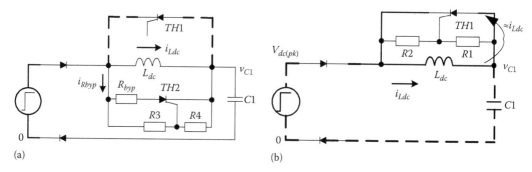

(a)

(b)

FIGURE 24.41 (a) Equivalent circuit for interval I; (b) equivalent circuit for interval II.

The operation of Figure 24.40 can be seen to have two distinct intervals of operation. Referring to Figure 24.41a, interval 1 of operation begins when the power is turned ON and the peak line-line voltage is applied to the inductor-resistor-capacitor combination and lasts till the voltage across the dc capacitor, v_{C1} goes above the peak input voltage. During interval 1, two current paths exist—one through the bypass resistor and the other through the dc link inductor. At the end of interval 1, current through R_{byp} is zero since TH2 is reverse biased.

The expression for capacitor current ($i_{Ldc} + i_{Rbyp}$) for zero initial capacitor voltage is

$$i_{C1} = \frac{V_{dc(pk)}}{\sqrt{L_{dc}/C_1}}\sin(\omega_1 t) + \frac{V_{dc(pk)}}{R_{byp}}e^{-t/R_{byp}C_1}; \quad \omega_1 = \frac{1}{\sqrt{L_{dc}*C_1}} \tag{24.24}$$

$$v_{C1} = V_{dc(pk)}(1-\cos(\omega_1 t)) + V_{dc(pk)}\left(1-e^{-t/R_{byp}C_1}\right) \tag{24.25}$$

$$v_{Ldc} = V_{dc(pk)} - v_{C1} = V_{dc(pk)}\left(\cos(\omega_1 t) + e^{-t/R_{byp}C_1} - 1\right) \tag{24.26}$$

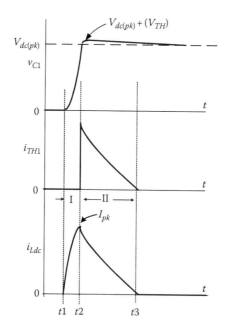

FIGURE 24.42 Theoretical waveforms.

Referring to Figure 24.41b, interval II begins when thyristor TH1 starts conducting and clamps the voltage at the capacitor to the rectifier output voltage. Interval II ends when i_{Ldc} decays to zero (Figure 24.42).

$$0 = L_{dc} * \frac{di_{Ldc}}{dt} + i_{Ldc} * R_{par} \tag{24.27}$$

$$i_{Ldc} = \frac{V_{dc(pk)}}{\sqrt{L_{dc}/C_1}} * \left(e^{-R_{par}t/L_{dc}} \right) \tag{24.28}$$

$$v_{Ldc} = -V_{TH}; \quad v_{C1} = V_{dc(pk)} + (V_{TH}) \tag{24.29}$$

24.3.6 HVDC Transmission Systems

One area where it is difficult to replace the use of high voltage, high current carrying thyristors is high voltage dc (HVDC) transmission systems. When large amount of power is to be transported over long distances, or under water, it has been found that high-voltage dc transmission is more economical. HVDC systems are in reality back to back rectifier systems. The sending end rectifier system consists typically of 12- or 24-pulse thyristor bridges while the receiving end consists of a similar configuration but in the opposite direction. The receiving end 12- or 24-pulse bridge operates in the inverter mode while the sending end operates in the rectifier mode. 12-pulse configuration is achieved by cascading two 6-pulse bridges in series while 24-pulse configuration needs four 6-pulse bridges cascaded in series. Typical advantages of high-voltage dc transmission over high-voltage ac transmission are listed as follows:

1. No stability problems due to transmission line length since no reactive power needs to be transmitted.
2. No limitation of cable lengths for underground cable or submarine cable transmission due to the fact that no charging power compensation need be done.
3. Ac power systems can be interconnected employing a dc tie without reference to system frequencies, short circuit power, etc.
4. High-speed control of dc power transmission is possible because the control angle, α, has a relatively short time constant.
5. Fault isolation between receiving end and sending end can be dynamically achieved due to fast efficient control of the high-voltage dc link.
6. Employing simple-control logic can change energy flow direction very fast. This can help in meeting peak demands at either the sending or the receiving station.
7. High reliability of thyristor converter and inverter stations makes this mode of transmission a viable solution for transmission lengths typically over 500 km.
8. The right-of-way needed for high-voltage dc transmission is much lower than that of ac transmission of the same power capacity.

The advantages of dc transmission over ac transmission should not be misunderstood. DC transmission should not be substituted for ac power transmission. In a power system, it is generally accepted that both ac and dc should be employed to compliment to each other. Integration of the two types of transmission enhances the salient features of each other and helps in realizing a power network that ensures high quality and reliability of power supply. A typical rectifier-inverter system employing a 12-pulse scheme is shown in Figure 24.43.

Typical dc link voltage can be as high as 400–600 kV. Higher voltage systems are also in use. Typical operating power levels are over 1000 MW. There are a few systems transmitting close to 3500 MW of power through two bipolar systems. Most thyristors employed in large HVDC transmission system are liquid cooled to improve their performance.

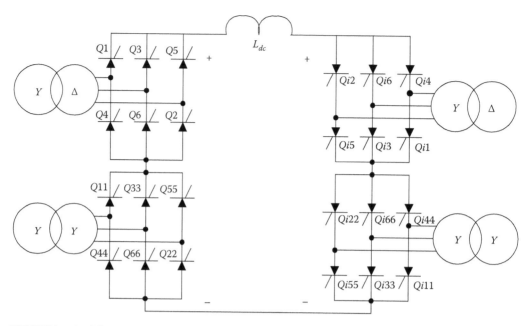

FIGURE 24.43 Schematic representation of a bipolar HVDC system employing 12-pulse rectification/inversion scheme.

24.3.7 Power System Interaction with Three-Phase Thyristor AC to DC Rectifier Systems

Similar to the diode rectifiers, the thyristor-based ac to dc rectifier is associated with low order current harmonics. In addition to current harmonics, there is voltage notching phenomenon occurring at the input terminals of an ac to dc thyristor-based rectifier system. The voltage notching is a very serious problem. Since thyristors are generally slower to turn ON and turn OFF compared to power semiconductor diodes, there are nontrivial durations during which an outgoing thyristor and an incoming thyristor remain in conduction thereby creating a short circuit across the power supply phases feeding the corresponding thyristors. Thyristors used in rectifiers are generally known as phase control type thyristors and have typical turn OFF times of 50–100 μs. Thyristors employed in inverter circuits typically are faster and have turn OFF times in the 10–50 μs ranges.

Notching can create major disturbances in sensitive electronic equipment that rely on the zero crossing of the voltage for satisfactory operation. Multiple, pseudo zero crossings of the voltage waveform can occur due to the notching effect of thyristor-based rectifier systems. Notching phenomenon can create large magnitudes of currents to flow into power-factor correcting capacitors, thereby potentially causing permanent damage to them. IEEE 519-1992 in the United States has strict regulations regarding the depth of the notch as well as the duration of the notch. AC line inductors in series with the supply feeding power to the three-phase bridge help to minimize the notching effect on the power system. The theory behind this phenomenon is discussed next.

When an external inductance is added in front of a three-phase ac to dc rectifier employing thyristors, the duration of commutation increases. In other words, the time duration for which the outgoing thyristor remains in conduction along with the incoming thyristor increases. This overlap duration causes the average output voltage to reduce because during this period, the output voltage is composed of two shorted phases and a healthy phase. The extent of reduction in the output voltage depends on the duration of overlap in electrical degrees. The duration of overlap in electrical degrees is commonly represented by μ. The overlap duration is directly proportional to the value of the external inductance used.

If no external line inductor is used, then this duration will depend on the existing inductance of the system including the wiring inductance. In order to compute the factors influencing the overlap duration, a simple model can be assumed. Assume that the line comprises inductance L in each phase. Let the dc load current be I_{dc} and let it be assumed that this current does not change during the overlap interval. The current in the incoming thyristor is zero at start and by the end of the overlap interval it increases to I_{dc}. Based on this assumption, the relationship between current and voltage can be expressed as

$$v_{ab} = \sqrt{2} * V_{L-L} * \sin(\omega t) = 2 * L * \left(\frac{di}{dt}\right)$$

$$\sqrt{2} * V_{L-L} * \int_{\alpha+(\pi/3)}^{\alpha+(\pi/3)+\mu} \sin(\omega t)\, d(t) = 2 * L * \int_{0}^{I_{dc}} di \tag{24.30}$$

$$I_{dc} = \frac{\sqrt{2} * V_{L-L} * (\cos(\alpha + \pi/3) - \cos(\alpha + \pi/3 + \mu))}{2\omega L} = \frac{\sqrt{2} * V_{L-L} * \sin(\alpha + \pi/3 + \mu/2) * \sin(\mu/2)}{\omega L}$$

For small values of overlap angle μ, $\sin(\mu/2) = \mu/2$ and $\sin(\alpha + \pi/3 + (\mu/2)) = \sin(\alpha + \pi/3)$. Rearranging the previous equation yields

$$\mu = \frac{2\omega L * I_{dc}}{\sqrt{2} * V_{L-L} * \sin(\alpha + \pi/3)} \tag{24.31}$$

From the preceding expression, it is interesting to note the following:

1. If the inductance L in the form of either external inductance or leakage inductance of transformer or lead length is large, the overlap duration will be large.
2. If the load current, I_{dc} is large, the overlap duration will be large.
3. If the delay angle is small, then the inductance will store more energy and so the duration of overlap will be large. The minimum value of delay angle α is 0° and the maximum value typically is 60°.

The average output voltage will reduce due to the overlap angle as mentioned before. In order to compute the average output voltage with a certain overlap angle, the limits of integration have to be changed. This exercise yields the following:

$$V_O = \frac{3}{\pi} \int_{\alpha+\mu+(\pi/3)}^{\alpha+\mu+(2\pi/3)} \sqrt{2}\, V_{L-L} \sin(\omega t)\, d(\omega t) \tag{24.32}$$

$$V_O = \frac{3 * \sqrt{2} * V_{L-L} * \cos(\alpha + \mu)}{\pi} = \frac{3 * \sqrt{2} * \sqrt{3} * V_{L-N} * \cos(\alpha + \mu)}{\pi}$$

Thus, it can be seen that the overlap angle has an equivalent effect of advancing the delay angle thereby reducing the average output voltage. From the discussions in the previous paragraphs on notching, it is interesting to note that adding external inductance increases the duration of the overlap and reduces the average value of the output dc voltage. However, when viewed from the ac source side, the notching effect is conspicuously reduced and in some cases not observable. Since all other electrical equipment in the system will be connected to the line side of the ac inductor (in front of a thyristor-based ac to dc rectifier), these equipments will not be affected by the notching phenomenon of thyristors. The external inductance also helps limit the circulating current between the two thyristors during the overlap duration.

24.4 Conclusion

Uncontrolled and controlled rectifier circuits have been discussed in this chapter. An introduction to the theory of diode and thyristor conduction has been presented to explain the important operating characteristics of these devices. Rectifier topologies employing both diodes and thyristors, their relative advantages and disadvantages have been discussed. The use of dual thyristor bridge converter to achieve four-quadrant operation of a dc motor has been discussed. The topic of HVDC transmission has been briefly introduced. Power quality issues relating to diode and thyristor-based rectifier topologies have also been addressed. To probe further into the various topics briefly discussed in this chapter, the reader is encouraged to refer to the references listed hereafter.

References

1. S. B. Dewan and A. Straughen, *Power Semiconductor Circuits*, John Wiley & Sons, New York, 1975. ISBN 0-471-21180-X.
2. R. G. Hoft, *Semiconductor Power Electronics*, Van Nostrand Reinhold Electrical/Computer Science and Engineering Series, Van Nostrand Reinhold Company, New York, 1986. ISBN 0-442-22543-1.
3. P. C. Sen, *Principles of Electric Machines and Power Electronics*, John Wiley & Sons, New York, 1997. ISBN 0-471-02295-0.
4. M. A. Laughton and M. G. Say (eds.), *Electrical Engineer's Reference Book – 14th Edition*, Butterworths, London, U.K., 1985. ISBN 0-408-00432-0.
5. *IEEE Recommended Practices and Requirements for Harmonic Control in Electrical Power Systems*, IEEE Std. 519-1992.
6. M. M. Swamy, Passive harmonic filter systems for variable frequency drives, U.S. Patent 5,444,609. August 1995.
7. H. Akagi, State of the art of active filters for power conditioning, Key note Speech KB 1, *EPE Conference 2005*, Dresden, Germany.
8. M. Swamy, T. J. Kume, and N. Takada, A hybrid 18 pulse rectification scheme for diode front END RECTIFIERS with large DC bus capacitor, IEEE Transactions on Industry Applications, 46(6), 2484–2494, November/December 2011.
9. M. M. Swamy, T. Kume, and N. Takada, Evaluation of an alternate soft charge circuit for diode front-end variable frequency drives, *IEEE Transactions on Industry Applications*, 46(5), 1999–2007, September/October 2010.

25

Inverters

Michael G.
Giesselmann
Texas Tech University

25.1 Introduction and Overview

Inverters are used to create single or poly-phase AC voltages from a DC supply. The DC supply is typically created by rectification of AC voltage from the utility power grid with due consideration of harmonics and input power factor. In the class of poly-phase inverters, three-phase inverters are by far the largest group. A very large number of inverters are used for adjustable speed motor drives for motors ranging from fractional hp (horsepower) to several 100 hp. Considering that more than 50% of all the electricity generated in the United States is used to drive electric motors [25, Figure 8.3], the importance of this application cannot be overstated. The typical inverter for motor drives is a "hard-switched" voltage source inverter producing pulse-width modulated (PWM) signals with a sinusoidal fundamental [11]. For large motors, multilevel inverters [19,24] are used, which are described in more detail later. Recently, research has confirmed and explained several detrimental effects such as electrical breakdown and excessive mechanical wear on motor windings and bearings respectively, resulting from unfiltered PWM waveforms from voltage source inverters. To avoid these detrimental effects, especially in the case of long cable runs between the inverter and the motor, voltage filters on the inverter outputs [8,21] can be used. Multilevel inverters besides their larger voltage and power ratings also inherently avoid this problem.

A very common application for single-phase inverters is a so-called uninterruptible power supply (UPS) for computers and other critical loads. Here the output waveforms range from square waves to almost ideal sinusoids. UPS designs are classified as either "off-line" or "online." An off-line UPS will connect the load to the utility for most of the time and quickly switch over to the inverter if the utility fails. An online UPS will always feed the load from the inverter and switch the supply of the DC bus instead. Since the DC bus is heavily buffered with capacitors, the load sees virtually no disturbance if the power fails.

In addition to the very common hard-switched inverters, active research is being conducted on "soft-switching" techniques [26]. Hard-switched inverters use controllable power semiconductors to connect an output terminal to a stable DC bus. On the other hand, soft-switching inverters have an oscillating intermediate circuit and attempt to open and close the power switches under zero-voltage and or zero-current conditions.

A separate class of inverters are the line-commutated inverters for multimegawatt power ratings that use thyristors (also called silicon controlled rectifiers [SCRs]). SCRs can only be turned "on" on command. After being turned on, the current in the device must approach zero in order for the device to turn off. All other inverters are self commutated, meaning that the power control devices can be turned on and off on command. Line-commutated inverters need the presence of a stable utility voltage to function. They are used for DC links between utilities, ultra-long distance energy transport, and very large motor drives including drives for large ships [1,2,14,16,18]. However, the technology for very large motor drives is more and more shifting to modern hard-switched inverters including multilevel inverters [7,19].

Modern inverters use isolated gate bipolar transistors (IGBTs) as the main power control devices [14]. Besides IGBTs, power MOSFETs are also used especially for lower voltages and power ratings and applications that require high efficiency and high switching frequency. In recent years, IGBTs, MOSFETs and their control and protection circuitry have made remarkable progress. IGBTs are now available with voltage ratings of 6500 V and current ratings up to 2400 A. MOSFETs have achieved on-state resistances approaching a few milliohms. In addition to the devices, manufacturers today offer customized control circuitry that provides for electrical isolation, proper operation of the devices under normal operating conditions and protection from a variety of fault conditions [14]. In addition, the industry provides good support for specialized passive devices such as capacitors and mechanical components such as low inductance bus-bar assemblies to facilitate the design of reliable inverters. In addition to the aforementioned inverters, a large number of special topologies are used. A good overview is given in Ref. [10].

25.2 Fundamental Issues

Inverters fall in the class of power electronics circuits. The most widely accepted definition of a power electronics circuit is that the circuit is actually processing electric energy rather than information. The actual power level is not very important for the classification of a circuit as a power electronics circuit. One of the most important performance considerations of power electronics circuits like inverters is their energy conversion efficiency. The most important reason for demanding high efficiency is the problem of removing large amounts of heat from the power devices. Of course, the judicious use of energy is also paramount, especially if the inverter is fed from batteries such as in electric cars. For these reasons, inverters operate the power devices, which control the flow of energy, as switches. In the case of an ideal switching event, there would be no power loss in the switch since either the current in the switch is zero (switch open) or the voltage across the switch is zero (switch closed) and the power loss is computed as the product of both. In reality, there are two mechanisms that do create some losses, however, which are on-state losses and switching losses [3,12,14,16]. On-state losses are due to the fact that the voltage across the switch in the on state is not zero, but typically in the range of 1–3 V for IGBTs. For power MOSFETs, the on-state voltage is often in the same range, but it can be substantially below 0.5 V due to the fact that these devices have a purely resistive conduction channel and no fixed minimum saturation voltage like bipolar junction devices (IGBTs). The switching losses are the second major loss mechanism and are due to the fact that, during the turn on and turn off transition, current is flowing while voltage is present across the device. In order to minimize the switching losses, the individual transitions have to be rapid (tens to hundreds of nanoseconds) and the maximum switching frequency (determining the frequency of transitions) needs to be carefully considered.

In order to avoid audible noise being radiated from motor windings or transformers, many modern inverters operate at switching frequencies substantially above 10 kHz [4,6].

25.3 Single-Phase Inverters

Figure 25.1 shows the basic topology of a full bridge inverter with single-phase output. This configuration is often called an H-bridge due to the arrangement of the power switches and the load. The inverter can deliver and accept both real and reactive power. The inverter has two legs, left and right. Each leg consists of two power control devices (here IGBTs) connected in series. The load is connected between the mid-points of the two-phase legs. Each power control device has a diode connected in anti-parallel

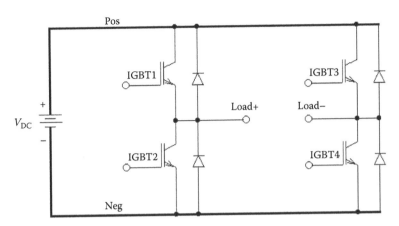

FIGURE 25.1   Topology of a single-phase, full bridge inverter.

to it. The diodes provide an alternate path for the load current if the power switches are turned off. For example, if the lower IGBT in the left leg is conducting and carrying current toward the negative DC bus, this current would "commutate" into the diode across the upper IGBT of the left leg, if the lower IGBT is turned off. Control of the circuit is accomplished by varying the turn on time of the upper and lower IGBT of each inverter leg, with the provision of never turning on both at the same time, to avoid a short circuit of the DC bus. In fact, modern drivers will not allow this to happen, even if the controller would erroneously command both devices to be turned on. The controller will therefore alternate the turn on commands for the upper and lower switch, that is, turn the upper switch on and the lower switch off and vice versa. The driver circuit will typically add some additional blanking time (typically 500–1000 ns) during the switch transitions to avoid any overlap in the conduction intervals.

The controller will hereby control the duty cycle of the conduction phase of the switches. The average potential of the center-point of each leg will be given by the DC bus voltage multiplied by the duty cycle of the upper switch, if the negative side of the DC bus is used as a reference. If this duty cycle is modulated with a sinusoidal signal with a frequency that is much smaller than the switching frequency, the short-term average of the center-point potential will follow the modulation signal. "Short-term" in this context means a small fraction of the period of the fundamental output frequency to be produced by the inverter. For the single-phase inverter, the modulation of the two legs is inverse of each other such that if the left leg has a large duty cycle for the upper switch, the right leg has a small one, etc. The output voltage is then given by Equation 25.1 in which m_a is the modulation factor. The boundaries for m_a are for linear modulation—values greater than 1 cause over-modulation and a noticeable increase in output voltage distortion:

$$V_{AC1}(t) = m_a \cdot V_{DC} \cdot \sin(\omega_1 \cdot t) \quad 0 \le m_a \le 1 \tag{25.1}$$

This voltage can be filtered using an LC low-pass filter. The voltage on the output of the filter will closely resemble the shape and frequency of the modulation signal. This means that the frequency, wave-shape, and amplitude of the inverter output voltage can all be controlled as long as the switching frequency is at least 25–100 times higher than the fundamental output frequency of the inverter [11]. The actual generation of the PWM signals is mostly done using micro-controllers and digital signal processors (DSPs) [5].

25.4 Three-Phase Inverters

Figure 25.2 shows a three-phase inverter, the most widely used topology in today's motor drives. The circuit is basically an extension of the H-bridge style single-phase inverter by an additional leg. The control strategy is similar to the control of the single-phase inverter, except that the reference signals

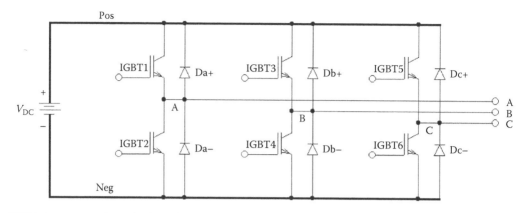

FIGURE 25.2   Topology of a three-phase inverter.

for the different legs have a phase shift of 120° instead of 180° for the single-phase inverter. Due to this phase shift, the odd triplen harmonics (3rd, 9th, 15th, etc.) of the reference waveform for each leg are eliminated from the line-to-line output voltage [14,17]. The even-numbered harmonics are canceled as well if the output voltages are purely AC, which is usually the case. For linear modulation, the amplitude of the AC output voltage is reduced with respect to the AC input voltage of a three-phase rectifier feeding the DC bus by a factor given by the following equation:

$$\frac{3}{(2\pi)} \cdot \sqrt{3} = 82.7\% \qquad (25.2)$$

To compensate for this voltage reduction, the amplitudes of the output voltages are sometimes boosted by intentionally injecting a third harmonic component into the reference waveform of each phase leg [14]. These third harmonic components cancel in the line-to-line output voltage that is applied to the motor.

Figure 25.3 shows the typical output of a three-phase inverter during a startup transient into a typical motor load. This figure was created using circuit simulation. The upper graph shows the PWM waveform between phases *A* and *B* whereas the lower graph shows the currents in all three phases. It is obvious that the motor acts a low-pass filter for the applied PWM voltage and the current assumes the wave-shape of the fundamental modulation signal with very small amounts of switching ripple.

Like the single-phase inverter based on the H-bridge topology, the inverter can deliver and accept both real and reactive power. In many cases the DC bus is fed by a diode rectifier from the utility, which cannot pass power back to the AC input. The topology of such a three-phase rectifier would be the same as shown in Figure 25.2 with all IGBTs deleted.

A reversal of power flow in an inverter with a rectifier front end would lead to a steady rise of the DC bus voltage beyond permissible levels. If the power flow to the load is only reversing for brief periods of time, such as to brake a motor occasionally, the DC bus voltage could be limited by dissipating the power in a so-called brake resistor. To accommodate a brake resistor, inverter modules with an additional seventh IGBT (called "brake chopper") are offered. This is shown in Figure 25.4. For long-term regeneration, the rectifier can be replaced by an additional three-phase inverter [14]. This additional inverter is often called a controlled synchronous rectifier. The additional inverter including its controller is of course much more expensive than a simple rectifier, but with this arrangement bidirectional power flow can be achieved. In addition, the interface toward the utility system can be managed such that the real and reactive power that is drawn from or delivered to the utility can be independently controlled. Also the harmonics content of the current in the utility link can be reduced to almost zero. The topology for an arrangement like this is shown in Figure 25.5.

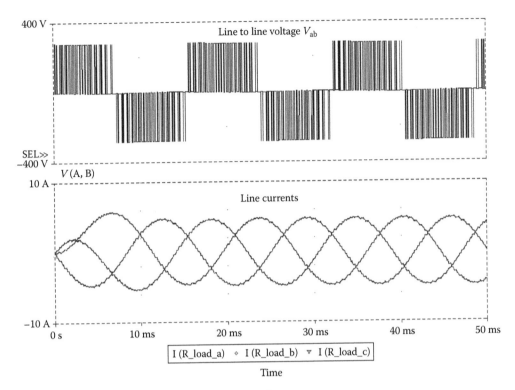

FIGURE 25.3 Typical waveforms of inverter voltages and currents.

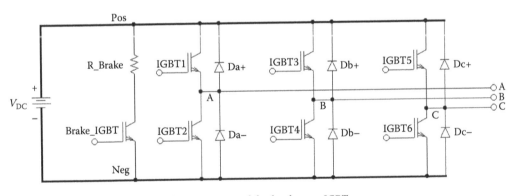

FIGURE 25.4 Topology of a three-phase inverter with brake chopper IGBT.

The inverter shown in Figure 25.2 provides a three-phase voltage without a neutral point. A fourth leg can be added to provide a four-wire system with a neutral point. Likewise 4-, 5-, or n-phase inverters can be realized by simply adding the appropriate number of phase legs.

Like in single-phase inverters, the generation of the PWM control signals is done using modern micro-controllers and DSPs. These digital controllers are typically not only controlling just the inverter, but through the controlled synthesis of the appropriate voltages, motors and attached loads are also controlled for high performance dynamic response. The most commonly used control principle for superior dynamic response is called field-oriented or vector control [5,6,9,13,20].

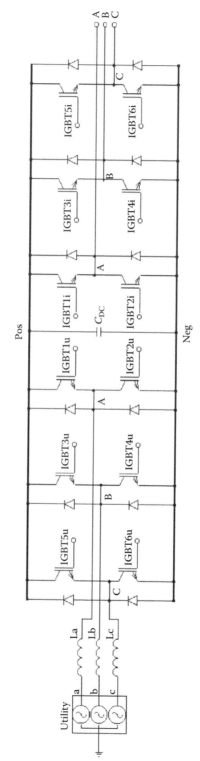

FIGURE 25.5 Topology of a three-phase inverter system for bidirectional power flow.

25.5 Multilevel Inverters

Multilevel inverters are a class of inverters where a DC source with several tabs between the positive and negative terminal or multiple isolated DC sources are present. The two main advantages of multilevel inverters are the much higher voltage and power capability and the reduced harmonics content of the output waveform due to the multiple DC levels. The higher voltage capability is due to the fact that clamping diodes are used to limit the nominal voltage stress on the IGBTs to the voltage differential between two tabs on the DC bus. In the case of cascaded H-Bridge converters, the voltage stress on each H-Bridge is limited to the level of the isolated DC source, which makes the topology completely modular with no theoretical voltage (and power) limit.

Figure 25.6 shows the topology of a three-level, three-phase inverter, which is also called "neutral point clamped" inverter. Here each phase leg consists of four IGBTs in series with additional anti-parallel and clamping diodes. The output is at the center-point of the phase leg. By switching on IGBT pairs 1 and 2, 2 and 3, or 3 and 4, respectively, the output of each phase can be connected to the top DC bus, the center connection of the DC supply, or the negative DC bus. This amounts to three distinct voltage levels for the voltage of each phase, which explains the name of the circuit. It turns out that the resulting line-to-line voltage has five distinct levels in a three-phase inverter. In each case, the nominal voltage stress on the IGBTs in the off state amounts to half the voltage between the positive and the negative rail.

Another inverter topology which inherently limits the voltage stress on the IGBTs to the DC-bus voltages of the individual H-Bridges is shown in Figure 25.7. Inverters that are based on this basic principle of cascaded H-bridges enable the synthesis of very high-quality output voltage waveforms with large numbers of individual voltage levels. Shown in the following are two possible examples of an almost limitless number of possible variations.

The schematic shown in Figure 25.7 represents one phase of a multiphase topology. The hardware configuration of the additional phases is identical. This inverter is called an asymmetric cascaded H-bridge inverter because the individual H-bridges have unequal DC bus voltages [22]. Symmetric cascaded inverters with identical DC bus voltages are also being used. The specific advantage of asymmetric inverters is the large number of voltage levels relative to the number of modules. For this example,

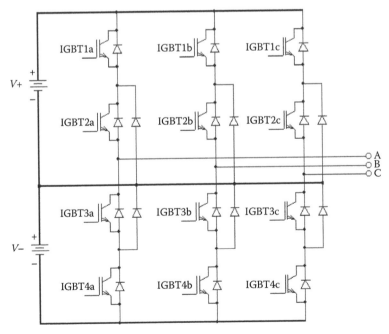

FIGURE 25.6 Topology of a three-level inverter.

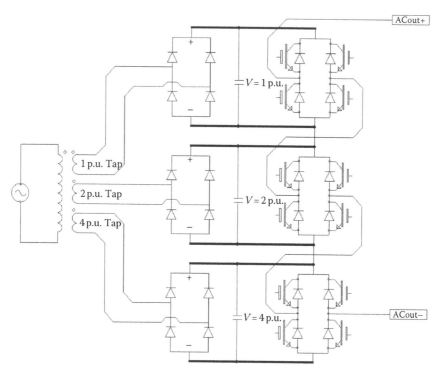

FIGURE 25.7   Topology of a three-stage, asymmetric cascaded H-bridge inverter.

the DC bus voltages of the stages have a ratio of 1:2:4 starting from the top stage. The output of each H-bridge can have three distinct values:

- Positive   DC bus voltage
- Negative   DC bus voltage
- Zero

Therefore, with the topology shown in Figure 25.7, the maximum positive and negative output voltage of the entire cascade 1 + 2 + 4 = 7 times the DC bus voltage of the upper stage. Furthermore, by appropriate selection of the switching states of the individual stages, all intermediate voltage levels including zero can be synthesized. Therefore, the total number of output levels is 2 * (1 + 2 + 4) + 1 = 15 for this inverter [22].

Figure 25.8 shows the reference waveforms and PWM output voltages for all three stages as well as the total output voltage of the entire cascade. The waveforms for the H-bridge with 4 p.u. DC bus voltage are shown in the uppermost diagram of Figure 25.8. It is apparent that this stage is only switching at the fundamental frequency which is appropriate for the high-voltage IGBTs that have to be used in this stage. The reference waveform of the H-bridge with 2 p.u. DC bus voltage is derived from the overall sinusoidal reference waveform by subtracting the output voltage of the 4 p.u. stage. In the same manner, the reference voltage for the 1 p.u. stage is obtained by subtracting the combined output of the lower stages from the sinusoidal reference. This last stage is essentially compensating for the difference between the reference and the output of the lower two stages, which provide the bulk of the voltage and power. As shown in Figure 25.8, each successive stage with lower DC bus voltage has a higher switching frequency. More details about the possible ratios of the DC bus voltages can be found in Ref. [22]. In Ref. [23] the value of multilevel inverters for utility applications is discussed in detail. The advantages of the cascaded H-bridge inverter in this field are its modularity, expandability, and superb output waveform quality.

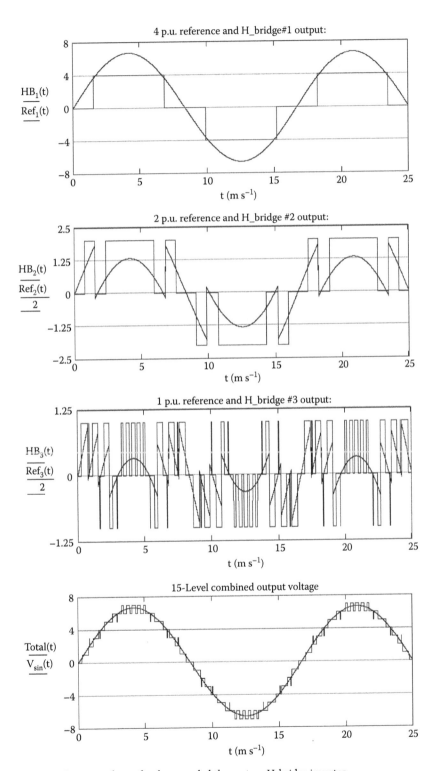

FIGURE 25.8   Switching waveforms for the cascaded three-stage H-bridge inverter.

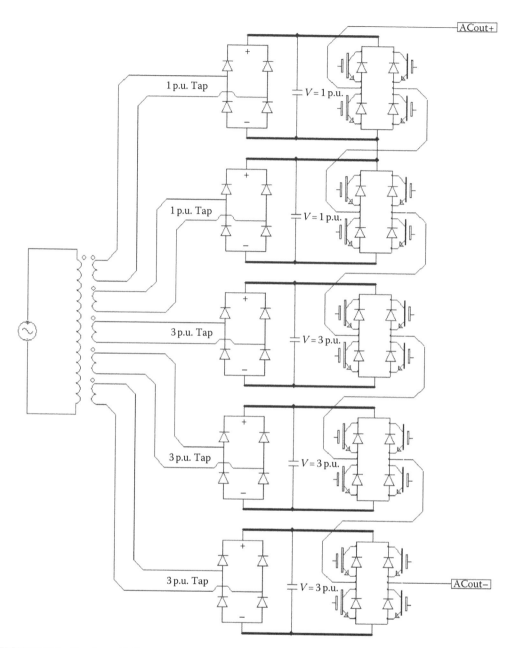

FIGURE 25.9 Topology of a five-stage, asymmetric cascaded H-bridge inverter.

While the inverter shown in Figure 25.7 illustrates an example that maximizes the number of output levels with a limited number of stages, the topology represented by Figure 25.9 (again for one phase) shows a more commonly used configuration that has only two different module elements. The inverter has three high-voltage modules which have a DC voltage that is three times higher than the DC bus voltage of the low-voltage stages. All the high-voltage stages are switched at the fundamental frequency, while the low-voltage stages are switched at 10 times the fundamental frequency. The total number of output levels is 2 * (1 + 1 + 3 + 3 + 3) + 1 = 23 for this inverter [22]. The upper graph in Figure 25.10 shows the typical stair-step output voltage of the three lower stages at 90% modulation level switched at the fundamental frequency. In order to equalize the heat load on the inverter stages, the modulation

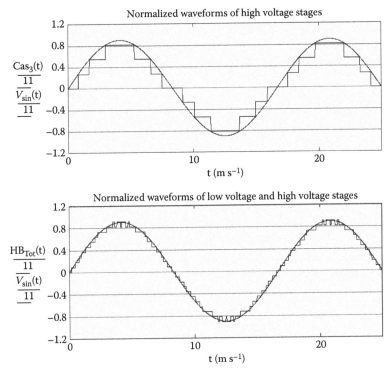

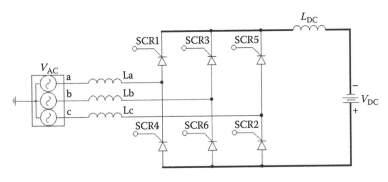

FIGURE 25.10 Switching waveforms for the cascaded five-stage H-bridge inverter.

patterns are typically cyclically rotated between modules of the same DC voltage level. The lower graph in Figure 25.10 shows the combined output waveform of the high- and low-voltage stages of the inverter phase leg along with an ideal sinusoidal reference for 90% modulation.

25.6 Line-Commutated Inverters

Figure 25.11 shows the topology of a line-commutated inverter. In Figure 25.11, the SCRs are numbered according to their firing sequence. The circuit can operate both as a rectifier and an inverter. The mode of operation is controlled by the firing angle of the SCRs in the circuit [1,2,14]. The reference value for the firing angle α is the instant when the voltage across each SCR becomes positive, that is, when an uncontrolled diode would turn on. This time corresponds to 30° past the positive going zero crossing of each phase. By delaying the turn-on angle α more than 90° past this instant, the polarity of the average DC bus voltage reverses and the circuit enters the inverter mode. The DC source in Figure 25.11 shows

FIGURE 25.11 Line-commutated converter in inverter mode.

the polarity of the DC voltage for inverter operation. The firing delay angle corresponds to the phase of the utility voltage. The maximum delay angle must be limited to less than 180°, to provide enough time for the next SCR in the sequence to acquire the load current. Equation 25.3 gives the value of the DC output voltage of the converter as a function of the delay angle α and the DC current I_{DC}, which is considered constant:

$$V_{DC} = \frac{3}{\pi} \cdot \left(\sqrt{2} \cdot V_{LL} \cdot \cos(\alpha) - \omega \cdot L_s \cdot I_{DC} \right) \tag{25.3}$$

where
  V_{LL} is the rms value of the AC line-to-line voltage
  ω is the radian frequency of the AC voltage

L_s is the value of the source impedance represented by inductors La, Lb, and Lc in Figure 25.11. Line-commutated inverters have a negative impact on the utility voltage and a relatively low total power factor. Equation 25.4 gives an estimate of the total power factor of the circuit shown in Figure 25.11 for constant DC current and negligible AC line reactors:

$$PE = \frac{3}{\pi} \cdot \cos(\alpha) \tag{25.4}$$

References

1. Ahmed, A., *Power Electronics for Technology*, Prentice Hall, Upper Saddle River, NJ, 1999, ISBN: 0-13-231069-4.
2. Barton, T.H., *Rectifiers, Cycloconverters, and AC Controllers*, Oxford University Press Inc., New York, 1994, ISBN: 0-19-856163-6.
3. Bird, B.M., King, K.G., and Pedder, D.A.G., *An Introduction to Power Electronics*, 2nd edn., John Wiley & Sons Inc., New York, 1993, ISBN: 0-471-92616-7.
4. Bose, B.K., *Modern Power Electronics, Evolution, Technology, and Applications*, IEEE Press, Piscataway, NJ, 1992, ISBN: 0-87942-282-3.
5. Bose, B.K., *Microcomputer Control of Power Electronics and Drives*, IEEE Press, Piscataway, NJ, 1987, ISBN: 0-87942-219-X.
6. Bose, B.K., *Power Electronics and Variable Frequency Drives*, IEEE Press, Piscataway, NJ, 1996.
7. Brumsickle, W.E., Divan, D.M., and Lipo, T.A., Reduced switching stress in high-voltage IGBT inverters via a three-level structure, *IEEE-APEC* 2, Anaheim, CA, February 1998, pp. 544–550.
8. Cash, M.A. and Habetler, T.G., Insulation failure prediction in induction machines using line-neutral voltages, *IEEE Transactions on Industry Applications*, 34(6), 1234–1239, November/December 1998.
9. De Donker, R. and Novotny, D.W., The universal field-oriented controller, *Conference Record, IEEE-IAS*, Minneapolis, MN, 1988, pp. 450–456.
10. Gottlieb, I.M., *Power Supplies, Switching Regulators, Inverters and Converters*, TAB Books Inc., Blue Ridge Summit, PA, 1984, ISBN: 0-8306-0665-3.
11. Holtz, J., Pulsewidth modulation—A survey, *IEEE Transactions on Industrial Electronics*, 39(5), 410–420, 1992.
12. Kassakian, J.G., Schlecht, M.F., and Verghese, G.C., *Principles of Power Electronics*, Addison-Wesley Publishing Company, New York, 1991, ISBN: 0-201-09689-7.

13. Lorenz, R.D. and Divan, D.M., Dynamic analysis and experimental evaluation of delta modulators for field oriented induction machines, *IEEE Transactions on Industry Applications*, 26(2), 296–301, 1990.
14. Mohan, N., Undeland, T., and Robbins, W., Eds., *Power Electronics, Converters, Applications, and Design*, 2nd edn., John Wiley & Sons, New York, 1995, ISBN: 0-471-58408-8.
15. Novotny, D.W. and Lipo, T.A., *Vector Control and Dynamics of AC Drives*, Oxford Science Publications, Oxford, U.K., 1996, ISBN: 0-19-856439-2.
16. Rashid, M.H., Ed., *Power Electronics, Circuits, Devices, and Applications*, 2nd edn., Prentice Hall, Englewood Cliffs, NJ, 1993, ISBN: 0-13-678996-X.
17. Shepherd, W. and Zand, P., *Energy Flow and Power Factor in Nonsinusoidal Circuits*, Cambridge University Press, London, U.K., 1979, ISBN: 0521-21990-6.
18. Tarter, R.E., *Solid State Power Conversion Handbook*, John Wiley & Sons Inc., New York, 1993, ISBN: 0-471-57243-8.
19. Tolbert, L.M., Peng, F.Z., and Habetler, T.G., Multilevel converters for large electric drives, *IEEE Transactions on Industry Applications*, 35(1), 36–44, January/February 1999.
20. Trzynadlowski, A.M., *The Field Orientation Principle in Control of Induction Motors*, Kluwer Academic, Dordrecht, the Netherlands, 1994, ISBN: 0-7923-9420-8.
21. Von Jouanne, A., Rendusara, D., Enjeti, P., and Gray, W., Filtering techniques to minimize the effect of long motor leads on PWM inverter fed AC motor drive systems, *IEEE Transactions on Industry Applications*, 32(4), 919–926, July/August 1996.
22. Song-Manguelle, J. and Rufer, A., Asymmetrical multilevel inverter for large induction machine drives, *Electric Drives and Power Electronics International Conference (EDPE 01), EDPE 2001: The 14th International Conference on Electrical Drives and Power Electronics*, The High Tatras, Slovakia, October 3–5, 2001.
23. Rudnick, H., Dixon, J., and Morán, L., Delivering clean and pure power, *IEEE Power and Energy Magazine*, 1(5), 32–40, September/October 2003.
24. Rodríguez, J., Lai, J.-S., and Peng, F.Z., Multilevel inverters: A survey of topologies, controls, and applications, *IEEE Transactions on Industrial Electronics*, 49(4), 724–738, August 2002.
25. Mohan, N., *First Course on Power Systems*, 2006 edn., MNPERE, Minneapolis, MN, 2006, ISBN 0-9715292-7-2.
26. Divan, D.M., Low stress switching for efficiency, *IEEE Spectrum*, 33(12), 33–39, December 1996.

26

Active Filters for Power Conditioning

Hirofumi Akagi
*Tokyo Institute of
Technology*

Much research has been performed on active filters for power conditioning and their practical applications since their basic principles of compensation were proposed around 1970 (Bird et al., 1969; Gyugyi and Strycula, 1976; Kawahira et al., 1983). In particular, recent remarkable progress in the capacity and switching speed of power semiconductor devices such as insulated-gate bipolar transistors (IGBTs) has spurred interest in active filters for power conditioning. In addition, state-of-the-art power electronics technology has enabled active filters to be put into practical use. More than one thousand sets of active filters consisting of voltage-fed pulse-width-modulation (PWM) inverters using IGBTs or gate-turn-off (GTO) thyristors are operating successfully in Japan.

Active filters for power conditioning provide the following functions:

- Reactive-power compensation
- Harmonic compensation, harmonic isolation, harmonic damping, and harmonic termination
- Negative-sequence current/voltage compensation
- Voltage regulation

The term "active filters" is also used in the field of signal processing. In order to distinguish active filters in power processing from active filters in signal processing, the term "active power filters" often appears in many technical papers or literature. However, the author prefers "active filters for power conditioning" to "active power filters," because the term "active power filters" is misleading to either "active filters for power" or "filters for active power." Therefore, this section takes the term "active filters for power conditioning" or simply uses the term "active filters" as long as no confusion occurs.

26.1 Harmonic-Producing Loads

26.1.1 Identified Loads and Unidentified Loads

Nonlinear loads drawing nonsinusoidal currents from utilities are classified into identified and unidentified loads. High-power diode/thyristor rectifiers, cycloconverters, and arc furnaces are typically characterized as identified harmonic-producing loads because utilities identify the individual nonlinear loads installed by high-power consumers on power distribution systems in many cases. The utilities determine the point of common coupling with high-power consumers who install their own harmonic-producing loads on power distribution systems, and also can determine the amount of harmonic current injected from an individual consumer.

A "single" low-power diode rectifier produces a negligible amount of harmonic current. However, multiple low-power diode rectifiers can inject a large amount of harmonics into power distribution systems. A low-power diode rectifier used as a utility interface in an electric appliance is typically considered as an unidentified harmonic-producing load. Attention should be paid to unidentified harmonic-producing loads as well as identified harmonic-producing loads.

26.1.2 Harmonic Current Sources and Harmonic Voltage Sources

In many cases, a harmonic-producing load can be represented by either a harmonic current source or a harmonic voltage source from a practical point of view. Figure 26.1a shows a three-phase diode rectifier with a DC link inductor L_d. When attention is paid to voltage and current harmonics, the rectifier can be considered as a harmonic current source shown in Figure 26.1b. The reason is that the load impedance is much larger than the supply impedance for harmonic frequency ω_h, as follows:

$$\sqrt{R_L^2 + (\omega_h L_d)^2} \gg \omega_h L_S.$$

Here, L_S is the sum of supply inductance existing upstream of the point of common coupling (PCC) and leakage inductance of a rectifier transformer. Note that the rectifier transformer is disregarded from Figure 26.1a. Figure 26.1b suggests that the supply harmonic current i_{Sh} is independent of L_S.

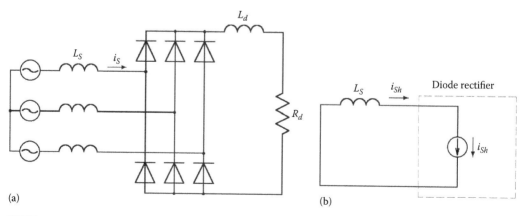

(a)   (b)

FIGURE 26.1 Diode rectifier with inductive load. (a) Power circuit. (b) Equivalent circuit for harmonic on a per-phase base.

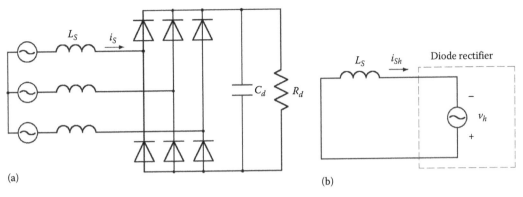

(a)   (b)

FIGURE 26.2 Diode rectifier with capacitive load. (a) Power circuit. (b) Equivalent circuit for harmonic on a per-phase base.

Figure 26.2a shows a three-phase diode rectifier with a DC link capacitor. The rectifier would be characterized as a harmonic voltage source shown in Figure 26.2b if it is seen from its AC terminals. The reason is that the following relation exists:

$$\frac{1}{\omega_h C_d} \ll \omega_h L_S.$$

This implies that i_{Sh} is strongly influenced by the inductance value of L_S.

26.2 Theoretical Approach to Active Filters for Power Conditioning

26.2.1 The Akagi-Nabae Theory

The theory of instantaneous power in three-phase circuits is referred to as the "Akagi-Nabae theory" (Akagi et al., 1983, 1984). Figure 26.3 shows a three-phase three-wire system on the *a-b-c* coordinates, where no zero-sequence voltage is included in the three-phase three-wire system. Applying the theory to Figure 26.3 can transform the three-phase voltages and currents on the *a-b-c* coordinates into the two-phase voltages and currents on the α-β coordinates, as follows:

$$\begin{bmatrix} e_\alpha \\ e_\beta \end{bmatrix} = \sqrt{\frac{2}{3}} \begin{bmatrix} 1 & -1/2 & -1/2 \\ 0 & \sqrt{3}/2 & -\sqrt{3}/2 \end{bmatrix} \begin{bmatrix} e_a \\ e_b \\ e_c \end{bmatrix} \qquad (26.1)$$

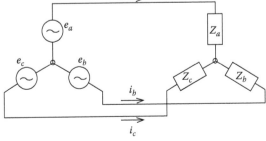

FIGURE 26.3 Three-phase three-wire system.

$$\begin{bmatrix} i_\alpha \\ i_\beta \end{bmatrix} = \sqrt{\frac{2}{3}} \begin{bmatrix} 1 & -1/2 & -1/2 \\ 0 & \sqrt{3}/2 & -\sqrt{3}/2 \end{bmatrix} \begin{bmatrix} i_a \\ i_b \\ i_c \end{bmatrix}. \tag{26.2}$$

As is well known, the instantaneous real power either on the *a-b-c* coordinates or on the α-β coordinates is defined by

$$p = e_a i_a + e_b i_b + e_c i_c = e_\alpha i_\alpha + e_\beta i_\beta. \tag{26.3}$$

To avoid confusion, *p* is referred to as three-phase instantaneous real power. According to the theory, the three-phase instantaneous imaginary power, *q*, is defined by

$$q = e_\alpha i_\beta - e_\beta i_\alpha. \tag{26.4}$$

The combination of the above two equations bears the following basic formulation:

$$\begin{bmatrix} p \\ q \end{bmatrix} = \begin{bmatrix} e_\alpha & e_\beta \\ -e_\beta & e_\alpha \end{bmatrix} \begin{bmatrix} i_\alpha \\ i_\beta \end{bmatrix}. \tag{26.5}$$

Here, $e_\alpha \dots i_\alpha$ or $e_\beta \dots i_\beta$ obviously means instantaneous power in the α-phase or the β-phase because either is defined by the product of the instantaneous voltage in one phase and the instantaneous current in the same phase. Therefore, *p* has a dimension of [W]. Conversely, neither $e_\alpha \cdot i_\beta$ nor $e_\beta \cdot i_\alpha$ means instantaneous power because either is defined by the product of the instantaneous voltage in one phase and the instantaneous current in the other phase. Accordingly, *q* is quite different from *p* in dimension and electric property although *q* looks similar in formulation to *p*. A common dimension for *q* should be introduced from both theoretical and practical points of view. A good candidate is [IW], that is, "imaginary watt."

Equation 26.5 is changed into the following equation:

$$\begin{bmatrix} i_\alpha \\ i_\beta \end{bmatrix} = \begin{bmatrix} e_\alpha & e_\beta \\ -e_\beta & e_\alpha \end{bmatrix} \begin{bmatrix} p \\ q \end{bmatrix} \tag{26.6}$$

Note that the determinant with respect to e_α and e_β in Equation 26.5 is not zero. The instantaneous currents on the α-β coordinates, i_α and i_β, are divided into two kinds of instantaneous current components, respectively:

$$\begin{bmatrix} i_\alpha \\ i_\beta \end{bmatrix} = \begin{bmatrix} e_\alpha & e_\beta \\ -e_\beta & e_\alpha \end{bmatrix}^{-1} \begin{bmatrix} p \\ 0 \end{bmatrix} + \begin{bmatrix} e_\alpha & e_\beta \\ -e_\beta & e_\alpha \end{bmatrix}^{-1} \begin{bmatrix} 0 \\ q \end{bmatrix}$$

$$\equiv \begin{bmatrix} i_{\alpha p} \\ i_{\beta p} \end{bmatrix} + \begin{bmatrix} i_{\alpha q} \\ i_{\beta q} \end{bmatrix} \tag{26.7}$$

Let the instantaneous powers in the α-phase and the β-phase be p_α and p_β, respectively. They are given by the conventional definition as follows:

$$\begin{bmatrix} p_\alpha \\ p_\beta \end{bmatrix} = \begin{bmatrix} e_\alpha i_\alpha \\ e_\beta i_\beta \end{bmatrix} = \begin{bmatrix} e_\alpha i_{\alpha p} \\ e_\beta i_{\beta p} \end{bmatrix} + \begin{bmatrix} e_\alpha i_{\alpha q} \\ e_\beta i_{\beta q} \end{bmatrix} \tag{26.8}$$

The three-phase instantaneous real power, p, is given as follows, by using Equations 26.7 and 26.8:

$$p = p_\alpha + p_\beta = e_\alpha i_{\alpha p} + e_\beta i_{\beta p} + e_\alpha i_{\alpha q} + e_\beta i_{\beta q}$$

$$= \frac{e_\alpha^2}{e_\alpha^2 + e_\beta^2} p + \frac{e_\beta^2}{e_\alpha^2 + e_\beta^2} p + \frac{-e_\alpha e_\beta}{e_\alpha^2 + e_\beta^2} q + \frac{e_\alpha e_\beta}{e_\alpha^2 + e_\beta^2} q \qquad (26.9)$$

The sum of the third and fourth terms on the right-hand side in Equation 26.9 is always zero. From Equations 26.8 and 26.9, the following equations are obtained:

$$p = e_\alpha i_{\alpha p} + e_\beta i_{\beta p} \equiv p_{\alpha p} + p_{\beta p} \qquad (26.10)$$

$$0 = e_\alpha i_{\alpha q} + e_\beta i_{\beta q} \equiv p_{\alpha q} + p_{\beta q}. \qquad (26.11)$$

Inspection of Equations 26.10 and 26.11 leads to the following essential conclusions:

- The sum of the power components, $p_{\alpha p}$ and $p_{\beta p}$, coincides with the three-phase instantaneous real power, p, which is given by Equation 26.3. Therefore, $p_{\alpha p}$ and $p_{\beta p}$ are referred to as the α-phase and β-phase instantaneous active powers.
- The other power components, $p_{\alpha q}$ and $p_{\beta q}$, cancel each other and make no contribution to the instantaneous power flow from the source to the load. Therefore, $p_{\alpha q}$ and $p_{\beta q}$ are referred to as the α-phase and β-phase instantaneous reactive powers.
- Thus, a shunt active filter without energy storage can achieve instantaneous compensation of the current components, $i_{\alpha q}$ and $i_{\beta q}$ or the power components, $p_{\alpha q}$ and $p_{\beta q}$. In other words, the Akagi-Nabae theory based on Equation 26.5 exactly reveals what components the active filter without energy storage can eliminate from the α-phase and β-phase instantaneous currents, i_α and i_β or the α-phase and β-phase instantaneous real powers, p_α and p_β.

26.2.2 Energy Storage Capacity

Figure 26.4 shows a system configuration of a shunt active filter for harmonic compensation of a diode rectifier, where the main circuit of the active filter consists of a three-phase voltage-fed PWM inverter and a DC capacitor, C_d. The active filter is controlled to draw the compensating current, i_{AF}, from the utility, so that the compensating current cancels the harmonic current flowing on the AC side of the diode rectifier with a DC link inductor.

Referring to Equation 26.6 yields the α-phase and β-phase compensating currents,

$$\begin{bmatrix} i_{AF\alpha} \\ i_{AF\beta} \end{bmatrix} = \begin{bmatrix} e_\alpha & e_\beta \\ -e_\beta & e_\alpha \end{bmatrix}^{-1} \begin{bmatrix} p_{AF} \\ q_{AF} \end{bmatrix}. \qquad (26.12)$$

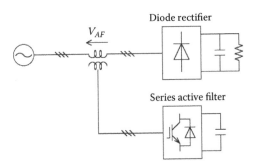

FIGURE 26.4 Shunt active filter.

Here, p_{AF} and q_{AF} are the three-phase instantaneous real and imaginary power on the AC side of the active filter, and they are usually extracted from p_L and q_L. Note that p_L and q_L are the three-phase instantaneous real and imaginary power on the AC side of a harmonic-producing load. For instance, when the active filter compensates for the harmonic current produced by the load, the following relationships exist:

$$p_{AF} = -\tilde{p}_L, \quad q_{AF} = -\tilde{q}_L. \tag{26.13}$$

Here, $\tilde{p}_L$ and $\tilde{q}_L$ are AC components of p_L and q_L, respectively. Note that the DC components of p_L and q_L correspond to the fundamental current present in i_L and the AC components to the harmonic current. In general, two high-pass filters in the control circuit extract $\tilde{p}_L$ from p_L and $\tilde{q}_L$ from q_L.

The active filter draws p_{AF} from the utility, and delivers it to the DC capacitor if no loss is dissipated in the active filter. Thus, p_{AF} induces voltage fluctuation of the DC capacitor. When the amplitude of p_{AF} is assumed to be constant, the lower the frequency of the AC component, the larger the voltage fluctuation (Akagi et al., 1984, 1986). If the period of the AC component is 1 h, the DC capacitor has to absorb or release electric energy given by integration of p_{AF} with respect to time. Thus, the following relationship exists between the instantaneous voltage across the DC capacitor, v_d and p_{AF}:

$$\frac{1}{2} C_d v_d^2(t) = \frac{1}{2} C_d v_d^2(0) + \int_0^t p_{AF}\, dt. \tag{26.14}$$

This implies that the active filter needs an extremely large-capacity DC capacitor to suppress the voltage fluctuation coming from achieving "harmonic" compensation of $\tilde{p}_L$. Hence, the active filter is no longer a harmonic compensator, and thereby it should be referred to as a "DC capacitor-based energy storage system," although it is impractical at present. In this case, the main purpose of the voltage-fed PWM inverter is to perform an interface between the utility and the bulky DC capacitor.

The active filter seems to "draw" q_{AF} from the utility, as shown in Figure 26.4. However, q_{AF} makes no contribution to energy transfer in the three-phase circuit. No energy storage, therefore, is required to the active filter, independent of q_{AF}, whenever $p_{AF} = 0$.

26.2.3 Classification of Active Filters

Various types of active filters have been proposed in technical literature (Moran, 1989; Grady et al., 1990; Akagi, 1994; Akagi and Fujita, 1995; Fujita and Akagi, 1997; Aredes et al., 1998). Classification of active filters is made from different points of view (Akagi, 1996). Active filters are divided into AC and DC filters. Active DC filters have been designed to compensate for current and/or voltage harmonics on the DC side of thyristor converters for high-voltage DC transmission systems (Watanabe, 1990; Zhang et al., 1993) and on the DC link of a PWM rectifier/inverter for traction systems. Emphasis, however, is put on active AC filter in the following because the term "active filters" refers to active AC filters in most cases.

26.2.4 Classification by Objectives: Who Is Responsible for Installing Active Filters?

The objective of "who is responsible for installing active filters" classifies them into the following two groups:

- Active filters installed by *individual consumers* on their own premises in the vicinity of one or more identified harmonic-producing loads.
- Active filters being installed by *electric power utilities* in substations and/or on distribution feeders.

Individual consumers should pay attention to current harmonics produced by their own harmonic-producing loads, and thereby the active filters installed by the individual consumers are aimed at compensating for current harmonics.

Utilities should concern themselves with voltage harmonics, and therefore active filters will be installed by utilities in the near future for the purpose of compensating for voltage harmonics and/or of achieving "harmonic damping" throughout power distribution systems or "harmonic termination" of a radial power distribution feeder. Section 26.4 describes a shunt active filter intended for installation by electric power utilities on the end bus of a power distribution line.

26.2.5 Classification by System Configuration

26.2.5.1 Shunt Active Filters and Series Active Filters

A standalone shunt active filter shown in Figure 26.4 is one of the most fundamental system configurations. The active filter is controlled to draw a compensating current, i_{AF}, from the utility, so that it cancels current harmonics on the AC side of a general-purpose diode/thyristor rectifier (Akagi et al., 1990; Peng et al., 1990; Bhattacharya et al., 1998) or a PWM rectifier for traction systems (Krah and Holtz, 1994). Generally, the shunt active filter is suitable for harmonic compensation of a current harmonic source such as diode/thyristor rectifier with a DC link inductor. The shunt active filter has the capability of damping harmonic resonance between an existing passive filter and the supply impedance.

Figure 26.5 shows a system configuration of a series active filter used alone. The series active filter is connected in series with the utility through a matching transformer, so that it is suitable for harmonic compensation of a voltage harmonic source such as a large-capacity diode rectifier with a DC link capacitor. The series active filter integrated into a diode rectifier with a DC common capacitor is discussed in section V. Table 26.1 shows comparisons between the shunt and series active filters.

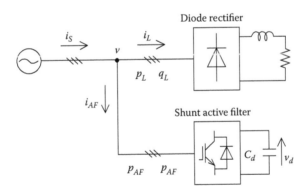

FIGURE 26.5 Series active filter.

TABLE 26.1 Comparison of Shunt Active Filters and Series Active Filters

	Shunt Active Filter	Series Active Filter
System configuration	Figure 26.4	Figure 26.5
Power circuit of active filter	Voltage-fed PWM inverter *with* current minor loop	Voltage-fed PWM inverter *without* current minor loop
Active filter acts as	Current source: i_{AF}	Voltage source: v_{AF}
Harmonic-producing load suitable	Diode/thyristor rectifiers with *inductive* loads, and cycloconverters	Large-capacity diode rectifiers with *capacitive* loads
Additional function	Reactive power compensation	AC voltage regulation
Present situation	Commercial stage	Laboratory stage

This concludes that the series active filter has a "dual" relationship in each item with the shunt active filter (Akagi, 1996; Peng, 1998).

26.2.5.2 Hybrid Active/Passive Filters

Figures 26.6 through 26.8 show three types of hybrid active/passive filters, the main purpose of which is to reduce initial costs and to improve efficiency. The shunt passive filter consists of one or more tuned LC filters and/or a high-pass filter. Table 26.2 shows comparisons among the three hybrid filters in which the active filters are different in function from the passive filters. Note that the hybrid filters are applicable to any current harmonic source, although a harmonic-producing load is represented by a thyristor rectifier with a DC link inductor in Figures 26.6 through 26.8.

Such a combination of a shunt active filter and a shunt passive filter as shown in Figure 26.6 has already been applied to harmonic compensation of naturally-commutated 12-pulse cycloconverters for steel mill drives (Takeda et al., 1987). The passive filters absorbs 11th and 13th harmonic currents while

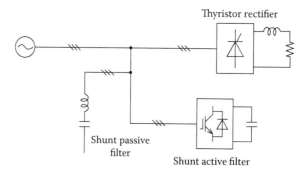

FIGURE 26.6   Combination of shunt active filter and shunt passive filter.

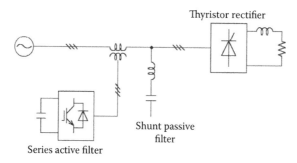

FIGURE 26.7   Combination of series active filter and shunt passive filter.

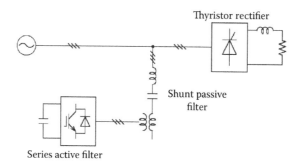

FIGURE 26.8   Series active filter connected in series with shunt passive filter.

TABLE 26.2 Comparison of Hybrid Active/Passive Filters

	Shunt Active Filter Plus Shunt Passive Filter	Series Active Filter Plus Shunt Passive Filter	Series Active Filter Connected in Series with Shunt Passive Filter
System configuration	Figure 26.6	Figure 26.7	Figure 26.8
Power circuit of active filter	• Voltage-fed PWM inverter *with* current minor loop	• Voltage-fed PWM inverter *without* current minor loop	• Voltage-fed PWM inverter *with* or *without* current minor loop
Function of active filter	• Harmonic compensation	• Harmonic isolation	• Harmonic isolation or harmonic compensation
Advantages	• General shunt active filters applicable • Reactive power 8 controllable	• Already existing shunt passive filters applicable • No harmonic current flowing through active filter	• Already existing shunt passive filters applicable • Easy protection of active filter
Problems or issues	• Share compensation in frequency domain between active filter and passive filter	• Difficult to protect active filter against overcurrent • No reactive power control	• No reactive power control
Present situation	• Commercial stage	• A few practical applications	• Commercial stage

the active filter compensates for 5th and 7th harmonic currents and achieves damping of harmonic resonance between the supply and the passive filter. One of the most important considerations in system design is to avoid competition for compensation between the passive filter and the active filter.

The hybrid active filters, shown in Figure 26.7 (Peng et al., 1990, 1993; Kawaguchi et al., 1997) and in Figure 26.8 (Fujita and Akagi, 1991; Balbo et al., 1994; van Zyl et al., 1995), are right now on the commercial stage, not only for harmonic compensation but also for harmonic isolation between supply and load, and for voltage regulation and imbalance compensation. They are considered prospective alternatives to pure active filters used alone. Other combined systems of active filters and passive filters or LC circuits have been proposed in Bhattacharya et al. (1997).

26.2.6 Classification by Power Circuit

There are two types of power circuits used for active filters; a voltage-fed PWM inverter (Akagi et al., 1986; Takeda et al., 1987) and a current-fed PWM inverter (Kawahira et al., 1983; van Schoor and van Wyk, 1987). These are similar to the power circuits used for AC motor drives. They are, however, different in their behavior because active filters act as nonsinusoidal current or voltage sources. The author prefers the voltage-fed to the current-fed PWM inverter because the voltage-fed PWM inverter is higher in efficiency and lower in initial costs than the current-fed PWM inverter (Akagi, 1994). In fact, almost all active filters that have been put into practical application in Japan have adopted the voltage-fed PWM inverter as the power circuit.

26.2.7 Classification by Control Strategy

The control strategy of active filters has a great impact not only on the compensation objective and required kVA rating of active filters, but also on the filtering characteristics in transient state as well as in steady state (Akagi et al., 1986).

26.2.7.1 Frequency-Domain and Time-Domain

There are mainly two kinds of control strategies for extracting current harmonics or voltage harmonics from the corresponding distorted current or voltage; one is based on the Fourier analysis in the frequency-domain (Grady et al., 1990), and the other is based on the Akagi-Nabae theory in the time-domain. The concept of the Akagi-Nabae theory in the time-domain has been applied to the control strategy of almost all the active filters installed by individual high-power consumers over the last 10 years in Japan.

26.2.7.2 Harmonic Detection Methods

Three kinds of harmonic detection methods in the time-domain have been proposed for shunt active filters acting as a current source i_{AF}. Taking into account the polarity of the currents i_S, i_L and i_{AF} in Figure 26.4 gives

$$\text{load-current detection: } i_{AF} = -i_{Lh}$$

$$\text{supply-current detection: } i_{AF} = K_S \cdot i_{Sh}$$

$$\text{voltage detection: } i_{AF} = K_V \cdot v_h.$$

Note that load-current detection is based on feedforward control, while supply-current detection and voltage detection are based on feedback control with gains of K_S and K_V, respectively. Load-current detection and supply-current detection are suitable for shunt active filters installed in the vicinity of one or more harmonic-producing loads by individual consumers. Voltage detection is suitable for shunt active filters that will be dispersed on power distribution systems by utilities, because the shunt active filter based on voltage detection is controlled in such a way to present infinite impedance to the external circuit for the fundamental frequency, and to present a resistor with low resistance of $1/K_V$ [Ω] for harmonic frequencies (Akagi et al., 1999).

Supply-current detection is the most basic harmonic detection method for series active filters acting as a voltage source v_{AF}. Referring to Figure 26.5 yields

$$\text{supply current detection: } v_{AF} = G \cdot i_{Sh}.$$

The series active filter based on supply-current detection is controlled in such a way to present zero impedance to the external circuit for the fundamental frequency and to present a resistor with high resistance of G [Ω] for the harmonic frequencies. The series active filters shown in Figure 26.5 (Fujita and Akagi, 1997) and Figure 26.7 (Peng et al., 1990) are based on supply current detection.

26.3 Integrated Series Active Filters

A small-rated series active filter integrated with a large-rated double-series diode rectifier has the following functions (Fujita and Akagi, 1997):

- Harmonic compensation of the diode rectifier
- Voltage regulation of the common DC bus
- Damping of harmonic resonance between the communication capacitors connected across individual diodes and the leakage inductors including the AC line inductors
- Reduction of current ripples flowing into the electrolytic capacitor on the common DC bus

26.3.1 System Configuration

Figure 26.9 shows a harmonic current-free AC/DC power conversion system described below. It consists of a combination of a double-series diode rectifier of 5 kW and a series active filter with a peak voltage and current rating of 0.38 kVA. The AC terminals of a single-phase H-bridge voltage-fed PWM

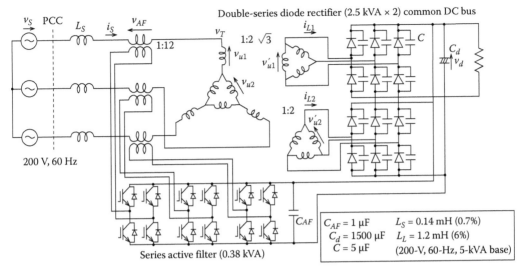

FIGURE 26.9 The harmonic current-free AC/DC power conversion system.

inverter are connected in "series" with a power line through a single-phase matching transformer, so that the combination of the matching transformers and the PWM inverters forms the "series" active filter. For small to medium-power systems, it is economically practical to replace the three single-phase inverters with a single three-phase inverter using six IGBTs. A small-rated high-pass filter for suppression of switching ripples is connected to the AC terminals of each inverter in the experimental system, although it is eliminated from Figure 26.9 for the sake of simplicity.

The primary windings of the Y-Δ and Δ-Δ connected transformers are connected in "series" with each other, so that the combination of the 3-phase transformers and two 3-phase diode rectifiers forms the "double-series" diode rectifier, which is characterized as a 3-phase 12-pulse rectifier. The DC terminals of the diode rectifier and the active filter form a common DC bus equipped with an electrolytic capacitor. This results not only in eliminating any electrolytic capacitor from the active filter, but also in reducing current ripples flowing into the electrolytic capacitor across the common DC bus.

Connecting only a commutation capacitor C in parallel with each diode plays an essential role in reducing the required peak voltage rating of the series active filter.

26.3.2 Operating Principle

Figure 26.10 shows an equivalent circuit for the power conversion system on a per-phase basis. The series active filter is represented as an AC voltage source v_{AF}, and the double-series diode rectifier as the series connection of a leakage inductor L_L of the transformers with an AC voltage source v_L. The reason

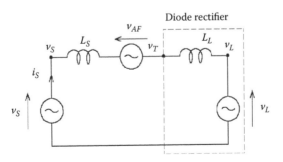

FIGURE 26.10 Single-phase equivalent circuit.

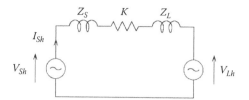

FIGURE 26.11 Single-phase equivalent circuit with respect to harmonics.

for providing the AC voltage source to the equivalent model of the diode rectifier is that the electrolytic capacitor C_d is directly connected to the DC terminal of the diode rectifier, as shown in Figure 26.9.

The active filter is controlled in such a way as to present zero impedance for the fundamental frequency and to act as a resistor with high resistance of $K [\Omega]$ for harmonic frequencies. The AC voltage of the active filter, which is applied to a power line through the matching transformer, is given by

$$v_{AF}^{*} = K \cdot i_{Sh} \tag{26.15}$$

where i_{Sh} is a supply harmonic current drawn from the utility. Note that v_{AF} and i_{Sh} are instantaneous values. Figure 26.11 shows an equivalent circuit with respect to current and voltage harmonics in Figure 26.10. Referring to Figure 26.11 enables derivation of the following basic equations:

$$I_{Sh} = \frac{V_{Sh} - V_{Lh}}{Z_S + Z_L + K} \tag{26.16}$$

$$V_{AF} = \frac{K}{Z_S + Z_L + K}(V_{Sh} - V_{Lh}) \tag{26.17}$$

where V_{AF} is equal to the harmonic voltage appearing across the resistor K in Figure 26.10.

If $K \gg Z_S + Z_L$, Equations 26.16 and 26.17 are changed into the following simple equations.

$$I_{Sh} \approx 0 \tag{26.18}$$

$$V_{AF} \approx V_{Sh} - V_{Lh}. \tag{26.19}$$

Equation 26.18 implies that an almost purely sinusoidal current is drawn from the utility. As a result, each diode in the diode rectifier continues conducting during a half cycle. Equation 26.19 suggests that the harmonic voltage V_{Lh}, which is produced by the diode rectifier, appears at the primary terminals of the transformers in Figure 26.9, although it does not appear upstream of the active filter or at the utility-consumer point of common coupling (PCC).

26.3.3 Control Circuit

Figure 26.12 shows a block diagram of a control circuit based on hybrid analog/digital hardware. The concept of the Akagi-Nabae theory (Akagi et al., 1983, 1984) is applied to the control circuit implementation. The p-q transformation circuit executes the following calculation to convert the three-phase supply current i_{Sv}, i_{Sv}, and i_{Sw} into the instantaneous active current i_p and the instantaneous reactive current i_q.

$$\begin{bmatrix} i_p \\ i_q \end{bmatrix} = \sqrt{\frac{2}{3}} \begin{bmatrix} \cos \omega t & \sin \omega t \\ -\sin \omega t & \cos \omega t \end{bmatrix} \cdot \begin{bmatrix} 1 & -1/2 & -1/2 \\ 0 & \sqrt{3}/2 & -\sqrt{3}/2 \end{bmatrix} \begin{bmatrix} i_{Su} \\ i_{Sv} \\ i_{Sw} \end{bmatrix}. \tag{26.20}$$

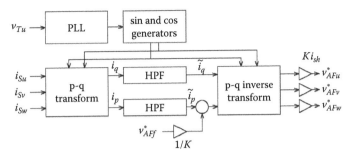

FIGURE 26.12 Control circuit for the series active filter.

The fundamental components in i_{Su}, i_{Sv}, and i_{Sw} correspond to DC components in i_p and i_q, and harmonic components to AC components. Two first-order high-pass-filters (HPFs) with the same cut-off frequency of 10 Hz as each other extract the AC components $\tilde{i}_p$ and $\tilde{i}_q$ from i_p and i_q, respectively. Then, the p-q transformation/inverse transformation of the extracted AC components produces the following supply harmonic currents:

$$
\begin{bmatrix} i_{Shu} \\ i_{Shv} \\ i_{Shw} \end{bmatrix} = \sqrt{\frac{2}{3}} \begin{bmatrix} 1 & 0 \\ -1/2 & \sqrt{3}/2 \\ -1/2 & -\sqrt{3}/2 \end{bmatrix} \cdot \begin{bmatrix} \cos \omega t & -\sin \omega t \\ \sin \omega t & \cos \omega t \end{bmatrix} \begin{bmatrix} \tilde{i}_p \\ \tilde{i}_q \end{bmatrix}.
\tag{26.21}
$$

Each harmonic current is amplified by a gain of K, and then it is applied to the gate control circuit of the active filter as a voltage reference v_{AF}^* in order to regulate the common DC bus voltage, v_{AFf}^* is divided by the gain of K, and then it is added to $\tilde{i}_p$.

The PLL (phase locked loop) circuit produces phase information ωt which is a 12-bit digital signal of 60×2^{12} samples per second. Digital signals, $\sin \omega t$ and $\cos \omega t$, are generated from the phase information, and then they are applied to the p-q (inverse) transformation circuits. Multifunction in the transformation circuits is achieved by means of eight multiplying D/A converters. Each voltage reference, v_{AF}^* is compared with two repetitive triangular waveforms of 10 kHz in order to generate the gate signals for the IGBTs. The two triangular waveforms have the same frequency, but one has polarity opposite to the other, so that the equivalent switching frequency of each inverter is 20 kHz, which is twice as high as that of the triangular waveforms.

26.3.4 Experimental Results

In the following experiment, the control gain of the active filter, K, is set to 27 Ω, which is equal to 3.3 p.u. on a 3ϕ 200-V, 15-A, 60-Hz basis. Equation 26.16 suggests that the higher the control gain, the better the performance of the active filter. An extremely high gain, however, may make the control system unstable, and thereby a trade-off between performance and stability exists in determining an optimal control gain. A constant load resistor is connected to the common DC bus, as shown in Figure 26.9.

Figures 26.13 and 26.14 show experimental waveforms, where a 5-μF commutation capacitor is connected in parallel with each diode used for the double-series diode rectifier. Table 26.3 shows the THD of i_S and the ratio of each harmonic current with respect to the fundamental current contained in i_S. Before starting the active filter, the supply 11th and 13th harmonic currents in Figure 26.13 are slightly magnified due to resonance between the commutation capacitors C and the AC line and leakage inductors, L_S and L_L. Nonnegligible amounts of third, fifth, and seventh harmonic currents, which are so-called non-characteristic current harmonics for the 3-phase 12-pulse diode rectifier, are drawn from the utility.

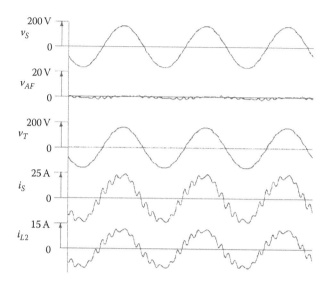

FIGURE 26.13 Experimental waveforms before starting the series active filter.

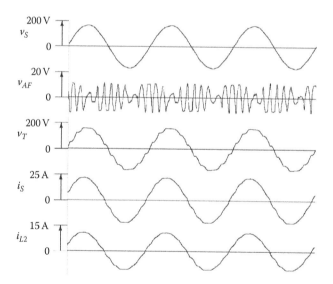

FIGURE 26.14 Experimental waveforms after starting the series active filter.

TABLE 26.3 Supply Current THD and Harmonics Expressed as the Harmonic-to-Fundamental Current Ratio %, Where Commutation Capacitors of 5 μF Are Connected

	THD	3rd	5th	7th	11th	13th
Before (Figure 26.13)	16.8	5.4	2.5	2.2	12.3	9.5
After (Figure 26.14)	1.6	0.7	0.2	0.4	0.8	1.0

Figure 26.14 shows experimental waveforms where the peak voltage of the series active filter is imposed on a limitation of ±12 V inside the control circuit based on hybrid analog/digital hardware. Note that the limitation of ±12 V to the peak voltage is equivalent to the use of three single-phase matching transformers with turn ratios of 1:20 under the common DC link voltage of 240 V. After starting the active filter, a sinusoidal current with a leading power factor of 0.96 is drawn because the active filter acts as a high resistor of 27 Ω, having the capability of compensating for both voltage harmonics V_{Sh} and V_{Lh}, as well as of damping the resonance. As shown in Figure 26.14, the waveforms of i_S and v_T are not affected by the voltage limitation, although the peak voltage v_{AF} frequently reaches the saturation or limitation voltage of ±12 V.

The required peak voltage and current rating of the series active filter in Figure 26.14 is given by

$$3 \times 12^V / \sqrt{2} \times 15^A = 0.38\,\text{kVA}, \tag{26.22}$$

which is only 7.6% of the kVA-rating of the diode rectifier.

The harmonic current-free AC-to-DC power conversion system has both practical and economical advantages. Hence, it is expected to be used as a utility interface with large industrial inverter-based loads such as multiple adjustable speed drives and uninterruptible power supplies in the range of 1–10 MW.

26.4 Practical Applications of Active Filters for Power Conditioning

26.4.1 Present Status and Future Trends

Shunt active filters have been put into practical applications mainly for harmonic compensation, with or without reactive-power compensation. Table 26.4 shows ratings and application examples of shunt active filters classified by compensation objectives.

Applications of shunt active filters are expanding, not only into industry and electric power utilities but also into office buildings, hospitals, water supply utilities, and rolling stock. At present, voltage-fed PWM inverters using IGBT modules are usually employed as the power circuits of active filters in a range of 10 kVA to 2 MVA, and DC capacitors are used as the energy storage components.

Since a combined system of a series active filter and a shunt passive filter was proposed in 1988 (Peng et al., 1990), much research has been done on hybrid active filters and their practical applications (Bhattacharya et al., 1997; Aredes et al., 1998). The reason is that hybrid active filters are attractive from both practical and economical points of view, in particular, for high-power applications. A hybrid active filter for harmonic damping has been installed at the Yamanashi test line for high-speed magnetically-levitated trains (Kawaguchi et al., 1997). The hybrid filter consists of a combination of a 5-MVA series active filter and a 25-MVA shunt passive filter. The series active filter makes a great contribution to damping of harmonic resonance between the supply inductor and the shunt passive filter.

TABLE 26.4 Shunt Active Filters on Commercial Base in Japan

Objective	Rating	Switching Devices	Applications
Harmonic compensation with or without reactive/negative-sequence current compensation	10 kVA ~ 2 MVA	IGBTs	Diode/thyristor rectifiers and cycloconverters for industrial loads
Voltage flicker compensation	5 MVA ~ 50 MVA	GTO thyristors	Arc furnaces
Voltage regulation	40 MVA ~ 60 MVA	GTO thyristors	Shinkansen (Japanese "bullet" trains)

26.4.2 Shunt Active Filters for Three-Phase Four-Wire Systems

Figure 26.15 depicts the system configuration of a shunt active filter for a three-phase four-wire system. The 300-kVA active filter developed by Meidensha has been installed in a broadcasting station (Yoshida et al., 1998). Electronic equipment for broadcasting requires single-phase 100-V AC power supply in Japan, and therefore the phase-neutral rms voltage is 100 V in Figure 26.15. A single-phase diode rectifier is used as an AC-to-DC power converter in an electronic device for broadcasting. The single-phase diode rectifier generates an amount of third-harmonic current that flows back to the supply through the neutral line. Unfortunately, the third-harmonic currents injected from all of the diode rectifiers are in phase, thus contributing to a large amount of third-harmonic current flowing in the neutral line. The current harmonics, which mainly contain the third, fifth, and seventh harmonic frequency components, may cause voltage harmonics at the secondary of a distribution transformer. The induced harmonic voltage may produce a serious effect on other harmonic-sensitive devices connected at the secondary of the transformer.

Figure 26.16 shows actually measured current waveform in Figure 26.15. The load currents, i_{La}, i_{Lb}, and i_{Lc}, and the neutral current flowing on the load side, i_{Ln}, are distorted waveforms including a large amount of harmonic current, while the supply currents, i_{Sa}, i_{Sb}, and i_{Sc}, and the neutral current flowing on the supply side, i_{Sn}, are almost sinusoidal waveforms with the help of the active filter.

26.4.3 The 48-MVA Shunt Active Filter for Compensation of Voltage Impact Drop, Variation, and Imbalance

Figure 26.17 shows a power system delivering electric power to the Japanese "bullet trains" on the Tokaido Shinkansen. Three shunt active filters for compensation of fluctuating reactive current/negative-sequence current have been installed in the Shintakatsuki substation by the Central Japan Railway Company (Iizuka et al., 1995). The shunt active filters, manufactured by Toshiba, consist of voltage-fed PWM inverters using GTO thyristors, each of which is rated at 16 MVA. A high-speed train with maximum output power of 12 MW draws unbalanced varying active and reactive power from the Scott transformer, the primary of which is connected to the 154-kV utility grid. More than 20 high-speed trains pass per hour during the daytime. This causes voltage impact drop, variation, and imbalance at the terminals of the 154-kV utility system, accompanied by a serious deterioration in the power quality of other consumers connected to the same power system. The purpose of the shunt active filters with

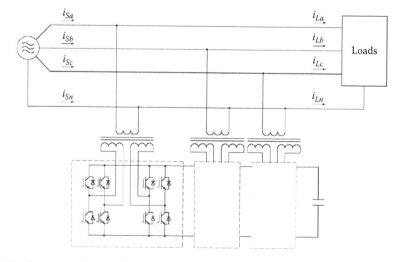

FIGURE 26.15   Shunt active filter for three-phase four-wire system.

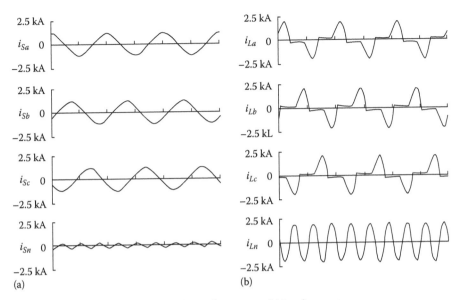

FIGURE 26.16   Actual current waveforms. (a) Supply currents. (b) Load currents.

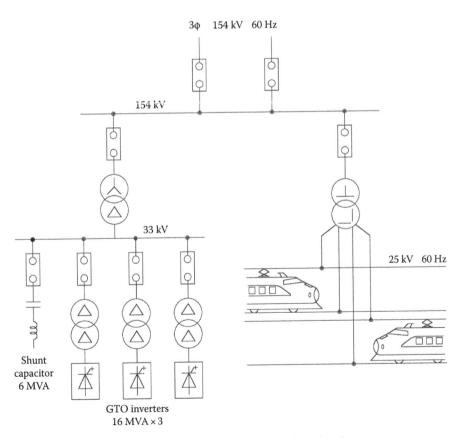

FIGURE 26.17   The 48-MVA shunt active filter installed in the Shintakatsuki substation.

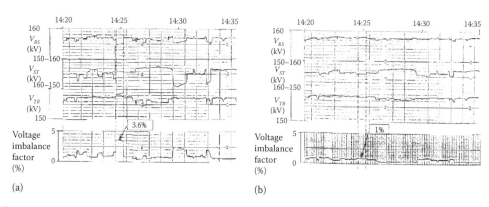

FIGURE 26.18   Installation effect (a) Before compensation. (b) After compensation.

a total rating of 48 MVA is to compensate for voltage impact drop, voltage variation, and imbalance at the terminals of the 154-kV power system, and to improve the power quality. The concept of the instantaneous power theory in the time-domain has been applied to the control strategy for the shunt active filter.

Figure 26.18 shows voltage waveforms on the 154-kV bus and the voltage imbalance factor before and after compensation, measured at 14:20–14:30 on July 27, 1994. The shunt active filters are effective not only in compensating for the voltage impact drop and variation, but also in reducing the voltage imbalance factor from 3.6% to 1%. Here, the voltage imbalance factor is the ratio of the negative to positive-sequence component in the three-phase voltages on the 154-kV bus. At present, several active filters in a range of 40–60 MVA have been installed in substations along the Tokaido Shinkansen (Takeda et al., 1995).

Acknowledgment

The author would like to thank Meidensha Corporation and Toshiba Corporation for providing helpful and valuable information of the 300-kVA active filter and the 48-MVA active filter.

References

Akagi, H., Trends in active power line conditioners, *IEEE Trans. Power Electron.*, 9, 3, 263–268, 1994.

Akagi, H., New trends in active filters for power conditioning, *IEEE Trans. Ind. Appl.*, 32, 6, 1312–1322, 1996.

Akagi, H. and Fujita, H., A new power line conditioner for harmonic compensation in power systems, *IEEE Trans. Power Delivery*, 10, 3, 1570–1575, 1995.

Akagi, H., Fujita, H., and Wada, K., A shunt active filter based on voltage detection for harmonic termination of a radial power distribution line, *IEEE Trans. Ind. Appl.*, 35, 3, 638–645, 1999.

Akagi, H., Kanazawa, Y., and Nabae, A., Generalized theory of the instantaneous reactive power in three-phase circuits, in *Proceedings of the 1983 International Power Electronics Conference*, Tokyo, Japan, 1983, pp. 1375–1386.

Akagi, H., Kanazawa, Y., and Nabae, A., Instantaneous reactive power compensators comprising switching devices without energy storage components, *IEEE Trans. Ind. Appl.*, 20, 3, 625–630, 1984.

Akagi, H., Nabae, A., and Atoh, S., Control strategy of active power filters using multiple voltage-source PWM converters, *IEEE Trans. Ind. Appl.*, 22, 3, 460–465, 1986.

Akagi, H., Tsukamoto, Y., and Nabae, A., Analysis and design of an active power filter using quad-series voltage-source PWM converters, *IEEE Trans. Ind. Appl.*, 26, 1, 93–98, 1990.

Aredes, M., Heumann, K., and Watanabe, E.H., A universal active power line conditioner, *IEEE Trans. Power Delivery*, 13, 2, 545–551, 1998.

Balbo, N., Penzo, R., Sella, D., Malesani, L., Mattavelli, P., and Zuccato, A., Simplified hybrid active filters for harmonic compensation in low voltage industrial applications, in *Proceedings of the 1994 IEEE/ PES International Conference on Harmonics in Power Systems*, Bologna, Italy, 1994, pp. 263–269.

Bhattacharya, S., Cheng, P., and Divan, D., Hybrid solutions for improving passive filter performance in high power applications, *IEEE Trans. Ind. Appl.*, 33, 3, 732–747, 1997.

Bhattacharya, S., Frank, T.M., Divan, D., and Banerjee, B., Active filter system implementation, *IEEE Ind. Appl. Mag.*, 4, 5, 47–63, 1998.

Bird, B.M., Marsh, J.F., and McLellan, P.R., Harmonic reduction in multiple converters by triple-frequency current injection, *IEE Proc.*, 116, 10, 1730–1734, 1969.

Fujita, H. and Akagi, H., A practical approach to harmonic compensation in power systems–series connection of passive and active filters, *IEEE Trans. Ind. Appl.*, 27, 6, 1020–1025, 1991.

Fujita, H. and Akagi, H., An approach to harmonic current-free AC/DC power conversion for large industrial loads: The integration of a series active filter and a double-series diode rectifier, *IEEE Trans. Ind. Appl.*, 33, 5, 1233–1240, 1997.

Fujita, H. and Akagi, H., The unified power quality conditioner: The integration of series- and shunt-active filters, *IEEE Trans. Power Electron.*, 13, 2, 315–322, 1998.

Grady, W.M., Samotyj, M.J., and Noyola, A.H., Survey of active power line conditioning methodologies, *IEEE Trans. Power Delivery*, 5, 3, 1536–1542, 1990.

Gyugyi, L. and Strycula, E.C., Active AC power filters, in *Proceedings of the 1976 IEEE/IAS Annual Meeting*, Orlando, FL, 1976, pp. 529–535.

Iizuka, A., Kishida, M., Mochinaga, Y., Uzuka, T., Hirakawa, K., Aoyama, F., and Masuyama, T., Self-commutated static var generators at Shintakatsuki substation, in *Proceedings of the 1995 International Power Electronics Conference*, Yokohama, Japan, 1995, pp. 609–614.

Kawaguchi, I., Ikeda, H., Ogihara, Y., Syogaki, M., and Morita, H., Novel active filter system composed of inverter bypass circuit for suppression of harmonic resonance at the Yamanashi maglev test line, in *Proceedings of the IEEE-IEEJ/IAS Power Conversion Conference*, 1997, pp. 175–180.

Kawahira, H., Nakamura, T., Nakazawa, S., and Nomura, M., Active power filters, in *Proceedings of the 1983 International Power Electronics Conference*, Tokyo, Japan, 1983, pp. 981–992.

Krah, J.O. and Holtz, J., Total compensation of line-side switching harmonics in converter-fed AC locomotives, in *Proceedings of the 1994 IEEE/IAS Annual Meeting*, Denver, CO, 1994, pp. 913–920.

Lêe, T.-N., Pereira, M., Renz, K., and Vaupel, G., Active damping of resonances in power systems, *IEEE Trans. Power Delivery*, 9, 2, 1001–1008, 1994.

Moran, S., A line voltage regulator/conditioner for harmonic-sensitive load isolation, in *Proceedings of the 1989 IEEE/IAS Annual Meeting*, Long Beach, CA, 1989, pp. 947–951.

Peng, F.Z., Application issues of active power filters, *IEEE Ind. Appl. Mag.*, 4, 5, 21–30, 1998.

Peng, F.Z., Akagi, H., and Nabae, A., A new approach to harmonic compensation in power systems—A combined system of shunt passive and series active filters, *IEEE Trans. Ind. Appl.*, 26, 6, 983–990, 1990a.

Peng, F.Z., Akagi, H., and Nabae, A., A study of active power filters using quad-series voltage-source PWM converters for harmonic compensation, *IEEE Trans. Power Electron.*, 5, 1, 9–15, 1990b.

Peng, F.Z., Akagi, H., and Nabae, A., Compensation characteristics of the combined system of shunt passive and series active filters, *IEEE Trans. Ind. Appl.*, 29, 1, 144–152, 1993.

Takeda, M., Ikeda, K., and Tominaga, Y., Harmonic current compensation with active filter, in *Proceedings of the 1987 IEEE/IAS Annual Meeting*, 1987, pp. 808–815, 1987.

Takeda, M., Murakami, S., Iizuka, A., Kishida, M., Mochinaga, Y., Hase, S., and Mochinaga, H., Development of an SVG series for voltage control over three-phase unbalance caused by railway load, in *Proceedings of the 1995 International Power Electronics Conference*, Yokohama, Japan, 1995, pp. 603–608.

Van Schoor, G. and van Wyk, J., A study of a system of a current-fed converters as an active three-phase filter, in *Proceedings of the 1987 IEEE/PELS Power Electronics Specialist Conference*, 1987, pp. 482–490.

Van Zyl, A., Enslin, J.H.R., and Spée, R., Converter based solution to power quality problems on radial distribution lines, in *Proceedings of the 1995 IEEE/IAS Annual Meeting*, 1995, pp. 2573–2580.

Watanabe, E.H., Series active filter for the DC side of HVDC transmission systems, in *Proceedings of the 1990 International Power Electronics Conference*, Tokyo, Japan, 1990, pp. 1024–1030.

Yoshida, T., Nakagawa, G., Kitamura, H., and Iwatani, K., Active filters (multi-functional harmonic suppressors) used to protect the quality of power supply from harmonics and reactive power generated in loads, *Meiden Rev.*, 262, 5, 13–17, 1998 (in Japanese).

Zhang, W., Asplund, G., Aberg, A., Jonsson, U., and Lööf, O., Active DC filter for HVDC system—A test installation in the Konti-Skan at Lindome converter station, *IEEE Trans. Power Delivery*, 8, 3, 1599–1605, 1993.

27

FACTS Controllers

Luis Morán
Universidad de Concepción

Juan Dixon
*Pontificia Universidad
Católica de Chile*

M. José Espinoza
Universidad de Concepción

José Rodríguez
*Universidad Téchnica
Federico Santa María*

27.1 Introduction

In the last decades power systems have faced new challenges due to deregulation, permanent increase in demand, the implementation of more stringent standard, and lack of investment in new transmission lines. The privatization of utilities has created a new electric power market scenario and a more complicated power system operation. Also, the increased dependence of modern society upon electricity has forced power systems to operate with very high reliability and with almost 100% availability. Moreover, power quality is now becoming a major concern among users and utilities, forcing the development and application of more stringent standards due to the connection of more sophisticated loads. These requirements have obliged to develop new technologies to improve power system operation and controllability. Based in these new technologies two concepts have been created: flexible ac transmission systems (FACTS) and flexible reliable and intelligent electrical energy delivery systems (FRIENDS). In these systems, compensation equipments based in static converters play an important role [1,2].

The FACTS concept was originally created in the 1980s to solve operation problems due to the restrictions on the construction of new transmission lines, to improve power system stability margins, and to facilitate power exchange between different generation companies and large power users. On the other hand, the FRIENDS concept was created in the 1990s and identifies the operation of utilities with new static compensators and communication systems whose goal is to develop a desirable structure for power delivery systems where distributed generations and distributed energy storage systems are located near the load side. Finally, custom power devices are special applications of FACTS, but oriented to satisfy requirements of power quality at the utility level [5].

The two main objectives of FACTS are to increase the power transfer capability of transmission lines and to maintain the power flow over designated routes. The first objective implies that power flow in a given line should be able to be increased up to the thermal limit by forcing the rated current through the series line impedance. This objective does not mean that the lines would normally be operated at their thermal limit (the transmission losses would be unacceptable), but this option would be available, if needed, to handle severe system contingencies. The second objective implies that, by being able to control the current in a line (for example by changing the effective line impedance) the power flow can be restricted to selected transmission corridors. The achievement of these two basic objectives significantly increases the utilization of existing (and new) transmission assets, and plays a major role in facilitating deregulation [9].

Therefore, the FACTS technology opens up new opportunities for controlling power and enhancing the usable capacity of present transmission systems. The possibility that power through a line can be controlled enables a large potential of increasing the capacity of existing lines. These opportunities arise through the ability of FACTS controllers to adjust the power system electrical parameters including series and shunt impedances, current, voltage, phase angle, and the damping oscillations. The implementation of such equipments requires the development of power electronics-based compensators and controllers. The coordination and overall control of these compensators to provide maximum system benefits and prevent undesirable interactions with different system configurations under normal and contingency conditions present a different technological challenge. This challenge is the development of appropriate system optimization control strategies, communication links, and security protocols.

The IEEE Power Engineering Society (PES) Task Force of the FACTS Working Group defined terms and definitions for FACTS and FACTS controllers. According with this IEEE Working Group, a FACTS controller is a power electronic-based system and other static equipment that provide control of one or more ac transmission system parameters [3].

Different types of FACTS controllers have been developed and implemented, for shunt and/or series compensation. Shunt compensation is used to influence the natural electrical characteristics of the transmission line to increase steady-state transmittable power and to control voltage profile along the line, while series compensation is used to change the transmission line impedance and is highly effective in controlling power flow through the line and in improving system stability. Most of FACTS controllers act over the reactive power flow in order to control voltage profile and to increase power system stability. The concept of VAR compensation using FACTS controllers embraces a wide and diverse field of both transmission system and customer problems, especially related with power quality issues, since most of power quality problems can be attenuated or solved with an adequate control of reactive power [4]. In general, the problem of reactive power compensation is viewed from two aspects: load compensation and voltage support. In load compensation the objectives are to increase the value of the system power factor, to balance the real power drawn from the ac supply, compensate voltage regulation, and to eliminate current harmonic components produced by large and fluctuating nonlinear industrial loads [6]. Voltage support is generally required to reduce voltage fluctuation at a given terminal of a transmission line. Reactive power compensation in transmission systems also improves the stability of the ac system by increasing the maximum active power that can be transmitted. It also helps to maintain a substantially flat voltage profile at all levels of power transmission, it improves HVDC conversion terminal performance, increases transmission efficiency, controls steady-state and temporary overvoltages, and can avoid disastrous blackouts [7,8].

Based on the use of reliable high-speed power electronics, powerful analytical tools, advanced control and microcomputer technologies, static compensators also known as FACTS controllers have been developed and represent a new concept for the operation of power transmission systems. This chapter presents the principles of operation and performance characteristics of different FACTS controllers. Static compensators implemented with thyristors and self-commutated converters are described. New static compensators such as static synchronous compensators (STATCOMs), unified power flow controllers (UPFC), dynamic voltage restorers (DVR), required to compensate modern power transmission and distribution systems are also presented and described [10].

27.2 Shunt Compensators

Shunt compensation is used basically to control the amount of reactive power that flows through the power system. In a linear circuit, the reactive power is defined as the ac component of the instantaneous power, with a frequency equal to 100/120 Hz in a 50 or 60 Hz system. The reactive power generated by the ac power source is stored in a capacitor or a reactor during a quarter of a cycle, and in the next quarter cycle is sent back to the power source. The reactive power oscillates between the ac source and the capacitor or reactor, and also between them, at a frequency equal to two times the rated value (50 or 60 Hz). For this reason it can be compensated using static equipments or VAR generators, avoiding its circulation between the load (inductive or capacitive) and the source, and therefore improving voltage regulation and stability of the power system. Reactive power compensation can be implemented with VAR generators connected in parallel or in series [11].

27.2.1 Shunt Compensation Principles

Figure 27.1 shows the principles and theoretical effects of shunt reactive power compensation in a basic ac system, which comprises a source V_1, a transmission line, and a typical inductive load. Figure 27.1a shows the system without compensation, and its associated phasor diagram. In the phasor diagram, the phase angle of the current has been related to the load side, which means that the active current I_P is in phase with the load voltage V_2. Since the load is assumed inductive, it requires reactive power for proper operation, which must be supplied by the source, increasing the current flow from the generator and through the lines. If reactive power is supplied near the load, the line current is minimized, reducing power losses and improving voltage regulation at the load terminals. This can be done with a capacitor, with a voltage source, or with a current source. In Figure 27.1b, a current-source device is being used to compensate the reactive component of the load current (I_Q). As a result, the system voltage regulation is improved and the reactive current component from the source is almost eliminated.

A current source or a voltage source can be used for reactive shunt compensation. The main advantages of using voltage- or current-source VAR generators (instead of inductors or capacitors) are that the reactive power generated is independent of the voltage at the point of connection and can be adjusted in a wide range.

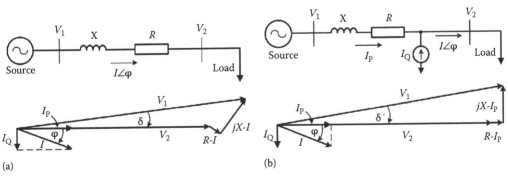

(a)

(b)

FIGURE 27.1 Principles of shunt compensation in a radial ac system. (a) System phasor diagram without reactive compensation. (b) Shunt compensation of the system with a current source.

Since shunt compensation is able to change the power flow in the system by varying the value of the applied shunt equivalent impedance, changing the reactive power flow in the system, during and following dynamic disturbances, the transient stability limit can be increased and effective power oscillation damping can be provided. Thereby, the voltage of the transmission line counteracts the accelerating swings of the disturbed machine and therefore damps the power oscillations.

Independent of the source type or system configuration, different requirements have to be taken into consideration for a successful operation of shunt compensators. Some of these requirements are simplicity, controllability, time response, cost, reliability, and harmonic distortion.

27.2.2 Traditional Shunt Compensation

In general, shunt compensators are classified depending on the technology used in their implementation. Rotating and static equipments were commonly used to compensate reactive power and to stabilize power systems. In the last decades, a large number of different static FACTS controllers, using power electronic technologies and digital control schemes have been proposed and developed [11]. There are two approaches to the realization of power electronics-based compensators: the one that employs thyristor-switched capacitors and reactors with tap-changing transformers, and the other group that uses self-commutated static converters. A brief description of the most commonly used shunt compensators is presented below.

27.2.2.1 Fixed or Mechanically Switched Capacitors

Shunt capacitors were first employed for power factor correction in the year 1914 [12]. The leading current drawn by the shunt capacitors compensates the lagging current drawn by the load. The selection of shunt capacitors depends on many factors, the most important of which is the amount of lagging reactive power taken by the load. In the case of widely fluctuating loads, the reactive power also varies over a wide range. Thus, a fixed capacitor bank may often lead to either over-compensation or under-compensation. Variable VAR compensation is achieved using switched capacitors [13]. Depending on the total VAR requirement, capacitor banks are switched into or switched out of the system. The smoothness of control is solely dependent on the number of capacitors switching units used. The switching is usually accomplished using relays and circuit breakers. However, these methods based on mechanical switches and relays have the disadvantage of being sluggish and unreliable. Also, they generate high inrush currents, and require frequent maintenance.

27.2.2.2 Synchronous Condensers

Synchronous condensers have played a major role in voltage and reactive power control for more than 50 years. Functionally, a synchronous condenser is simply a synchronous machine connected to the power system. After the unit is synchronized, the field current is adjusted to either generate or absorb reactive power as required by the ac system. The machine can provide continuous reactive power control when used with the proper automatic exciter circuit. Synchronous condensers have been used at both distribution and transmission voltage levels to improve stability and to maintain voltages within desired limits under varying load conditions and contingency situations. However, synchronous condensers are rarely used today because they require substantial foundations and a significant amount of starting and protective equipment. They also contribute to the short circuit current and they cannot be controlled fast enough to compensate rapid load changes. Moreover, their losses are much higher than those associated with static compensators, and the cost is much higher as well. Their advantage lies in their high temporary overload capability [4].

27.2.2.3 Thyristorized VAR Compensators

As in the case of the synchronous condenser, the aim of achieving fine control over the entire VAR range, has been fulfilled with the development of static compensators but with the advantage of faster

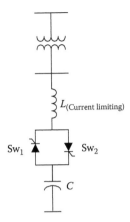

FIGURE 27.2 The thyristor-switched capacitor configuration.

response times [11]. Thyristorized VAR compensators consist of standard reactive power shunt elements (reactors and capacitors) which are controlled to provide rapid and variable reactive power. They can be grouped into two basic categories, the thyristor-switched capacitor (TSC) and the thyristor-controlled reactor (TCR).

27.2.2.3.1 Thyristor-Switched Capacitors

Figure 27.2 shows the basic scheme of a static compensator of the TSC type. First introduced by ASEA in 1971 [12], the shunt capacitor bank is split up into appropriately small steps, which are individually switched in and out using bidirectional thyristor switches. Each single-phase branch consists of two major parts, the capacitor C and the thyristor switches Sw_1 and Sw_2. In addition, there is a minor component, the inductor L, whose purpose is to limit the rate of rise of the current through the thyristors and to prevent resonance with the network (normally 6% with respect to X_C). The capacitor may be switched with a minimum of transients if the thyristor is turned on at the instant when the capacitor voltage and the network voltage have the same value. Static compensators of the TSC type have the following properties: stepwise control, average delay of one half a cycle (maximum one cycle), and no generation of harmonics since current transient component can be attenuated effectively [12,13].

The current that flows through the capacitor at a given time t is defined by the following expression:

$$i(t) = \frac{V_m}{X_C - X_L}\cos(\omega t + \alpha) - \frac{V_m}{X_C - X_L}\cos\theta\cos(\omega_r t) + \left[\frac{X_C V_m \sin(\alpha)}{\omega_r L(X_C - X_L)} - \frac{V_{C0}}{\omega_r L}\right]\sin(\omega_r t) \qquad (27.1)$$

where
X_C and X_L are the compensator capacitive and inductive reactances
V_m the source maximum instantaneous voltage
α the voltage phase-shift angle at which the capacitor is connected
ω_r the system resonant frequency $\left(\omega_r = 1/\sqrt{LC}\right)$
V_{C0} is the capacitor voltage at $t = 0^-$

Expression (27.1) has been obtained assuming that the system equivalent resistance is negligible as compared with the system reactance. This assumption is valid in high-voltage transmission lines. If the capacitor is connected at the moment that the source voltage is maximum and V_{C0} is equal to the source voltage peak value, V_m ($\alpha = \pm90°$) the current transient component is zero.

Despite the attractive theoretical simplicity of the switched capacitor scheme, its popularity has been hindered by a number of practical disadvantages: the VAR compensation is not continuous, each capacitor

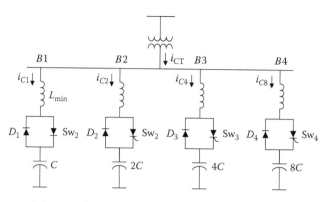

FIGURE 27.3   Binary rated thyristor-diode-switched capacitor configuration.

bank requires a separate thyristor switch and therefore the construction is not economical, the steady-state voltage across the nonconducting thyristor switch is twice the peak supply voltage, and the thyristor must be rated for or protected by external means against line voltage transients and fault currents.

An attractive solution to the disadvantages of using TSC is to replace one of the thyristor switches by a diode. In this case, inrush currents are eliminated when thyristors are fired at the right time, and a more continuous reactive power control can be achieved if the rated power of each capacitor bank is selected following a binary combination, as described in Ref. [14]. This configuration is shown in Figure 27.3. In this figure, the inductor L_{min} is used to prevent any inrush current produced by a firing pulse out of time.

To connect each branch, a firing pulse is applied at the thyristor gate, but only when the voltage supply reaches its maximum negative value. In this way, a soft connection is obtained [1]. The current will increase starting from zero without distortion, following a sinusoidal waveform, and after the cycle is completed, the capacitor voltage will have the voltage $-V_m$, and the thyristor automatically will block. In this form of operation, both connection and disconnection of the branch will be soft, and without transient component. If the firing pulses and the voltage $-V_m$ are properly adjusted, neither harmonics nor inrush currents are generated, since two important conditions are achieved: dv/dt at $v = -V_m$ is zero, and anode-to-cathode thyristor voltage is equal to zero. Assuming that $v(t) = V_m \sin \omega t$ is the source voltage, $V_{C;0}$ the initial capacitor voltage, and $v^{Th}(t)$ the thyristor anode-to-cathode voltage, the right connection of the branch will be when $v^{Th}(t) = 0$, that is

$$v^{Th}(t) = v(t) - V_{C0} = V_m \sin \omega t - V_{C0} \tag{27.2}$$

since $V_{C0} = -V_m$:

$$v^{Th}(t) = V_m \sin \omega t + V_m = V_m(1 + \sin \omega t) \tag{27.3}$$

then, $v^{Th}(t) = 0$ when $\sin \omega t = -1 \Rightarrow \omega t = 270°$.

At $\omega t = 270°$, the thyristor is switched on, and the capacitor C begins to discharge. At this point, $\sin(270°) = -\cos(0°)$, and hence $v_C(t)$ for $\omega t \geq 270°$ will be: $V_C(t_0) = -V_m \cos \omega t_0$. The compensating capacitor current starting at t_0 will be

$$i_c = C \frac{dv_c}{dt} = C \cdot V_m \frac{d}{dt}(-\cos \omega \cdot t_o) = C \cdot V_m \sin \omega \cdot t_o \tag{27.4}$$

Equation 27.4 shows that the current starts from zero as a sinusoidal waveform without distortion and/or inrush component. If the above switching conditions are satisfied, the inductor L may be minimized or even eliminated.

The experimental oscillograms of Figure 27.4 show how the binary connection of many branches allows an almost continuous compensating current variation. These experimental current waveforms were obtained in a 5 kVAr laboratory prototype. The advantages of this topology are that many

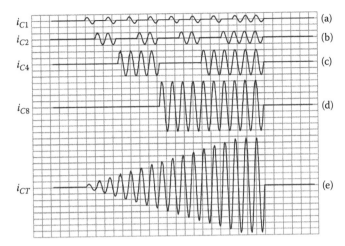

FIGURE 27.4 Experimental compensating phase current of the thyristor-diode switched capacitor: (a) current through B1; (b) current through B2; (c) current through B3; (d) current through B4, and (e) total system compensating current.

compensation levels can be implemented with few branches allowing continuous variations without distortion. Moreover, the topology is simpler and more economical as compared with TSC. The main drawback is that it has a time delay of one complete cycle compared with the half cycle of TSC.

27.2.2.3.2 Thyristor-Controlled Reactor

Figure 27.5 shows the power scheme of a static compensator of the TCR type. In most cases, the compensator also includes a fixed capacitor or a filter for low order harmonics, which is not shown in this figure. Each of the three phase branches includes an inductor L, and the thyristor switches Sw_1 and Sw_2. Reactors may be both switched and phase-angle controlled [15–17].

When phase-angle control is used, a continuous range of reactive power consumption is obtained. It results, however, in the generation of odd harmonic current components during the control process. Full conduction is achieved with a gating angle of 90°. Partial conduction is obtained with gating angles between 90° and 180°, as shown in Figure 27.6. By increasing the thyristor gating angle, the fundamental component of the current reactor is reduced. This is equivalent to increase the inductance, reducing the reactive power absorbed by the static compensator.

However, it should be pointed out that the change in the reactor current may only take place at discrete points of time, which means that adjustments cannot be made quicker than once per half cycle.

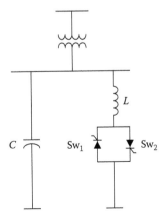

FIGURE 27.5 The thyristor-controlled reactor configuration.

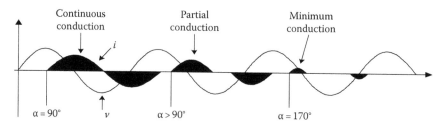

FIGURE 27.6 Simulated voltage and current waveforms in a TCR for different thyristor phase-shift angles, α.

Static compensators of the TCR type are characterized by the ability to perform continuous control, with maximum delay of one-half cycle and practically no transients. The principal disadvantages of this configuration are the generation of low frequency harmonic current components, and higher losses when working in the inductive region (i.e., absorbing reactive power). The relation between the fundamental component of the reactor current and the phase-shift angle α is given by

$$I_1 = \frac{V_{rms}}{\pi\omega L}(2\pi - 2\alpha + \sin(2\alpha)) \tag{27.5}$$

In a single-phase unit, with balanced phase-shift angles, only odd harmonic components are presented in the reactor current. The amplitude of each harmonic component is defined by

$$I_k = \frac{4V_{rms}}{\pi X_L}\left[\frac{\sin(k+1)\alpha}{2(k+1)} + \frac{\sin(k-1)\alpha}{2(k-1)} - \cos(\alpha)\frac{\sin(k\alpha)}{k}\right] \tag{27.6}$$

In order to eliminate low frequency current harmonics (third, fifth, seventh), delta configurations (for zero sequence harmonics) and passive filters may be used, as shown in Figure 27.7a. Twelve pulse configurations are also used as shown in Figure 27.7b. In this case, passive filters are not required, since the fifth and seventh current harmonics are eliminated by the phase-shift introduced by the transformer.

27.2.2.3.3 VAR Compensation Characteristics

One of the main characteristics of static VAR compensators is that the amount of reactive power interchanged with the system depends on the applied voltage, as shown in Figure 27.8. This figure displays the steady-state $V_T - Q$ characteristics of a combination of fixed capacitor–thyristor-controlled reactor

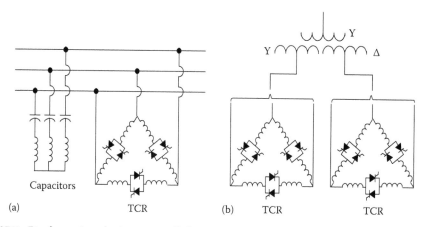

FIGURE 27.7 Fixed capacitor–thyristor-controlled reactor configuration. (a) Six pulse topology. (b) Twelve-pulse topology.

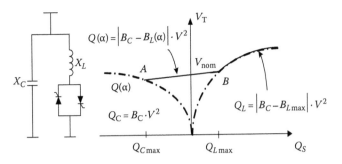

FIGURE 27.8 Voltage–reactive power characteristic of a FC–TCR.

(FC–TCR) compensator. This characteristic shows the amount of reactive power generated or absorbed by the FC–TCR, as a function of the applied voltage. At rated voltage, the FC–TCR presents a linear characteristic, which is limited by the rated power of the capacitor and reactor respectively. Beyond these limits, the $V_T – Q$ characteristic is not linear, which is one of the main disadvantages of this type of VAR compensator.

27.2.2.3.4 Combined Thyristor-Switched Capacitor and Thyristor-Controlled Reactor

Irrespective of the reactive power control range required, any static compensator can be built up from one or both of the above-mentioned schemes (i.e., TSC and TCR), as shown in Figure 27.9. In those cases where the system with switched capacitors is used, the reactive power is divided into a suitable number of steps and the variation will therefore take place stepwise. Continuous control may be obtained with the addition of a TCR. If it is required to absorb reactive power, the entire capacitor bank is disconnected and the equalizing reactor becomes responsible for the absorption. By coordinating the control between the reactor and the capacitor steps, it is possible to obtain fully stepless control. Static compensators of the combined TSC and TCR type are characterized by a continuous control, practically no transients, low generation of harmonics (because the controlled reactor rating is small compared to the total reactive power), and flexibility in control and operation. An obvious disadvantage of the TSC–TCR as compared with TCR- and TSC-type compensators is the higher cost. A smaller TCR rating results in some savings, but these savings are more than absorbed by the cost of the capacitor switches and the more complex control system [12].

The voltage–current characteristic of this compensator is shown in Figure 27.10.

To reduce transient phenomena and harmonic distortion, and to improve the dynamics of the compensator, some researchers have applied self-commutation to TSC and TCR. Some examples of this can be found in Refs. [18,19]. However, best results have been obtained using self-commutated compensators based on conventional two-level and three-level inverters [20].

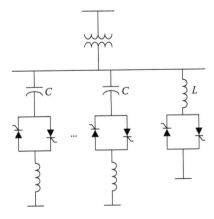

FIGURE 27.9 Combined TSC and TCR configuration.

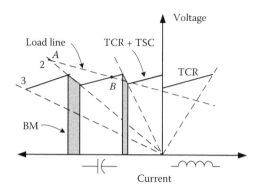

FIGURE 27.10   Steady-state voltage–current characteristic of a combined TSC–TCR compensator.

27.2.3 Self-Commutated Shunt Compensators

The application of self-commutated converters as a means of compensating reactive power has demonstrated to be an effective solution. This technology has been used to implement more sophisticated compensator equipment such as STATCOM, UPFC, and dynamic voltage compensators [3,21]. The STATCOM is based on a solid-state voltage source, implemented with an inverter, and connected in parallel to the power system through a coupling reactor, in analogy with a synchronous machine, generating balanced set of three sinusoidal voltages at the fundamental frequency, with controllable amplitude and phase-shift angle. A STATCOM is a controlled reactive power source. This equipment, however, has no inertia and limited overload capability. Examples of these topologies are shown in Figures 27.11 and 27.12 [10,21].

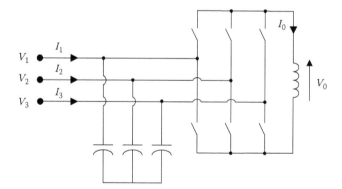

FIGURE 27.11   A VAR compensator topology implemented with a current-source converter.

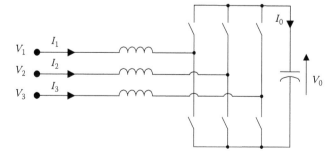

FIGURE 27.12   A VAR compensator topology implemented with a voltage-source converter.

27.2.3.1 Principles of Operation

With the remarkable progress of gate commutated semiconductor devices, attention has been focused on self-commutated FACTS controllers capable of generating or absorbing reactive power without requiring large banks of capacitors or reactors. Several approaches are possible including current-source and voltage-source converters. The current-source approach shown in Figure 27.11 uses a reactor supplied with a regulated dc current, while the voltage-source inverter, displayed in Figure 27.12, uses a capacitor with a regulated dc voltage.

The principal advantages of self-commutated FACTS controllers are the significant reduction of size and the potential reduction in cost achieved from the elimination of a large number of passive components and lower relative capacity requirement for the semiconductor switches [20,21]. Because of its smaller size, self-commutated VAR compensators are well suited for applications where space is a premium.

Self-commutated compensators are used to stabilize transmission systems, improve voltage regulation, correct power factor, and also correct load unbalances [22]. Moreover, they can be used for the implementation of shunt and series compensators. Figure 27.13 shows a shunt STATCOM, implemented with a boost type voltage-source converter. Neglecting the internal power losses of the overall converter, the control of the reactive power is done by adjusting the amplitude of the fundamental component of the output voltage V_{MOD}, which can be modified with the PWM pattern as shown in Figure 27.14. When V_{MOD} is larger than the voltage V_{COMP}, the VAR compensator generates reactive power (Figure 27.13b) and when V_{MOD} is smaller than V_{COMP}, the compensator absorbs reactive power (Figure 27.13c).

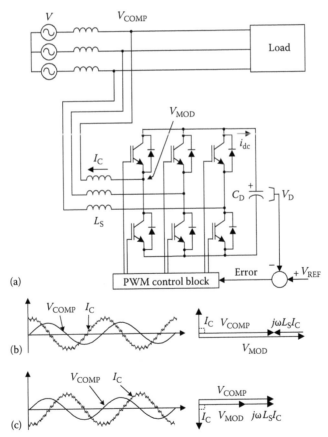

FIGURE 27.13 Simulated current and voltage waveforms of a voltage-source self-commutated shunt VAR compensator. (a) STATCOM topology. (b) Simulated current and voltage waveforms for leading compensation ($V_{MOD} > V_{COMP}$). (c) Simulated current and voltage waveforms for lagging compensation ($V_{MOD} < V_{COMP}$).

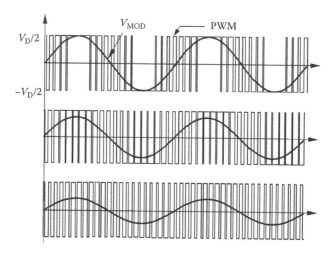

FIGURE 27.14 Simulated compensator output voltage waveform for different modulation index (amplitude of the fundamental voltage component).

Its principle of operation is similar to the synchronous machine. The compensation current can be leading or lagging, depending on the relative amplitudes of V_{COMP} and V_{MOD}. The capacitor voltage V_D, connected to the dc link of the converter, is kept equal to a reference value V_{REF} through of a special feedback control loop, which controls the phase-shift angle, δ, between V_{COMP} and V_{MOD}.

The amplitude of the compensator output voltage (V_{MOD}) can be controlled by changing the switching pattern modulation index (Figure 27.14), or by changing the amplitude of the converter dc voltage V_D. Faster time response is achieved by changing the switching pattern modulation index instead of V_D. The converter dc voltage V_D, is changed by adjusting the small amount of active power absorbed by the converter and defined by

$$P = \frac{V_{COMP} \cdot V_{MOD}}{X_S} \sin \delta \tag{27.7}$$

where
   X_S is the converter link reactor
   δ is the phase-shift angle between voltages V_{COMP} and V_{MOD}

To increase the amplitude of V_D a small positive average value of current must circulate through the dc capacitor so that V_D will increase until it reaches the required value. In the same way, if it is necessary to decrease the amplitude of V_D then a small negative average value of current must flow through the dc capacitor. The active power flow of the converter is defined by δ. If δ is positive the converter absorbs active power (increasing V_D), and if δ is negative the converter generates active power, and therefore V_D decreases.

One of the major problems that must be solved when self-commutated converters are used in high-voltage systems is the limited capacity of the gate-controlled semiconductors available in the market (IGBTs and IGCTs). Actual semiconductors can handle a few thousands of amperes and 6–10 kV reverse voltage blocking capabilities, which is clearly not enough for high-voltage applications. This problem can be overcome by using more sophisticated converters topologies, as described below.

27.2.3.2 Multilevel Converters

Multilevel converters are being investigated and some topologies are used today as STATCOM. The main advantages of multilevel converters are less harmonic generation and higher voltage capability because of serial connection of bridges or semiconductors. The most popular arrangement today is the three-level neutral-point clamped (NPC) topology.

27.2.3.2.1 Three-Level Neutral-Point Clamped Topology

Figure 27.15 shows a STATCOM implemented with a three-level NPC converter. Three-level converters [23] are becoming the standard topology for medium voltage converter applications, such as machine drives and active front-end rectifiers. The advantage of three-level converters is that they can reduce the generated harmonic content, for the same switching frequency, since they produce a voltage waveform with more levels than the conventional two-level topology. Another advantage is that they can reduce the semiconductor voltage rating and the associated switching frequency. Three-level converters consist of 12 self-commutated semiconductors such as IGBTs or IGCTs, each of them shunted by a reverse parallel connected power diode, and 6 diode branches connected between the midpoint of the dc link bus and the midpoint of each pair of switches as shown in Figure 27.15. By connecting the dc source sequentially to the output terminals, the converter can produce a set of PWM signals in which the frequency, amplitude, and phase of the ac voltage can be modified with adequate control signals.

27.2.3.2.2 Multilevel Converters with Single-Phase Inverters

Another exciting technology that has been successfully proven uses basic "H" bridges as shown in Figure 27.16, connected to line through power transformers. The system uses sinusoidal pulse width modulation (SPWM) with triangular carriers shifted and depending on the number of converters connected in the chain of bridges, the voltage waveform becomes more and more sinusoidal. Figure 27.16a shows one phase of this topology implemented with eight "H" bridges and Figure 27.16b shows the voltage waveforms obtained as a function of number of "H" bridges.

An interesting result with this converter is that the ac voltages become modulated by pulse width and by amplitude (PWM and AM). This is because when the pulse modulation changes, the steps of the amplitude also change. The maximum number of steps of the resultant voltage is equal to two times the number of converters plus the zero level. Then, four bridges will result in a nine-level converter per phase.

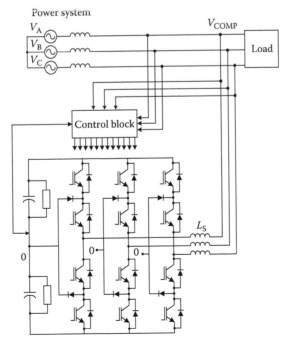

FIGURE 27.15 A STATCOM implemented with a three-level NPC inverter.

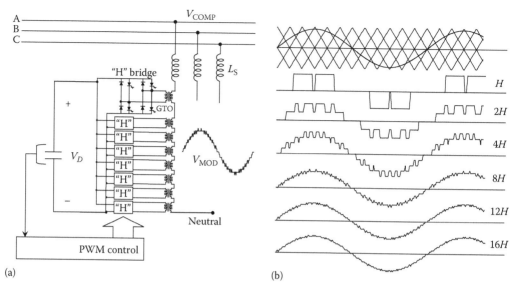

FIGURE 27.16 Multilevel STATCOM implemented with "H" bridge converters and associated voltage waveforms. (a) Multilevel converter with eight "H" bridges per phase and triangular carriers shifted. (b) Voltage waveforms as a function of number of bridges.

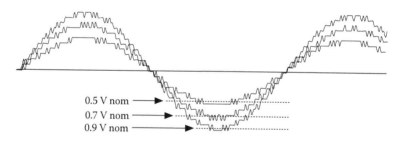

FIGURE 27.17 Amplitude modulation in the voltage waveform topology of Figure 27.16.

Figure 27.17 shows the principles of AM operation. When the voltage decreases, some steps disappear, and then the amplitude modulation becomes a discrete function.

27.2.3.2.3 Optimized Multilevel Converter

The number of levels can increase rapidly with few converters when voltage escalation is applied. In a similar way of converter in Figure 27.16, the topology of Figure 27.18 has a common dc link with voltage isolation through output transformers, connected in series at the line side. However, the voltages at the line side are scaled in power of three. By using this strategy, the number of voltage steps is maximized and few converters are required to obtain almost sinusoidal voltage waveforms. In the topology of Figure 27.18, amplitude modulation with 81 levels of voltage is obtained using only 4 "H" converters per phase (4-stage inverter). In this way, STATCOMs with "harmonic-free" characteristics can be implemented.

It is important to remark that the bridge with the higher voltage is being commutated at the line frequency, which is a major advantage of this topology for high power applications. Another interesting characteristic of this converter, compared with the multilevel strategy with carriers shifted, is that only 4 "H" bridges per phase are required to get 81 levels of voltage. In the previous multilevel converter with carriers shifted, 40 "H" bridges instead of four are required. For high power applications, probably a less complicated three-stage (3 "H" bridges per phase) is enough. In this case, 27 levels or steps of voltage are obtained, which will provide good enough voltage and current waveforms for high-quality operation [26].

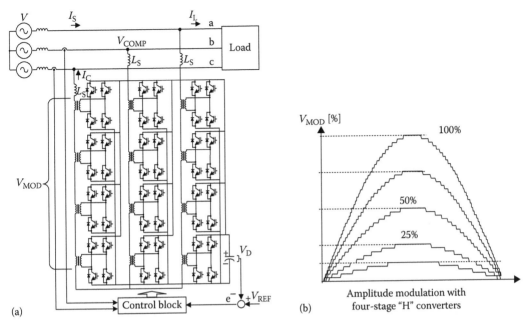

FIGURE 27.18 Optimized STATCOM using "H" bridges converters. (a) Four-stage, 81-level STATCOM, using "H" bridges scaled in power of three; (b) converter output voltage using amplitude modulation.

27.2.3.3 STATCOM Design Principles

The principal components of the STATCOM are the PWM converter, the coupling reactor, and the dc capacitor. The converter topology depends on the rated voltage and power, and can be implemented with a voltage- or current-source approach, although most of the STATCOM have been implemented using voltage-source converters. The design of the coupling reactor and dc capacitor is based on the functions they perform. The main purpose of the coupling reactor is to absorb the instantaneous voltage difference that appears between the utility voltage and the converter input voltage. Also, in case of using a voltage-source converter, the reactor helps to reduce the harmonic distortion of the line current, and can contribute to stabilize the converter switching frequency.

27.2.3.3.1 Coupling Reactor Design

Two criteria have been used to design the coupling reactor. The first one is based in the maximum current harmonic allowed in the input current and the second one is related with the control scheme implemented in the current loop. In the first case, the reactor value is given by expression

$$X_1 = \frac{1}{THD_i} \sqrt{\sum_{k \neq 1}^{\infty} \frac{(V_{ak})^2}{k^2}} \tag{27.8}$$

where
THD_i is the maximum harmonic distortion allowed in the converter input current (normally 5%)
V_{ak} the converter input voltage harmonic component
k the order of the harmonic component

The converter input voltage, V_{ak}, depends on the PWM technique implemented in the converter.

In case a current control loop is implemented, the reactor value must be rated in order to allow intersections between the current reference and the converter input current. The block diagram of a typical

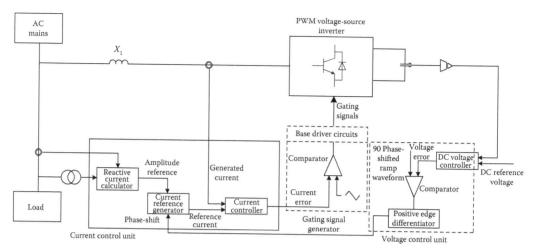

FIGURE 27.19 The block diagram of a current controlled STATCOM.

current control scheme used in STATCOM is shown in Figure 27.19. The principal advantage of this current control scheme is that it forces the converter to operate with constant switching frequency. The constant switching frequency is achieved by comparing the current error signal with a triangular reference waveform. The purpose of introducing the triangular waveform is to stabilize the inverter switching frequency by forcing it to be constant and equal to the frequency of the triangular reference.

The current control method can be explained by considering the hysteresis technique plus the addition of a fixed frequency triangular waveform inside the imaginary hysteresis window (Figure 27.20). If the current reference, i_s^*, is higher than the generated current, i_s, the error waveform is positive and when compared with the triangular waveform it results in a positive pulse. This pulse will then turn on an inverter bottom switch that will increase the corresponding output line current. In the same way, if i_s^* is lower than i_s, the error waveform is negative and the gating signals are adjusted so that the line current decreases (a top switch is turned on). The slope of the error signal is chosen to be always smaller

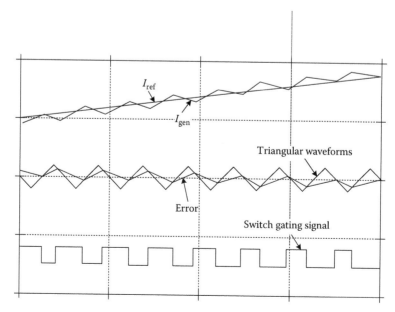

FIGURE 27.20 Principles of operation of the fixed switching frequency current control technique.

than the slope of the triangular waveform in order to ensure that an intersection between the two signals exists. Thus, the current error signal is forced to remain between the maximum and the minimum of the triangular waveform and as a result, the line current follows the reference closely. Moreover, since the error between i_s^* and i_s is always kept within the positive and negative peaks of the triangular waveform, the system has an inherent overcurrent protection. However, a large variation in the reference current will generate a large error signal, which can be higher than the amplitude of the triangular waveform. In this case there will not be an intersection between the error and the triangular waveform, thus the switching pattern will not change until the error is reduced and a new intersection occurs.

The minimum amplitude of the triangular waveform is determined by the maximum slope of the error, which is in turn fixed by the maximum slope of the line current. The maximum slope of the line current is determined by the maximum instantaneous voltage drop across the coupling reactor and the value of the inductance. Reducing the amplitude of the triangular waveform below this value will generate multiple crossing between the current error and the triangular signals thus disturbing the operation of the inverter. The value of the coupling reactor inductance is obtained from expression

$$L = \Delta V_{\text{reactor}} \frac{1}{4 A_{\text{triangular}} f_t} \tag{27.9}$$

where
$\Delta V_{\text{reactor}}$ is the maximum expected voltage drop across the coupling reactor
$A_{\text{triangular}}$ the amplitude of the triangular control waveform
f_t the triangular waveform frequency (equals to the converter switching frequency)

27.2.3.3.2 dc Capacitor Design

The function of the dc capacitor is to maintain a dc voltage constant and equal to the reference value imposed by the control scheme. In theory, the dc capacitor is not used as an energy storage element, although it may be necessary to compensate active power in case of outage, especially in industrial applications. Two design criteria are used to calculate the capacitance value. The first is related with the maximum voltage harmonic distortion factor allowed in the dc voltage, and the second one with the maximum voltage fluctuation that can be tolerated in the dc voltage in case of load rejection or a step increase in the load power. In the first case the capacitance is obtained with expression (27.9), and in the second one (maximum voltage fluctuation) with expression (27.10). If the dc capacitor is designed with the constraint that with rated leading var compensation, the ripple factor for V_D is less than a given value, the capacitance is defined by expression

$$X_C = \frac{R_V V_D}{\left\{ \sum_k \frac{I_{ck}^2}{k^2} \right\}^{1/2}} \tag{27.10}$$

where
R_V is the maximum voltage ripple factor across the dc capacitor
I_{ck} the capacitor current harmonic component

Independently of the STATCOM control scheme, an overvoltage is created when the converter line currents are forced to reduce their amplitude, or when they are forced to change from a 90° leading to a 90° lagging phase-shift. On the other hand, an undervoltage is generated when the line currents are forced to increase their amplitude, or when they are forced to change from a 90° lagging to a 90° leading phase-shift. These dc voltage fluctuations are created because transiently the dc capacitor supplies (or absorbs) the extra power required by the system. The amplitude of this voltage fluctuation depends on the instant at which the transient occurs and on the amount of change in the line current amplitude.

The maximum overvoltage is generated when one of the line currents is forced to change from 1 per unit leading to 1 per unit lagging, and can be calculated using expression

$$V_{\mathrm{D\,max}} = \frac{1}{C} \int_{\theta/\omega} i_{\mathrm{c}}(t)\mathrm{d}t + V_{\mathrm{D0}} \tag{27.11}$$

where

$V_{\mathrm{D\,max}}$ is the maximum overvoltage across the dc capacitor
V_{D0} is the steady-state dc voltage
i_{c} is the instantaneous capacitor current

The dc current $i_{\mathrm{c}}(t)$ is defined by the product of the converter line currents with the respective switching functions. The principal design constraint for C is to keep the voltage fluctuation below a defined value, ΔV, that is

$$C = \frac{1}{\Delta V} \int_{t} i_{\mathrm{c}}(t)\mathrm{d}t \tag{27.12}$$

Comparing the two design criteria, the second one provides a larger value of C, which contributes to increase the STATCOM stability, making its operation safer.

27.2.3.4 Semiconductor Devices Used for Self-Commutated STATCOMs and FACTS Controllers

Three are the most relevant devices for applications in self-commutated STATCOMs or other FACTS controllers: thyristors, insulated gate bipolar transistors (IGBTs), and integrated gate controlled thyristors (IGCTs). This field of application requires that the semiconductor must be able to block high voltages in the kV range. High-voltage IGBTs required to apply self-commutated converters in FACTS controllers reach now the level of 6.5 kV, allowing for the construction of circuits with a power of several MVA. Also IGCTs are reaching levels higher than 6 kV. Perhaps, the most important development in semiconductors for high-voltage applications is the light triggered thyristor (LTT). This device is the most important for ultrahigh power applications. Recently, LTT devices have been developed with a capability of up to 13.5 kV and a current of up to 6 kA. These new devices reduce the number of elements in series and in parallel, reducing consequently the number of gate and protection circuits. With these elements, it is possible to reduce cost and increase rated power in FACTS installations of up to several hundreds of MVARs [24].

27.2.4 Comparison between Thyristorized and Self-Commutated Compensators

Thyristorized and self-commutated FACTS controllers are very similar in their functional compensation capability, but the basic operating principles, as shown, are fundamentally different. A STATCOM functions as a shunt-connected synchronous voltage source whereas a thyristorized compensator operates as a shunt-connected, controlled reactive admittance. This difference accounts for the STATCOM's superior functional characteristics, better performance, and greater application flexibility.

In the linear operating range of the *V–I* characteristic, the functional compensation capability of the STATCOM and Static VAR Compensator (SVC) is similar. Concerning the nonlinear operating range, the STATCOM is able to control its output current over the rated maximum capacitive or inductive range independently of the ac system voltage, whereas the maximum attainable compensating current of the SVC decreases linearly with ac voltage. Thus, the STATCOM is more effective than the SVC in providing voltage support under large system disturbances during which the voltage

TABLE 27.1 Comparison of Basic Types of Shunt Compensators

		Static Compensator		
	Synchronous Condenser	TCR (with Shunt Capacitors if Necessary)	TSC (with TCR if Necessary)	Self-Commutated Compensator
Accuracy of compensation	Good	Very good	Good, very good with TCR	Excellent
Control flexibility	Good	Very good	Good, very good with TCR	Excellent
Reactive power capability	Leading/lagging	Lagging/leading indirect	Leading/lagging indirect	Leading/lagging
Control	Continuous	Continuous	Discontinuous (continuous with TCR)	Continuous
Response time	Slow	Fast, 0.5 to 2 cycles	Fast, 0.5 to 2 cycles	Very fast but depends on the control system and switching frequency
Harmonics	Very good	Very high (large-size filters are needed)	Good, filters are necessary with TCR	Good, but depends on switching pattern
Losses	Moderate	Good, but increase in lagging mode	Good, but increase in leading mode	Very good, but increase with switching frequency
Phase balancing ability	Limited	Good	Limited	Very good with 1-ϕ units, limited with 3-ϕ units
Cost	High	Moderate	Moderate	Low to moderate

excursions would be well outside of the linear operating range of the compensator. The ability of the STATCOM to maintain full capacitive output current at low system voltage also makes it more effective than the SVC in improving the transient stability limit. The attainable response time and the bandwidth of the closed voltage regulation loop of the STATCOM are also significantly better than those of the SVC.

Table 27.1 summarizes the comparative merits of the main types of shunt FACTS compensation. The significant advantages of self-commutated compensators make them an interesting alternative to improve compensation characteristics and also to increase the performance of ac power systems.

As compared with thyristor-controlled capacitor and reactor banks, self-commutated VAR compensators have the following advantages:

1. They can provide both leading and lagging reactive power, thus enabling a considerable saving in capacitors and reactors. This in turn reduces the possibility of resonances at some critical operating conditions.
2. Since the time response of self-commutated converter can be faster than the fundamental power network cycle, reactive power can be controlled continuously and precisely.
3. High frequency modulation of self-commutated converter results in a low harmonic content of the supply current, thus reducing the size of passive filter components.
4. They do not generate inrush current.
5. The dynamic performance under voltage variations and transients is improved.
6. Self-commutated VAR compensators are capable of generating 1 p.u. reactive current even when the line voltages are very low. This ability to support the power system is better than that obtained with thyristor-controlled VAR compensators because the current in shunt capacitors and reactors is proportional to the voltage.
7. Self-commutated compensators with appropriate control can also act as active line harmonic filters, DVR, or UPFC.

Figure 27.21 shows the voltage/current characteristic of a self-commutated VAR compensator compared with that of thyristor-controlled SVC. This figure illustrates that the self-commutated compensator offers

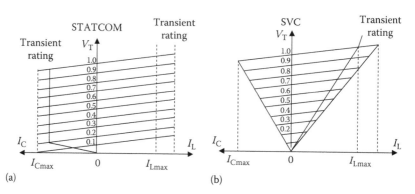

FIGURE 27.21 Voltage–current characteristics of shunt VAR compensators. (a) Compensator implemented with self-commutated converter (STATCOM). (b) Compensator implemented with back to back thyristors.

better voltage support and improved transient stability margin by providing more reactive power at lower voltages. Because no large capacitors and reactors are used to generate reactive power, the self-commutated compensator provides faster time response and better stability to variations in system impedances.

27.2.5 Superconducting Magnetic Energy Storage

The principal limitation of STATCOM is that it cannot provide active power, since they have limited energy storage components (dc capacitor or reactor). This limitation is overcome, if the traditional electrolytic capacitor used in the dc link is replaced by an SMES, as shown in Figure 27.22. This device is capable to store and instantaneously discharge large quantities of power [27,28]. It stores energy in the magnetic field created by the flow of dc current in a coil of superconducting material that has been cryogenically cooled. These systems have been in use for several years to improve power quality and to provide a premium-quality service for individual customers vulnerable to voltage fluctuations. The SMES recharges within minutes and can repeat the charge/discharge sequence thousands of times without any degradation of the magnet. Recharge time can be accelerated to meet specific requirements, depending on system capacity. It is claimed that SMES is 97%–98% efficient and it is much better at providing reactive power on demand. Figure 27.23 shows a different SMES topology using three-level converters.

The first commercial application of SMES was in 1981 [28] along the 500-kV Pacific Intertie, which interconnects California and the Northwest. The device's purpose was to demonstrate the feasibility of SMES to improve transmission capacity by damping inter-area modal oscillations. Since that time, many studies have been performed and prototypes developed for installing SMES to enhance transmission line capacity and performance. A major cost driver for SMES is the amount of stored energy.

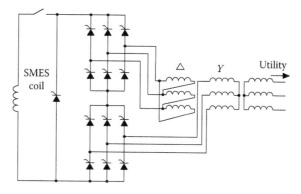

FIGURE 27.22 SMES implemented with a 12-pulse thyristor converter.

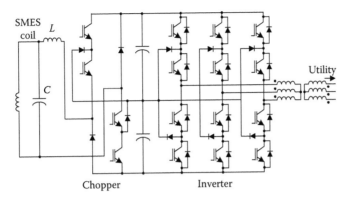

FIGURE 27.23 SMES implemented with a three-level converter.

Previous studies have shown that SMES can substantially increase transmission line capacity when utilities apply relatively small amounts of stored energy and a large power rating (greater than 50 MW).

Another interesting application of SMES for frequency stabilization is in combination with static synchronous series compensator (SSSC) [29].

27.3 Series Compensation

FACTS controllers can also be of the series type. Typical series compensators use capacitors to reduce the equivalent reactance of a power line at rated frequency, thus increasing the voltage at the load terminals. The connection of a series compensator generates reactive power that, in a self-regulated manner, balances a fraction of the line's reactance.

27.3.1 Series Compensation Principles

Like shunt compensation, series compensation may also be implemented with current- or voltage-source converters. In this type of compensation, the compensator injects a voltage in series with the load, eliminating voltage unbalance at the load terminals, and supplying the voltage component required to operate with rated, balanced, and constant value. Figure 27.24 shows the principle of series compensation.

Figure 27.25 shows a radial power system, with the reference angle in V_2, and the results obtained with the series compensation using a voltage source, which has been adjusted to have unity power factor operation at V_2. However, the compensation strategy is different when compared with shunt compensation. In this case, the voltage V_{COMP} has been added between the line and the load to change the angle of V_2, which becomes the voltage at the load terminals. With the appropriate magnitude and phase-shift adjustment of V_{COMP}, unity power factor can be reached at V_2. As can be seen from the phasor diagram of Figure 27.25b, V_{COMP} generates a voltage phasor with opposite direction to the voltage drop in the line inductance. In this type of compensation, the load current cannot be changed, which means that the compensator voltage

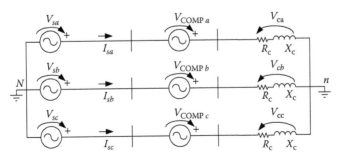

FIGURE 27.24 Three-phase equivalent circuit of a series compensated system.

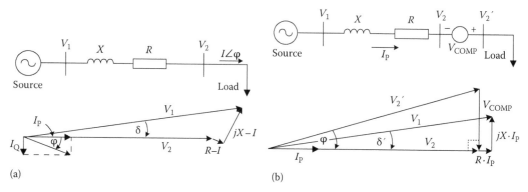

FIGURE 27.25 Principles of series compensation. (a) The radial system without compensation. (b) Series compensation with a voltage source.

must be adjusted to reach the required voltage at the load terminals. Moreover, the amount of apparent power interchanged with the series compensator in this case depends on the V_{COMP} phase and magnitude. Real and reactive power can be generated from the series compensator, depending on the load and the system requirements. In most cases, series compensation is used to control voltage at the load terminal, so that only reactive power needs to be exchanged. In this case, series capacitors represent a good alternative.

Series compensation with capacitors is the most common strategy. Series capacitors are connected with a transmission line as shown in Figure 27.26, which means that all the equipment must be installed on a platform that is fully insulated for the system voltage (both the terminals are at the line voltage). On this platform, the capacitor is located together with overvoltage protection circuits. The overvoltage protection is a key design factor as the capacitor bank has to withstand the throughput fault current, even at a severe fault. The primary overvoltage protection typically involves nonlinear metal-oxide varistors, a spark gap, and a fast bypass switch.

Series compensation is used basically to improve voltage regulation in transmission and distribution power systems. Different compensation equipments using power electronics have been developed to operate as series compensators, as described below.

27.3.2 Static Synchronous Series Compensator

A voltage-source converter can also be used as a series compensator as shown in Figure 27.27. The SSSC injects a voltage in series to the line, 90° phase-shifted with the load current, operating as a controllable series capacitor. The basic difference, as compared with series capacitor, is that the voltage injected by an SSSC is not related to the line current and can be independently controlled [3]. If the phase-shift angle between the injected voltage and the line current is not 90°, active power flows through or from the static compensator. In this case, an energy storage element must be included. This element normally is connected in the dc bus, and modifies the compensator control scheme.

In case of medium voltage application, the SSSC is also known as voltage dynamic restorer, and is used to compensate voltage sag, swells, and unbalance. The principle of operation is similar to the SSSC,

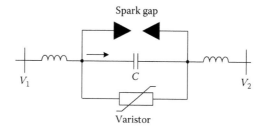

FIGURE 27.26 Series capacitor compensator and associated protection system.

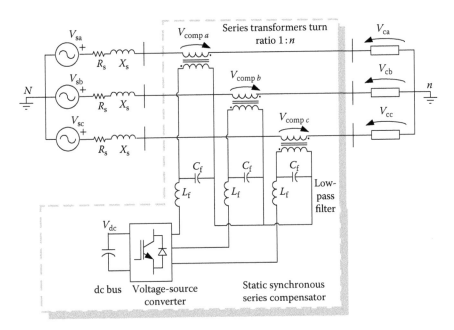

FIGURE 27.27 Static synchronous series compensator (SSSC).

but the main difference is that a DVR can compensate active power, since it may be implemented with an energy storage element in the dc bus.

When voltage sags or swells are present at the load terminals, the SSSC responds by injecting three ac voltages in series with the incoming three-phase network voltages, compensating for the difference between faulted and prefault voltages (Figure 27.28). Each phase of the injected voltages can be controlled separately (their magnitude and angle). Active and reactive power required to generate these voltages is supplied by the voltage-source converter, fed from a dc link as shown in Figure 27.28. In order to be able to mitigate voltage sag, the SSSC must present fast control response (Figure 27.28).

Voltage fluctuation at the load terminals can be in magnitude or in phase (or both) as shown in Figure 27.29. When the power supply conditions remain normal the SSSC operate in low-loss standby mode, with the converter side of the booster transformer shorted. Since no voltage-source converter (VSC) modulation takes place, the SSSC produces only conduction losses. In case active power needs to be compensated, an energy storage element must be supplied. This energy storage element is normally connected to the dc bus (supper capacitors, batteries), or the other alternative is to keep the dc capacitor loaded at rated voltage through a noncontrolled rectifier connected to the power line, as shown in Figure 27.30. This is the typical power circuit configuration of a DVR.

27.3.2.1 Compensation Strategies

The magnitude of the compensated voltage can be obtained from the equivalent circuit shown in Figure 27.24, and the phasor diagram shown in Figure 27.20, and is equal to

$$\dot{V}_{compj} = \dot{V}_{cj} - \dot{V}_{sj} \tag{27.13}$$

$$V_{compj}\angle\beta = V_{cj}\angle\alpha - V_{sj}\angle\delta \tag{27.14}$$

Using complex variables

$$V_{compj}\angle\beta = V_{cj}(\cos\alpha + j\sin\alpha) - V_{sj}(\cos\delta + j\sin\delta) \tag{27.15}$$

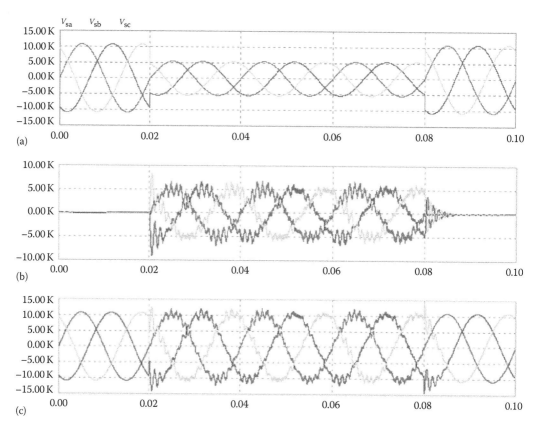

FIGURE 27.28 Voltage waveforms using series compensation. (a) Source voltage. (b) Series compensator voltage. (c) Compensated voltage at the load terminals.

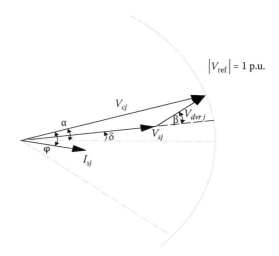

FIGURE 27.29 Phasor diagram for series voltage compensation.

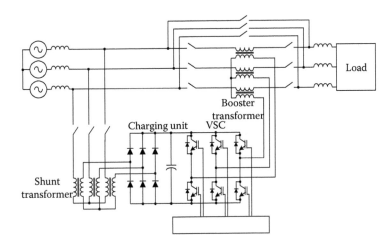

FIGURE 27.30 Power circuit topology of a dynamic voltage restorer (DVR).

The magnitude of the voltage component that must be generated is equal to

$$\left|V_{comp}\right| = \sqrt{V_{cj}^2 + V_{sj}^2 - 2 \cdot V_{cj} \cdot V_{sj} \cdot \cos(\alpha - \delta)} \tag{27.16}$$

From this expression, different compensation strategies are derived.

27.3.2.1.1 Boosting or In-Phase Compensation

The first compensation strategy is called boosting or in-phase compensation, since the compensated voltage is generated in phase with the load voltage as shown in Figure 27.31. In this case the compensator does not correct the phase-shift introduced at the voltage terminals.

The active and reactive power delivered by the series compensator is given by

$$P_{compj} = V_{compj} \cdot I_{sj} \cdot \cos(\varphi) \tag{27.17}$$

$$Q_{compj} = V_{compj} \cdot I_{sj} \cdot \sin(\varphi) \tag{27.18}$$

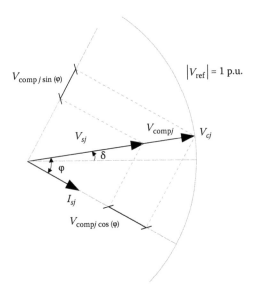

FIGURE 27.31 Phasor diagram of the compensated voltage scheme.

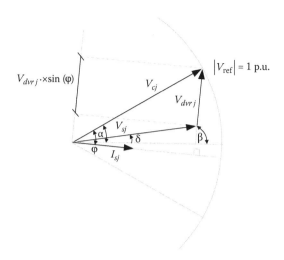

FIGURE 27.32 Phasor diagram of the minimum power injection method.

The in-phase compensation strategy requires the injection of both active and reactive power due to the angle between the compensation voltage and the system current (φ). The active power is taken from the energy storage unit, which defines the ride-through capability of the static series compensator (SSSC or DVR).

27.3.2.1.2 Minimum Power Injection

Unlike the boosting method, the minimum power injection strategy defines an optimal angle for which the compensation voltage V_{comp} is 90° phase-shift with the system current, canceling the injection of active power from the series compensator. The limiting factor to compensate the sag only with reactive power is the load power factor. Figure 27.32 illustrates the phasor diagram of the compensation voltage with minimum power injection.

The active power delivered by the series compensator can be obtained from Figure 27.32.

$$P_{compj} = P_{sj} - P_{cj} \tag{27.19}$$

$$P_{compj} = (V_{cj} \cdot I_{sj} \cdot \cos(\varphi)) - (V_{sj} \cdot I_{sj} \cdot (\cos\varphi - \alpha + \delta_j)) \tag{27.20}$$

$$P_{compj} = \cos(\varphi_c) - V_{sj} \cdot \cos(\varphi_s) \tag{27.21}$$

The active power injected by the series compensator, P_{compj}, is minimum when the load power factor is equal to 1 (Equation 27.21).

27.3.2.2 Reference Signal Generation

The reference signal is used by the power converter control scheme to generate the required output voltage necessary to keep the load voltage at a given value. Different methods can be used, and the differences are based on the simplicity of implementation, number of calculations, and execution time.

27.3.2.2.1 Pythagoras Method

Based on the Pythagoras theorem, it is possible to generate the signal required to compensate the system voltage drop at the load terminals. This scheme is based on the synchronization of the compensator control signals with the system voltage. The equation used is

$$\sin^2\theta + \cos^2\theta = 1 \tag{27.22}$$

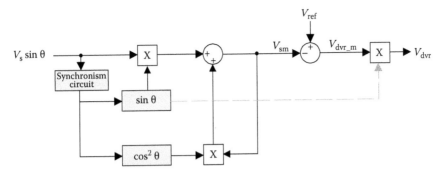

FIGURE 27.33 The block diagram of the Pitagoras algorithm.

The block diagram for the generation of the compensation voltage reference signal for single-phase application is shown in Figure 27.33.

One of the problems that must be taken into consideration during the generation of the compensated voltage reference signal is the sign, which must indicate if the voltage perturbation is a sag or a swell. The type of perturbation defines the polarity of the compensated voltage. If the perturbation is a sag, the compensated voltage must be in phase with the system voltage, but, if the perturbation is a swell, the compensated voltage must be phase-shifted 180° with respect to the equivalent system voltage. The synchronization circuit used to generate the sin θ signal defines the control scheme time response. Different circuits and methods can be used to synchronize the reference signal with the system voltage: zero voltage detector, fast Fourier transform, or phase locked loop. The fastest algorithm is the zero crossing detection, with a time response equal to half a cycle of the system frequency. The Pitagoras method allows the generation of a positive reference signal in case of sag and a negative signal in case of a swell.

27.3.2.2.2 Synchronous Reference Frame

The synchronous reference frame method uses the transformation from the stationary reference frame *abc* to a *dq* axes, which are synchronized with the system voltage. The advantage of transforming the voltages from *abc* to *dq* is that sinusoidal signals generate continuous signals in *dq* axis, allowing the use of linear control algorithms. The block diagram of the synchronous reference frame algorithm is shown in Figure 27.34. This scheme generates dc error signals, which are easier to process as compared with the Pitagoras algorithm. Since the system voltages are line to line, in order to transform them in phase to neutral, they must be multiplied by $1/(1 - a^2)$ term.

Another alternative to go from *abc* to *dq* axis is to use αβ transformation, as shown in Figure 27.35. In this case, the phase-shift angle θ, used to synchronize the *dq* axis, is obtained from Equation 27.22.

The principal disadvantage of the indirect algorithm is the presence of a second order harmonic in Vα and Vβ, used to calculate θ. The amplitude of this second order harmonic depends on the voltage unbalance in the *abc* reference frame. The use of a passive low-pass filter to eliminate the second order

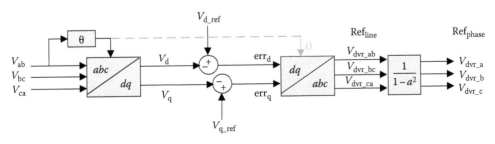

FIGURE 27.34 The block diagram of the synchronous reference frame direct algorithm.

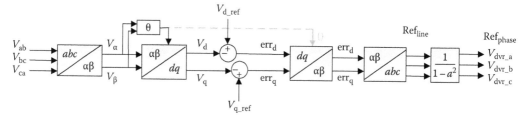

FIGURE 27.35 The block diagram of the synchronous reference frame indirect algorithm.

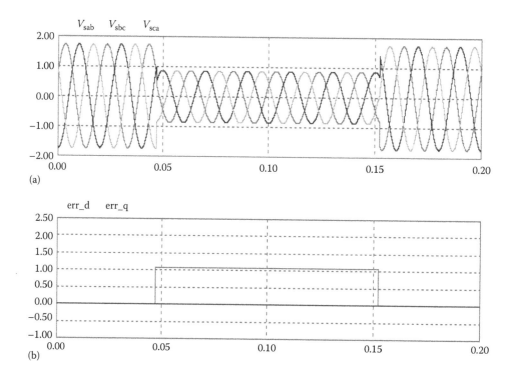

FIGURE 27.36 Simulated voltage waveforms for a balance sag. (a) Line to line system voltage in *abc* reference frame. (b) Voltage signals in *dq* axes.

harmonic introduces a significant time delay, affecting the transient response of the SSSC. Simulated waveforms showing the characteristics of synchronous reference frame method are shown in Figures 27.36 and 27.37, for a balance and unbalance voltage perturbation.

27.3.2.2.3 Sequence Components in Synchronous Reference Frame

Another alternative to generate the reference signal is to use sequence components. From the line voltages in *abc* frame, the negative, zero, and positive sequence voltage components are calculated:

$$
\begin{bmatrix} V_{ab}^0 \\ V_{ab}^1 \\ V_{ab}^2 \end{bmatrix} = \frac{1}{3} \begin{bmatrix} 1 & 1 & 1 \\ 1 & a & a^2 \\ 1 & a^2 & a \end{bmatrix} * \begin{bmatrix} V_{ab} \\ V_{bc} \\ V_{ca} \end{bmatrix}
$$
(27.23)

The block diagram used to generate the reference signals with the sequence components algorithm is shown in Figure 27.38. The reference voltage signals are phase to neutral, and are calculated from the

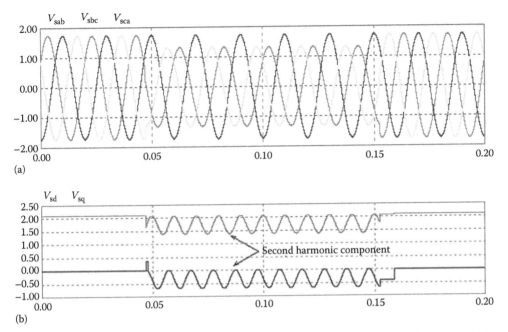

(a)

(b)

FIGURE 27.37 Simulated voltage waveforms for an unbalance sag. (a) Line to line system voltage in *abc* reference frame. (b) Voltage signals in *dq* axes.

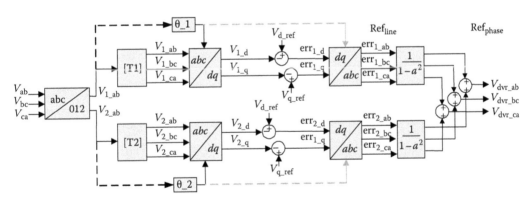

FIGURE 27.38 The block diagram of the sequence components algorithm.

line to line voltages. By transforming the sequence components from *abc* to *dq* reference frame, dc quantities are obtained.

If the voltage perturbation is balanced, the reference signals have only positive sequence component, but if the perturbation is unbalanced, positive and negative sequence components are obtained as shown in Figure 27.39.

27.3.2.3 Power Circuit Design

The power circuit topology of the SSSC is composed by the three-phase PWM voltage-source inverter, the second order resonant LC filters, the coupling transformers, and the secondary ripple frequency (Figure 27.27). The main design characteristics for each of the power components are described below.

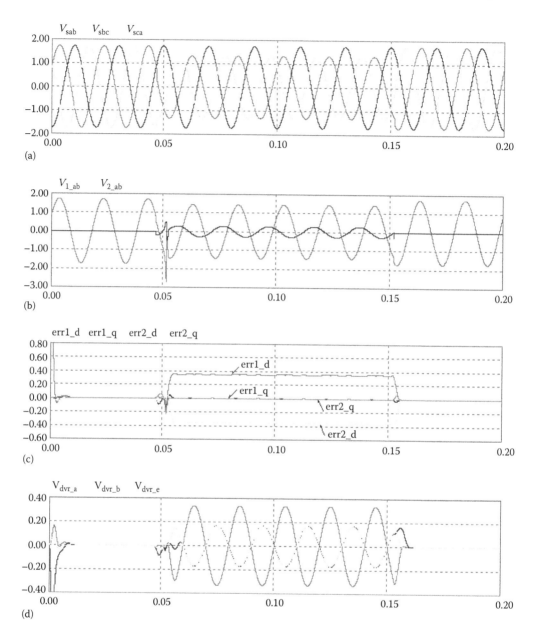

FIGURE 27.39 Simulated waveforms for an unbalance voltage sag. (a) Unbalanced line to line system voltage in *abc* reference frame. (b) Positive and negative voltage sequence components in *abc* reference. (c) Positive and negative sequence reference signal voltages in *dq* reference frame. (d) Positive and negative sequence reference signal voltages in *abc* reference frame.

27.3.2.3.1 PWM Voltage-Source Inverter

The rated apparent power required by the inverter can be obtained by calculating the apparent power generated in the primary of the series transformer. The voltage reflected across the primary winding of the coupling transformer is defined by the following expression:

$$V_{series} = \sqrt{K_2^2(V_1 + V_2)^2} \tag{27.24}$$

where

V_{series} is the rms voltage across the primary winding of the coupling transformer
K_2 is equal to 1

The fundamental component of the primary voltage depends on the amplitude of the negative and positive sequence component of the source-voltage defined by the system voltage regulation and unbalance. The transformer primary current depends on the load.

27.3.2.3.2 Coupling Transformer

The primary windings of the coupling transformers are connected in series to the power system, and as such, they behave as current transformers. The total apparent power required by each series transformer is one-third the total apparent power of the inverter. The turn ratio of the coupling transformer is specified according with the inverter dc bus voltage. The correct value of the turn ratio a must be specified according with the overall SSSC performance. The turn ratio of the coupling transformer must be optimized through the simulation of the overall SSSC, since it depends on the values of different related parameters. In general, the transformer turn ratio must be high in order to reduce the amplitude of the inverter output current and to reduce the voltage induced across the primary winding. Also, the selection of the transformer turn ratio influences the performance of the ripple filter connected at the output of the PWM inverter.

27.3.2.3.3 Secondary Ripple Filter

It is important to note that the design of the secondary ripple filter depends mainly on the coupling transformer turn ratio and on the frequency of the triangular waveform used to generate the inverter gating signals. The ripple filter connected at the output of the inverter avoids the induction of the high frequency ripple voltage generated by the PWM inverter switching pattern at the terminals of the primary windings of the coupling transformers. In this way, the voltage applied in series to the power system corresponds to the components required to compensate voltage unbalance and regulation. The single-phase equivalent circuit of the ripple filter is shown in Figure 27.40.

The voltage reflected to the primary winding of the series transformer has the same waveform as the voltage across the filter capacitor. For low frequency components, the inverter output voltage must be almost equal to the voltage across C_f. However, for high frequency components, most of the inverter output voltage must drop across L_f, in which case the voltage at the capacitor terminals is almost zero. Moreover, C_f and L_f must be selected in order to not exceed the burden of the series transformer. The ripple filter must be designed for the carrier frequency of the PWM voltage-source inverter. To calculate C_f and L_f the system equivalent impedance at the carrier frequency, Z_{sys}, reflected to the secondary must be known. This impedance is equal to

$$Z_{\text{sys(secondary)}} = a^2 Z_{\text{sys(primary)}} \qquad (27.25)$$

For the carrier frequency, the following design criteria must be satisfied:

1. $X_{Cf} \ll X_{Lf}$ to ensure that at the carrier frequency most of the inverter output voltage will drop across L_f.
2. X_{Cf} and $X_{Lf} \ll Z_{\text{sys}}$ to ensure that the voltage divider is between L_f and C_f.

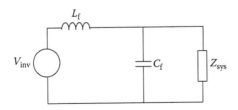

FIGURE 27.40 The single-phase equivalent circuit of the inverter ripple filter.

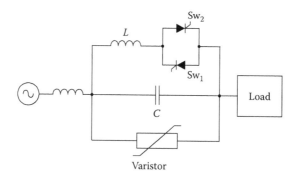

FIGURE 27.41   Power circuit topology of a thyristor-controlled series compensator.

27.3.3 Thyristor-Controlled Series Compensation

Figure 27.41 shows a single line diagram of a thyristor-controlled series compensator (TCSC). TCSC provides a proven technology that addresses specific dynamic problems in transmission systems. TCSCs are an excellent tool to introduce if increased damping is required when interconnecting large electrical systems. Additionally, they can overcome the problem of subsynchronous resonance (SSR), a phenomenon that involves an interaction between large thermal generating units and series compensated transmission systems.

There are two bearing principles of the TCSC concept. First, the TCSC provides electromechanical damping between large electrical systems by changing the reactance of a specific interconnecting power line, i.e., the TCSC will provide a variable capacitive reactance. Second, the TCSC shall change its apparent impedance (as seen by the line current) for subsynchronous frequencies such that a prospective SSR is avoided. Both these objectives are achieved with the TCSC using control algorithms that operate concurrently. The controls will function on the thyristor circuit (in parallel to the main capacitor bank) such that controlled charges are added to the main capacitor, making it a variable capacitor at fundamental frequency but a "virtual inductor" at subsynchronous frequencies.

For power oscillation damping, the TCSC scheme introduces a component of modulation of the effective reactance of the power transmission corridor. By suitable system control, this modulation of the reactance is made to counteract the oscillations of the active power transfer, in order to damp these out.

27.4 Hybrid Compensation

Static synchronous VAR compensators (STATCOM) and SSSC or DVR can be integrated to get a system capable of controlling the power flow of a transmission line during steady-state conditions and providing dynamic voltage compensation and short circuit current limitation during system disturbances [31].

27.4.1 Unified Power Flow Controller

The UPFC, shown in Figure 27.42, consists of two switching converters operated from a common dc link provided by a dc capacitor, one connected in series with the line and the other in parallel. This arrangement functions as an ideal ac to ac power converter in which the real power can freely flow in either direction between the ac terminals of the two inverters, and each one can independently generate (or absorb) reactive power at its own ac output terminal. The series converter of the UPFC injects via series transformer, an ac voltage with controllable magnitude and phase angle in series with the transmission line. The shunt converter supplies or absorbs the real power demanded by the series converter through the common dc link. The inverter connected in series provides the main function of the UPFC by injecting an ac voltage V_{pq} with controllable magnitude ($0 \leq V_{pq} \leq V_{pqmax}$) and phase angle $\rho(0 \leq \rho \leq 360°)$, at the power frequency, in series with the line via a transformer. The transmission line current flows

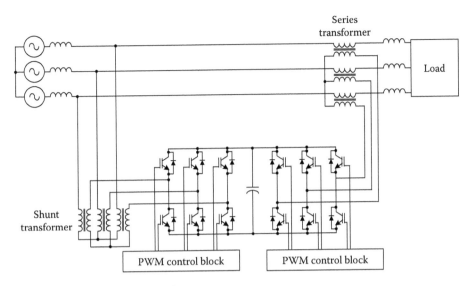

FIGURE 27.42   UPFC power circuit topology.

through the series voltage source resulting in real and reactive power exchange between the UPFC and the ac system. The real power exchanged at the ac terminal, i.e., the terminal of the coupling transformer, is converted by the inverter into dc power, which appears at the dc link as positive or negative real power demand. The reactive power exchanged at the ac terminal is generated internally by the inverter.

The basic function of the inverter connected in parallel (inverter 1) is to supply or absorb the real power demanded by the inverter connected in series to the ac system (inverter 2), at the common dc link. Inverter 1 can also generate or absorb controllable reactive power, if it is desired, and thereby it can provide independent shunt reactive compensation for the line. It is important to note that whereas there is a closed "direct" path for the real power negotiated by the action of series voltage injection through inverter 1 and back to the line, the corresponding reactive power exchanged is supplied or absorbed locally by inverter 2 and therefore it does not flow through the line. Thus, inverter 1 can be operated at a unity power factor or be controlled to have a reactive power exchange with the line independently of the reactive power exchanged by inverter 2. This means that there is no continuous reactive power flow through the UPFC.

27.4.2 Interline Power Flow Controller

An interline power flow controller (IPFC), shown in Figure 27.43, consists of two series voltage-source converters whose dc capacitors are coupled, allowing active power to circulate between different power lines [33]. The IPFC addresses the problem of compensating a number of transmission lines at a given substation. Series capacitive compensators are used to increase the transmittable active power over a given line but they are unable to control the reactive power flow in, and thus the proper load balancing of the line. With the IPFC active power can be transferred between different lines, and therefore it is possible to

1. Equalize both active and reactive power flows between different lines
2. Reduce the burden of the overloaded lines by active power transfer
3. Compensate against resistive line voltage drops and the corresponding reactive power demand
4. Increase the effectiveness of the overall compensating system for dynamic disturbances

In the IPFC the converters do not only provide series reactive compensation but can also be controlled to supply active power to a common dc link from its own transmission line. Like this, active power can be provided from the overloaded lines for active power compensation in other lines. This scheme requires a rigorous maintenance of the overall power balance at the common dc terminal by appropriate

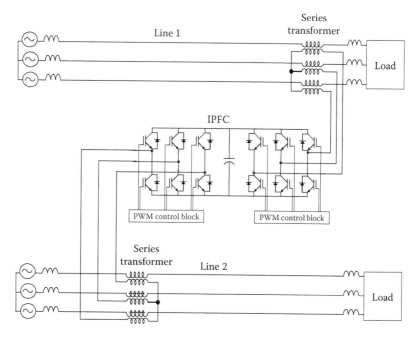

FIGURE 27.43   IPFC power circuit topology.

control action, allowing that underloaded lines provide appropriate active power transfer to the over-loaded ones. When operating below its rated capacity, the IPFC is in regulation mode, allowing the regulation of the active and reactive power flows on one line, and the active power flow on the other line. In addition, the net active power generation by the two coupled converters is zero, neglecting power losses.

27.4.3 Unified Power Quality Conditioner

Unlike the previous alternatives, the unified power quality conditioner (UPQC), shown in Figure 27.44, has two main objectives. First, to provide a near sinusoidal and constant voltage to a critical load regardless of the variations (sag, swells, distortion) of the voltage at the ac main terminals in the point of common coupling (PCC). Second, compensate the reactive and/or distorted current taken by the load achieving a near sinusoidal and unity power factor at the PCC. For these reasons, the UPQC is more oriented for load compensation in industrial power distribution systems.

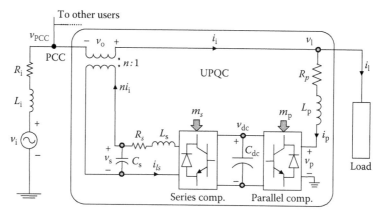

FIGURE 27.44   UPQC power circuit topology.

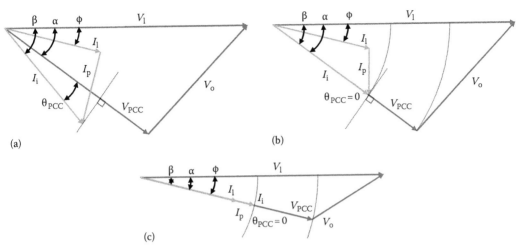

FIGURE 27.45 UPQC phasor's diagrams. (a) For a general application. (b) For unity power load power factor.
(c) For unity system power factor.

The UPQC consists of two PWM converters operated from a common dc voltage link, which must be kept constant. One converter connected in series with the line and the other connected in parallel [31]. The series converter task is to achieve the first objective (clean load voltage) and the parallel converter the second one (sinusoidal current at the system terminals).

The ac voltages of the two converters can be generated with an arbitrary amplitude and phase, thus providing four degrees of freedom. On the other hand, the objectives of achieving a given power factor at the system terminals, a given rms load voltage, and a constant dc link voltage leave one degree of freedom. From the control point of view, this extra degree of freedom could be used to set a given phase-shift between the load voltage and the system voltage. This arbitrary phase angle forces active power to circulate through the dc link of the UPQC. This approach extends the compensating capabilities of the UPQC as compared to null active power circulating through the dc link.

Figure 27.45a shows the general phasor diagram of the UPQC connected to a load with $\cos \phi$ power factor, a system power factor $\cos \theta$, a phase-shift angle α between the load and system voltages, a voltage system V_{PCC} ($V_{PCC} < V_1$), and rated voltage V_1 at the load side. Figure 27.45b shows that with the UPFC unity power factor can be achieved at the PCC ($\theta_{PCC} = 0$). Finally, Figure 27.45c shows that in the case of unity power factor at the PCC, the current I_p can be minimized, therefore the losses at the UPFC are reduced. This operating condition can be achieved if the currents I_i, I_1, and I_p are in phase, which is achieved for an $\alpha \neq 0$ (unless no sag/swell is present). It is important to note in this case that the active power flowing through the dc link is not zero since the phase-shift angle between the series injected voltage V_0 and the supply current I_i is not 90°.

27.5 Facts Controller's Applications

The implementation of high-performance FACTS controllers enables power grid owners to increase existing transmission network capacity while maintaining or improving the operating margins necessary for grid stability. As a result, more power can reach consumers with a minimum impact on the environment, after substantially shorter project implementation times, and at lower investment costs—all compared to the alternative of building new transmission lines or power generation facilities. Some of the examples of high-performance power system controllers that have been installed and are operating in power systems are described below. Some of these projects have been sponsored by the Electric Power Research Institute (EPRI), based on a research program implemented to develop and promote FACTS.

27.5.1 500 kV Winnipeg–Minnesota Interconnection (Canada–USA) [33]

Northern States Power Co. (NSP) of Minnesota is operating an SVC in its 500 kV power transmission network between Winnipeg and Minnesota. This device is located at Forbes substation, in the state of Minnesota, and is shown in Figure 27.46. The purpose is to increase the power interchange capability on existing transmission lines. This solution was chosen instead of building a new line as it was found superior with respect to increased advantage utilization as well as reduced environmental impact. With the SVC in operation, the power transmission capability was increased in about 200 MW.

The system has a dynamic range of 450 MVAr inductive to 1000 MVAr capacitive at 500 kV, making it one of the largest of its kind in the world. It consists of an SVC and two 500 kV, 300 MVAr mechanically switched capacitor banks (MSC). The large inductive capability of the SVC is required to control the overvoltage during loss of power from the incoming HVDC at the northern end of the 500 kV line.

The SVC consists of two thyristor-switched reactors (TSR) and three TSC. Additionally, the SVC has been designed to withstand brief (<200 ms) overvoltages up to 150% of rated voltage. Without the SVC, power transmission capacity of the NSP network would be severely limited, either due to excessive voltage fluctuations following certain fault situations in the underlying 345 kV system, or due to severe overvoltages at loss of feeding power from HVDC lines coming from Manitoba.

27.5.2 Channel Tunnel Rail Link [31]

Today, it is possible to travel between London and Paris in just over 2 h, at a maximum speed of 300 km/h. The railway power system is designed for power loads in the range of 10 MW. The traction feeding system is a modern 50-Hz, 2–25-kV supply incorporating an autotransformer scheme to keep the voltage drop along the traction lines low. Power step-down from the grid is direct, via transformers connected between two phases. A major feature of this power system, shown in Figure 27.47, is the SVC support. The primary purpose of VAR is to balance the unsymmetrical load and to support the railway voltage in the case of a feeder station trip—when two sections have to be fed from one station. The second purpose of the SVCs is to ensure a low tariff for the active power by maintaining unity power factor during

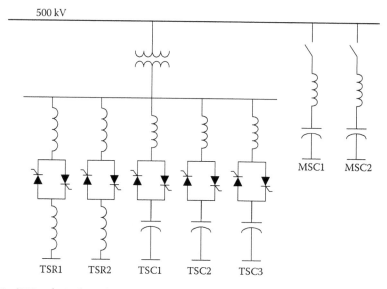

FIGURE 27.46 SVC at the Forbes substation.

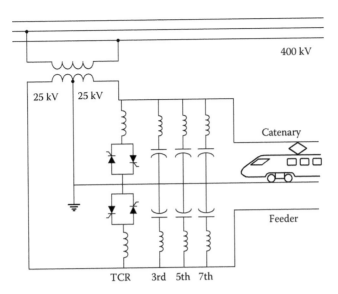

FIGURE 27.47 VAR compensation system for the channel tunnel.

normal operation. Finally, the SVCs alleviate harmonic pollution by filtering the harmonics from the traction load.

Harmonic compensation is important because strict limits apply to the traction system's contribution to the harmonic level at the supergrid connection points. The SVCs for voltage support only are connected on the traction side of the interconnecting power transformers. The supergrid transformers for the traction supply have two series-connected medium-voltage windings, each with its midpoint grounded. This results in two voltages, 180° apart, between the winding terminals and ground. The SVCs are connected across these windings; consequently, there are identical single-phase SVCs connected feeder to ground and catenary to ground. The traction load of up to 120 MW is connected between two phases. Without compensation, this would result in an approximately 2% negative phase sequence voltage. To counteract the unbalanced load, a load balancer (an asymmetrically controlled SVC) has been installed in the Sellindge substation. This has a three-phase connection to the grid. The load balancer transfers active power between the phases in order to create a balanced load (as seen by the supergrid).

27.5.3 STATCOM "Voltage Controller" ±100 MVAr STATCOM at Sullivan Substation (TVA) in Northeastern Tennessee [25]

The Sullivan substation is supplied by a 500 kV bulk power network and by four 161 kV lines that are interconnected through a 1200 MVA transformer bank. Seven distributors and one large industrial customer are served from this substation. The STATCOM, shown in Figure 27.48, is implemented with a 48 pulse, two-level voltage-source inverter that combines eight six pulse three-phase inverter bridges, each with a nominal rating of 12.5 MVA, a single step-down transformer having a wye and delta secondary to couple the inverter to the 161 kV transmission line, and a closed-loop liquid cooling system that contains a pumping skid and a fan-cooled, liquid to air heat exchanger unit, a central control system with operator interface. The STATCOM system is housed in one building that is a standard commercial design with metal walls and roof and measured 27.4 × 15.2 m.

The statcom regulates the 161 kV bus voltage during daily load increases to minimize the activation of the tap changing mechanism on the transformer bank, which interconnects the two power systems. The use of this VAR compensator to regulate the bus voltage has resulted in the reduction of the use tap

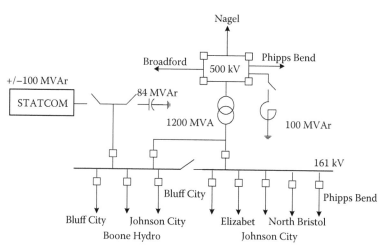

FIGURE 27.48 The 100 MVAr STATCOM at Sullivan substation.

changer from about 250 times per month to 2–5 times per month. Tap changing mechanisms are prone to failure, and the estimated cost of each failure is about $1 million. Without the STATCOM, the transmission company would be compelled either to install a second transformer bank or to construct a fifth 161 kV line into the area; both are costly alternatives.

27.5.4 Unified Power Flow Controller "All Transmission Parameters Controller": ±160 MVA Shunt and ±160 MVA Series at Inez Substation (AEP), Northeastern Virginia, USA [25]

The Inez load area has a power demand of approximately 2000 MW and is served by a long and heavily loaded 138 kV transmission lines. This means that, during normal power delivery, there is a very small voltage stability margin for system contingencies. Single contingency outages in the area will adversely affect the underlying 138 kV system, and in certain cases, a second contingency would be intolerable, resulting in a wide-area blackout. A reliable power supply to the Inez area requires effective voltage support and added real power supply facilities. System studies have identified a reinforcement plan that includes, among other things, the following system upgrades:

1. Construction of a new double-circuit high-capacity 138-kV transmission line from Big Sandy to Inez substation
2. Installation of FACTS controller to provide dynamic voltage support at the Inez substation and to ensure full utilization of the new high-capacity transmission line

The UPFC satisfies all these needs, providing independent dynamic control of transmission voltage as well as real and reactive power flow. The UPFC installation (see Figure 27.49) comprises two identical 3-phase 48-pulse, 160 MVA voltage-source inverters couple to two sets of dc capacitor banks. The two inverters are interfaced with the ac system via two transformers, a set of magnetically coupled windings configured to construct a 48-pulse sinusoidal waveshape. With this arrangement, the following operation modes are possible.

Inverter 1 (connected in parallel) can operate as a STATCOM, with either one of the two main shunt transformers, while inverter 2 (connected in series) operates as an SSSC. Alternatively, inverter 2 can be connected to the spare shunt transformer and operates as an additional STATCOM. With the later configuration, a formidable shunt reactive capability of ±320 MVA would be available, necessary for voltage

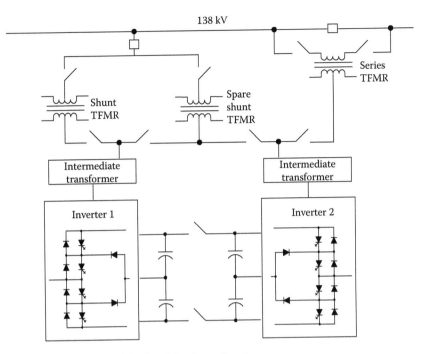

FIGURE 27.49 Inverter pole assembly of UPFC at Inez substation.

support at some transmission contingencies in the Inez area. The expected benefits of the installed UPFC are the following:

1. Dynamic voltage support at the Inez substation to prevent voltage collapse under double transmission contingency conditions
2. Flexible and independent control of real and reactive power flow on the new high capacity (950 MVA thermal rating) of the 138 kV transmission line
3. Reduction of real power losses by more than 24 MW, which is equivalent to a reduction of CO_2 emissions by about 85,000 tons per year
4. More than 100 MW increase in the power transfer and excellent voltage support at the Inez bus

27.5.5 Convertible Static Compensator in the New York 345 kV Transmission System [35]

Convertible Static Compensator (CSC), a versatile and reconfigurable device based on FACTS technology, was designed, developed, tested, and commissioned in the New York 345 kV transmission system. The CSC, shown in Figure 27.50, consists of two 100 MVA voltage-source converters which can be reconfigured and operated as either STATCOM, SSSC, UPFC, or IPFC. The CSC installation at the New York Power Authority's (NYPA) Marcy 345 kV substation consists of a 200 MVA shunt transformer with two identical secondary windings, and two 100 MVA series coupling transformers for series devices in two 345 kV lines. The CSC provides voltage control on the 345 kV Marcy bus, improved power flow transfers, and superior power flow control on the two 345 kV lines leaving the Marcy substation: Marcy–New Scotland line and Marcy–Coopers Corner line.

Each voltage-source inverter of Figure 27.50 has 12 three-level NPC poles connected to a common dc bus. Inverter pole outputs are connected to an intermediate transformer, which synthesize the 3-phase near-sinusoidal 48-pulse voltage waveform that is coupled into the transmission system.

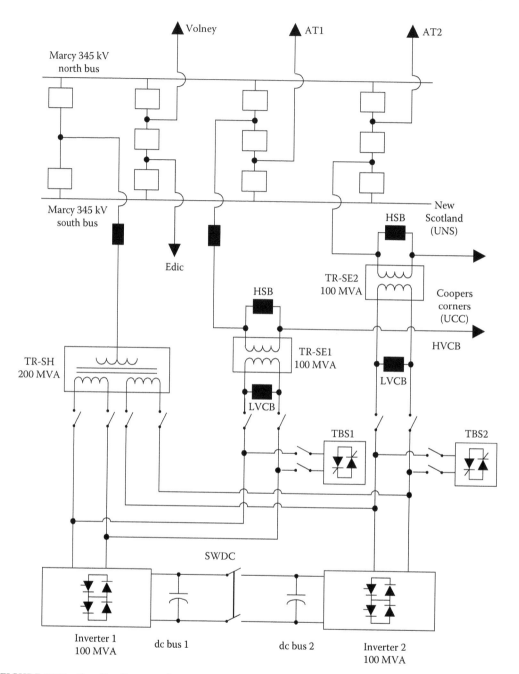

FIGURE 27.50 One-line diagram of 2 × 100 MVA CSC.

27.6 Conclusions

The principles of operation and the actual technological development of different FACTS controllers have been presented and analyzed. Starting from the principles of active, reactive, and voltage compensation, including classical solutions using phase-controlled semiconductors has been reviewed. The introduction of self-commutated topologies based on IGBT and IGCT semiconductors produced a dramatic improvement in the performance of FACTS controllers, since they have a faster dynamic behavior, they can control more variables, and introduce less pollution to power systems. The introduction of new self-commutated topologies at even higher voltage levels will increase the impact of FACTS controllers in future applications.

Some relevant examples of FACTS controllers have been described, where it can be observed that modern compensators improve power systems performance, helping to increase reliability and the quality of power delivered to the customers. These examples show that static power compensators will be used on a much wider scale in the future as grid performance and reliability becomes an even more important factor. Having better grid controllability will allow utilities to reduce investment in the transmission lines themselves. The combination of modern control with real-time information and power electronic technologies will move them very close to their physical limits. Besides, the development of faster and more powerful semiconductor valves will increase in the near future the applicability of FACTS controllers to higher limits.

Acknowledgments

The authors would like to thank Fondecyt (the Chilean Research Council) for the financial support given through the Projects #105 0067 and #105 0958. The support obtained from Universidad de Concepción, Pontificia Universidad Católica de Chile, and Universidad Federico Santa María is also acknowledged.

References

1. K. Stahlkopf and M. Wilhelm, Tighter controls for busier systems, *IEEE Spectr.*, 1997, 34(4), 48–52.
2. R. Grünbaum, Å. Petersson, and B. Thorvaldsson, FACTS, improving the performance of electrical grids, *ABB Rev.*, March 2003, 11–18.
3. N. Hingorani and L. Gyugyi, *Understanding FACTS, Concepts and Technology of Flexible AC Transmission Systems*, IEEE Press, New York, 2000.
4. T. Miller, *Reactive Power Control in Electric Systems*, John Wiley & Sons, Chichester, U.K., 1982.
5. E. Wanner, R. Mathys, and M. Hausler, Compensation systems for industry, *Brown Boveri Rev.*, 1983, 70, 330–340.
6. G. Bonnard, The problems posed by electrical power supply to industrial installations, *Proc. IEE Part B*, 1985, 132, 335–340.
7. A. Hammad and B. Roesle, New roles for static VAR compensators in transmission systems, *Brown Boveri Rev.*, 1986, 73, 314–320.
8. N. Grudinin and I. Roytelman, Heading off emergencies in large electric grids, *IEEE Spectr.*, 1997, 34(4), 43–47.
9. C. Taylor, Improving grid behavior, *IEEE Spectr.*, 1999, 36(6), 40–45.
10. R. Grünbaum, M. Noroozian, and B. Thorvaldsson, FACTS—Powerful systems for flexible power transmission, *ABB Rev.*, May 1999, 5, 4–17.
11. Canadian Electrical Association, *Static Compensators for Reactive Power Control*, Cantext Publications, Winnipeg, Manitoba, Canada, 1984.
12. H. Frank and S. Ivner, Thyristor-controlled shunt compensation in power networks, *ASEA J.*, 1981, 54, 121–127.
13. H. Frank and B. Landstrom, Power factor correction with thyristor-controlled capacitors, *ASEA J.*, 1971, 45(6), 180–184.
14. J. Dixon, Y. del Valle, M. Orchard, M. Ortúzar, L. Morán, and C. Maffrand, A full compensating system for general loads, based on a combination of thyristor binary compensator, and a PWM-IGBT active power filter, *IEEE Trans. Ind. Electron.*, 2003, 50(5), 982–989.
15. S. Torseng, Shunt-connected reactors and capacitors controlled by thyristors, *IEE Proc. Part C*, 1981, 128(6), 366–373.
16. L. Gyugyi, Reactive power generation and control by thyristor circuits, *IEEE Trans. Ind. Appl.*, 1979, IA-15(5), 521–532.

17. L. Gyugyi, R. Otto, and T. Putman, Principles and applications of static, thyristor-controlled shunt compensators, *IEEE Trans. PAS*, 1980, PAS-97(5), 1935–1945.

18. A. Chakravorti and A. Emanuel, A current regulated switched capacitor static volt ampere reactive compensator, *IEEE Trans. Ind. Appl.*, 1994, 30(4), 986–997.

19. H. Jin, G. Goós, and L. Lopes, An efficient switched-reactor-based static var compensator, *IEEE Trans. Ind. Appl.*, 1994, 30(4), 997–1005.

20. Y. Sumi, Y. Harumoto, T. Hasegawa, M. Yano, K. Ikeda, and T. Mansura, New static var control using force-commutated inverters, *IEEE Trans. PAS*, 1981, PAS-100(9), 4216–4223.

21. L. Morán, P. Ziogas, and G. Joos, Analysis and design of a synchronous solid-state VAR compensator, *IEEE Trans. Ind. Appl.*, 1989, IA-25(4), 598–608.

22. J. Dixon, J. García, and L. Morán, Control system for a three-phase active power filter which simultaneously compensates power factor and unbalanced loads, *IEEE Trans. Ind. Electron.*, 1995, 42(6), 636–641.

23. J. Lai and F. Peng, Multilevel converters a new breed of power converters, *IEEE Trans. Ind. Appl.*, 1996, IA-32(3), 509–517.

24. L. Lorenz, Power semiconductors: State of the art and future developments. Keynote Speech at the *International Power Electronics Conference, IPEC 2005*, Niigata, Japan, April 2005, CD ROM.

25. A. Edris, FACTS technology development: An update, *IEEE Power Eng. Rev.*, March 2000, 20(3), 4–9.

26. S. Bhattacharya, B. Fardenesh, B. Shperling, and S. Zelingher, Convertible static compensator: Voltage source converter based FACTS application in the New York 345 kV transmission system, *International Power Electronics Conference, IPEC 2005*, Niigata, Japan, April 2005, pp. 2286–2294.

27. M. Superczynski, Analysis of the power conditioning system for a superconducting magnetic energy storage unit, Master's thesis, Virginia Polytechnic Institute and State University, Blacksburg, VA, August 2000.

28. Reassessment of superconducting magnetic energy storage (SMES) transmission system benefits, Report number 01006795, March 2002.

29. I. Ngamroo, Robust frequency stabilization by coordinated superconducting magnetic energy storage with static synchronous series compensator, *Int. J. Emerg. Electr. Power Syst.*, 2005, 3(1), 16–25.

30. C. Sepúlveda, J. Espinoza, and R. Ortega, Analysis and design of a linear control strategy for three-phase UPQCs, Conference Record of *IECON'04*, Busan, Korea, November 3–6, 2004.

31. R. Grünbaum, Å. Petersson, and B. Thorvaldsson, FACTS improving the performance of electrical grids, *ABB Rev.*, Special Report on Power Technologies, 2003, 13–18.

32. J. Dixon and L. Morán, A clean four-quadrant sinusoidal power rectifier, using multistage converters for subway applications, *IEEE Trans. Ind. Electron.*, 2005, 52(3), 653–661.

33. X. Wei, J. Chow, B. Fardanesh, and A. Edris, A dispatch strategy for an interline power flow controller operating at rated capacity, PSCE 2004, *2004 IEEE/PES Power Systems Conference and Exposition*, New York, October, pp. 10–13, 2004.

34. C. Luongo, Superconducting storage systems: An overview, *IEEE Trans. Magn.*, 1996, 32(4), 2214–2223.

35. K. Sen, Recent developments in electric power transmission technology, The Carnegie Mellon Electricity Industry Center, EPP Conference room, April 15, 2003 (http://wpweb2.tepper.cmu.edu/ceic/SeminarPDFs/Sen_CEIC_Seminar_4_15_03.pdf).

28

Power Electronics for Renewable Energy

Wei Qiao
*University of
Nebraska-Lincoln*

28.1 Introduction

Power electronics plays an important role in wind and photovoltaic (PV) power systems. In these systems, power electronic converters are used to convert the electrical power from the form generated by wind turbine generators and PV cells to the form required by electric grids and loads in terms of frequency, voltage, power factor, harmonics, etc. This chapter reviews power electronics for control and grid integration of various wind and PV power systems.

28.2 Power Electronics for Wind Power Systems

28.2.1 Basic Concepts of Wind Power Systems

The main components of a wind power system are illustrated in Figure 28.1, which include a turbine rotor and blades, a yaw mechanism, a gearbox, a generator, a power electronic converter system, a transformer to connect the wind power system to a power grid, and a wind turbine generator control system. The wind turbine converts kinetic power in wind (i.e., aerodynamic power) to mechanical power by means of rotation of turbine rotor and blades. The mechanical power is transmitted from the turbine shaft directly or through a gearbox to the generator shaft, depending on the number of poles of the generator. If the generator has a low number of poles (e.g., four poles), a gearbox is commonly used to connect the low-speed turbine shaft and the high-speed generator shaft. If a generator with a high number of poles is used, the gearbox may not be necessary. The generator converts mechanical power to electrical power, which is fed into a power grid or used to supply local loads through optional power electronic converters and a power transformer with circuit breakers. The power transformer is normally located close to the wind turbine to avoid high currents flowing in long low-voltage cables. The use of power electronic converters enables the wind turbine generator to operate at variable speed to generate the maximum power and to have many other operational benefits, such as reactive power and power factor

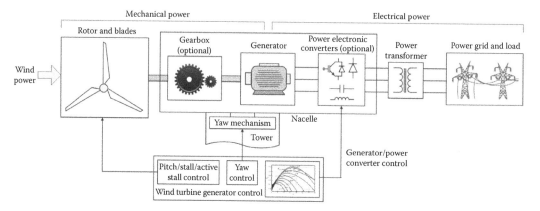

FIGURE 28.1   Main components of a wind power system.

control, reduced mechanical stresses of the drive-train system, and enhanced grid fault ride through capability. The power transformer may be mounted in the nacelle to minimize electrical losses to the grid or at the base of the tower on the foundation. Grid connection is usually made at the foundation. The yaw mechanism rotates the rotor plane of the wind turbine to be perpendicular to the wind direction in order to extract the maximum power from wind.

Wind power to electrical power conversion of the wind turbine generator is regulated by an electronic control system, which consists of the controllers for the generator and power converters, the turbine blades, and the yaw mechanism. The generator/power converter controller regulates the generator and power converters to generate a certain amount of electrical power with the voltage and frequency required by the power grid and loads. The turbine blade-angle controller optimizes the mechanical power output of the wind turbine and limits the mechanical power at the rated value during strong wind speed conditions. The power limitation may be done by stall, active stall, or pitch control (Akhmatov 2003). The yaw controller regulates the yaw mechanism to turn the rotor plane of the wind turbine to face the prevailing wind in order to generate the maximum power. If multiple wind turbine generators are connected to form a wind power plant, the control system of each wind turbine generator is usually coordinated by a wind plant central control system through a Supervisory Control and Data Acquisition (SCADA) System.

28.2.2 Power Electronics for Control and Grid Integration of Wind Turbine Generators

The voltage magnitude and frequency of the AC electrical power generated by a wind turbine generator are usually variable due to the variation of the wind sources. Therefore, power electronic converters are commonly employed to convert the electrical power from the form generated by the wind turbine generator into the form required by the power grid or load. Depending on the generator and the power electronics system, wind turbine generators can be divided into four types, including type 1: a fixed-speed wind turbine with a squirrel-cage induction generator (SCIG); type 2: a partial variable-speed wind turbine with a wound-rotor induction generator (WRIG), which has adjustable rotor resistances; type 3: a variable-speed wind turbine with a doubly fed induction generator (DFIG); and type 4: a variable-speed wind turbine generator with full-scale power electronic converters.

28.2.2.1 Power Electronics for Wind Turbine Type 1

Wind turbine type 1 represents one of the oldest wind power conversion technologies. It consists of an SCIG connected to the turbine rotor blades through a gearbox, as shown in Figure 28.2. The SCIG can

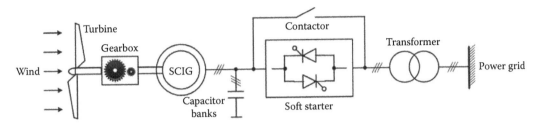

FIGURE 28.2 Configuration of wind turbine type 1.

only operate in a narrow speed range slightly higher (e.g., 0%–1% higher) than the synchronous speed. Consequently, wind turbine type 1 is commonly called a fixed-speed wind turbine. Many type 1 wind turbines use dual-speed induction generators where two sets of windings are used within the same stator frame. The first set is designed to operate in a low rotational speed (corresponding to low-wind speed operation); while the second set is designed to operate in a high rotational speed (corresponding to high-wind speed operation). The reactive power necessary to energize the magnetic circuits of the SCIG must be supplied from the power grid or a switched capacitor bank in parallel with each phase of the SCIG's stator windings. The mechanical power generated by the wind turbine can be limited aerodynamically by stall control, active stall, or pitch control. The advantages of type 1 wind turbine are the simple and cheap construction and no need of synchronization device. Some disadvantages include (1) the wind turbine usually does not operate at optimal power points to generate the maximum power due to the fixed rotating speed; (2) the wind turbine often suffers from high mechanical stresses, since wind gusts may cause torque pulsations on the drive train; and (3) it requires a stiff power grid to enable stable operation, because the SCIG consumes reactive power during operation.

Connecting an SCIG to a power grid produces transients that are short duration with high inrush currents, which cause severe voltage disturbances to the grid and high torque spikes in the drive train of the wind turbine. To mitigate such adverse effects, many type 1 wind turbines employ a phase-controlled soft starter to limit the rms values of the inrush currents to a level below two times of the SCIG rated current (Chen et al. 2009). The soft starter, as shown in Figure 28.2, consists of two antiparallel-connected thyristors in series with each phase of the SCIG to allow the phase current to follow in both directions. The firing angles of the thyristors are properly controlled to limit the stator currents of the SCIG by building up the magnetic flux slowly in the SCIG during the transient start-up period. The soft starter operates until the voltages at both sides of the soft starter are the same. Since the soft starter has a limited thermal capacity, at this moment, a contactor that electrically connects the wind turbine generator and the low-voltage terminals of the power transformer is energized, thus bypassing the soft starter. The contactor carries the full-load current when the connection to the grid has been completed. Finally, the capacitor banks are connected for reactive power compensation. To facilitate the excitation of the SCIG, it is desirable to connect the capacitor banks during the start-up period. However, the soft starter produces harmonic currents that can damage the capacitors, and, therefore, the connection of the capacitor banks will not be initialized until the grid connection process of the SCIG has finished.

28.2.2.2 Power Electronics for Wind Turbine Type 2

Wind turbine type 2 consists of a WRIG with an adjustable external rotor resistance connected to the turbine rotor blades through a gearbox, as shown in Figure 28.3. The adjustable external rotor resistance is implemented by a combination of external three-phase resistors connected in parallel with a power electronic circuit, which consists of a B6 diode bridge and an insulated gate bipolar transistor (IGBT) module. Both the resistors and the power electronic circuit are connected to the rotor windings via brushes and slip rings. The duty ratio of the IGBT module is controlled to dynamically adjust the

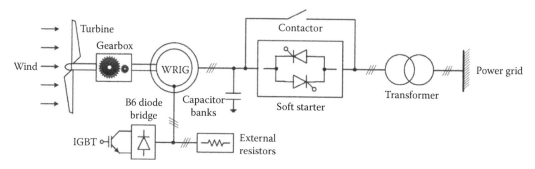

FIGURE 28.3 Configuration of wind turbine type 2.

effective value of the external rotor resistance of the WRIG. It is well known that the electrical power, P_e, generated by a WRIG depends on the rotor resistance and slip as follows:

$$P_e = n_p I_2^2 \left(\frac{R_2}{s} \right) \approx n_p V_1^2 \left(\frac{s}{R_2} \right) \text{ (If } |s| \text{ is small, e.g., } |s| \leq 0.02) \tag{28.1}$$

where

n_p and V_1 are the number of phases and per-phase terminal voltage of the stator windings, respectively
I_2 and R_2 are the per-phase rotor current and resistance referring to the stator side, respectively
s is the slip

The wind turbine generator starts to generate electrical power when the rotor speed is above the synchronous speed. As the wind speed increases, the input aerodynamic power increases, the rotor slip increases; as a consequence, the electrical output power increases. If the electrical output power is lower than the rated value, the external rotor resistors are short circuited by setting the duty ratio of the IGBT module to be unity. Once the electrical output power reaches its rated value, the external rotor resistance is adjusted to keep the output of the wind turbine generator constant. This is done by keeping the ratio of the total rotor resistance to the slip to be constant as follows:

$$\frac{R_{2total}}{s} = \frac{R_2}{s_{rated}} \tag{28.2}$$

where

s_{rated} is the rated slip when the rotor resistance is R_2
R_{2total} is the sum of R_2 and the effective external rotor resistance

Therefore, by adjusting the external rotor resistance using the power electronic circuit in Figure 28.3, the wind turbine generator can be operated in a wider speed range. Vestas uses this concept to achieve a speed variation of 0%–10% above the synchronous speed for their so-called *OptiSlip* wind turbine generators. To limit the rotor speed to its maximum value and to reduce the mechanical loads on the blades and the turbine structures, the aerodynamic power is also controlled by controlling the pitch angle of the blades in the high wind speed regions.

Figure 28.4 illustrates the detailed topology and control for the power electronic circuit to adjust the external rotor resistance of a WRIG. A surge arrester may be connected in parallel with the IGBT module to protect it against overvoltage that is created in the DC circuit due to current pulsing. The control system regulates the rotor currents flowing through the external rotor resistors. Therefore, the effect of varying rotor resistance on the rotor terminal of the WRIG is created. The control system consists of an

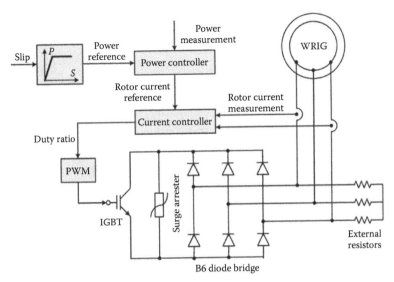

FIGURE 28.4 Typology and control of the power electronic circuit to adjust the external rotor resistance of a WRIG.

inner-loop current controller and an outer-loop power controller. The inputs for the current controller are the measured rotor current and the rotor current reference value, which is received from the power controller. The output of the current controller is the duty ratio for switching the IGBT. When the wind turbine is connected to the grid, the power controller is activated. The inputs for the power controller are the measured electrical output power and the power reference obtained from the power-slip relationship of the WRIG. If the wind speed is not high enough to produce enough torque on the turbine for running on the rated power, the power is controlled to increase with the generator slip up to 2%. If the wind speed rises to a point where the rated power can be produced, the wind turbine will be controlled to output a constant rated power; while the slip will be controlled up to 4% by using the pitch control. Short time speed changes at rated power output are controlled by possible slip changes between the rated slip which is approximate 0.5% with no external resistance connected and the maximum allowable slip of 10%.

As illustrated in Figure 28.3, type 2 wind turbine generators still need a soft starter to limit the inrush currents during the start-up process and switched capacitor banks for reactive power compensation. Compared to wind turbine type 1, wind turbine type 2 has some advantages. It provides a partial variable-speed operation with a small power electronic converter, and, therefore, energy capture efficiency is increased; the mechanical loads to the turbine structures at high wind speeds are reduced; the flicker is mitigated and the quality of output power is improved; the action frequency of pitch control system is reduced; the noise emission is reduced in weak wind conditions because the turbine is rotating with lower speed; and the reliability of the wind turbine is improved and its life is extended. However, the connection of the external rotor resistances to the rotor terminal is usually done with brushes and slip rings, which is a drawback in comparison with SCIG due to the need of additional parts and increased maintenance requirements.

28.2.2.3 Power Electronics for Wind Turbine Type 3

Figure 28.5 illustrates the basic configuration of a type 3 wind turbine generator. It consists of a low-speed wind turbine driving a high-speed WRIG through a gearbox. The WRIG is connected to a power grid at both stator and rotor terminals. The stator is directly connected to the power grid, while the rotor is connected to the grid by a variable-frequency AC–DC–AC power electronic converter through slip rings. As a consequence, the generator in this configuration is commonly called a DFIG. In order to produce electrical power at constant voltage and frequency to the power grid over a wide operating range from subsynchronous to supersynchronous speeds, the power flow between the rotor circuit and the

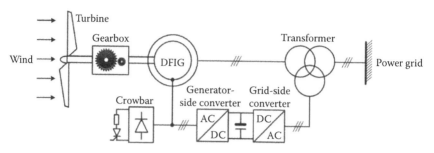

FIGURE 28.5 Configuration of wind turbine type 3.

grid must be controlled both in magnitude and in direction. Therefore, the variable-frequency converter typically consists of two four-quadrant AC–DC voltage sources converters (VSCs), that is, a generator-side converter and a grid-side converter, connected back-to-back by a common DC link. The generator-side converter and grid-side converter usually have a rating of a fraction (typically 30%) of the generator nominal power to carry the slip power. As a consequence, the wind turbine generator can operate with the rotational speed in a range of ±30% around the synchronous speed. Below the synchronous speed, the rotor power flows from the grid to the rotor winding; above the synchronous speed, the rotor power flows from the rotor winding to the grid. In this configuration, the active and reactive power of the generator and power converter can be controlled independently. By controlling the active power of the converter, the rotational speed of the generator, and thus the speed of the rotor of the wind turbine, can be regulated. The acoustical noise from type 3 wind turbines can be effectively reduced, since the system can operate at a lower speed when the wind becomes weak. The dynamic response and controllability are excellent in comparison with type 1 and type 2 wind turbine systems. This type of wind turbines needs neither a soft-starter nor a reactive power compensator. They are typically equipped with a blade pitch control to limit the aerodynamic power during conditions of high wind speeds.

Power electronic converters are constructed by power semiconductor devices, inductors, and capacitors with driving, protection, and control circuits to perform voltage magnitude and frequency conversion and control. There are two different types of power electronic converters: naturally commutated and forced-commutated converters. The naturally commutated converters are mainly thyristor converters, which use the line voltages of the power grid present at one side of the converters to facilitate the turn-off of the power semiconductor devices. A thyristor converter consumes inductive reactive power, and it is not able to control the reactive power. Thyristor converters are mainly used for high voltage and high power applications, such as conventional high-voltage direct-current (HVDC) systems and some flexible AC transmission system (FACTS) devices.

The forced-commutated converters are constructed by controllable power semiconductor devices, such as IGBTs, metal–oxide–semiconductor field-effect transistors (MOSFETs), integrated gate commutated thyristors (IGCTs), MOS-gate thyristors, and silicon carbide FETs, which are turned on and off at frequencies that are higher than the line frequency. Forced-commutated converters, such as IGBT-based pulsewidth modulated (PWM) VSCs, are normally used in type 3 wind turbine generators. As shown in Figure 28.6, the DFIG normally uses two bidirectional back-to-back PWM-VSCs sharing a common

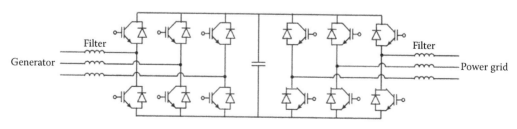

FIGURE 28.6 Topology of back-to-back PWM-VSCs used in wind power systems.

DC link. This type of converters has the ability to control both the active and reactive power delivered to the grid. The reactive power to the grid from the DFIG and converter can be controlled as zero or to a value required by the grid operator within the converter's rating limit. These features offer potential for optimizing the grid integration with respect to active and reactive power control, power quality, and voltage and angular stability. The high-frequency switching of a PWM-VSC may produce harmonics and interharmonics, which are generally in the range of a few kHz. Due to the high switching frequencies, the harmonics are relatively easy to be removed by small-size filters.

In order to reduce the cost per megawatt and increase the efficiency of wind energy conversion, the nominal power of wind turbines has been continuously growing in the past years. As a consequence, there is an increasing interest in multilevel power converters especially for medium to high-power, high-voltage wind turbine applications (Carrasco et al. 2006; Portillo et al. 2006). The increase of voltage rating allows for connection of the converters of the wind turbine systems directly to the wind plant distribution grid, avoiding the use of a bulky transformer. The general idea behind the multilevel converter technology is to create a sinusoidal voltage from stepped voltage waveforms, typically obtained from an array of power semiconductors and capacitor voltage sources. The commonly used multilevel converter topologies can be classified in three categories (Rodriguez et al. 2002), which are diode-clamped multilevel converters, capacitor-clamped multilevel converters, and cascaded multilevel converters. Figure 28.7 illustrates commonly used three-level converters using these topologies. Other topologies of multilevel converters can be found in (Hansen et al. 2002; Rodriguez et al. 2002; Carrasco et al. 2006).

Initially, the motivation of using multilevel converters was to achieve a higher voltage and power capability. As the ratings of the components increase and the switching and conducting properties improve, other advantages of multilevel converters become more and more attractive. Multilevel converters can generate output voltages with lower distortion and lower dv/dt (Rodriguez et al. 2002). Consequently, the size of the output filters is reduced. For the same harmonic performance, multilevel converters can be operated with a lower switching frequency when compared with two-level converters. Therefore, the switching losses of multilevel converters are reduced.

The most commonly reported disadvantage of multilevel converters with split DC link is the voltage imbalance between the DC-link capacitors. Nevertheless, for a three-level converter, this problem is not serious, and the problem in the three-level converter is mainly caused by differences in the real capacitance of each capacitor as well as the inaccuracy in the deadtime implementation or an unbalanced load (Shen and Butterworth 1997). By a proper modulation control of the switches, the imbalance problem can be solved (Lim et al. 1999).

The three-level diode-clamped multilevel converter and the three-level capacitor-clamped multilevel converter exhibit an unequal current stress on the semiconductors. It appears that the upper and lower switches in a converter leg might be derated compared to the switches in the middle. For an appropriate

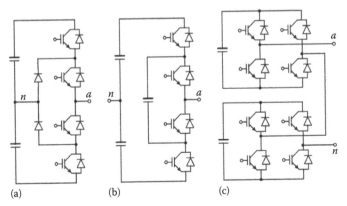

(a) (b) (c)

FIGURE 28.7 Multilevel converter topologies: (a) one leg of a diode-clamped three-level converter; (b) one leg of a capacitor-clamped three-level converter; and (c) one leg of an H-bridge cascaded three-level converter.

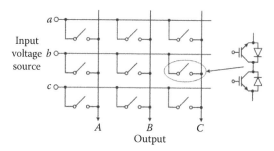

FIGURE 28.8   Topology of a matrix converter.

design of the converter, different devices are required (Rodriguez et al. 2002). The unequal current stress and the unequal voltage stress might constitute a design problem for the multilevel converter with bidirectional switch interconnection (Hansen et al. 2002).

The cascaded H-bridge multilevel converter is heavy, bulky, and complex. Moreover, connecting separated DC sources between two converters in a back-to-back fashion is difficult, because a short circuit will occur when two back-to-back converters are not switching synchronously (Lai and Peng 1996).

Another type of circuit configuration is the matrix converter, as shown in Figure 28.8. A matrix converter is a one-stage AC–AC converter that is composed of an array of nine bidirectional semiconductor switches, connecting each phase of the input to each phase of the output. The basic idea behind the matrix converter is that a desired input current, a desired output voltage, and a desired output frequency can be obtained by properly operating the switches that connect the output terminals of the converter to its input terminals. In order to protect the converter, the following two control rules must be complied with. First, only one switch in an output leg is allowed to be on at any instant of time. Second, all of the three output phases must be connected to an input phase at any instant of time. The actual combination of the switches depends on the modulation strategy.

Grid faults, even far away from the location of a wind turbine, can cause voltage sags at the connection point of the wind turbine. Such voltage sags result in an imbalance between the turbine input power and the generator output power, which initiates the machine stator and rotor current transients, the converter current transient, the DC-link voltage fluctuations, and a change in speed. One of the major problems of the type 3 wind turbines operating during grid faults is that the voltage sags may cause overvoltage in the DC link and overcurrent in the DFIG rotor circuit and the generator-side converter, which in turn may destroy the generator-side converter. To protect the generator-side converter from overvoltage or overcurrent during grid faults, a crowbar circuit is usually connected between the rotor circuit of the DFIG and the generator-side converter to short-circuit the rotor windings. During this time, the generator-side converter is blocked from switching (Qiao et al. 2009), and its controllability is naturally lost. Consequently, there is no longer the independent control of active and reactive power in the DFIG. The DFIG becomes a conventional SCIG. It produces an amount of active power and starts to absorb an amount of reactive power. The grid-side converter can be operated to regulate the reactive power exchanged with the grid.

The crowbar circuit is connected between the rotor of the DFIG and the generator-side converter. The crowbar circuit may have various topologies. Figure 28.9a shows a passive crowbar (Petersson et al. 2005)

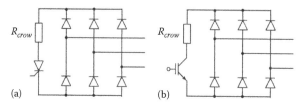

FIGURE 28.9   Topologies of crowbar circuits: (a) passive crowbar; (b) active crowbar.

consisting of a diode bridge that rectifies the rotor phase currents and a single thyristor in series with a resistor. The thyristor is turned on when the DC-link voltage reaches its limit value or the rotor current reaches its limit value. Simultaneously, the rotor circuit of the DFIG is disconnected from the generator-side converter and connected to the crowbar. When the grid fault is cleared, the generator-side converter is restarted, and after synchronization, the rotor circuit of the DFIG is connected back to the generator-side converter (Qiao et al. 2009).

Figure 28.9b shows an active crowbar topology (Seman et al. 2006), which replaces the thyristor in the passive crowbar with a fully controllable semiconductor switch, such as an IGBT. This type of crowbar may be able to cut the short-circuit rotor current at anytime. If either the rotor current or the DC-link voltage exceeds the limit values, the IGBTs of the generator-side converter are blocked and the active crowbar is turned on. The crowbar resistor voltage and DC-link voltage are monitored during the operation of the crowbar. When both voltages reduce below certain values, the crowbar is turned off. After a short delay for the decay of the rotor currents, the generator-side converter is restarted and connected back to the rotor circuit of the DFIG. In both topologies, the value of the crowbar resistance has significant effects on the dynamic performance of the DFIG, such as the maximum short-circuit current of the DFIG and reactive power control capability.

28.2.2.4 Power Electronics for Wind Turbine Type 4

Wind turbine type 4 may have a variety of configurations, as illustrated in Figure 28.10. It could use an SCIG (Figure 28.10a) or a wound-rotor synchronous generator (SG) (Figure 28.10b) connected to the turbine shaft through a gearbox. It could use a wound-rotor SG (Figure 28.10c) or a permanent magnet SG (PMSG) (Figure 28.10d) connected directly to the turbine shaft without gearbox. The wound-rotor SGs in Figure 28.10b and c need an extra small AC–DC power converter, which feeds the excitation winding for field excitation. The generator is connected to the power grid through an AC–DC–AC power electronic converter, whose rating is the same as that of the electric generator used. Since the generator is decoupled from the grid, the generator can operate in a wide variable frequency range for optimal operation. The grid-side PWM converter can be used to control the active and reactive power delivered to the grid independently and to provide grid support features, such as power factor or voltage regulation. Therefore, compared to other types of wind turbines, the dynamic response of type 4 wind turbines is improved.

The AC–DC–AC converters in Figure 28.10 can be implemented by using the bidirectional back-to-back PWM-VSCs as shown in Figure 28.6 to achieve full control of the active and reactive power for the generator. A wound-rotor SG or a PMSG requires only a simple diode bridge rectifier for the generator-side converter, as shown in Figure 28.11. For a three-phase system, the diode rectifier consists of six diodes. The diode rectifier is simple and has a low cost. However, it can only be used in one quadrant. Therefore, it is not possible to control the active or reactive power of the generator by controlling the diode rectifier. In order to achieve variable-speed operation, the wind turbine equipped with a SG will require a boost DC–DC converter inserted between the diode rectifier and the DC-link, as shown in Figure 28.11.

The type 4 wind turbines with a PMSG are the most popular configuration of small wind turbines for residential and other nonutility applications, in which the grid-side converter may be a single-phase full-bridge instead of a three-phase PWM inverter. In this case, the AC terminals of the inverter may be connected between line and line or line and neutral of the power grid. Moreover, it is possible to use a matrix converter shown in Figure 28.8 to replace the AC–DC–AC converter for type 4 wind turbine systems.

28.2.3 Control of Power Electronic Converters for Variable-Speed Wind Turbine Generators

The control system of a variable-speed wind turbine generator generally consists of two parts: the electrical control on the generator and power converters and the mechanical control on the wind turbine blade

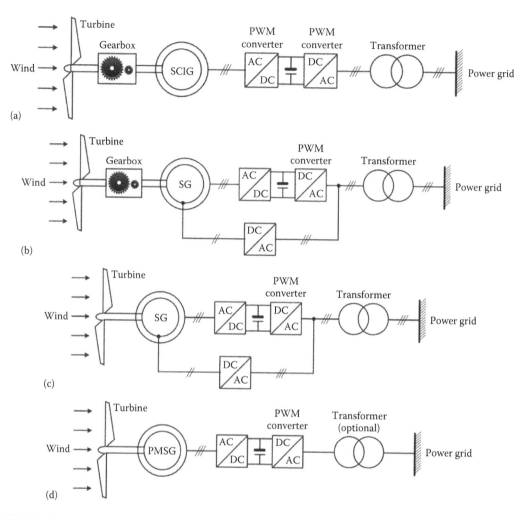

FIGURE 28.10 Configurations of wind turbine type 4 equipped with: (a) SCIG and gearbox; (b) wound-rotor SG and gearbox; (c) wound-rotor SG with a high number of poles but no gearbox; (d) PMSG but no gearbox.

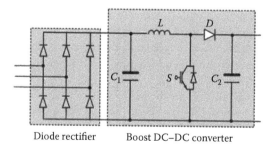

FIGURE 28.11 Topology of diode rectifier and boost DC–DC converter used in type 4 wind turbine systems.

pitch angle and yaw mechanism. As shown in Figure 28.12 for a type 3 wind turbine generator, the control of power converters includes the control for the generator-side converter and the control for the grid-side converter. If multiple wind turbine generators are connected to form a wind power plant, the control system of each wind turbine generator is usually coordinated by a wind power plant supervisory controller, which can generate the active and reactive power references for each wind turbine generator.

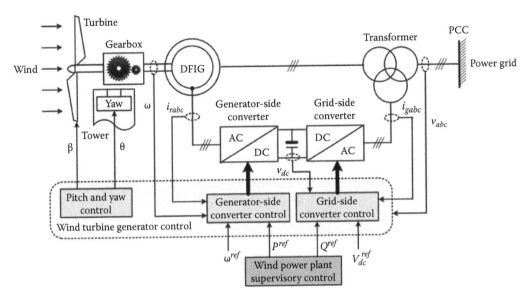

FIGURE 28.12   Control of a type 3 wind turbine generator.

28.2.3.1 Control of Power Electronic Converters for Wind Turbine Type 3

28.2.3.1.1 Control of Generator-Side Converter

Two control schemes originated from motor drives applications, that is, field oriented control and direct torque control, are commonly used for controlling the generator-side converter. In the field-oriented control, the generator-side converter can be controlled to govern the stator active and reactive power independently. This is usually achieved by rotor current regulation in an orthogonal dq reference frame aligned to one of the fluxes in the generator. The most commonly used one is the stator-flux-oriented dq reference frame, where the d-axis is aligned with the stator flux linkage vector λ_s, namely, $\lambda_{ds} = \lambda_s$ and $\lambda_{qs} = 0$ (Qiao et al. 2008). In the stator-flux-oriented dq reference frame, the q-axis and d-axis rotor currents are decoupled to control the stator active power (or rotor speed) and the reactive power, respectively. Consequently, the overall vector control scheme for the generator-side converter consists of two cascaded control loops, as shown in Figure 28.13. The inner current control loops regulate independently the q-axis and d-axis rotor current components according to the following dynamical model of the generator (Liang et al. 2010):

$$\begin{bmatrix} v_{dr} \\ v_{qr} \end{bmatrix} = \begin{bmatrix} R_r + L'_r p & -(\omega_{\lambda s} - \omega_r)L'_r \\ (\omega_{\lambda s} - \omega_r)L'_r & R_r + L'_r p \end{bmatrix} \cdot \begin{bmatrix} i_{dr} \\ i_{qr} \end{bmatrix} + \frac{L_m}{L_s} \begin{bmatrix} v_{ds} \\ v_{qs} - \omega_r \lambda_{ds} \end{bmatrix} \tag{28.3}$$

where
    $\omega_{\lambda s}$ is the rotating speed of the stator flux space vector
    ω_r is the rotor rotating speed of the generator
    R_r is the rotor resistance
    L_m is the mutual inductance
    L'_r is the rotor transient inductance, $L'_r = \sigma L_r$ with $\sigma = 1 - L_m^2/(L_s L_r)$
    L_s and L_r are the stator and rotor inductances, respectively
    p is the derivative operator

In steady states, $\omega_{\lambda s}$ is equal to the synchronous speed ω_s; $v_{ds} = 0$ and $v_{qs} = \omega_s \lambda_{ds}$ with the stator resistance neglected.

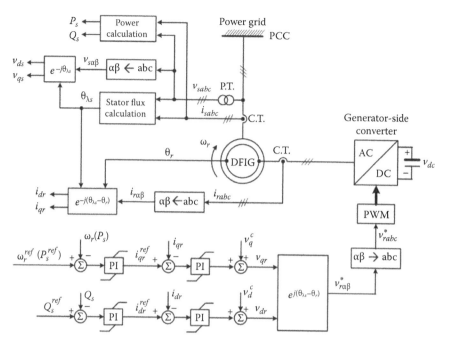

FIGURE 28.13 Stator-flux-oriented vector control scheme for the generator-side converter, where $v_q^c = (\omega_{\lambda s} - \omega_r)L_r' i_{dr} + L_m(v_{qs} - \omega_r \lambda_{ds})/L_s$ and $v_d^c = -(\omega_{\lambda s} - \omega_r)L_r' i_{qr} + L_m v_{ds}/L_s$.

A similar control scheme is used for the DFIG systems, where the electromagnetic torque and the stator flux of the generator are controlled by using decoupled q-axis and d-axis rotor current regulation, respectively. The actual stator flux and torque as well as the stator flux angle are determined based on the generator equations using measured voltages and currents.

The direct torque control proposed by Depenbrock (1988) eliminates the inner current loops and the needs of transformations between different references frames. In the direct torque control, the stator flux is estimated by integrating the stator voltages, and the electromagnetic torque is estimated as a cross product of the estimated stator flux vector and the measured stator current vector. The magnitudes of the stator flux and the electromagnetic torque of the generator are then controlled directly by using hysteresis comparators, as shown in Figure 28.14. The outputs of the hysteresis comparators as well as the flux angle are used directly to determine the switching states of the converter. If either the estimated flux or torque deviates from the reference more than allowed tolerance, the power switches of the converter are turned off and on in such a way that the flux and torque will return in their tolerance bands as fast as possible.

28.2.3.1.2 Maximum Power Point Tracking

The aerodynamic model of a wind turbine can be characterized by the well-known C_p-λ-β curves. C_p is called the power coefficient, which is a function of both tip speed ratio (TSR) λ and the blade pitch angle β. The TSR is defined by

$$\lambda = \frac{\omega_t R}{v_w} \tag{28.4}$$

where
 R is the blade length
 ω_t is the wind turbine rotational speed
 v_w is the wind speed

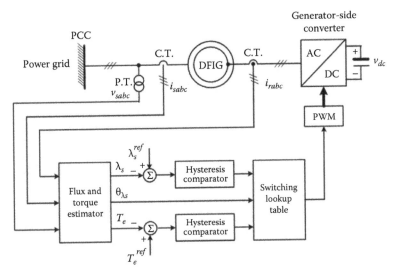

FIGURE 28.14 Direct torque control scheme for the generator-side converter.

Given the power coefficient C_P, the mechanical power extracted by the turbine from the wind is calculated by

$$P_m = \frac{1}{2}\rho A_r v_w^3 C_P(\lambda,\beta) \tag{28.5}$$

where
  ρ is the air density
  $A_r = \pi R^2$ is the area swept by the rotor blades

At a specific wind speed, there is a unique value of ω_t to achieve the maximum power coefficient C_P, as shown in Figure 28.15 and thereby extract the maximum mechanical (wind) power. Figure 28.16 illustrates a typical power-wind speed curve of a wind turbine. If the wind speed is between the cut-in and nominal values, the wind turbine is usually operated in variable speed mode, in which the rotational

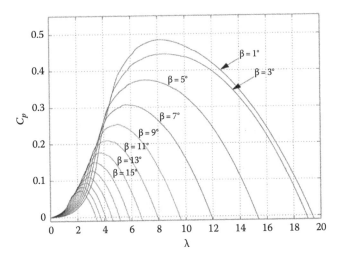

FIGURE 28.15 Typical C_P-λ-β curves of a wind turbine.

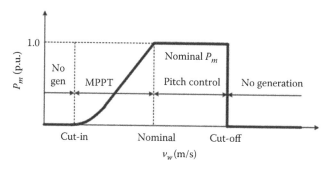

FIGURE 28.16 A typical power-wind speed curve of a wind turbine.

speed is adjusted (by means of speed or active power control in the DFIG shown in Figure 28.13), such that the maximum value of C_p is achieved. This is called the maximum power point tracking (MPPT). The wind turbine pitch control is deactivated, and the pitch angle β is fixed. If the wind speed is above the nominal value but below the cut-off value, the rotor speed can no longer be controlled within the limits by increasing the generated power, as this would lead to overloading of the generator and/or the converter. In such a situation, the pitch control is activated to increase the pitch angle to maintain nominal mechanical power extracted from the wind.

There are some methods to perform MPPT control for wind turbines (Qiao et al. 2008).

1. *TSR control:* The wind speed is measured by an anemometer. The controller regulates the rotating speed of the wind turbine to maintain an optimal TSR. However, the accurate wind speed may be difficult to obtain. In addition, the use of an external anemometer increases the complexity and cost of the system.
2. *Power signal feedback (PSF) control:* This control requires the knowledge of the maximum power curves of the turbine, which may be obtained through simulations and practical tests. The measured speed and output power of the wind turbine generator are used to determine the target speed/power for the MPPT control by utilizing the power-wind turbine speed-wind speed characteristics of the wind turbine.
3. *Hill climbing searching (HCS) control:* At a certain wind speed, when the wind turbine speed increases, the output power should normally increase as well; otherwise, the wind turbine speed should be decreased. In the HCS method, based on the comparison of the wind turbine output power at the present and previous time steps, the controller incrementally increases, decreases, or fixes the control variables to achieve the maximum power point of the system. However, this method could be ineffective for large wind turbines, since large turbines are difficult to adjust the speed fast.

In practice, MPPT controllers may use combinations of the three techniques.

28.2.3.1.3 Control of Grid-Side Converter

The objective of the grid-side converter control is to maintain a constant DC-link voltage regardless of the magnitude and direction of the rotor power. The grid-side converter can also be arranged to control the reactive power exchanged between the converter and the grid. This is usually achieved by current regulation in an orthogonal synchronously rotating *dq* reference frame aligned to the stator terminal voltage vector of the generator, that is, a synchronous voltage-oriented control. In this control scheme, the *d*-axis and *q*-axis components of the converter's AC-terminal currents are decoupled to control the DC-link voltage and the reactive power, respectively. Similar to the generator-side converter control, the overall vector control scheme for the grid-side converter consists of two cascaded control loops, as shown in Figure 28.17. A phase-locked loop (PLL) is used for the transformation between the stationary αβ reference frame and the synchronously rotating *dq* reference frame.

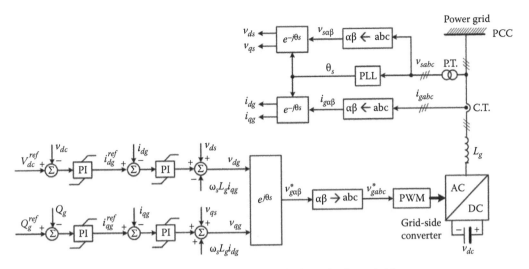

FIGURE 28.17 Synchronous voltage-oriented vector control scheme for the grid-side converter.

28.2.3.2 Control of Power Electronic Converters for Wind Turbine Type 4

The control of the generator-side converter for type 4 wind turbines is achieved by stator current regulation in the rotor *dq* reference frame. The *q*-axis and *d*-axis stator currents are decoupled, where the *q*-axis stator current controls the stator active power (or rotor speed) and the *d*-axis stator current is controlled to be zero. Consequently, the overall vector control scheme for the generator-side converter consists of two cascaded control loops, as shown in Figure 28.18. The reference value of rotor speed or active power of the PMSG is generated by an MPPT algorithm, which is the same as that for type 3 wind turbines. In addition, the control of the grid-side converter for type 4 wind turbines is the same as that for type 3 wind turbines as well.

Figure 28.19 illustrates the control scheme for a type 4 wind turbine with a three-phase diode rectifier and a boost DC–DC converter. The input voltage and inductor current of the DC–DC converter are sensed to calculate the power delivered by the converter. The calculated power, the input voltage, and the inductor current are used by a MPPT algorithm to determine the optimal voltage reference of the

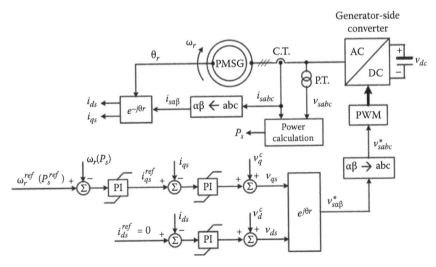

FIGURE 28.18 Vector control scheme for the generator-side converter of a type 4 wind turbine.

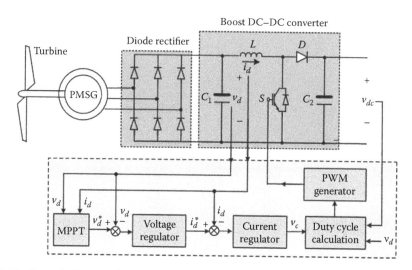

FIGURE 28.19 Control of boost DC/DC converter for a type 4 wind turbine.

DC–DC converter. The MPPT algorithm checks the power against the previous step to adjust the voltage reference of the DC–DC converter with a certain step that will cause the wind turbine to output a current and voltage so as to be operating at the maximum power under the particular conditions at that time. The reference voltage is compared to the measured voltage, and the error is passed through a voltage regulator to generate the current reference signal. The current reference is used by the inner-loop current regulator to generate a control signal, v_c, which is then used to generate the appropriate duty cycle for PWM switching of the semiconductor switch, *S*.

28.2.4 Power Electronics for Wind Power Plants

A wind power plant typically consists of many individual wind turbine generators. The electrical powers generated by individual wind turbine generators are usually collected to a plant substation through three-phase AC power lines or cables and then delivered to the utility power grid through a voltage-step-up power transformer. With the increasing penetration of wind power into the electric power grids, the utilities in many countries have established grid codes for operation and grid connection of wind power plants. Many of these grid codes require wind power plants to provide frequency and voltage control, active and reactive power regulation, and quick responses under power grid transient and dynamic situations. The aim of these grid codes is to ensure that the continued growth of wind power does not compromise the power quality as well as the security and reliability of the electric power grids. The power electronic technology plays an important role in both system configurations and control of the wind power plant to fulfill the grid code requirements.

For example, the type 1 wind turbines consume reactive power during operation due to the use of SCIGs. To minimize the power losses of delivering the electrical power, mitigate the voltage fluctuations caused by wind power fluctuations, and increase voltage stability, dynamic reactive compensators, such as the thyristor-based static var compensator (SVC; Hingorani and Gyugyi 2000) or the VSC-based static synchronous compensator (STATCOM; Qiao and Harley 2007), may be used for the wind power plants consisting of type 1 wind turbines. Figure 28.20 illustrates such a system with a STATCOM. The STATCOM is a shunt-connected FACTS device. It uses a VSC to inject reactive power to or absorb reactive power from the power grid through an inductor. The VSC uses power electronic devices, such as IGBTs, IGCTs, or gate turn-OFF thyristors (GTOs) and can be configured as a multilevel bidirectional converter. Compared to the SVC, the STATCOM is able to provide faster and smoother dynamic reactive compensation and voltage control because of its rapid and continuous response characteristics.

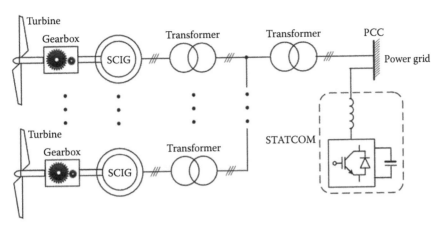

FIGURE 28.20 The use of STATCOM to assist with grid connection of type 1 wind turbines.

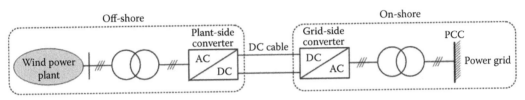

FIGURE 28.21 VSC-based HVDC system for grid connection of off-shore wind power plants.

Therefore, the STATCOM is more suitable for voltage fluctuation mitigation at the PCC and grid fault ride through of the wind power plant.

HVDC is an interesting alternative for long distance power transmission from an offshore wind power plant. In an HVDC transmission system, the low or medium AC voltage at the wind power plant is converted into a high DC voltage on the transmission side, and the DC power is transferred to the onshore system where the DC voltage is converted back into AC voltage, as shown in Figure 28.21. For certain power levels, an HVDC system based on the VSC technology may be used instead of the conventional thyristor-based HVDC technology. Another possible DC transmission system configuration is shown in Figure 28.22 for a wind power plant consisting of type 4 wind turbines. In this configuration, individual wind turbines are connected through a common DC grid in the off-shore wind power plant, while a grid-side power converter terminal realizes the on-shore grid connection.

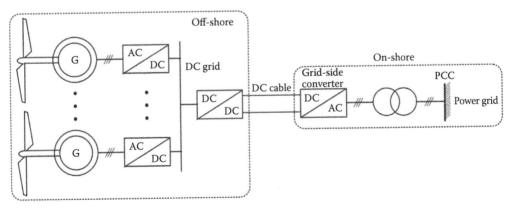

FIGURE 28.22 An alternative HVDC system for grid connection of off-shore wind power plants.

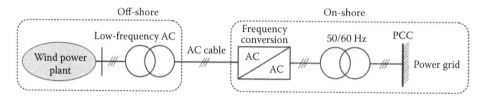

FIGURE 28.23   Low-frequency HVAC transmission for off-shore wind power plants.

Because of the physical limitation of submarine cables and associated power loss and transmission capabilities, 50/60 Hz high-voltage AC (HAVC) may not be efficient or economically beneficial for long distance (e.g., ≥50 km) power transmission from an offshore wind power plant. HVDC is capable of extending the power transmission capability and distance. However, it can also increase the capital costs (i.e., equipment and installation costs) as well as operation and maintenance costs in certain perspectives. First, HVDC requires an offshore power conversion substation. Unlike HVAC where the platform mainly accommodates transformers and switchgears, a HVDC offshore platform will include more equipment and apparatus, such as high-voltage power converters. It requires a larger and heavier foundation and platform structure that leads to more expensive installations. Second, HVDC faces more challenges for fault mitigation and protection. A HVDC system needs to rely on solid-state devices, or upstream/downstream AC breakers to clear a fault. Contributions of DC link to fault currents, shutdown sequences, and limited valve short-circuit current capability can all complicate the protection and control requirements. Finally, semiconductor devices operating at a very high voltage level can raise more reliability issues than conventional AC equipment. Maintenance and repair of HVDC power converters requires special skills compared to conventional AC transformers and switchgears. These issues, together with installations at offshore sites, make the system reliability and availability performance more challenging and maintenance more expensive.

Another interesting option for long distance power transmission from an offshore wind power plant is to transmit power at a low frequency, such as 16.7 or 20 Hz. Submarine cables have strong capacitance that causes high reactive current through the cable. The reactive current is proportional to voltage and frequency. In other words, reduction of the frequency will lower the reactive current and, therefore, increase the cable capacity to allow it carrying more real power. For example, by reducing the frequency to 16.7 Hz, reactive current will decrease by 67%–72% compared to 50/60 Hz. Therefore, it enables the cable to transmit more real power over longer distance. Figure 28.23 illustrates the use of low-frequency HVAC to transmit offshore wind plant. The low-frequency AC output from individual wind turbines is stepped up to high voltage by a power transformer at the offshore substation and then sent to the shore. At the onshore frequency conversion substation, the low-frequency output is finally converted to 50/60 Hz utility voltage. Such a frequency conversion substation can use thyristor converters, which is more reliable and less expensive than IGBT-based HVDC substations.

28.3 Power Electronics for PV Power Systems

28.3.1 Basic Concepts of PV Power Systems

A PV cell uses the photoelectric effect to produce DC electrical power when exposed to sunlight and connected to a suitable load. A typical PV module (or panel) consists of multiple (e.g., 36 or 72) cells connected in series, encapsulated in a structure made of, for example, aluminum and tedlar. The series connection of cells produces a high voltage (around 25–45 V) across the terminals of the module, but the weakest cell determines the current seen at the terminals. This causes reduction in the available power, which, to some extent, can be mitigated by the use of bypass diodes, in parallel with the cells. The parallel connection of the bypass diodes with the cells solves the weakest-link problem, but the voltage seen at the terminals is reduced. Without any moving parts inside the PV module, the

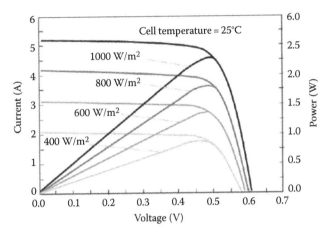

FIGURE 28.24 Typical current-voltage and power-voltage characteristic curves of a PV cell. (Courtesy of Ecowave, Aspendale, VIC, Australia.)

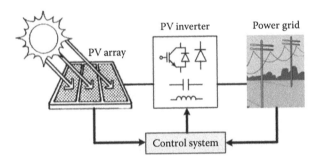

FIGURE 28.25 Main components of a grid-connected PV power system.

tear-and-wear is very low. Therefore, lifetimes of more than 25 years are easily reached for PV modules. However, the power generation capability may be reduced to 75%–80% of the nominal value due to aging. Figure 28.24 shows typical curves of the current-voltage and power-voltage characteristics of a PV cell, with insolation as a parameter. It should be pointed out that the characteristic curves also depend on cell temperature. These curves reveal that the captured power is determined by loading conditions (terminal voltage and current). At a certain temperature and insolation condition, there exists a unique terminal voltage at which the PV cell captures the maximum power from sunlight.

Figure 28.25 illustrates the main components of a grid-connected PV power system, which consists of a PV array, a PV inverter, and a control system connected to the power grid. The PV array can be a single module, a string of PV modules, or multiple parallel strings of PV modules. The PV inverter converts the DC electrical power generated by the PV array into AC electrical power with the frequency and the voltage level required by the power grid. Moreover, the use of the PV inverter enables the PV array to operate at the optimal voltage or current level to extract the maximum power from sunlight and to have other operational benefits, such as reactive power and power factor control. The control system regulates the PV inverter to perform these functions by using the information acquired from the PV array and the power grid.

28.3.2 Topologies of PV Power Systems

Three PV power system topologies are commonly used, which are the centralized inverter topology, string inverter topology, and module-integrated inverter topology (Kjaer et al. 2005; Blaabjerg et al. 2006; Carrasco et al. 2006).

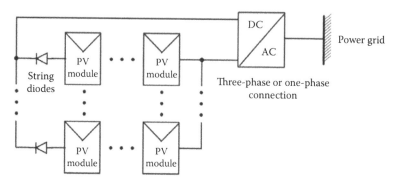

FIGURE 28.26   PV grid integration using a central inverter.

28.3.2.1 Centralized Inverter Topology

In this topology, the PV power system (typical >10 kW) consists of many parallel strings that are connected to a single central inverter on the DC-side, as shown in Figure 28.26. This inverter system is characterized by high efficiency and low cost per kilowatt. However, the energy produced by the PV array cannot be maximized due to module mismatching and potential partial shading conditions. Moreover, a failure of the central inverter will result in the out of operation of the whole PV power system. Therefore, the reliability of the PV power system is limited due to the dependence of power conversion on a single inverter.

28.3.2.2 String Inverter Topology

Similar to the central inverter, the PV array in this topology is divided into several parallel strings. Each PV string is connected to a designated inverter, the so-called string inverter, as shown in Figure 28.27. String inverters have the capability of separate MPPT for each PV string. This increases the energy production by reducing module mismatching and partial shading losses as well as the reliability of the PV power system. String inverters have been a standard in PV system technology for grid-connected PV power systems.

An evolution of the string inverter topology applicable for higher power levels is the multistring inverter topology, as shown in Figure 28.28. It allows the connection of multiple PV strings with separate MPPT control via DC/DC converters to a common DC/AC inverter. Consequently, a compact and cost-effective solution, which combines the advantages of central and string topologies, is achieved. This multistring topology allows the integration of PV strings of different technologies and of various orientations. These characteristics allow time-shifted solar power, which optimizes the operation efficiencies of each string separately.

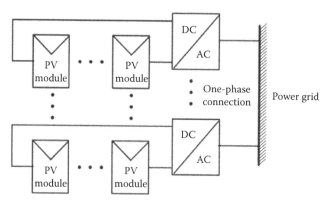

FIGURE 28.27   PV grid integration using string inverters.

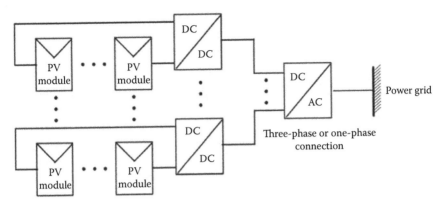

FIGURE 28.28 PV grid integration using a multistring inverter.

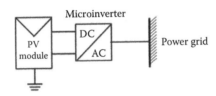

FIGURE 28.29 PV grid integration using a module-integrated inverter.

28.3.2.3 Module-Integrated Inverter Topology

This system uses one inverter (also called a microinverter) for each module, as shown in Figure 28.29. This topology optimizes the energy production of each module by adapting the inverter to the PV module characteristics using the so-called MPPT control. Although the module-integrated inverter optimizes the energy production of each module, it has a lower efficiency than the string inverter-based PV system. Module-integrated inverters are characterized by a more extended AC-side cabling, since each module of the PV system has to be connected to the available AC grid (e.g., 120 V/60 Hz). Moreover, the maintenance processes are quite complicated, especially for facade-integrated PV systems. This concept is suitable for PV systems of 50–400 W peak.

28.3.3 Topologies and Control for PV Inverters

The topologies of PV inverters are categorized on the basis of the number of power processing stages, whether an isolation is used or not, and location of the isolation. Based on the number of power processing stages, PV inverters can be classified into two major categories: single-stage inverters without DC–DC converter and dual-stage inverters with a DC–DC converter. Whether a DC–DC converter is used or not depends on PV string configuration and the voltage level of the power grid. Having more modules in series and a lower grid voltage, such as in the United States and Japan, it is possible to avoid using a DC–DC converter to boost the terminal voltage of the PV string. Thus, a single-stage PV inverter can be used, leading to a higher efficiency. The single-stage inverter must handle all tasks, such as MPPT control, grid-current control, and voltage amplification. This is the typical configuration for a centralized inverter, which must be designed to handle a peak power of twice the nominal power. In a dual-stage inverter, the DC–DC converter performs the MPPT (and perhaps voltage amplification). Depending on the control of the DC–AC inverter, the output of the DC–DC converter is either a pure DC voltage (and the DC–DC converter is only designed to handle the nominal power), or the output current of the DC–DC converter is modulated to follow a rectified sine wave (the DC–DC converter should now handle a peak power of twice the nominal power). The DC–AC inverter in the former solution is used to control the grid current by means of PWM or bang-bang operation. In the latter solution, the DC–AC inverter is switched at line frequency, unfolding the rectified current to a full-wave sine, while the DC–DC converter takes care of the current control. A high efficiency can be reached for the latter solution if the nominal power is low. On the other hand, it is advised to operate the DC–AC inverter in PWM mode if the nominal power is high.

The issue of isolation is mainly related to safety standards. Isolation is typically acquired by using a transformer (a requirement in the United States). In a single-stage inverter, a line-frequency transformer is placed between the AC terminals of the inverter and the power grid. In a dual-stage inverter, the transformer can be placed on either the grid or low-frequency side or on the high-frequency side in the DC–DC converter. The high-frequency transformer is more compact, but special attention must be paid in design to reduce losses. Modern inverters tend to use a high-frequency transformer. Commonly used high-frequency DC–DC converter topologies with isolation include full-bridge, single-inductor push–pull, double-inductor push–pull, and flyback. A full-bridge converter is usually used at power levels above 750 W due to its good transformer utilization. The main disadvantages of full-bridge topology in comparison with push–pull topology are the higher active part count and the higher transformer ratio needed for boosting the DC voltage to the grid level. In some countries where grid isolation is not mandatory, PV inverters with a DC–DC converter without isolation are usually used to simplify the PV inverter design. In this case, simple DC–DC boost, buck, buck–boost converters can be used.

A common DC–AC inverter topology is the half-bridge two-level voltage source inverter (VSI), which can create two different voltage levels and requires double DC-link voltage and double switching frequency in order to obtain the same performance as the full bridge. A variant of this topology is the standard full-bridge three-level VSI, which can create a sinusoidal grid current by applying the positive or negative DC-link or zero voltage to the grid. The semiconductor devices of the VSI can be switched at the line frequency or using PWM control.

Another interesting PV inverter topology without boost and isolation can be achieved by using multi-level concepts. This is beneficial for the power grid and results in an improvement in the total harmonic distortion (THD) performance of the output voltages and current. However, other problems such as commutation and conduction losses appear.

An MPPT control is usually implemented either in the DC–DC converter or in the DC–AC converter of a PV power system to capture the maximum power. Several algorithms can be used to implement the MPPT, for example, perturb and observe method, incremental conductance method, parasitic capacitance method, and constant voltage method (Blaabjerg et al. 2006).

References

Akhmatov, V. 2003. Analysis of dynamic behavior of electric power systems with large amount of wind power. PhD dissertation, Technical University of Denmark, Lyngby, Denmark.

Blaabjerg, F., Iov, F., Teodorescu, R., and Chen, Z. 2006. Power electronics in renewable energy systems. In *Proceedings of the 2006 Power Electronics and Motion Control Conference*, Portoroz, Slovenia, pp. 1–17.

Carrasco, J. M., Franquelo, L. G., Bialasiewicz, J. T. et al. 2006. Power-electronic systems for the grid integration of renewable energy sources: A survey. *IEEE Transactions on Industrial Electronics* 53: 1002–1016.

Chen, Z., Guerrero, J. M., and Blaabjerg, F. 2009. A review of the state of the art of power electronics for wind turbines. *IEEE Transactions on Power Electronics* 24: 1859–1875.

Depenbrock, M. 1988. Direct self-control (DSC) of inverter-fed induction machine. *IEEE Transactions on Power Electronics* 3: 420–429.

Hansen, L. H., Helle, L., Blaabjerg, F. et al. 2002. Conceptual survey of generators and power electronics for wind turbines. Risø-R-1205(EN), Pitney Bowes Management Services, Denmark.

Hingorani, N. G. and Gyugyi, L. 2000. *Understanding FACTS: Concepts and Technology of Flexible AC Transmission Systems*. IEEE Press, New York.

Kjaer, S. B., Pedersen, J. K., and Blaabjerg, F. 2005. A review of single-phase grid-connected inverters for photovoltaic modules. *IEEE Transactions on Industry Applications* 41: 1292–1306.

Lai, J. S. and Peng, F. Z. 1996. Multilevel converters-a new breed of power converters. *IEEE Transactions on Industry Applications* 32: 509–517.

Liang, J., Qiao, W., and Harley, R. G. 2010. Feed-forward transient current control for low-voltage ride-through enhancement of DFIG wind turbines. *IEEE Transactions on Energy Conversion* 25: 836–843.

Lim, S.-K., Kim, J.-H., and Nam, K. 1999. A DC-link voltage balancing algorithm for 3-level converter using the zero sequence current. In *Proceedings of IEEE PESC'99*, Charleston, SC, pp. 1083–1088.

Petersson, A., Lundberg, S., and Thiringer, T. 2005. A DFIG wind turbine ride-through system. Influence on the energy production. *Wind Energy* 8: 251–263.

Portillo, R., Prats, M., Leon, J. I. et al. 2006. Modelling strategy for back-to-back three-level converters applied to high-power wind turbines. *IEEE Transactions on Industrial Electronics* 53: 1483–1491.

Qiao, W. and Harley, R. G. 2007. Power quality and dynamic performance improvement of wind farms using a STATCOM. In *Proceedings of the 38th IEEE Power Electronics Specialists Conference*, Orlando, FL, pp. 1832–1838.

Qiao, W., Venayagamoorthy, G. K., and Harley, R. G. 2009. Real-time implementation of a STATCOM on a wind farm equipped with doubly fed induction generators. *IEEE Transactions on Industry Applications* 45: 98–107.

Qiao, W., Zhou, W., Aller, J. M., and Harley, R. G. 2008. Wind speed estimation based sensorless output maximization control for a wind turbine driving a DFIG. *IEEE Transactions on Power Electronics* 23: 1156–1169.

Rodriguez, J., Lai, J.-S., and Peng, F. Z. 2002. Multilevel inverters: A survey of topologies, controls, and applications. *IEEE Transactions on Industrial Electronics* 49: 724–738.

Seman, S., Niiranen, J., and Arkkio, A. 2006. Ride-through analysis of doubly fed induction wind-power generator under unsymmetrical network disturbance. *IEEE Transactions on Power Systems* 21: 1782–1789.

Shen, J. and Butterworth, N. 1997. Analysis and design of a three-level PWM converter system for railway-traction applications. *IEE Proceedings on Electronic Power Application* 144: 357–371.

Index

Index page.